FRIDAYS: Class meets in MS310

Introduction to
PROBABILITY
Second Revised Edition

Charles M. Grinstead
Swarthmore College

J. Laurie Snell
Dartmouth College

 American Mathematical Society

1991 *Mathematics Subject Classification.* Primary 60–01.

Library of Congress Cataloging-in-Publication Data
Grinstead, Charles M. (Charles Miller), 1952–
  Introduction to probability. — 2nd rev. ed. / Charles M. Grinstead, J. Laurie Snell.
    p. cm.
  Rev. ed. of: Introduction to probability / J. Laurie Snell. 1st ed. c1988.
  Includes bibliographical references and index.
  ISBN 0-8218-0749-8 (alk. paper)
  1. Probabilities. I. Snell, J. Laurie (James Laurie), 1925– . II. Snell, J. Laurie (James Laurie), 1925– Introduction to probability. III. Title.
QA273.S668 1997
519.2—dc21
                                                                                            97-8126
                                                                                            CIP

Copying and reprinting. Individual readers of this publication, and nonprofit libraries acting for them, are permitted to make fair use of the material, such as to copy a chapter for use in teaching or research. Permission is granted to quote brief passages from this publication in reviews, provided the customary acknowledgment of the source is given.

Republication, systematic copying, or multiple reproduction of any material in this publication (including abstracts) is permitted only under license from the American Mathematical Society. Requests for such permission should be addressed to the Assistant to the Publisher, American Mathematical Society, P. O. Box 6248, Providence, Rhode Island 02940-6248. Requests can also be made by e-mail to reprint-permission@ams.org.

© 1997 by the authors.
Printed in the United States of America.
Reprinted with corrections 1998.

∞ The paper used in this book is acid-free and falls within the guidelines
established to ensure permanence and durability.
Visit the AMS home page at URL: http://www.ams.org/.

10 9 8 7 6 5 4 3 2   02 01 00 99 98

To our wives
and in memory of
Reese T. Prosser

Contents

1 Discrete Probability Distributions   **1**
    1.1 Simulation of Discrete Probabilities 1
    1.2 Discrete Probability Distributions 18

2 Continuous Probability Densities   **41**
    2.1 Simulation of Continuous Probabilities 41
    2.2 Continuous Density Functions 55

3 Combinatorics *The Art of Counting!*   **75**
    3.1 Permutations 75
    3.2 Combinations 92
    3.3 Card Shuffling 120

4 Conditional Probability   **133**
    4.1 Discrete Conditional Probability 133
    4.2 Continuous Conditional Probability 162
    4.3 Paradoxes 175

5 Distributions and Densities   **183**
    5.1 Important Distributions 183
    5.2 Important Densities 205

6 Expected Value and Variance   **225**
    6.1 Expected Value 225
    6.2 Variance of Discrete Random Variables 257
    6.3 Continuous Random Variables 268

7 Sums of Random Variables   **285**
    7.1 Sums of Discrete Random Variables 285
    7.2 Sums of Continuous Random Variables 291

8 Law of Large Numbers   **305**
    8.1 Discrete Random Variables 305
    8.2 Continuous Random Variables 316

9 Central Limit Theorem — 325
9.1 Bernoulli Trials 325
9.2 Discrete Independent Trials 340
9.3 Continuous Independent Trials 355

10 Generating Functions — 365
10.1 Discrete Distributions 365
10.2 Branching Processes 377
10.3 Continuous Densities 394

11 Markov Chains — 405
11.1 Introduction 405
11.2 Absorbing Markov Chains 415
11.3 Ergodic Markov Chains 433
11.4 Fundamental Limit Theorem 447
11.5 Mean First Passage Time 452

12 Random Walks — 471
12.1 Random Walks in Euclidean Space 471
12.2 Gambler's Ruin 486
12.3 Arc Sine Laws 493

Appendices — 499
A Normal Distribution Table 499
B Galton's Data 500
C Life Table 501

Index — 503

Preface

Probability theory began in seventeenth century France when the two great French mathematicians, Blaise Pascal and Pierre de Fermat, corresponded over two problems from games of chance. Problems like those Pascal and Fermat solved continued to influence such early researchers as Huygens, Bernoulli, and DeMoivre in establishing a mathematical theory of probability. Today, probability theory is a well-established branch of mathematics that finds applications in every area of scholarly activity from music to physics, and in daily experience from weather prediction to predicting the risks of new medical treatments.

This text is designed for an introductory probability course taken by sophomores, juniors, and seniors in mathematics, the physical and social sciences, engineering, and computer science. It presents a thorough treatment of probability ideas and techniques necessary for a firm understanding of the subject. The text can be used in a variety of course lengths, levels, and areas of emphasis.

For use in a standard one-term course, in which both discrete and continuous probability is covered, students should have taken as a prerequisite two terms of calculus, including an introduction to multiple integrals. In order to cover Chapter 11, which contains material on Markov chains, some knowledge of matrix theory is necessary.

The text can also be used in a discrete probability course. The material has been organized in such a way that the discrete and continuous probability discussions are presented in a separate, but parallel, manner. This organization dispels an overly rigorous or formal view of probability and offers some strong pedagogical value in that the discrete discussions can sometimes serve to motivate the more abstract continuous probability discussions. For use in a discrete probability course, students should have taken one term of calculus as a prerequisite.

Very little computing background is assumed or necessary in order to obtain full benefits from the use of the computing material and examples in the text. All of the programs that are used in the text have been written in each of the languages TrueBASIC, Maple, and Mathematica.

This book is on the Web at http://www.dartmouth.edu/~chance, and is part of the Chance project, which is devoted to providing materials for beginning courses in probability and statistics. The computer programs, solutions to the odd-numbered exercises, and current errata are also available at this site. Instructors may obtain all of the solutions by writing to either of the authors, at jlsnell@dartmouth.edu and cgrinst1@swarthmore.edu. It is our intention to place items related to this book at this site, and we invite our readers to submit their contributions.

FEATURES

Level of rigor and emphasis: Probability is a wonderfully intuitive and applicable field of mathematics. We have tried not to spoil its beauty by presenting too much formal mathematics. Rather, we have tried to develop the key ideas in a somewhat leisurely style, to provide a variety of interesting applications to probability, and to show some of the nonintuitive examples that make probability such a lively subject.

Exercises: There are over 600 exercises in the text providing plenty of opportunity for practicing skills and developing a sound understanding of the ideas. In the exercise sets are routine exercises to be done with and without the use of a computer and more theoretical exercises to improve the understanding of basic concepts. More difficult exercises are indicated by an asterisk. A solution manual for all of the exercises is available to instructors.

Historical remarks: Introductory probability is a subject in which the fundamental ideas are still closely tied to those of the founders of the subject. For this reason, there are numerous historical comments in the text, especially as they deal with the development of discrete probability.

Pedagogical use of computer programs: Probability theory makes predictions about experiments whose outcomes depend upon chance. Consequently, it lends itself beautifully to the use of computers as a mathematical tool to simulate and analyze chance experiments.

In the text the computer is utilized in several ways. First, it provides a laboratory where chance experiments can be simulated and the students can get a feeling for the variety of such experiments. This use of the computer in probability has been already beautifully illustrated by William Feller in the second edition of his famous text *An Introduction to Probability Theory and Its Applications* (New York: Wiley, 1950). In the preface, Feller wrote about his treatment of fluctuation in coin tossing: "The results are so amazing and so at variance with common intuition that even sophisticated colleagues doubted that coins actually misbehave as theory predicts. The record of a simulated experiment is therefore included."

In addition to providing a laboratory for the student, the computer is a powerful aid in understanding basic results of probability theory. For example, the graphical illustration of the approximation of the standardized binomial distributions to the normal curve is a more convincing demonstration of the Central Limit Theorem than many of the formal proofs of this fundamental result.

Finally, the computer allows the student to solve problems that do not lend themselves to closed-form formulas such as waiting times in queues. Indeed, the introduction of the computer changes the way in which we look at many problems in probability. For example, being able to calculate exact binomial probabilities for experiments up to 1000 trials changes the way we view the normal and Poisson approximations.

ACKNOWLEDGMENTS FOR FIRST EDITION

Anyone writing a probability text today owes a great debt to William Feller, who taught us all how to make probability come alive as a subject matter. If you find an example, an application, or an exercise that you really like, it probably had its origin in Feller's classic text, An Introduction to Probability Theory and Its Applications.

This book had its start with a course given jointly at Dartmouth College with Professor John Kemeny. I am indebted to Professor Kemeny for convincing me that it is both useful and fun to use the computer in the study of probability. He has continuously and generously shared his ideas on probability and computing with me. No less impressive has been the help of John Finn in making the computing an integral part of the text and in writing the programs so that they not only can be easily used, but they also can be understood and modified by the student to explore further problems. Some of the programs in the text were developed through collaborative efforts with John Kemeny and Thomas Kurtz on a Sloan Foundation project and with John Finn on a Keck Foundation project. I am grateful to both foundations for their support.

I am indebted to many other colleagues, students, and friends for valuable comments and suggestions. A few whose names stand out are: Eric and Jim Baumgartner, Tom Bickel, Bob Beck, Ed Brown, Christine Burnley, Richard Crowell, David Griffeath, John Lamperti, Beverly Nickerson, Reese Prosser, Cathy Smith, and Chris Thron.

The following individuals were kind enough to review various drafts of the manuscript. Their encouragement, criticisms, and suggestions were very helpful.

Ron Barnes	University of Houston, Downtown College
Thomas Fischer	Texas A & M University
Richard Groeneveld	Iowa State University
James Kuelbs	University of Wisconsin, Madison
Greg Lawler	Duke University
Sidney Resnick	Colorado State University
Malcom Sherman	SUNY Albany
Olaf Stackelberg	Kent State University
Murad Taqqu	Boston University
Abraham Wender	University of North Carolina

In addition, I would especially like to thank James Kuelbs, Sidney Resnick, and their students for using the manuscript in their courses and sharing their experience and invaluable suggestions with me.

The versatility of Dartmouth's mathematical word processor PREPPY, written by Professor James Baumgartner, has made it much easier to make revisions, but has made the job of typist extraordinaire Marie Slack correspondingly more challenging. Her high standards and willingness always to try the next more difficult task have made it all possible.

Finally, I must thank all the people at Random House who helped during the development and production of this project. First, among these was my editor Wayne Yuhasz, whose continued encouragement and commitment were very helpful during the development of the manuscript. The entire production team provided efficient and professional support: Margaret Pinette, project manager; Michael Weinstein, production manager; and Kate Bradfor of Editing, Design, and Production, Inc.

ACKNOWLEDGMENTS FOR SECOND EDITION

The debt to William Feller has not diminished in the years between the two editions of this book. His book on probability is likely to remain the classic book in this field for many years.

The process of revising the first edition of this book began with some high-level discussions involving the two present co-authors together with Reese Prosser and John Finn. It was during these discussions that, among other things, the first co-author was made aware of the concept of "negative royalties" by Professor Prosser.

We are indebted to many people for their help in this undertaking. First and foremost, we thank Mark Kernighan for his almost 40 pages of single-spaced comments on the first edition. Many of these comments were very thought-provoking; in addition, they provided a student's perspective on the book. Most of the major changes in the second edition have their genesis in these notes.

We would also like to thank Fuxing Hou, who provided extensive help with the typesetting and the figures. Her incessant good humor in the face of many trials, both big ("we need to change the entire book from Lamstex to Latex") and small ("could you please move this subscript down just a bit?"), was truly remarkable.

We would also like to thank Lee Nave, who typed the entire first edition of the book into the computer. Lee corrected most of the typographical errors in the first edition during this process, making our job easier.

Karl Knaub and Jessica Sklar are responsible for the implementations of the computer programs in Mathematica and Maple, and we thank them for their efforts. We also thank Jessica for her work on the solution manual for the exercises, building on the work done by Gang Wang for the first edition.

Tom Shemanske and Dana Williams provided much TeX-nical assistance. Their patience and willingness to help, even to the extent of writing intricate TeX macros, are very much appreciated.

The following people used various versions of the second edition in their probability courses, and provided valuable comments and criticisms.

Marty Arkowitz	*Dartmouth College*
Aimee Johnson	*Swarthmore College*
Bill Peterson	*Middlebury College*
Dan Rockmore	*Dartmouth College*
Shunhui Zhu	*Dartmouth College*

Reese Prosser and John Finn provided much in the way of moral support and camaraderie throughout this project. Certainly, one of the high points of this entire

PREFACE   xi

endeavour was Professor Prosser's telephone call to a casino in Monte Carlo, in an attempt to find out the rules involving the "prison" in roulette.

Peter Doyle motivated us to make this book part of a larger project on the Web, to which others can contribute. He also spent many hours actually carrying out the operation of putting the book on the Web.

Finally, we thank Sergei Gelfand and the American Mathematical Society for their interest in our book, their help in its production, and their willingness to let us put the book on the Web.

Chapter 1

Discrete Probability Distributions

1.1 Simulation of Discrete Probabilities

Probability

In this chapter, we shall first consider chance experiments with a finite number of possible outcomes $\omega_1, \omega_2, \ldots, \omega_n$. For example, we roll a die and the possible outcomes are 1, 2, 3, 4, 5, 6 corresponding to the side that turns up. We toss a coin with possible outcomes H (heads) and T (tails).

It is frequently useful to be able to refer to an outcome of an experiment. For example, we might want to write the mathematical expression which gives the sum of four rolls of a die. To do this, we could let X_i, $i = 1, 2, 3, 4$, represent the values of the outcomes of the four rolls, and then we could write the expression

$$X_1 + X_2 + X_3 + X_4$$

for the sum of the four rolls. The X_i's are called *random variables*. A random variable is simply an expression whose value is the outcome of a particular experiment. Just as in the case of other types of variables in mathematics, random variables can take on different values.

Let X be the random variable which represents the roll of one die. We shall assign probabilities to the possible outcomes of this experiment. We do this by assigning to each outcome ω_j a nonnegative number $m(\omega_j)$ in such a way that

$$m(\omega_1) + m(\omega_2) + \cdots + m(\omega_6) = 1 \ .$$

The function $m(\omega_j)$ is called the *distribution function* of the random variable X. For the case of the roll of the die we would assign equal probabilities or probabilities 1/6 to each of the outcomes. With this assignment of probabilities, one could write

$$P(X \leq 4) = \frac{2}{3}$$

to mean that the probability is 2/3 that a roll of a die will have a value which does not exceed 4.

Let Y be the random variable which represents the toss of a coin. In this case, there are two possible outcomes, which we can label as H and T. Unless we have reason to suspect that the coin comes up one way more often than the other way, it is natural to assign the probability of 1/2 to each of the two outcomes.

In both of the above experiments, each outcome is assigned an equal probability. This would certainly not be the case in general. For example, if a drug is found to be effective 30 percent of the time it is used, we might assign a probability .3 that the drug is effective the next time it is used and .7 that it is not effective. This last example illustrates the intuitive *frequency concept of probability*. That is, if we have a probability p that an experiment will result in outcome A, then if we repeat this experiment a large number of times we should expect that the fraction of times that A will occur is about p. To check intuitive ideas like this, we shall find it helpful to look at some of these problems experimentally. We could, for example, toss a coin a large number of times and see if the fraction of times heads turns up is about 1/2. We could also simulate this experiment on a computer.

Simulation

We want to be able to perform an experiment that corresponds to a given set of probabilities; for example, $m(\omega_1) = 1/2$, $m(\omega_2) = 1/3$, and $m(\omega_3) = 1/6$. In this case, one could mark three faces of a six-sided die with an ω_1, two faces with an ω_2, and one face with an ω_3.

In the general case we assume that $m(\omega_1)$, $m(\omega_2)$, ..., $m(\omega_n)$ are all rational numbers, with least common denominator n. If $n > 2$, we can imagine a long cylindrical die with a cross-section that is a regular n-gon. If $m(\omega_j) = n_j/n$, then we can label n_j of the long faces of the cylinder with an ω_j, and if one of the end faces comes up, we can just roll the die again. If $n = 2$, a coin could be used to perform the experiment.

We will be particularly interested in repeating a chance experiment a large number of times. Although the cylindrical die would be a convenient way to carry out a few repetitions, it would be difficult to carry out a large number of experiments. Since the modern computer can do a large number of operations in a very short time, it is natural to turn to the computer for this task.

Random Numbers

We must first find a computer analog of rolling a die. This is done on the computer by means of a *random number generator*. Depending upon the particular software package, the computer can be asked for a real number between 0 and 1, or an integer in a given set of consecutive integers. In the first case, the real numbers are chosen in such a way that the probability that the number lies in any particular subinterval of this unit interval is equal to the length of the subinterval. In the second case, each integer has the same probability of being chosen.

1.1. SIMULATION OF DISCRETE PROBABILITIES

.203309	.762057	.151121	.623868
.932052	.415178	.716719	.967412
.069664	.670982	.352320	.049723
.750216	.784810	.089734	.966730
.946708	.380365	.027381	.900794

Table 1.1: Sample output of the program **RandomNumbers**.

Let X be a random variable with distribution function $m(\omega)$, where ω is in the set $\{\omega_1, \omega_2, \omega_3\}$, and $m(\omega_1) = 1/2$, $m(\omega_2) = 1/3$, and $m(\omega_3) = 1/6$. If our computer package can return a random integer in the set $\{1, 2, ..., 6\}$, then we simply ask it to do so, and make 1, 2, and 3 correspond to ω_1, 4 and 5 correspond to ω_2, and 6 correspond to ω_3. If our computer package returns a random real number r in the interval (0, 1), then the expression

$$\lfloor 6r \rfloor + 1$$

will be a random integer between 1 and 6. (The notation $\lfloor x \rfloor$ means the greatest integer not exceeding x, and is read "floor of x.")

The method by which random real numbers are generated on a computer is described in the historical discussion at the end of this section. The following example gives sample output of the program **RandomNumbers**.

Example 1.1 (Random Number Generation) The program **RandomNumbers** generates n random real numbers in the interval $[0, 1]$, where n is chosen by the user. When we ran the program with $n = 20$, we obtained the data shown in Table 1.1. □

Example 1.2 (Coin Tossing) As we have noted, our intuition suggests that the probability of obtaining a head on a single toss of a coin is 1/2. To have the computer toss a coin, we can ask it to pick a random real number in the interval $[0, 1]$ and test to see if this number is less than 1/2. If so, we shall call the outcome *heads*; if not we call it *tails*. Another way to proceed would be to ask the computer to pick a random integer from the set $\{0, 1\}$. The program **CoinTosses** carries out the experiment of tossing a coin n times. Running this program, with $n = 20$, resulted in:

THTTTHTTTTHTTTTTHHTT.

Note that in 20 tosses, we obtained 5 heads and 15 tails. Let us toss a coin n times, where n is much larger than 20, and see if we obtain a proportion of heads closer to our intuitive guess of 1/2. The program **CoinTosses** keeps track of the number of heads. When we ran this program with $n = 1000$, we obtained 494 heads. When we ran it with $n = 10000$, we obtained 5039 heads.

We notice that when we tossed the coin 10,000 times, the proportion of heads was close to the "true value" .5 for obtaining a head when a coin is tossed. A mathematical model for this experiment is called Bernoulli Trials (see Chapter 3). The *Law of Large Numbers,* which we shall study later (see Chapter 8), will show that in the Bernoulli Trials model, the proportion of heads should be near .5, consistent with our intuitive idea of the frequency interpretation of probability.

Of course, our program could be easily modified to simulate coins for which the probability of a head is p, where p is a real number between 0 and 1. □

In the case of coin tossing, we already knew the probability of the event occurring on each experiment. The real power of simulation comes from the ability to estimate probabilities when they are not known ahead of time. This method has been used in the recent discoveries of strategies that make the casino game of blackjack favorable to the player. We illustrate this idea in a simple situation in which we can compute the true probability and see how effective the simulation is.

Example 1.3 (Dice Rolling) We consider a dice game that played an important role in the historical development of probability. The famous letters between Pascal and Fermat, which many believe started a serious study of probability, were instigated by a request for help from a French nobleman and gambler, Chevalier de Méré. It is said that de Méré had been betting that, in four rolls of a die, at least one six would turn up. He was winning consistently and, to get more people to play, he changed the game to bet that, in 24 rolls of two dice, a pair of sixes would turn up. It is claimed that de Méré lost with 24 and felt that 25 rolls were necessary to make the game favorable. It was *un grand scandale* that mathematics was wrong.

We shall try to see if de Méré is correct by simulating his various bets. The program **DeMere1** simulates a large number of experiments, seeing, in each one, if a six turns up in four rolls of a die. When we ran this program for 1000 plays, a six came up in the first four rolls 48.6 percent of the time. When we ran it for 10,000 plays this happened 51.98 percent of the time.

We note that the result of the second run suggests that de Méré was correct in believing that his bet with one die was favorable; however, if we had based our conclusion on the first run, we would have decided that he was wrong. *Accurate results by simulation require a large number of experiments.* □

The program **DeMere2** simulates de Méré's second bet that a pair of sixes will occur in n rolls of a pair of dice. The previous simulation shows that it is important to know how many trials we should simulate in order to expect a certain degree of accuracy in our approximation. We shall see later that in these types of experiments, a rough rule of thumb is that, at least 95% of the time, the error does not exceed the reciprocal of the square root of the number of trials. Fortunately, for this dice game, it will be easy to compute the exact probabilities. We shall show in the next section that for the first bet the probability that de Méré wins is $1 - (5/6)^4 = .518$.

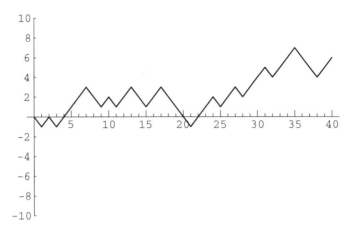

Figure 1.1: Peter's winnings in 40 plays of heads or tails.

One can understand this calculation as follows: The probability that no 6 turns up on the first toss is $(5/6)$. The probability that no 6 turns up on either of the first two tosses is $(5/6)^2$. Reasoning in the same way, the probability that no 6 turns up on any of the first four tosses is $(5/6)^4$. Thus, the probability of at least one 6 in the first four tosses is $1 - (5/6)^4$. Similarly, for the second bet, with 24 rolls, the probability that de Méré wins is $1 - (35/36)^{24} = .491$, and for 25 rolls it is $1 - (35/36)^{25} = .506$.

Using the rule of thumb mentioned above, it would require 27,000 rolls to have a reasonable chance to determine these probabilities with sufficient accuracy to assert that they lie on opposite sides of .5. It is interesting to ponder whether a gambler can detect such probabilities with the required accuracy from gambling experience. Some writers on the history of probability suggest that de Méré was, in fact, just interested in these problems as intriguing probability problems.

Example 1.4 (Heads or Tails) For our next example, we consider a problem where the exact answer is difficult to obtain but for which simulation easily gives the qualitative results. Peter and Paul play a game called *heads or tails*. In this game, a fair coin is tossed a sequence of times—we choose 40. Each time a head comes up Peter wins 1 penny from Paul, and each time a tail comes up Peter loses 1 penny to Paul. For example, if the results of the 40 tosses are

THTHHHHTTHTHHTTHHTTTTHHHTHHTHHHTHHHTTTHH.

Peter's winnings may be graphed as in Figure 1.1.

Peter has won 6 pennies in this particular game. It is natural to ask for the probability that he will win j pennies; here j could be any even number from -40 to 40. It is reasonable to guess that the value of j with the highest probability is $j = 0$, since this occurs when the number of heads equals the number of tails. Similarly, we would guess that the values of j with the lowest probabilities are $j = \pm 40$.

A second interesting question about this game is the following: How many times in the 40 tosses will Peter be in the lead? Looking at the graph of his winnings (Figure 1.1), we see that Peter is in the lead when his winnings are positive, but we have to make some convention when his winnings are 0 if we want all tosses to contribute to the number of times in the lead. We adopt the convention that, when Peter's winnings are 0, he is in the lead if he was ahead at the previous toss and not if he was behind at the previous toss. With this convention, Peter is in the lead 34 times in our example. Again, our intuition might suggest that the most likely number of times to be in the lead is 1/2 of 40, or 20, and the least likely numbers are the extreme cases of 40 or 0.

It is easy to settle this by simulating the game a large number of times and keeping track of the number of times that Peter's final winnings are j, and the number of times that Peter ends up being in the lead by k. The proportions over all games then give estimates for the corresponding probabilities. The program **HTSimulation** carries out this simulation. Note that when there are an even number of tosses in the game, it is possible to be in the lead only an even number of times. We have simulated this game 10,000 times. The results are shown in Figures 1.2 and 1.3. These graphs, which we call spike graphs, were generated using the program **Spikegraph**. The vertical line, or spike, at position x on the horizontal axis, has a height equal to the proportion of outcomes which equal x. Our intuition about Peter's final winnings was quite correct, but our intuition about the number of times Peter was in the lead was completely wrong. The simulation suggests that the least likely number of times in the lead is 20 and the most likely is 0 or 40. This is indeed correct, and the explanation for it is suggested by playing the game of heads or tails with a large number of tosses and looking at a graph of Peter's winnings. In Figure 1.4 we show the results of a simulation of the game, for 1000 tosses and in Figure 1.5 for 10,000 tosses.

In the second example Peter was ahead most of the time. It is a remarkable fact, however, that, if play is continued long enough, Peter's winnings will continue to come back to 0, but there will be very long times between the times that this happens. These and related results will be discussed in Chapter 12. □

In all of our examples so far, we have simulated equiprobable outcomes. We illustrate next an example where the outcomes are not equiprobable.

Example 1.5 (Horse Races) Four horses (Acorn, Balky, Chestnut, and Dolby) have raced many times. It is estimated that Acorn wins 30 percent of the time, Balky 40 percent of the time, Chestnut 20 percent of the time, and Dolby 10 percent of the time.

We can have our computer carry out one race as follows: Choose a random number x. If $x < .3$ then we say that Acorn won. If $.3 \leq x < .7$ then Balky wins. If $.7 \leq x < .9$ then Chestnut wins. Finally, if $.9 \leq x$ then Dolby wins.

The program **HorseRace** uses this method to simulate the outcomes of n races. Running this program for $n = 10$ we found that Acorn won 40 percent of the time, Balky 20 percent of the time, Chestnut 10 percent of the time, and Dolby 30 percent

1.1. SIMULATION OF DISCRETE PROBABILITIES 7

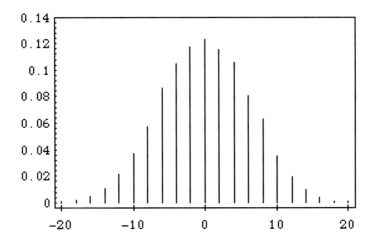

Figure 1.2: Distribution of winnings.

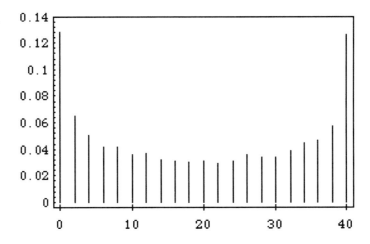

Figure 1.3: Distribution of number of times in the lead.

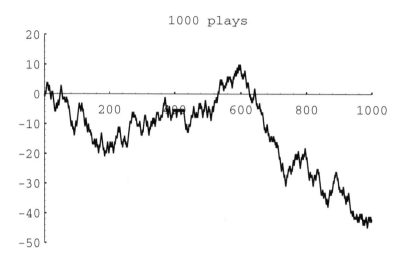

Figure 1.4: Peter's winnings in 1000 plays of heads or tails.

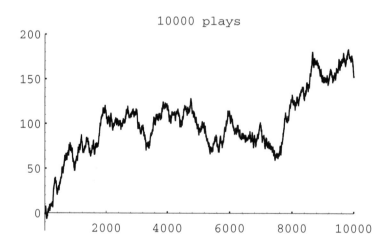

Figure 1.5: Peter's winnings in 10,000 plays of heads or tails.

of the time. A larger number of races would be necessary to have better agreement with the past experience. Therefore we ran the program to simulate 1000 races with our four horses. Although very tired after all these races, they performed in a manner quite consistent with our estimates of their abilities. Acorn won 29.8 percent of the time, Balky 39.4 percent, Chestnut 19.5 percent, and Dolby 11.3 percent of the time.

The program **GeneralSimulation** uses this method to simulate repetitions of an arbitrary experiment with a finite number of outcomes occurring with known probabilities. □

Historical Remarks

Anyone who plays the same chance game over and over is really carrying out a simulation, and in this sense the process of simulation has been going on for centuries. As we have remarked, many of the early problems of probability might well have been suggested by gamblers' experiences.

It is natural for anyone trying to understand probability theory to try simple experiments by tossing coins, rolling dice, and so forth. The naturalist Buffon tossed a coin 4040 times, resulting in 2048 heads and 1992 tails. He also estimated the number π by throwing needles on a ruled surface and recording how many times the needles crossed a line (see Section 2.1). The English biologist W. F. R. Weldon[1] recorded 26,306 throws of 12 dice, and the Swiss scientist Rudolf Wolf[2] recorded 100,000 throws of a single die without a computer. Such experiments are very time-consuming and may not accurately represent the chance phenomena being studied. For example, for the dice experiments of Weldon and Wolf, further analysis of the recorded data showed a suspected bias in the dice. The statistician Karl Pearson analyzed a large number of outcomes at certain roulette tables and suggested that the wheels were biased. He wrote in 1894:

> Clearly, since the Casino does not serve the valuable end of huge laboratory for the preparation of probability statistics, it has no scientific *raison d'être*. Men of science cannot have their most refined theories disregarded in this shameless manner! The French Government must be urged by the hierarchy of science to close the gaming-saloons; it would be, of course, a graceful act to hand over the remaining resources of the Casino to the Académie des Sciences for the endowment of a laboratory of orthodox probability; in particular, of the new branch of that study, the application of the theory of chance to the biological problems of evolution, which is likely to occupy so much of men's thoughts in the near future.[3]

However, these early experiments were suggestive and led to important discoveries in probability and statistics. They led Pearson to the *chi-squared test*, which

---

[1] T. C. Fry, *Probability and Its Engineering Uses,* 2nd ed. (Princeton: Van Nostrand, 1965).
[2] E. Czuber, *Wahrscheinlichkeitsrechnung,* 3rd ed. (Berlin: Teubner, 1914).
[3] K. Pearson, "Science and Monte Carlo," *Fortnightly Review,* vol. 55 (1894), p. 193; cited in S. M. Stigler, *The History of Statistics* (Cambridge: Harvard University Press, 1986).

is of great importance in testing whether observed data fit a given probability distribution.

By the early 1900s it was clear that a better way to generate random numbers was needed. In 1927, L. H. C. Tippett published a list of 41,600 digits obtained by selecting numbers haphazardly from census reports. In 1955, RAND Corporation printed a table of 1,000,000 random numbers generated from electronic noise. The advent of the high-speed computer raised the possibility of generating random numbers directly on the computer, and in the late 1940s John von Neumann suggested that this be done as follows: Suppose that you want a random sequence of four-digit numbers. Choose any four-digit number, say 6235, to start. Square this number to obtain 38,875,225. For the second number choose the middle four digits of this square (i.e., 8752). Do the same process starting with 8752 to get the third number, and so forth.

More modern methods involve the concept of modular arithmetic. If a is an integer and m is a positive integer, then by $a \pmod{m}$ we mean the remainder when a is divided by m. For example, 10 (mod 4) = 2, 8 (mod 2) = 0, and so forth. To generate a random sequence $X_0, X_1, X_2, \ldots$ of numbers choose a starting number X_0 and then obtain the numbers X_{n+1} from X_n by the formula

$$X_{n+1} = (aX_n + c) \pmod{m},$$

where a, c, and m are carefully chosen constants. The sequence $X_0, X_1, X_2, \ldots$ is then a sequence of integers between 0 and $m-1$. To obtain a sequence of real numbers in $[0,1)$, we divide each X_j by m. The resulting sequence consists of rational numbers of the form j/m, where $0 \leq j \leq m-1$. Since m is usually a very large integer, we think of the numbers in the sequence as being random real numbers in $[0,1)$.

For both von Neumann's squaring method and the modular arithmetic technique the sequence of numbers is actually completely determined by the first number. Thus, there is nothing really random about these sequences. However, they produce numbers that behave very much as theory would predict for random experiments. To obtain different sequences for different experiments the initial number X_0 is chosen by some other procedure that might involve, for example, the time of day.[4]

During the Second World War, physicists at the Los Alamos Scientific Laboratory needed to know, for purposes of shielding, how far neutrons travel through various materials. This question was beyond the reach of theoretical calculations. Daniel McCracken, writing in the *Scientific American*, states:

> The physicists had most of the necessary data: they knew the average distance a neutron of a given speed would travel in a given substance before it collided with an atomic nucleus, what the probabilities were that the neutron would bounce off instead of being absorbed by the nucleus, how much energy the neutron was likely to lose after a given

---

[4]For a detailed discussion of random numbers, see D. E. Knuth, *The Art of Computer Programming*, vol. II (Reading: Addison-Wesley, 1969).

collision and so on.[5]

John von Neumann and Stanislas Ulam suggested that the problem be solved by modeling the experiment by chance devices on a computer. Their work being secret, it was necessary to give it a code name. Von Neumann chose the name "Monte Carlo." Since that time, this method of simulation has been called the *Monte Carlo Method*.

William Feller indicated the possibilities of using computer simulations to illustrate basic concepts in probability in his book *An Introduction to Probability Theory and Its Applications*. In discussing the problem about the number of times in the lead in the game of "heads or tails" Feller writes:

> The results concerning fluctuations in coin tossing show that widely held beliefs about the law of large numbers are fallacious. These results are so amazing and so at variance with common intuition that even sophisticated colleagues doubted that coins actually misbehave as theory predicts. The record of a simulated experiment is therefore included.[6]

Feller provides a plot showing the result of 10,000 plays of *heads or tails* similar to that in Figure 1.5.

The martingale betting system described in Exercise 10 has a long and interesting history. Russell Barnhart pointed out to the authors that its use can be traced back at least to 1754, when Casanova, writing in his memoirs, *History of My Life*, writes

> She [Casanova's mistress] made me promise to go to the casino [the Ridotto in Venice] for money to play in partnership with her. I went there and took all the gold I found, and, determinedly doubling my stakes according to the system known as the martingale, I won three or four times a day during the rest of the Carnival. I never lost the sixth card. If I had lost it, I should have been out of funds, which amounted to two thousand zecchini.[7]

Even if there were no zeros on the roulette wheel so the game was perfectly fair, the martingale system, or any other system for that matter, cannot make the game into a favorable game. The idea that a fair game remains fair and unfair games remain unfair under gambling systems has been exploited by mathematicians to obtain important results in the study of probability. We will introduce the general concept of a martingale in Chapter 6.

The word *martingale* itself also has an interesting history. The origin of the word is obscure. The *Oxford English Dictionary* gives examples of its use in the

---

[5]D. D. McCracken, "The Monte Carlo Method," *Scientific American*, vol. 192 (May 1955), p. 90.

[6]W. Feller, *Introduction to Probability Theory and its Applications*, vol. 1, 3rd ed. (New York: John Wiley & Sons, 1968), p. xi.

[7]G. Casanova, *History of My Life,* vol. IV, Chap. 7, trans. W. R. Trask (New York: Harcourt-Brace, 1968), p. 124.

early 1600s and says that its probable origin is the reference in Rabelais's Book One, Chapter 19:

> Everything was done as planned, the only thing being that Gargantua doubted if they would be able to find, right away, breeches suitable to the old fellow's legs; he was doubtful, also, as to what cut would be most becoming to the orator—the martingale, which has a draw-bridge effect in the seat, to permit doing one's business more easily; the sailor-style, which affords more comfort for the kidneys; the Swiss, which is warmer on the belly; or the codfish-tail, which is cooler on the loins.[8]

In modern uses martingale has several different meanings, all related to *holding down*, in addition to the gambling use. For example, it is a strap on a horse's harness used to hold down the horse's head, and also part of a sailing rig used to hold down the bowsprit.

The Labouchere system described in Exercise 9 is named after Henry du Pre Labouchere (1831–1912), an English journalist and member of Parliament. Labouchere attributed the system to Condorcet. Condorcet (1743–1794) was a political leader during the time of the French revolution who was interested in applying probability theory to economics and politics. For example, he calculated the probability that a jury using majority vote will give a correct decision if each juror has the same probability of deciding correctly. His writings provided a wealth of ideas on how probability might be applied to human affairs.[9]

Exercises

1. Modify the program **CoinTosses** to toss a coin n times and print out after every 100 tosses the proportion of heads minus 1/2. Do these numbers appear to approach 0 as n increases? Modify the program again to print out, every 100 times, both of the following quantities: the proportion of heads minus 1/2, and the number of heads minus half the number of tosses. Do these numbers appear to approach 0 as n increases?

2. Modify the program **CoinTosses** so that it tosses a coin n times and records whether or not the proportion of heads is within .1 of .5 (i.e., between .4 and .6). Have your program repeat this experiment 100 times. About how large must n be so that approximately 95 out of 100 times the proportion of heads is between .4 and .6?

3. In the early 1600s, Galileo was asked to explain the fact that, although the number of triples of integers from 1 to 6 with sum 9 is the same as the number of such triples with sum 10, when three dice are rolled, a 9 seemed to come up less often than a 10—supposedly in the experience of gamblers.

---

[8]Quoted in the *Portable Rabelais*, ed. S. Putnam (New York: Viking, 1946), p. 113.
[9]Le Marquise de Condorcet, *Essai sur l'Application de l'Analyse à la Probabilité dès Décisions Rendues a la Pluralité des Voix* (Paris: Imprimerie Royale, 1785).

(a) Write a program to simulate the roll of three dice a large number of times and keep track of the proportion of times that the sum is 9 and the proportion of times it is 10.

(b) Can you conclude from your simulations that the gamblers were correct?

4 In raquetball, a player continues to serve as long as she is winning; a point is scored only when a player is serving and wins the volley. The first player to win 21 points wins the game. Assume that you serve first and have a probability .6 of winning a volley when you serve and probability .5 when your opponent serves. Estimate, by simulation, the probability that you will win a game.

5 Consider the bet that all three dice will turn up sixes at least once in n rolls of three dice. Calculate $f(n)$, the probability of at least one triple-six when three dice are rolled n times. Determine the smallest value of n necessary for a favorable bet that a triple-six will occur when three dice are rolled n times. (DeMoivre would say it should be about $216 \log 2 = 149.7$ and so would answer 150—see Exercise 1.2.17. Do you agree with him?)

6 In Las Vegas, a roulette wheel has 38 slots numbered 0, 00, 1, 2, ..., 36. The 0 and 00 slots are green and half of the remaining 36 slots are red and half are black. A croupier spins the wheel and throws in an ivory ball. If you bet 1 dollar on red, you win 1 dollar if the ball stops in a red slot and otherwise you lose 1 dollar. Write a program to find the total winnings for a player who makes 1000 bets on red.

7 Another form of bet for roulette is to bet that a specific number (say 17) will turn up. If the ball stops on your number, you get your dollar back plus 35 dollars. If not, you lose your dollar. Write a program that will plot your winnings when you make 500 plays of roulette at Las Vegas, first when you bet each time on red (see Exercise 6), and then for a second visit to Las Vegas when you make 500 plays betting each time on the number 17. What differences do you see in the graphs of your winnings on these two occasions?

8 An astute student noticed that, in our simulation of the game of heads or tails (see Example 1.4), the proportion of times the player is always in the lead is very close to the proportion of times that the player's total winnings end up 0. Work out these probabilities by enumeration of all cases for two tosses and for four tosses, and see if you think that these probabilities are, in fact, the same.

9 The *Labouchere system* for roulette is played as follows. Write down a list of numbers, usually 1, 2, 3, 4. Bet the sum of the first and last, $1 + 4 = 5$, on red. If you win, delete the first and last numbers from your list. If you lose, add the amount that you last bet to the end of your list. Then use the new list and bet the sum of the first and last numbers (if there is only one number, bet that amount). Continue until your list becomes empty. Show that, if this happens, you win the sum, $1 + 2 + 3 + 4 = 10$, of your original list. Simulate this system and see if you do always stop and, hence, always win. If so, why is this not a foolproof gambling system?

10 Another well-known gambling system is the *martingale doubling system*. Suppose that you are betting on red to turn up in roulette. Every time you win, bet 1 dollar next time. Every time you lose, double your previous bet. Continue to play until you have won at least 5 dollars or you have lost more than 100 dollars. Write a program to simulate this system and play it a number of times and see how you do. In his book *The Newcomes*, W. M. Thackeray remarks "You have not played as yet? Do not do so; above all avoid a martingale if you do."[10] Was this good advice?

11 Modify the program **HTSimulation** so that it keeps track of the maximum of Peter's winnings in each game of 40 tosses. Have your program print out the proportion of times that your total winnings take on values 0, 2, 4, ..., 40. Calculate the corresponding exact probabilities for games of two tosses and four tosses.

12 In an upcoming national election for the President of the United States, a pollster plans to predict the winner of the popular vote by taking a random sample of 1000 voters and declaring that the winner will be the one obtaining the most votes in his sample. Suppose that 48 percent of the voters plan to vote for the Republican candidate and 52 percent plan to vote for the Democratic candidate. To get some idea of how reasonable the pollster's plan is, write a program to make this prediction by simulation. Repeat the simulation 100 times and see how many times the pollster's prediction would come true. Repeat your experiment, assuming now that 49 percent of the population plan to vote for the Republican candidate; first with a sample of 1000 and then with a sample of 3000. (The Gallup Poll uses about 3000.) (This idea is discussed further in Chapter 9, Section 9.1.)

13 The psychologist Tversky and his colleagues[11] say that about four out of five people will answer (a) to the following question:

A certain town is served by two hospitals. In the larger hospital about 45 babies are born each day, and in the smaller hospital 15 babies are born each day. Although the overall proportion of boys is about 50 percent, the actual proportion at either hospital may be more or less than 50 percent on any day. At the end of a year, which hospital will have the greater number of days on which more than 60 percent of the babies born were boys?

(a) the large hospital

---

[10]W. M. Thackerey, *The Newcomes* (London: Bradbury and Evans, 1854–55).
[11]See K. McKean, "Decisions, Decisions," *Discover*, June 1985, pp. 22–31. Kevin McKean, Discover Magazine, ©1987 Family Media, Inc. Reprinted with permission. This popular article reports on the work of Tverksy et. al. in *Judgement Under Uncertainty: Heuristics and Biases* (Cambridge: Cambridge University Press, 1982).

(b) the small hospital

(c) neither—the number of days will be about the same.

Assume that the probability that a baby is a boy is .5 (actual estimates make this more like .513). Decide, by simulation, what the right answer is to the question. Can you suggest why so many people go wrong?

14 You are offered the following game. A fair coin will be tossed until the first time it comes up heads. If this occurs on the jth toss you are paid 2^j dollars. You are sure to win at least 2 dollars so you should be willing to pay to play this game—but how much? Few people would pay as much as 10 dollars to play this game. See if you can decide, by simulation, a reasonable amount that you would be willing to pay, per game, if you will be allowed to make a large number of plays of the game. Does the amount that you would be willing to pay per game depend upon the number of plays that you will be allowed?

15 Tversky and his colleagues[12] studied the records of 48 of the Philadelphia 76ers basketball games in the 1980–81 season to see if a player had times when he was hot and every shot went in, and other times when he was cold and barely able to hit the backboard. The players estimated that they were about 25 percent more likely to make a shot after a hit than after a miss. In fact, the opposite was true—the 76ers were 6 percent more likely to score after a miss than after a hit. Tversky reports that the number of hot and cold streaks was about what one would expect by purely random effects. Assuming that a player has a fifty-fifty chance of making a shot and makes 20 shots a game, estimate by simulation the proportion of the games in which the player will have a streak of 5 or more hits.

16 Estimate, by simulation, the average number of children there would be in a family if all people had children until they had a boy. Do the same if all people had children until they had at least one boy and at least one girl. How many more children would you expect to find under the second scheme than under the first in 100,000 families? (Assume that boys and girls are equally likely.)

17 Mathematicians have been known to get some of the best ideas while sitting in a cafe, riding on a bus, or strolling in the park. In the early 1900s the famous mathematician George Pólya lived in a hotel near the woods in Zurich. He liked to walk in the woods and think about mathematics. Pólya describes the following incident:

> At the hotel there lived also some students with whom I usually took my meals and had friendly relations. On a certain day one of them expected the visit of his fiancée, what (sic) I knew, but I did not foresee that he and his fiancée would also set out for a

---

[12]ibid.

16   CHAPTER 1. DISCRETE PROBABILITY DISTRIBUTIONS

a. Random walk in one dimension.

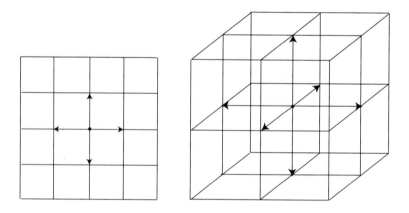

b. Random walk in two dimensions.   c. Random walk in three dimensions.

Figure 1.6: Random walk.

1.1. SIMULATION OF DISCRETE PROBABILITIES

stroll in the woods, and then suddenly I met them there. And then I met them the same morning repeatedly, I don't remember how many times, but certainly much too often and I felt embarrassed: It looked as if I was snooping around which was, I assure you, not the case.[13]

This set him to thinking about whether random walkers were destined to meet.

Pólya considered random walkers in one, two, and three dimensions. In one dimension, he envisioned the walker on a very long street. At each intersection the walker flips a fair coin to decide which direction to walk next (see Figure 1.6a). In two dimensions, the walker is walking on a grid of streets, and at each intersection he chooses one of the four possible directions with equal probability (see Figure 1.6b). In three dimensions (we might better speak of a random climber), the walker moves on a three-dimensional grid, and at each intersection there are now six different directions that the walker may choose, each with equal probability (see Figure 1.6c).

The reader is referred to Section 12.1, where this and related problems are discussed.

(a) Write a program to simulate a random walk in one dimension starting at 0. Have your program print out the lengths of the times between returns to the starting point (returns to 0). See if you can guess from this simulation the answer to the following question: Will the walker always return to his starting point eventually or might he drift away forever?

(b) The paths of two walkers in two dimensions who meet after n steps can be considered to be a single path that starts at $(0,0)$ and returns to $(0,0)$ after $2n$ steps. This means that the probability that two random walkers in two dimensions meet is the same as the probability that a single walker in two dimensions ever returns to the starting point. Thus the question of whether two walkers are sure to meet is the same as the question of whether a single walker is sure to return to the starting point.

Write a program to simulate a random walk in two dimensions and see if you think that the walker is sure to return to $(0,0)$. If so, Pólya would be sure to keep meeting his friends in the park. Perhaps by now you have conjectured the answer to the question: Is a random walker in one or two dimensions sure to return to the starting point? Pólya answered this question for dimensions one, two, and three. He established the remarkable result that the answer is *yes* in one and two dimensions and *no* in three dimensions.

---

[13]G. Pólya, "Two Incidents," *Scientists at Work: Festschrift in Honour of Herman Wold,* ed. T. Dalenius, G. Karlsson, and S. Malmquist (Uppsala: Almquist & Wiksells Boktryckeri AB, 1970).

(c) Write a program to simulate a random walk in three dimensions and see whether, from this simulation and the results of (a) and (b), you could have guessed Pólya's result.

1.2 Discrete Probability Distributions

In this book we shall study many different experiments from a probabilistic point of view. What is involved in this study will become evident as the theory is developed and examples are analyzed. However, the overall idea can be described and illustrated as follows: to each experiment that we consider there will be associated a random variable, which represents the outcome of any particular experiment. The set of possible outcomes is called the *sample space*. In the first part of this section, we will consider the case where the experiment has only finitely many possible outcomes, i.e., the sample space is finite. We will then generalize to the case that the sample space is either finite or countably infinite. This leads us to the following definition.

Random Variables and Sample Spaces

Definition 1.1 Suppose we have an experiment whose outcome depends on chance. We represent the outcome of the experiment by a capital Roman letter, such as X, called a *random variable*. The *sample space* of the experiment is the set of all possible outcomes. If the sample space is either finite or countably infinite, the random variable is said to be *discrete*. □

We generally denote a sample space by the capital Greek letter Ω. As stated above, in the correspondence between an experiment and the mathematical theory by which it is studied, the sample space Ω corresponds to the set of possible outcomes of the experiment.

We now make two additional definitions. These are subsidiary to the definition of sample space and serve to make precise some of the common terminology used in conjunction with sample spaces. First of all, we define the elements of a sample space to be *outcomes*. Second, each subset of a sample space is defined to be an *event*. Normally, we shall denote outcomes by lower case letters and events by capital letters.

Example 1.6 A die is rolled once. We let X denote the outcome of this experiment. Then the sample space for this experiment is the 6-element set

$$\Omega = \{1, 2, 3, 4, 5, 6\},$$

where each outcome i, for $i = 1, \ldots, 6$, corresponds to the number of dots on the face which turns up. The event

$$E = \{2, 4, 6\}$$

1.2. DISCRETE PROBABILITY DISTRIBUTIONS

corresponds to the statement that the result of the roll is an even number. The event E can also be described by saying that X is even. Unless there is reason to believe the die is loaded, the natural assumption is that every outcome is equally likely. Adopting this convention means that we assign a probability of 1/6 to each of the six outcomes, i.e., $m(i) = 1/6$, for $1 \leq i \leq 6$. □

Distribution Functions

We next describe the assignment of probabilities. The definitions are motivated by the example above, in which we assigned to each outcome of the sample space a nonnegative number such that the sum of the numbers assigned is equal to 1.

Definition 1.2 Let X be a random variable which denotes the value of the outcome of a certain experiment, and assume that this experiment has only finitely many possible outcomes. Let Ω be the sample space of the experiment (i.e., the set of all possible values of X, or equivalently, the set of all possible outcomes of the experiment.) A *distribution function* for X is a real-valued function m whose domain is Ω and which satisfies:

1. $m(\omega) \geq 0$, for all $\omega \in \Omega$, and

2. $\sum_{\omega \in \Omega} m(\omega) = 1$.

For any subset E of Ω, we define the *probability* of E to be the number $P(E)$ given by

$$P(E) = \sum_{\omega \in E} m(\omega).$$

□

Example 1.7 Consider an experiment in which a coin is tossed twice. Let X be the random variable which corresponds to this experiment. We note that there are several ways to record the outcomes of this experiment. We could, for example, record the two tosses, in the order in which they occurred. In this case, we have $\Omega = \{HH, HT, TH, TT\}$. We could also record the outcomes by simply noting the number of heads that appeared. In this case, we have $\Omega = \{0, 1, 2\}$. Finally, we could record the two outcomes, without regard to the order in which they occurred. In this case, we have $\Omega = \{HH, HT, TT\}$.

We will use, for the moment, the first of the sample spaces given above. We will assume that all four outcomes are equally likely, and define the distribution function $m(\omega)$ by

$$m(\text{HH}) = m(\text{HT}) = m(\text{TH}) = m(\text{TT}) = \frac{1}{4}.$$

Let $E = \{$HH,HT,TH$\}$ be the event that at least one head comes up. Then, the probability of E can be calculated as follows:

$$\begin{aligned} P(E) &= m(\text{HH}) + m(\text{HT}) + m(\text{TH}) \\ &= \frac{1}{4} + \frac{1}{4} + \frac{1}{4} = \frac{3}{4} . \end{aligned}$$

Similarly, if $F = \{$HH,HT$\}$ is the event that heads comes up on the first toss, then we have

$$\begin{aligned} P(F) &= m(\text{HH}) + m(\text{HT}) \\ &= \frac{1}{4} + \frac{1}{4} = \frac{1}{2} . \end{aligned}$$

$\square$

Example 1.8 (Example 1.6 continued) The sample space for the experiment in which the die is rolled is the 6-element set $\Omega = \{1, 2, 3, 4, 5, 6\}$. We assumed that the die was fair, and we chose the distribution function defined by

$$m(i) = \frac{1}{6}, \quad \text{for } i = 1, \ldots, 6 .$$

If E is the event that the result of the roll is an even number, then $E = \{2, 4, 6\}$ and

$$\begin{aligned} P(E) &= m(2) + m(4) + m(6) \\ &= \frac{1}{6} + \frac{1}{6} + \frac{1}{6} = \frac{1}{2} . \end{aligned}$$

$\square$

Notice that it is an immediate consequence of the above definitions that, for every $\omega \in \Omega$,

$$P(\{\omega\}) = m(\omega) .$$

That is, the probability of the elementary event $\{\omega\}$, consisting of a single outcome ω, is equal to the value $m(\omega)$ assigned to the outcome ω by the distribution function.

Example 1.9 Three people, A, B, and C, are running for the same office, and we assume that one and only one of them wins. The sample space may be taken as the 3-element set $\Omega = \{$A,B,C$\}$ where each element corresponds to the outcome of that candidate's winning. Suppose that A and B have the same chance of winning, but that C has only 1/2 the chance of A or B. Then we assign

$$m(\text{A}) = m(\text{B}) = 2m(\text{C}) .$$

Since

$$m(\text{A}) + m(\text{B}) + m(\text{C}) = 1 ,$$

1.2. DISCRETE PROBABILITY DISTRIBUTIONS

we see that
$$2m(\text{C}) + 2m(\text{C}) + m(\text{C}) = 1 \ ,$$
which implies that $5m(\text{C}) = 1$. Hence,
$$m(\text{A}) = \frac{2}{5} \ , \qquad m(\text{B}) = \frac{2}{5} \ , \qquad m(\text{C}) = \frac{1}{5} \ .$$

Let E be the event that either A or C wins. Then $E = \{\text{A},\text{C}\}$, and
$$P(E) = m(\text{A}) + m(\text{C}) = \frac{2}{5} + \frac{1}{5} = \frac{3}{5} \ .$$
$\square$

In many cases, events can be described in terms of other events through the use of the standard constructions of set theory. We will briefly review the definitions of these constructions. The reader is referred to Figure 1.7 for Venn diagrams which illustrate these constructions.

Let A and B be two sets. Then the union of A and B is the set
$$A \cup B = \{x \,|\, x \in A \text{ or } x \in B\} \ .$$

The intersection of A and B is the set
$$A \cap B = \{x \,|\, x \in A \text{ and } x \in B\} \ .$$

The difference of A and B is the set
$$A - B = \{x \,|\, x \in A \text{ and } x \notin B\} \ .$$

The set A is a subset of B, written $A \subset B$, if every element of A is also an element of B. Finally, the complement of A is the set
$$\tilde{A} = \{x \,|\, x \in \Omega \text{ and } x \notin A\} \ .$$

The reason that these constructions are important is that it is typically the case that complicated events described in English can be broken down into simpler events using these constructions. For example, if A is the event that "it will snow tomorrow and it will rain the next day," B is the event that "it will snow tomorrow," and C is the event that "it will rain two days from now," then A is the intersection of the events B and C. Similarly, if D is the event that "it will snow tomorrow or it will rain the next day," then $D = B \cup C$. (Note that care must be taken here, because sometimes the word "or" in English means that exactly one of the two alternatives will occur. The meaning is usually clear from context. In this book, we will always use the word "or" in the inclusive sense, i.e., A or B means that at least one of the two events A, B is true.) The event $\tilde{B}$ is the event that "it will not snow tomorrow." Finally, if E is the event that "it will snow tomorrow but it will not rain the next day," then $E = B - C$.

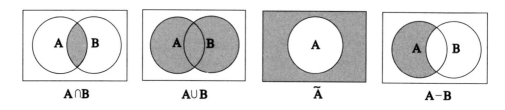

Figure 1.7: Basic set operations.

Properties

Theorem 1.1 The probabilities assigned to events by a distribution function on a sample space Ω satisfy the following properties:

1. $P(E) \geq 0$ for every $E \subset \Omega$.

2. $P(\Omega) = 1$.

3. If $E \subset F \subset \Omega$, then $P(E) \leq P(F)$.

4. If A and B are *disjoint* subsets of Ω, then $P(A \cup B) = P(A) + P(B)$.

5. $P(\tilde{A}) = 1 - P(A)$ for every $A \subset \Omega$.

Proof. For any event E the probability $P(E)$ is determined from the distribution m by
$$P(E) = \sum_{\omega \in E} m(\omega),$$
for every $E \subset \Omega$. Since the function m is nonnegative, it follows that $P(E)$ is also nonnegative. Thus, Property 1 is true.

Property 2 is proved by the equations
$$P(\Omega) = \sum_{\omega \in \Omega} m(\omega) = 1.$$

Suppose that $E \subset F \subset \Omega$. Then every element ω that belongs to E also belongs to F. Therefore,
$$\sum_{\omega \in E} m(\omega) \leq \sum_{\omega \in F} m(\omega),$$
since each term in the left-hand sum is in the right-hand sum, and all the terms in both sums are non-negative. This implies that
$$P(E) \leq P(F),$$
and Property 3 is proved.

1.2. DISCRETE PROBABILITY DISTRIBUTIONS

Suppose next that A and B are disjoint subsets of Ω. Then every element ω of $A \cup B$ lies either in A and not in B or in B and not in A. It follows that

$$\begin{aligned} P(A \cup B) &= \sum_{\omega \in A \cup B} m(\omega) = \sum_{\omega \in A} m(\omega) + \sum_{\omega \in B} m(\omega) \\ &= P(A) + P(B) , \end{aligned}$$

and Property 4 is proved.

Finally, to prove Property 5, consider the disjoint union

$$\Omega = A \cup \tilde{A} .$$

Since $P(\Omega) = 1$, the property of disjoint additivity (Property 4) implies that

$$1 = P(A) + P(\tilde{A}) ,$$

whence $P(\tilde{A}) = 1 - P(A)$. □

It is important to realize that Property 4 in Theorem 1.1 can be extended to more than two sets. The general finite additivity property is given by the following theorem.

Theorem 1.2 If $A_1, \ldots, A_n$ are pairwise disjoint subsets of Ω (i.e., no two of the A_i's have an element in common), then

$$P(A_1 \cup \cdots \cup A_n) = \sum_{i=1}^{n} P(A_i) .$$

Proof. Let ω be any element in the union

$$A_1 \cup \cdots \cup A_n .$$

Then $m(\omega)$ occurs exactly once on each side of the equality in the statement of the theorem. □

We shall often use the following consequence of the above theorem.

Theorem 1.3 Let $A_1, \ldots, A_n$ be pairwise disjoint events with $\Omega = A_1 \cup \cdots \cup A_n$, and let E be any event. Then

$$P(E) = \sum_{i=1}^{n} P(E \cap A_i) .$$

Proof. The sets $E \cap A_1, \ldots, E \cap A_n$ are pairwise disjoint, and their union is the set E. The result now follows from Theorem 1.2. □

Corollary 1.1 For any two events A and B,

$$P(A) = P(A \cap B) + P(A \cap \tilde{B}) \ .$$

$\square$

Property 4 can be generalized in another way. Suppose that A and B are subsets of Ω which are not necessarily disjoint. Then:

Theorem 1.4 If A and B are subsets of Ω, then

$$P(A \cup B) = P(A) + P(B) - P(A \cap B) \ . \tag{1.1}$$

Proof. The left side of Equation 1.1 is the sum of $m(\omega)$ for ω in either A or B. We must show that the right side of Equation 1.1 also adds $m(\omega)$ for ω in A or B. If ω is in exactly one of the two sets, then it is counted in only one of the three terms on the right side of Equation 1.1. If it is in both A and B, it is added twice from the calculations of $P(A)$ and $P(B)$ and subtracted once for $P(A \cap B)$. Thus it is counted exactly once by the right side. Of course, if $A \cap B = \emptyset$, then Equation 1.1 reduces to Property 4. (Equation 1.1 can also be generalized; see Theorem 3.8.) $\square$

Tree Diagrams

Example 1.10 Let us illustrate the properties of probabilities of events in terms of three tosses of a coin. When we have an experiment which takes place in stages such as this, we often find it convenient to represent the outcomes by a *tree diagram* as shown in Figure 1.8.

A *path* through the tree corresponds to a possible outcome of the experiment. For the case of three tosses of a coin, we have eight paths $\omega_1, \omega_2, \ldots, \omega_8$ and, assuming each outcome to be equally likely, we assign equal weight, $1/8$, to each path. Let E be the event "at least one head turns up." Then $\tilde{E}$ is the event "no heads turn up." This event occurs for only one outcome, namely, $\omega_8 = \text{TTT}$. Thus, $\tilde{E} = \{\text{TTT}\}$ and we have

$$P(\tilde{E}) = P(\{\text{TTT}\}) = m(\text{TTT}) = \frac{1}{8} \ .$$

By Property 5 of Theorem 1.1,

$$P(E) = 1 - P(\tilde{E}) = 1 - \frac{1}{8} = \frac{7}{8} \ .$$

Note that we shall often find it is easier to compute the probability that an event does not happen rather than the probability that it does. We then use Property 5 to obtain the desired probability.

1.2. DISCRETE PROBABILITY DISTRIBUTIONS

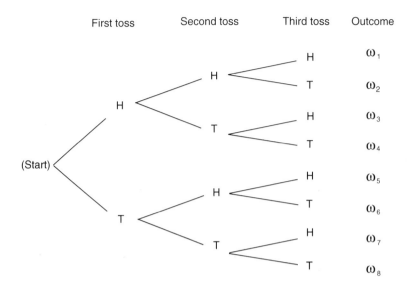

Figure 1.8: Tree diagram for three tosses of a coin.

Let A be the event "the first outcome is a head," and B the event "the second outcome is a tail." By looking at the paths in Figure 1.8, we see that

$$P(A) = P(B) = \frac{1}{2} \ .$$

Moreover, $A \cap B = \{\omega_3, \omega_4\}$, and so $P(A \cap B) = 1/4$. Using Theorem 1.4, we obtain

$$\begin{aligned} P(A \cup B) &= P(A) + P(B) - P(A \cap B) \\ &= \frac{1}{2} + \frac{1}{2} - \frac{1}{4} = \frac{3}{4} \ . \end{aligned}$$

Since $A \cup B$ is the 6-element set,

$$A \cup B = \{\text{HHH,HHT,HTH,HTT,TTH,TTT}\} \ ,$$

we see that we obtain the same result by direct enumeration. $\square$

In our coin tossing examples and in the die rolling example, we have assigned an equal probability to each possible outcome of the experiment. Corresponding to this method of assigning probabilities, we have the following definitions.

Uniform Distribution

Definition 1.3 The *uniform distribution* on a sample space Ω containing n elements is the function m defined by

$$m(\omega) = \frac{1}{n} \ ,$$

for every $\omega \in \Omega$. $\square$

It is important to realize that when an experiment is analyzed to describe its possible outcomes, there is no single correct choice of sample space. For the experiment of tossing a coin twice in Example 1.2, we selected the 4-element set $\Omega = \{\text{HH}, \text{HT}, \text{TH}, \text{TT}\}$ as a sample space and assigned the uniform distribution function. These choices are certainly intuitively natural. On the other hand, for some purposes it may be more useful to consider the 3-element sample space $\bar{\Omega} = \{0, 1, 2\}$ in which 0 is the outcome "no heads turn up," 1 is the outcome "exactly one head turns up," and 2 is the outcome "two heads turn up." The distribution function $\bar{m}$ on $\bar{\Omega}$ defined by the equations

$$\bar{m}(0) = \frac{1}{4}, \qquad \bar{m}(1) = \frac{1}{2}, \qquad \bar{m}(2) = \frac{1}{4}$$

is the one corresponding to the uniform probability density on the original sample space Ω. Notice that it is perfectly possible to choose a different distribution function. For example, we may consider the uniform distribution function on $\bar{\Omega}$, which is the function $\bar{q}$ defined by

$$\bar{q}(0) = \bar{q}(1) = \bar{q}(2) = \frac{1}{3} \ .$$

Although $\bar{q}$ is a perfectly good distribution function, it is not consistent with observed data on coin tossing.

Example 1.11 Consider the experiment that consists of rolling a pair of dice. We take as the sample space Ω the set of all ordered pairs (i, j) of integers with $1 \leq i \leq 6$ and $1 \leq j \leq 6$. Thus,

$$\Omega = \{\, (i, j) : 1 \leq i, j \leq 6 \,\} \ .$$

(There is at least one other "reasonable" choice for a sample space, namely the set of all unordered pairs of integers, each between 1 and 6. For a discussion of why we do not use this set, see Example 3.14.) To determine the size of Ω, we note that there are six choices for i, and for each choice of i there are six choices for j, leading to 36 different outcomes. Let us assume that the dice are not loaded. In mathematical terms, this means that we assume that each of the 36 outcomes is equally likely, or equivalently, that we adopt the uniform distribution function on Ω by setting

$$m((i,j)) = \frac{1}{36}, \qquad 1 \leq i, j \leq 6 \ .$$

What is the probability of getting a sum of 7 on the roll of two dice—or getting a sum of 11? The first event, denoted by E, is the subset

$$E = \{(1,6), (6,1), (2,5), (5,2), (3,4), (4,3)\} \ .$$

A sum of 11 is the subset F given by

$$F = \{(5,6), (6,5)\} \ .$$

Consequently,

$$P(E) = \sum_{\omega \in E} m(\omega) = 6 \cdot \tfrac{1}{36} = \tfrac{1}{6} \ ,$$

$$P(F) = \sum_{\omega \in F} m(\omega) = 2 \cdot \tfrac{1}{36} = \tfrac{1}{18} \ .$$

1.2. DISCRETE PROBABILITY DISTRIBUTIONS

What is the probability of getting neither *snakeeyes* (double ones) nor *boxcars* (double sixes)? The event of getting either one of these two outcomes is the set

$$E = \{(1,1), (6,6)\} \ .$$

Hence, the probability of obtaining neither is given by

$$P(\tilde{E}) = 1 - P(E) = 1 - \frac{2}{36} = \frac{17}{18} \ .$$

□

In the above coin tossing and the dice rolling experiments, we have assigned an equal probability to each outcome. That is, in each example, we have chosen the uniform distribution function. These are the natural choices provided the coin is a fair one and the dice are not loaded. However, the decision as to which distribution function to select to describe an experiment is *not* a part of the basic mathematical theory of probability. The latter begins only when the sample space and the distribution function have already been defined.

Determination of Probabilities

It is important to consider ways in which probability distributions are determined in practice. One way is by *symmetry*. For the case of the toss of a coin, we do not see any physical difference between the two sides of a coin that should affect the chance of one side or the other turning up. Similarly, with an ordinary die there is no essential difference between any two sides of the die, and so by symmetry we assign the same probability for any possible outcome. In general, considerations of symmetry often suggest the uniform distribution function. Care must be used here. We should not always assume that, just because we do not know any reason to suggest that one outcome is more likely than another, it is appropriate to assign equal probabilities. For example, consider the experiment of guessing the sex of a newborn child. It has been observed that the proportion of newborn children who are boys is about .513. Thus, it is more appropriate to assign a distribution function which assigns probability .513 to the outcome *boy* and probability .487 to the outcome *girl* than to assign probability 1/2 to each outcome. This is an example where we use statistical observations to determine probabilities. Note that these probabilities may change with new studies and may vary from country to country. Genetic engineering might even allow an individual to influence this probability for a particular case.

Odds

Statistical estimates for probabilities are fine if the experiment under consideration can be repeated a number of times under similar circumstances. However, assume that, at the beginning of a football season, you want to assign a probability to the event that Dartmouth will beat Harvard. You really do not have data that relates to this year's football team. However, you can determine your own personal probability by seeing what kind of a bet you would be willing to make. For example, suppose that you are willing to make a 1 dollar bet giving 2 to 1 odds that Dartmouth will win. Then you are willing to pay 2 dollars if Dartmouth loses in return for receiving 1 dollar if Dartmouth wins. This means that you think the appropriate probability for Dartmouth winning is 2/3.

Let us look more carefully at the relation between odds and probabilities. Suppose that we make a bet at r to 1 odds that an event E occurs. This means that we think that it is r times as likely that E will occur as that E will not occur. In general, r to s odds will be taken to mean the same thing as r/s to 1, i.e., the ratio between the two numbers is the only quantity of importance when stating odds.

Now if it is r times as likely that E will occur as that E will not occur, then the probability that E occurs must be $r/(r+1)$, since we have

$$P(E) = r\, P(\tilde{E})$$

and

$$P(E) + P(\tilde{E}) = 1\ .$$

In general, the statement that the odds are r to s in favor of an event E occurring is equivalent to the statement that

$$\begin{aligned} P(E) &= \frac{r/s}{(r/s)+1} \\ &= \frac{r}{r+s}\ . \end{aligned}$$

If we let $P(E) = p$, then the above equation can easily be solved for r/s in terms of p; we obtain $r/s = p/(1-p)$. We summarize the above discussion in the following definition.

Definition 1.4 If $P(E) = p$, the *odds* in favor of the event E occurring are $r : s$ (r to s) where $r/s = p/(1-p)$. If r and s are given, then p can be found by using the equation $p = r/(r+s)$. □

Example 1.12 (Example 1.9 continued) In Example 1.9 we assigned probability 1/5 to the event that candidate C wins the race. Thus the odds in favor of C winning are 1/5 : 4/5. These odds could equally well have been written as 1 : 4, 2 : 8, and so forth. A bet that C wins is fair if we receive 4 dollars if C wins and pay 1 dollar if C loses. □

Infinite Sample Spaces

If a sample space has an infinite number of points, then the way that a distribution function is defined depends upon whether or not the sample space is countable. A sample space is *countably infinite* if the elements can be counted, i.e., can be put in one-to-one correspondence with the positive integers, and *uncountably infinite*

1.2. DISCRETE PROBABILITY DISTRIBUTIONS

otherwise. Infinite sample spaces require new concepts in general (see Chapter 2), but countably infinite spaces do not. If

$$\Omega = \{\omega_1, \omega_2, \omega_3, \ldots\}$$

is a countably infinite sample space, then a distribution function is defined exactly as in Definition 1.2, except that the sum must now be a *convergent* infinite sum. Theorem 1.1 is still true, as are its extensions Theorems 1.2 and 1.4. One thing we cannot do on a countably infinite sample space that we could do on a finite sample space is to define a *uniform* distribution function as in Definition 1.3. You are asked in Exercise 20 to explain why this is not possible.

Example 1.13 A coin is tossed until the first time that a head turns up. Let the outcome of the experiment, ω, be the first time that a head turns up. Then the possible outcomes of our experiment are

$$\Omega = \{1, 2, 3, \ldots\} \ .$$

Note that even though the coin could come up tails every time we have not allowed for this possibility. We will explain why in a moment. The probability that heads comes up on the first toss is $1/2$. The probability that tails comes up on the first toss and heads on the second is $1/4$. The probability that we have two tails followed by a head is $1/8$, and so forth. This suggests assigning the distribution function $m(n) = 1/2^n$ for $n = 1, 2, 3, \ldots$. To see that this is a distribution function we must show that

$$\sum_\omega m(\omega) = \frac{1}{2} + \frac{1}{4} + \frac{1}{8} + \cdots = 1 \ .$$

That this is true follows from the formula for the sum of a geometric series,

$$1 + r + r^2 + r^3 + \cdots = \frac{1}{1-r} \ ,$$

or

$$r + r^2 + r^3 + r^4 + \cdots = \frac{r}{1-r} \ , \tag{1.2}$$

for $-1 < r < 1$.

Putting $r = 1/2$, we see that we have a probability of 1 that the coin eventually turns up heads. The possible outcome of tails every time has to be assigned probability 0, so we omit it from our sample space of possible outcomes.

Let E be the event that the first time a head turns up is after an even number of tosses. Then

$$E = \{2, 4, 6, 8, \ldots\} \ ,$$

and

$$P(E) = \frac{1}{4} + \frac{1}{16} + \frac{1}{64} + \cdots \ .$$

Putting $r = 1/4$ in Equation 1.2 see that

$$P(E) = \frac{1/4}{1 - 1/4} = \frac{1}{3} \ .$$

Thus the probability that a head turns up for the first time after an even number of tosses is $1/3$ and after an odd number of tosses is $2/3$. $\square$

Historical Remarks

An interesting question in the history of science is: Why was probability not developed until the sixteenth century? We know that in the sixteenth century problems in gambling and games of chance made people start to think about probability. But gambling and games of chance are almost as old as civilization itself. In ancient Egypt (at the time of the First Dynasty, ca. 3500 B.C.) a game now called "Hounds and Jackals" was played. In this game the movement of the hounds and jackals was based on the outcome of the roll of four-sided dice made out of animal bones called astragali. Six-sided dice made of a variety of materials date back to the sixteenth century B.C. Gambling was widespread in ancient Greece and Rome. Indeed, in the Roman Empire it was sometimes found necessary to invoke laws against gambling. Why, then, were probabilities not calculated until the sixteenth century?

Several explanations have been advanced for this late development. One is that the relevant mathematics was not developed and was not easy to develop. The ancient mathematical notation made numerical calculation complicated, and our familiar algebraic notation was not developed until the sixteenth century. However, as we shall see, many of the combinatorial ideas needed to calculate probabilities were discussed long before the sixteenth century. Since many of the chance events of those times had to do with lotteries relating to religious affairs, it has been suggested that there may have been religious barriers to the study of chance and gambling. Another suggestion is that a stronger incentive, such as the development of commerce, was necessary. However, none of these explanations seems completely satisfactory, and people still wonder why it took so long for probability to be studied seriously. An interesting discussion of this problem can be found in Hacking.[14]

The first person to calculate probabilities systematically was Gerolamo Cardano (1501–1576) in his book *Liber de Ludo Aleae*. This was translated from the Latin by Gould and appears in the book *Cardano: The Gambling Scholar* by Ore.[15] Ore provides a fascinating discussion of the life of this colorful scholar with accounts of his interests in many different fields, including medicine, astrology, and mathematics. You will also find there a detailed account of Cardano's famous battle with Tartaglia over the solution to the cubic equation.

In his book on probability Cardano dealt only with the special case that we have called the uniform distribution function. This restriction to equiprobable outcomes was to continue for a long time. In this case Cardano realized that the probability that an event occurs is the ratio of the number of favorable outcomes to the total number of outcomes.

Many of Cardano's examples dealt with rolling dice. Here he realized that the outcomes for two rolls should be taken to be the 36 ordered pairs (i, j) rather than the 21 unordered pairs. This is a subtle point that was still causing problems much later for other writers on probability. For example, in the eighteenth century the famous French mathematician d'Alembert, author of several works on probability, claimed that when a coin is tossed twice the number of heads that turn up would

---

[14] I. Hacking, *The Emergence of Probability* (Cambridge: Cambridge University Press, 1975).

[15] O. Ore, *Cardano: The Gambling Scholar* (Princeton: Princeton University Press, 1953).

1.2. DISCRETE PROBABILITY DISTRIBUTIONS

be 0, 1, or 2, and hence we should assign equal probabilities for these three possible outcomes.[16] Cardano chose the correct sample space for his dice problems and calculated the correct probabilities for a variety of events.

Cardano's mathematical work is interspersed with a lot of advice to the potential gambler in short paragraphs, entitled, for example: "Who Should Play and When," "Why Gambling Was Condemned by Aristotle," "Do Those Who Teach Also Play Well?" and so forth. In a paragraph entitled "The Fundamental Principle of Gambling," Cardano writes:

> The most fundamental principle of all in gambling is simply equal conditions, e.g., of opponents, of bystanders, of money, of situation, of the dice box, and of the die itself. To the extent to which you depart from that equality, if it is in your opponent's favor, you are a fool, and if in your own, you are unjust.[17]

Cardano did make mistakes, and if he realized it later he did not go back and change his error. For example, for an event that is favorable in three out of four cases, Cardano assigned the correct odds 3 : 1 that the event will occur. But then he assigned odds by squaring these numbers (i.e., 9 : 1) for the event to happen twice in a row. Later, by considering the case where the odds are 1 : 1, he realized that this cannot be correct and was led to the correct result that when f out of n outcomes are favorable, the odds for a favorable outcome twice in a row are $f^2 : n^2 - f^2$. Ore points out that this is equivalent to the realization that if the probability that an event happens in one experiment is p, the probability that it happens twice is p^2. Cardano proceeded to establish that for three successes the formula should be p^3 and for four successes p^4, making it clear that he understood that the probability is p^n for n successes in n independent repetitions of such an experiment. This will follow from the concept of independence that we introduce in Section 4.1.

Cardano's work was a remarkable first attempt at writing down the laws of probability, but it was not the spark that started a systematic study of the subject. This came from a famous series of letters between Pascal and Fermat. This correspondence was initiated by Pascal to consult Fermat about problems he had been given by Chevalier de Méré, a well-known writer, a prominent figure at the court of Louis XIV, and an ardent gambler.

The first problem de Méré posed was a dice problem. The story goes that he had been betting that at least one six would turn up in four rolls of a die and winning too often, so he then bet that a pair of sixes would turn up in 24 rolls of a pair of dice. The probability of a six with one die is 1/6 and, by the product law for independent experiments, the probability of two sixes when a pair of dice is thrown is $(1/6)(1/6) = 1/36$. Ore[18] claims that a gambling rule of the time suggested that, since four repetitions was favorable for the occurrence of an event with probability 1/6, for an event six times as unlikely, $6 \cdot 4 = 24$ repetitions would be sufficient for

---

[16] J. d'Alembert, "Croix ou Pile," in *L'Encyclopédie*, ed. Diderot, vol. 4 (Paris, 1754).
[17] O. Ore, op. cit., p. 189.
[18] O. Ore, "Pascal and the Invention of Probability Theory," *American Mathematics Monthly*, vol. 67 (1960), pp. 409–419.

a favorable bet. Pascal showed, by exact calculation, that 25 rolls are required for a favorable bet for a pair of sixes.

The second problem was a much harder one: it was an old problem and concerned the determination of a fair division of the stakes in a tournament when the series, for some reason, is interrupted before it is completed. This problem is now referred to as the problem of points. The problem had been a standard problem in mathematical texts; it appeared in Fra Luca Paccioli's book *summa de Arithmetica, Geometria, Proportioni et Proportionalità*, printed in Venice in 1494,[19] in the form:

> A team plays ball such that a total of 60 points are required to win the game, and each inning counts 10 points. The stakes are 10 ducats. By some incident they cannot finish the game and one side has 50 points and the other 20. One wants to know what share of the prize money belongs to each side. In this case I have found that opinions differ from one to another but all seem to me insufficient in their arguments, but I shall state the truth and give the correct way.

Reasonable solutions, such as dividing the stakes according to the ratio of games won by each player, had been proposed, but no correct solution had been found at the time of the Pascal-Fermat correspondence. The letters deal mainly with the attempts of Pascal and Fermat to solve this problem. Blaise Pascal (1623–1662) was a child prodigy, having published his treatise on conic sections at age sixteen, and having invented a calculating machine at age eighteen. At the time of the letters, his demonstration of the weight of the atmosphere had already established his position at the forefront of contemporary physicists. Pierre de Fermat (1601–1665) was a learned jurist in Toulouse, who studied mathematics in his spare time. He has been called by some the prince of amateurs and one of the greatest pure mathematicians of all times.

The letters, translated by Maxine Merrington, appear in Florence David's fascinating historical account of probability, *Games, Gods and Gambling*.[20] In a letter dated Wednesday, 29th July, 1654, Pascal writes to Fermat:

> Sir,
>
> Like you, I am equally impatient, and although I am again ill in bed, I cannot help telling you that yesterday evening I received from M. de Carcavi your letter on the problem of points, which I admire more than I can possibly say. I have not the leisure to write at length, but, in a word, you have solved the two problems of points, one with dice and the other with sets of games with perfect justness; I am entirely satisfied with it for I do not doubt that I was in the wrong, seeing the admirable agreement in which I find myself with you now...
>
> Your method is very sound and is the one which first came to my mind in this research; but because the labour of the combination is excessive, I have found a short cut and indeed another method which is much

---
[19]ibid., p. 414.
[20]F. N. David, *Games, Gods and Gambling* (London: G. Griffin, 1962), p. 230 ff.

1.2. DISCRETE PROBABILITY DISTRIBUTIONS

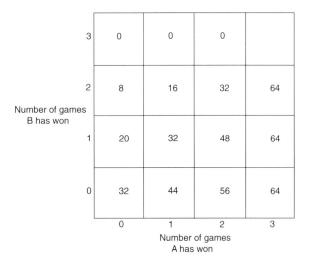

Figure 1.9: Pascal's table.

quicker and neater, which I would like to tell you here in a few words: for henceforth I would like to open my heart to you, if I may, as I am so overjoyed with our agreement. I see that truth is the same in Toulouse as in Paris.

Here, more or less, is what I do to show the fair value of each game, when two opponents play, for example, in three games and each person has staked 32 pistoles.

Let us say that the first man had won twice and the other once; now they play another game, in which the conditions are that, if the first wins, he takes all the stakes; that is 64 pistoles; if the other wins it, then they have each won two games, and therefore, if they wish to stop playing, they must each take back their own stake, that is, 32 pistoles each.

Then consider, Sir, if the first man wins, he gets 64 pistoles; if he loses he gets 32. Thus if they do not wish to risk this last game but wish to separate without playing it, the first man must say: 'I am certain to get 32 pistoles, even if I lost I still get them; but as for the other 32, perhaps I will get them, perhaps you will get them, the chances are equal. Let us then divide these 32 pistoles in half and give one half to me as well as my 32 which are mine for sure.' He will then have 48 pistoles and the other 16...

Pascal's argument produces the table illustrated in Figure 1.9 for the amount due player A at any quitting point.

Each entry in the table is the average of the numbers just above and to the right of the number. This fact, together with the known values when the tournament is completed, determines all the values in this table. If player A wins the first game, then he needs two games to win and B needs three games to win; and so, if the tounament is called off, A should receive 44 pistoles.

The letter in which Fermat presented his solution has been lost; but fortunately, Pascal describes Fermat's method in a letter dated Monday, 24th August, 1654. From Pascal's letter:[21]

> This is your procedure when there are two players: If two players, playing several games, find themselves in that position when the first man needs *two* games and second needs *three*, then to find the fair division of stakes, you say that one must know in how many games the play will be absolutely decided.
>
> It is easy to calculate that this will be in *four* games, from which you can conclude that it is necessary to see in how many ways four games can be arranged between two players, and one must see how many combinations would make the first man win and how many the second and to share out the stakes in this proportion. I would have found it difficult to understand this if I had not known it myself already; in fact you had explained it with this idea in mind.

Fermat realized that the number of ways that the game might be finished may not be equally likely. For example, if A needs two more games and B needs three to win, two possible ways that the tournament might go for A to win are WLW and LWLW. These two sequences do not have the same chance of occurring. To avoid this difficulty, Fermat extended the play, adding fictitious plays, so that all the ways that the games might go have the same length, namely four. He was shrewd enough to realize that this extension would not change the winner and that he now could simply count the number of sequences favorable to each player since he had made them all equally likely. If we list all possible ways that the extended game of four plays might go, we obtain the following 16 possible outcomes of the play:

WWWW	WLWW	LWWW	LLWW
WWWL	WLWL	LWWL	LLWL
WWLW	WLLW	LWLW	LLLW
WWLL	WLLL	LWLL	LLLL .

Player A wins in the cases where there are at least two wins (the 11 underlined cases), and B wins in the cases where there are at least three losses (the other 5 cases). Since A wins in 11 of the 16 possible cases Fermat argued that the probability that A wins is 11/16. If the stakes are 64 pistoles, A should receive 44 pistoles in agreement with Pascal's result. Pascal and Fermat developed more systematic methods for counting the number of favorable outcomes for problems like this, and this will be one of our central problems. Such counting methods fall under the subject of *combinatorics*, which is the topic of Chapter 3.

---

[21]ibid., p. 239ff.

1.2. DISCRETE PROBABILITY DISTRIBUTIONS

We see that these two mathematicians arrived at two very different ways to solve the problem of points. Pascal's method was to develop an algorithm and use it to calculate the fair division. This method is easy to implement on a computer and easy to generalize. Fermat's method, on the other hand, was to change the problem into an equivalent problem for which he could use counting or combinatorial methods. We will see in Chapter 3 that, in fact, Fermat used what has become known as Pascal's triangle! In our study of probability today we shall find that both the algorithmic approach and the combinatorial approach share equal billing, just as they did 300 years ago when probability got its start.

Exercises

1 Let $\Omega = \{a, b, c\}$ be a sample space. Let $m(a) = 1/2$, $m(b) = 1/3$, and $m(c) = 1/6$. Find the probabilities for all eight subsets of Ω.

2 Give a possible sample space Ω for each of the following experiments:

   (a) An election decides between two candidates A and B.

   (b) A two-sided coin is tossed.

   (c) A student is asked for the month of the year and the day of the week on which her birthday falls.

   (d) A student is chosen at random from a class of ten students.

   (e) You receive a grade in this course.

3 For which of the cases in Exercise 2 would it be reasonable to assign the uniform distribution function?

4 Describe in words the events specified by the following subsets of

$$\Omega = \{HHH, HHT, HTH, HTT, THH, THT, TTH, TTT\}$$

(see Example 1.6).

   (a) $E = \{\text{HHH,HHT,HTH,HTT}\}$.
   (b) $E = \{\text{HHH,TTT}\}$.
   (c) $E = \{\text{HHT,HTH,THH}\}$.
   (d) $E = \{\text{HHT,HTH,HTT,THH,THT,TTH,TTT}\}$.

5 What are the probabilities of the events described in Exercise 4?

6 A die is loaded in such a way that the probability of each face turning up is proportional to the number of dots on that face. (For example, a six is three times as probable as a two.) What is the probability of getting an even number in one throw?

7 Let A and B be events such that $P(A \cap B) = 1/4$, $P(\tilde{A}) = 1/3$, and $P(B) = 1/2$. What is $P(A \cup B)$?

8 A student must choose one of the subjects, art, geology, or psychology, as an elective. She is equally likely to choose art or psychology and twice as likely to choose geology. What are the respective probabilities that she chooses art, geology, and psychology?

9 A student must choose exactly two out of three electives: art, French, and mathematics. He chooses art with probability 5/8, French with probability 5/8, and art and French together with probability 1/4. What is the probability that he chooses mathematics? What is the probability that he chooses either art or French?

10 For a bill to come before the president of the United States, it must be passed by both the House of Representatives and the Senate. Assume that, of the bills presented to these two bodies, 60 percent pass the House, 80 percent pass the Senate, and 90 percent pass at least one of the two. Calculate the probability that the next bill presented to the two groups will come before the president.

11 What odds should a person give in favor of the following events?

(a) A card chosen at random from a 52-card deck is an ace.

(b) Two heads will turn up when a coin is tossed twice.

(c) Boxcars (two sixes) will turn up when two dice are rolled.

12 You offer 3 : 1 odds that your friend Smith will be elected mayor of your city. What probability are you assigning to the event that Smith wins?

13 In a horse race, the odds that Romance will win are listed as 2 : 3 and that Downhill will win are 1 : 2. What odds should be given for the event that either Romance or Downhill wins?

14 Let X be a random variable with distribution function $m_X(x)$ defined by

$$m_X(-1) = 1/5, \quad m_X(0) = 1/5, \quad m_X(1) = 2/5, \quad m_X(2) = 1/5 \ .$$

(a) Let Y be the random variable defined by the equation $Y = X + 3$. Find the distribution function $m_Y(y)$ of Y.

(b) Let Z be the random variable defined by the equation $Z = X^2$. Find the distribution function $m_Z(z)$ of Z.

*15 John and Mary are taking a mathematics course. The course has only three grades: A, B, and C. The probability that John gets a B is .3. The probability that Mary gets a B is .4. The probability that neither gets an A but at least one gets a B is .1. What is the probability that at least one gets a B but neither gets a C?

16 In a fierce battle, not less than 70 percent of the soldiers lost one eye, not less than 75 percent lost one ear, not less than 80 percent lost one hand, and not less than 85 percent lost one leg. What is the minimal possible percentage of those who simultaneously lost one ear, one eye, one hand, and one leg?[22]

*17 Assume that the probability of a "success" on a single experiment with n outcomes is $1/n$. Let m be the number of experiments necessary to make it a favorable bet that at least one success will occur (see Exercise 1.1.5).

(a) Show that the probability that, in m trials, there are no successes is $(1 - 1/n)^m$.

(b) (de Moivre) Show that if $m = n \log 2$ then
$$\lim_{n \to \infty} \left(1 - \frac{1}{n}\right)^m = \frac{1}{2} \, .$$

Hint:
$$\lim_{n \to \infty} \left(1 - \frac{1}{n}\right)^n = e^{-1} \, .$$

Hence for large n we should choose m to be about $n \log 2$.

(c) Would DeMoivre have been led to the correct answer for de Méré's two bets if he had used his approximation?

18 (a) For events $A_1, \ldots, A_n$, prove that
$$P(A_1 \cup \cdots \cup A_n) \leq P(A_1) + \cdots + P(A_n) \, .$$

(b) For events A and B, prove that
$$P(A \cap B) \geq P(A) + P(B) - 1.$$

19 If A, B, and C are any three events, show that
$$\begin{aligned} P(A \cup B \cup C) &= P(A) + P(B) + P(C) \\ &\quad - P(A \cap B) - P(B \cap C) - P(C \cap A) \\ &\quad + P(A \cap B \cap C) \, . \end{aligned}$$

20 Explain why it is not possible to define a uniform distribution function (see Definition 1.3) on a countably infinite sample space. *Hint*: Assume $m(\omega) = a$ for all ω, where $0 \leq a \leq 1$. Does $m(\omega)$ have all the properties of a distribution function?

21 In Example 1.13 find the probability that the coin turns up heads for the first time on the tenth, eleventh, or twelfth toss.

22 A die is rolled until the first time that a six turns up. We shall see that the probability that this occurs on the nth roll is $(5/6)^{n-1} \cdot (1/6)$. Using this fact, describe the appropriate infinite sample space and distribution function for the experiment of rolling a die until a six turns up for the first time. Verify that for your distribution function $\sum_\omega m(\omega) = 1$.

---

[22]See Knot X, in Lewis Carroll, *Mathematical Recreations*, vol. 2 (Dover, 1958).

23 Let Ω be the sample space

$$\Omega = \{0, 1, 2, \ldots\},$$

and define a distribution function by

$$m(j) = (1-r)^j r,$$

for some fixed r, $0 < r < 1$, and for $j = 0, 1, 2, \ldots$. Show that this is a distribution function for Ω.

24 Our calendar has a 400-year cycle. B. H. Brown noticed that the number of times the thirteenth of the month falls on each of the days of the week in the 4800 months of a cycle is as follows:

Sunday 687

Monday 685

Tuesday 685

Wednesday 687

Thursday 684

Friday 688

Saturday 684

From this he deduced that the thirteenth was more likely to fall on Friday than on any other day. Explain what he meant by this.

25 Tversky and Kahneman[23] asked a group of subjects to carry out the following task. They are told that:

> Linda is 31, single, outspoken, and very bright. She majored in philosophy in college. As a student, she was deeply concerned with racial discrimination and other social issues, and participated in anti-nuclear demonstrations.

The subjects are then asked to rank the likelihood of various alternatives, such as:

(1) Linda is active in the feminist movement.
(2) Linda is a bank teller.
(3) Linda is a bank teller and active in the feminist movement.

Tversky and Kahneman found that between 85 and 90 percent of the subjects rated alternative (1) most likely, but alternative (3) more likely than alternative (2). Is it? They call this phenomenon the *conjunction fallacy,* and note that it appears to be unaffected by prior training in probability or statistics. Explain why this is a fallacy. Can you give a possible explanation for the subjects' choices?

---

[23] K. McKean, "Decisions, Decisions," pp. 22–31.

1.2. DISCRETE PROBABILITY DISTRIBUTIONS

26 Two cards are drawn successively from a deck of 52 cards. Find the probability that the second card is higher in rank than the first card. *Hint*: Show that $1 = P(\text{higher}) + P(\text{lower}) + P(\text{same})$ and use the fact that $P(\text{higher}) = P(\text{lower})$.

27 A *life table* is a table that lists for a given number of births the estimated number of people who will live to a given age. In Appendix C we give a life table based upon 100,000 births for ages from 0 to 85, both for women and for men. Show how from this table you can estimate the probability $m(x)$ that a person born in 1981 would live to age x. Write a program to plot $m(x)$ both for men and for women, and comment on the differences that you see in the two cases.

*28 Here is an attempt to get around the fact that we cannot choose a "random integer."

   (a) What, intuitively, is the probability that a "randomly chosen" positive integer is a multiple of 3?

   (b) Let $P_3(N)$ be the probability that an integer, chosen at random between 1 and N, is a multiple of 3 (since the sample space is finite, this is a legitimate probability). Show that the limit

$$P_3 = \lim_{N \to \infty} P_3(N)$$

exists and equals 1/3. This formalizes the intuition in (a), and gives us a way to assign "probabilities" to certain events that are infinite subsets of the positive integers.

   (c) If A is any set of positive integers, let $A(N)$ mean the number of elements of A which are less than or equal to N. Then define the "probability" of A as

$$P(A) = \lim_{N \to \infty} A(N)/N \ ,$$

provided this limit exists. Show that this definition would assign probability 0 to any finite set and probability 1 to the set of all positive integers. Thus, the probability of the set of all integers is not the sum of the probabilities of the individual integers in this set. This means that the definition of probability given here is not a completely satisfactory definition.

   (d) Let A be the set of all positive integers with an odd number of digits. Show that $P(A)$ does not exist. This shows that under the above definition of probability, not all sets have probabilities.

29 (from Sholander[24]) In a standard clover-leaf interchange, there are four ramps for making right-hand turns, and inside these four ramps, there are four more ramps for making left-hand turns. Your car approaches the interchange from the south. A mechanism has been installed so that at each point where there exists a choice of directions, the car turns to the right with fixed probability r.

---
[24]M. Sholander, Problem #1034, *Mathematics Magazine,* vol. 52, no. 3 (May 1979), p. 183.

(a) If $r = 1/2$, what is your chance of emerging from the interchange going west?

(b) Find the value of r that maximizes your chance of a westward departure from the interchange.

30 (from Benkoski[25]) Consider a "pure" cloverleaf interchange in which there are no ramps for right-hand turns, but only the two intersecting straight highways with cloverleaves for left-hand turns. (Thus, to turn right in such an interchange, one must make three left-hand turns.) As in the preceding problem, your car approaches the interchange from the south. What is the value of r that maximizes your chances of an eastward departure from the interchange?

31 (from vos Savant[26]) A reader of Marilyn vos Savant's column wrote in with the following question:

> My dad heard this story on the radio. At Duke University, two students had received A's in chemistry all semester. But on the night before the final exam, they were partying in another state and didn't get back to Duke until it was over. Their excuse to the professor was that they had a flat tire, and they asked if they could take a make-up test. The professor agreed, wrote out a test and sent the two to separate rooms to take it. The first question (on one side of the paper) was worth 5 points, and they answered it easily. Then they flipped the paper over and found the second question, worth 95 points: 'Which tire was it?' What was the probability that both students would say the same thing? My dad and I think it's 1 in 16. Is that right?"

(a) Is the answer 1/16?

(b) The following question was asked of a class of students. "I was driving to school today, and one of my tires went flat. Which tire do you think it was?" The responses were as follows: right front, 58%, left front, 11%, right rear, 18%, left rear, 13%. Suppose that this distribution holds in the general population, and assume that the two test-takers are randomly chosen from the general population. What is the probability that they will give the same answer to the second question?

---

[25]S. Benkoski, Comment on Problem #1034, *Mathematics Magazine,* vol. 52, no. 3 (May 1979), pp. 183-184.

[26]M. vos Savant, *Parade Magazine,* 3 March 1996, p. 14.

Chapter 2

Continuous Probability Densities

2.1 Simulation of Continuous Probabilities

In this section we shall show how we can use computer simulations for experiments that have a whole continuum of possible outcomes.

Probabilities

Example 2.1 We begin by constructing a spinner, which consists of a circle of *unit circumference* and a pointer as shown in Figure 2.1. We pick a point on the circle and label it 0, and then label every other point on the circle with the distance, say x, from 0 to that point, measured counterclockwise. The experiment consists of spinning the pointer and recording the label of the point at the tip of the pointer. We let the random variable X denote the value of this outcome. The sample space is clearly the interval $[0, 1)$. We would like to construct a probability model in which each outcome is equally likely to occur.

If we proceed as we did in Chapter 1 for experiments with a finite number of possible outcomes, then we must assign the probability 0 to each outcome, since otherwise, the sum of the probabilities, over all of the possible outcomes, would not equal 1. (In fact, summing an uncountable number of real numbers is a tricky business; in particular, in order for such a sum to have any meaning, at most countably many of the summands can be different than 0.) However, if all of the assigned probabilities are 0, then the sum is 0, not 1, as it should be.

In the next section, we will show how to construct a probability model in this situation. At present, we will assume that such a model can be constructed. We will also assume that in this model, if E is an arc of the circle, and E is of length p, then the model will assign the probability p to E. This means that if the pointer is spun, the probability that it ends up pointing to a point in E equals p, which is certainly a reasonable thing to expect.

42   CHAPTER 2. CONTINUOUS PROBABILITY DENSITIES

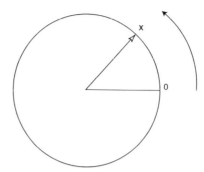

Figure 2.1: A spinner.

To simulate this experiment on a computer is an easy matter. Many computer software packages have a function which returns a random real number in the interval $[0, 1]$. Actually, the returned value is always a rational number, and the values are determined by an algorithm, so a sequence of such values is not truly random. Nevertheless, the sequences produced by such algorithms behave much like theoretically random sequences, so we can use such sequences in the simulation of experiments. On occasion, we will need to refer to such a function. We will call this function *rnd*. □

Monte Carlo Procedure and Areas

It is sometimes desirable to estimate quantities whose exact values are difficult or impossible to calculate exactly. In some of these cases, a procedure involving chance, called a *Monte Carlo procedure*, can be used to provide such an estimate.

Example 2.2 In this example we show how simulation can be used to estimate areas of plane figures. Suppose that we program our computer to provide a pair (x, y) or numbers, each chosen independently at random from the interval $[0, 1]$. Then we can interpret this pair (x, y) as the coordinates of a point chosen *at random* from the unit square. Events are subsets of the unit square. Our experience with Example 2.1 suggests that the point is equally likely to fall in subsets of equal area. Since the total area of the square is 1, the probability of the point falling in a specific subset E of the unit square should be equal to its area. Thus, we can estimate the area of any subset of the unit square by estimating the probability that a point chosen at random from this square falls in the subset.

We can use this method to estimate the area of the region E under the curve $y = x^2$ in the unit square (see Figure 2.2). We choose a large number of points (x, y) at random and record what fraction of them fall in the region $E = \{\, (x, y) : y \leq x^2 \,\}$.

The program **MonteCarlo** will carry out this experiment for us. Running this program for 10,000 experiments gives an estimate of .325 (see Figure 2.3).

From these experiments we would estimate the area to be about $1/3$. Of course,

2.1. SIMULATION OF CONTINUOUS PROBABILITIES

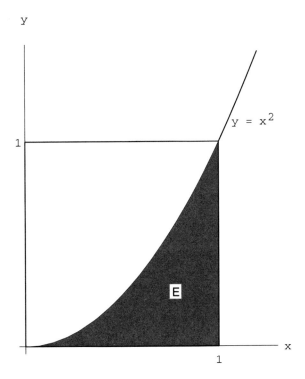

Figure 2.2: Area under $y = x^2$.

for this simple region we can find the exact area by calculus. In fact,

$$\text{Area of } E = \int_0^1 x^2 \, dx = \frac{1}{3} \ .$$

We have remarked in Chapter 1 that, when we simulate an experiment of this type n times to estimate a probability, we can expect the answer to be in error by at most $1/\sqrt{n}$ at least 95 percent of the time. For 10,000 experiments we can expect an accuracy of 0.01, and our simulation did achieve this accuracy.

This same argument works for any region E of the unit square. For example, suppose E is the circle with center $(1/2, 1/2)$ and radius $1/2$. Then the probability that our random point (x, y) lies inside the circle is equal to the area of the circle, that is,

$$P(E) = \pi \left(\frac{1}{2}\right)^2 = \frac{\pi}{4} \ .$$

If we did not know the value of π, we could estimate the value by performing this experiment a large number of times! □

The above example is not the only way of estimating the value of π by a chance experiment. Here is another way, discovered by Buffon.[1]

---

[1] G. L. Buffon, in "Essai d'Arithmétique Morale," *Oeuvres Complètes de Buffon avec Supplements,* tome iv, ed. Duménil (Paris, 1836).

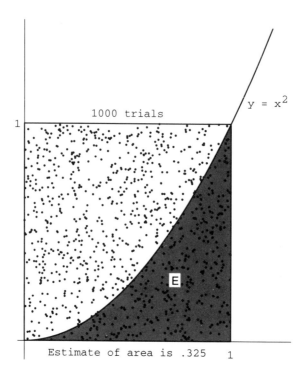

Figure 2.3: Computing the area by simulation.

Buffon's Needle

Example 2.3 Suppose that we take a card table and draw across the top surface a set of parallel lines a unit distance apart. We then drop a common needle of unit length at random on this surface and observe whether or not the needle lies across one of the lines. We can describe the possible outcomes of this experiment by coordinates as follows: Let d be the distance from the center of the needle to the nearest line. Next, let L be the line determined by the needle, and define θ as the acute angle that the line L makes with the set of parallel lines. (The reader should certainly be wary of this description of the sample space. We are attempting to coordinatize a set of line segments. To see why one must be careful in the choice of coordinates, see Example 2.6.) Using this description, we have $0 \le d \le 1/2$, and $0 \le \theta \le \pi/2$. Moreover, we see that the needle lies across the nearest line if and only if the hypotenuse of the triangle (see Figure 2.4) is less than half the length of the needle, that is,

$$\frac{d}{\sin \theta} < \frac{1}{2}.$$

Now we assume that when the needle drops, the pair (θ, d) is chosen at random from the rectangle $0 \le \theta \le \pi/2$, $0 \le d \le 1/2$. We observe whether the needle lies across the nearest line (i.e., whether $d \le (1/2) \sin \theta$). The probability of this event E is the fraction of the area of the rectangle which lies inside E (see Figure 2.5).

2.1. SIMULATION OF CONTINUOUS PROBABILITIES

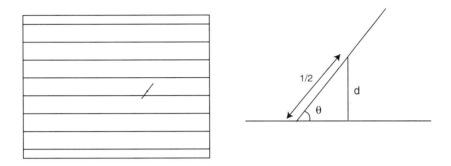

Figure 2.4: Buffon's experiment.

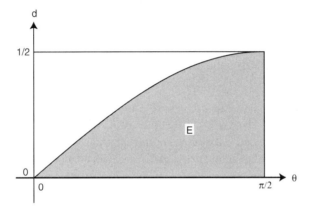

Figure 2.5: Set E of pairs (θ, d) with $d < \frac{1}{2}\sin\theta$.

Now the area of the rectangle is $\pi/4$, while the area of E is

$$\text{Area} = \int_0^{\pi/2} \frac{1}{2}\sin\theta\, d\theta = \frac{1}{2}\ .$$

Hence, we get

$$P(E) = \frac{1/2}{\pi/4} = \frac{2}{\pi}\ .$$

The program **BuffonsNeedle** simulates this experiment. In Figure 2.6, we show the position of every 100th needle in a run of the program in which 10,000 needles were "dropped." Our final estimate for π is 3.139. While this was within 0.003 of the true value for π we had no right to expect such accuracy. The reason for this is that our simulation estimates $P(E)$. While we can expect this estimate to be in error by at most 0.001, a small error in $P(E)$ gets magnified when we use this to compute $\pi = 2/P(E)$. Perlman and Wichura, in their article "Sharpening Buffon's

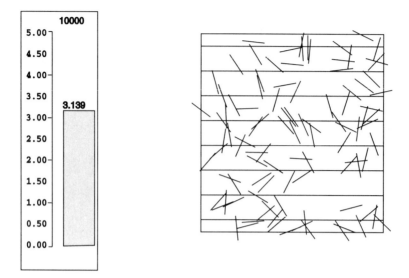

Figure 2.6: Simulation of Buffon's needle experiment.

Needle,"[2] show that we can expect to have an error of not more than $5/\sqrt{n}$ about 95 percent of the time. Here n is the number of needles dropped. Thus for 10,000 needles we should expect an error of no more than 0.05, and that was the case here. We see that a large number of experiments is necessary to get a decent estimate for π. □

In each of our examples so far, events of the same size are equally likely. Here is an example where they are not. We will see many other such examples later.

Example 2.4 Suppose that we choose two random real numbers in $[0, 1]$ and add them together. Let X be the sum. How is X distributed?

To help understand the answer to this question, we can use the program **Areabargraph**. This program produces a bar graph with the property that on each interval, the *area*, rather than the height, of the bar is equal to the fraction of outcomes that fell in the corresponding interval. We have carried out this experiment 1000 times; the data is shown in Figure 2.7. It appears that the function defined by

$$f(x) = \begin{cases} x, & \text{if } 0 \leq x \leq 1, \\ 2 - x, & \text{if } 1 < x \leq 2 \end{cases}$$

fits the data very well. (It is shown in the figure.) In the next section, we will see that this function is the "right" function. By this we mean that if a and b are any two real numbers between 0 and 2, with $a \leq b$, then we can use this function to calculate the probability that $a \leq X \leq b$. To understand how this calculation might be performed, we again consider Figure 2.7. Because of the way the bars were constructed, the sum of the areas of the bars corresponding to the interval

---

[2]M. D. Perlman and M. J. Wichura, "Sharpening Buffon's Needle," *The American Statistician*, vol. 29, no. 4 (1975), pp. 157–163.

2.1. SIMULATION OF CONTINUOUS PROBABILITIES   47

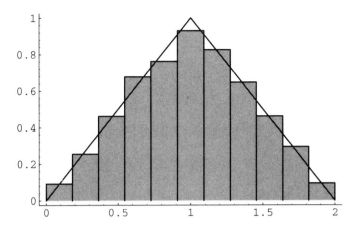

Figure 2.7: Sum of two random numbers.

$[a, b]$ approximates the probability that $a \leq X \leq b$. But the sum of the areas of these bars also approximates the integral

$$\int_a^b f(x)\, dx \ .$$

This suggests that for an experiment with a continuum of possible outcomes, if we find a function with the above property, then we will be able to use it to calculate probabilities. In the next section, we will show how to determine the function $f(x)$. □

Example 2.5 Suppose that we choose 100 random numbers in $[0, 1]$, and let X represent their sum. How is X distributed? We have carried out this experiment 10000 times; the results are shown in Figure 2.8. It is not so clear what function fits the bars in this case. It turns out that the type of function which does the job is called a *normal density* function. This type of function is sometimes referred to as a "bell-shaped" curve. It is among the most important functions in the subject of probability, and will be formally defined in Section 5.2 of Chapter 4.3. □

Our last example explores the fundamental question of how probabilities are assigned.

Bertrand's Paradox

Example 2.6 A chord of a circle is a line segment both of whose endpoints lie on the circle. Suppose that a chord is drawn *at random* in a unit circle. What is the probability that its length exceeds $\sqrt{3}$?

Our answer will depend on what we mean by *random*, which will depend, in turn, on what we choose for coordinates. The sample space Ω is the set of all possible chords in the circle. To find coordinates for these chords, we first introduce a

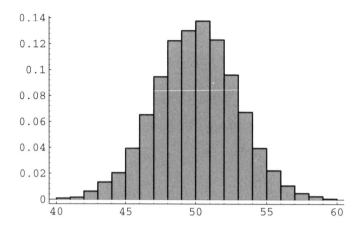

Figure 2.8: Sum of 100 random numbers.

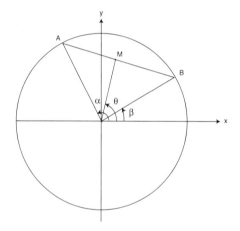

Figure 2.9: Random chord.

rectangular coordinate system with origin at the center of the circle (see Figure 2.9). We note that a chord of a circle is perpendicular to the radial line containing the midpoint of the chord. We can describe each chord by giving:

1. The rectangular coordinates (x, y) of the midpoint M, or

2. The polar coordinates (r, θ) of the midpoint M, or

3. The polar coordinates $(1, \alpha)$ and $(1, \beta)$ of the endpoints A and B.

In each case we shall interpret *at random* to mean: choose these coordinates at random.

We can easily estimate this probability by computer simulation. In programming this simulation, it is convenient to include certain simplifications, which we describe in turn:

1. To simulate this case, we choose values for x and y from $[-1, 1]$ at random. Then we check whether $x^2 + y^2 \leq 1$. If not, the point $M = (x, y)$ lies outside the circle and cannot be the midpoint of any chord, and we ignore it. Otherwise, M lies inside the circle and is the midpoint of a unique chord, whose length L is given by the formula:

$$L = 2\sqrt{1 - (x^2 + y^2)} \ .$$

2. To simulate this case, we take account of the fact that any rotation of the circle does not change the length of the chord, so we might as well assume in advance that the chord is horizontal. Then we choose r from $[-1, 1]$ at random, and compute the length of the resulting chord with midpoint $(r, \pi/2)$ by the formula:

$$L = 2\sqrt{1 - r^2} \ .$$

3. To simulate this case, we assume that one endpoint, say B, lies at $(1, 0)$ (i.e., that $\beta = 0$). Then we choose a value for α from $[0, 2\pi]$ at random and compute the length of the resulting chord, using the Law of Cosines, by the formula:

$$L = \sqrt{2 - 2\cos\alpha} \ .$$

The program **BertrandsParadox** carries out this simulation. Running this program produces the results shown in Figure 2.10. In the first circle in this figure, a smaller circle has been drawn. Those chords which intersect this smaller circle have length at least $\sqrt{3}$. In the second circle in the figure, the vertical line intersects all chords of length at least $\sqrt{3}$. In the third circle, again the vertical line intersects all chords of length at least $\sqrt{3}$.

In each case we run the experiment a large number of times and record the fraction of these lengths that exceed $\sqrt{3}$. We have printed the results of every 100th trial up to 10,000 trials.

It is interesting to observe that these fractions are *not* the same in the three cases; they depend on our choice of coordinates. This phenomenon was first observed by Bertrand, and is now known as *Bertrand's paradox*.[3] It is actually not a paradox at all; it is merely a reflection of the fact that different choices of coordinates will lead to different assignments of probabilities. Which assignment is "correct" depends on what application or interpretation of the model one has in mind.

One can imagine a real experiment involving throwing long straws at a circle drawn on a card table. A "correct" assignment of coordinates should not depend on where the circle lies on the card table, or where the card table sits in the room. Jaynes[4] has shown that the only assignment which meets this requirement is (2). In this sense, the assignment (2) is the natural, or "correct" one (see Exercise 11).

We can easily see in each case what the true probabilities are if we note that $\sqrt{3}$ is the length of the side of an inscribed equilateral triangle. Hence, a chord has

---

[3] J. Bertrand, *Calcul des Probabilités* (Paris: Gauthier-Villars, 1889).

[4] E. T. Jaynes, "The Well-Posed Problem," in *Papers on Probability, Statistics and Statistical Physics*, R. D. Rosencrantz, ed. (Dordrecht: D. Reidel, 1983), pp. 133–148.

50   CHAPTER 2. CONTINUOUS PROBABILITY DENSITIES

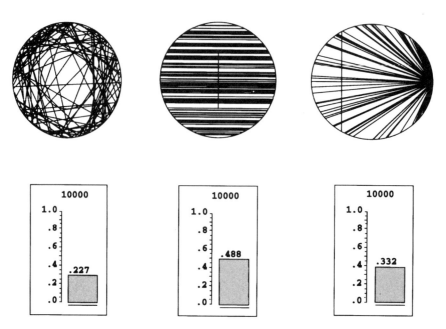

Figure 2.10: Bertrand's paradox.

length $L > \sqrt{3}$ if its midpoint has distance $d < 1/2$ from the origin (see Figure 2.9). The following calculations determine the probability that $L > \sqrt{3}$ in each of the three cases.

1. $L > \sqrt{3}$ if (x, y) lies inside a circle of radius $1/2$, which occurs with probability
$$p = \frac{\pi(1/2)^2}{\pi(1)^2} = \frac{1}{4} \ .$$

2. $L > \sqrt{3}$ if $|r| < 1/2$, which occurs with probability
$$\frac{1/2 - (-1/2)}{1 - (-1)} = \frac{1}{2} \ .$$

3. $L > \sqrt{3}$ if $2\pi/3 < \alpha < 4\pi/3$, which occurs with probability
$$\frac{4\pi/3 - 2\pi/3}{2\pi - 0} = \frac{1}{3} \ .$$

We see that our simulations agree quite well with these theoretical values. □

Historical Remarks

G. L. Buffon (1707–1788) was a natural scientist in the eighteenth century who applied probability to a number of his investigations. His work is found in his monumental 44-volume *Histoire Naturelle* and its supplements.[5] For example, he

---

[5]G. L. Buffon, *Histoire Naturelle, Generali et Particular avec le Descriptión du Cabinet du Roy,* 44 vols. (Paris: L'Imprimerie Royale, 1749–1803).

2.1. SIMULATION OF CONTINUOUS PROBABILITIES

Experimenter	Length of needle	Number of casts	Number of crossings	Estimate for π
Wolf, 1850	.8	5000	2532	3.1596
Smith, 1855	.6	3204	1218.5	3.1553
De Morgan, c.1860	1.0	600	382.5	3.137
Fox, 1864	.75	1030	489	3.1595
Lazzerini, 1901	.83	3408	1808	3.1415929
Reina, 1925	.5419	2520	869	3.1795

Table 2.1: Buffon needle experiments to estimate π.

presented a number of mortality tables and used them to compute, for each age group, the expected remaining lifetime. From his table he observed: the expected remaining lifetime of an infant of one year is 33 years, while that of a man of 21 years is also approximately 33 years. Thus, a father who is not yet 21 can hope to live longer than his one year old son, but if the father is 40, the odds are already 3 to 2 that his son will outlive him.[6]

Buffon wanted to show that not all probability calculations rely only on algebra, but that some rely on geometrical calculations. One such problem was his famous "needle problem" as discussed in this chapter.[7] In his original formulation, Buffon describes a game in which two gamblers drop a loaf of French bread on a wide-board floor and bet on whether or not the loaf falls across a crack in the floor. Buffon asked: what length L should the bread loaf be, relative to the width W of the floorboards, so that the game is fair. He found the correct answer ($L = (\pi/4)W$) using essentially the methods described in this chapter. He also considered the case of a checkerboard floor, but gave the wrong answer in this case. The correct answer was given later by Laplace.

The literature contains descriptions of a number of experiments that were actually carried out to estimate π by this method of dropping needles. N. T. Gridgeman[8] discusses the experiments shown in Table 2.1. (The halves for the number of crossing comes from a compromise when it could not be decided if a crossing had actually occurred.) He observes, as we have, that 10,000 casts could do no more than establish the first decimal place of π with reasonable confidence. Gridgeman points out that, although none of the experiments used even 10,000 casts, they are surprisingly good, and in some cases, too good. The fact that the number of casts is not always a round number would suggest that the authors might have resorted to clever stopping to get a good answer. Gridgeman comments that Lazzerini's estimate turned out to agree with a well-known approximation to π, $355/113 = 3.1415929$, discovered by the fifth-century Chinese mathematician, Tsu Ch'ungchih. Gridgeman says that he did not have Lazzerini's original report, and while waiting for it (knowing

---
[6]G. L. Buffon, "Essai d'Arithmétique Morale," p. 301.
[7]ibid., pp. 277–278.
[8]N. T. Gridgeman, "Geometric Probability and the Number π" Scripta Mathematika, vol. 25, no. 3, (1960), pp. 183–195.

only the needle crossed a line 1808 times in 3408 casts) deduced that the length of the needle must have been 5/6. He calculated this from Buffon's formula, assuming $\pi = 355/113$:

$$L = \frac{\pi P(E)}{2} = \frac{1}{2}\left(\frac{355}{113}\right)\left(\frac{1808}{3408}\right) = \frac{5}{6} = .8333 \ .$$

Even with careful planning one would have to be extremely lucky to be able to stop so cleverly.

The second author likes to trace his interest in probability theory to the Chicago World's Fair of 1933 where he observed a mechanical device dropping needles and displaying the ever-changing estimates for the value of π. (The first author likes to trace his interest in probability theory to the second author.)

Exercises

*1 In the spinner problem (see Example 2.1) divide the unit circumference into three arcs of length 1/2, 1/3, and 1/6. Write a program to simulate the spinner experiment 1000 times and print out what fraction of the outcomes fall in each of the three arcs. Now plot a bar graph whose bars have width 1/2, 1/3, and 1/6, and areas equal to the corresponding fractions as determined by your simulation. Show that the heights of the bars are all nearly the same.

2 Do the same as in Exercise 1, but divide the unit circumference into five arcs of length 1/3, 1/4, 1/5, 1/6, and 1/20.

3 Alter the program **MonteCarlo** to estimate the area of the circle of radius 1/2 with center at $(1/2, 1/2)$ inside the unit square by choosing 1000 points at random. Compare your results with the true value of $\pi/4$. Use your results to estimate the value of π. How accurate is your estimate?

4 Alter the program **MonteCarlo** to estimate the area under the graph of $y = \sin \pi x$ inside the unit square by choosing 10,000 points at random. Now calculate the true value of this area and use your results to estimate the value of π. How accurate is your estimate?

5 Alter the program **MonteCarlo** to estimate the area under the graph of $y = 1/(x+1)$ in the unit square in the same way as in Exercise 4. Calculate the true value of this area and use your simulation results to estimate the value of log 2. How accurate is your estimate?

6 To simulate the Buffon's needle problem we choose independently the distance d and the angle θ at random, with $0 \leq d \leq 1/2$ and $0 \leq \theta \leq \pi/2$, and check whether $d \leq (1/2)\sin\theta$. Doing this a large number of times, we estimate π as $2/a$, where a is the fraction of the times that $d \leq (1/2)\sin\theta$. Write a program to estimate π by this method. Run your program several times for each of 100, 1000, and 10,000 experiments. Does the accuracy of the experimental approximation for π improve as the number of experiments increases?

7 For Buffon's needle problem, Laplace[9] considered a grid with *horizontal* and *vertical* lines one unit apart. He showed that the probability that a needle of length $L \leq 1$ crosses at least one line is

$$p = \frac{4L - L^2}{\pi}.$$

To simulate this experiment we choose at random an angle θ between 0 and $\pi/2$ and independently two numbers d_1 and d_2 between 0 and $L/2$. (The two numbers represent the distance from the center of the needle to the nearest horizontal and vertical line.) The needle crosses a line if either $d_1 \leq (L/2)\sin\theta$ or $d_2 \leq (L/2)\cos\theta$. We do this a large number of times and estimate π as

$$\bar{\pi} = \frac{4L - L^2}{a},$$

where a is the proportion of times that the needle crosses at least one line. Write a program to estimate π by this method, run your program for 100, 1000, and 10,000 experiments, and compare your results with Buffon's method described in Exercise 6. (Take $L = 1$.)

8 A long needle of length L much bigger than 1 is dropped on a grid with horizontal and vertical lines one unit apart. We will see (in Exercise 6.3.28) that the average number a of lines crossed is approximately

$$a = \frac{4L}{\pi}.$$

To estimate π by simulation, pick an angle θ at random between 0 and $\pi/2$ and compute $L\sin\theta + L\cos\theta$. This may be used for the number of lines crossed. Repeat this many times and estimate π by

$$\bar{\pi} = \frac{4L}{a},$$

where a is the average number of lines crossed per experiment. Write a program to simulate this experiment and run your program for the number of experiments equal to 100, 1000, and 10,000. Compare your results with the methods of Laplace or Buffon for the same number of experiments. (Use $L = 100$.)

The following exercises involve experiments in which not all outcomes are equally likely. We shall consider such experiments in detail in the next section, but we invite you to explore a few simple cases here.

9 A large number of waiting time problems have an *exponential distribution* of outcomes. We shall see (in Section 5.2) that such outcomes are simulated by computing $(-1/\lambda)\log(\text{rnd})$, where $\lambda > 0$. For waiting times produced in this way, the average waiting time is $1/\lambda$. For example, the times spent waiting for

---
[9]P. S. Laplace, *Théorie Analytique des Probabilités* (Paris: Courcier, 1812).

a car to pass on a highway, or the times between emissions of particles from a radioactive source, are simulated by a sequence of random numbers, each of which is chosen by computing $(-1/\lambda)\log(\text{rnd})$, where $1/\lambda$ is the average time between cars or emissions. Write a program to simulate the times between cars when the average time between cars is 30 seconds. Have your program compute an area bar graph for these times by breaking the time interval from 0 to 120 into 24 subintervals. On the same pair of axes, plot the function $f(x) = (1/30)e^{-(1/30)x}$. Does the function fit the bar graph well?

10 In Exercise 9, the distribution came "out of a hat." In this problem, we will again consider an experiment whose outcomes are not equally likely. We will determine a function $f(x)$ which can be used to determine the probability of certain events. Let T be the right triangle in the plane with vertices at the points $(0,0)$, $(1,0)$, and $(0,1)$. The experiment consists of picking a point at random in the interior of T, and recording only the x-coordinate of the point. Thus, the sample space is the set $[0,1]$, but the outcomes do not seem to be equally likely. We can simulate this experiment by asking a computer to return two random real numbers in $[0,1]$, and recording the first of these two numbers if their sum is less than 1. Write this program and run it for 10,000 trials. Then make a bar graph of the result, breaking the interval $[0,1]$ into 10 intervals. Compare the bar graph with the function $f(x) = 2 - 2x$. Now show that there is a constant c such that the height of T at the x-coordinate value x is c times $f(x)$ for every x in $[0,1]$. Finally, show that

$$\int_0^1 f(x)\,dx = 1 \ .$$

How might one use the function $f(x)$ to determine the probability that the outcome is between .2 and .5?

11 Here is another way to pick a chord *at random* on the circle of unit radius. Imagine that we have a card table whose sides are of length 100. We place coordinate axes on the table in such a way that each side of the table is parallel to one of the axes, and so that the center of the table is the origin. We now place a circle of unit radius on the table so that the center of the circle is the origin. Now pick out a point (x_0, y_0) at random in the square, and an angle θ at random in the interval $(-\pi/2, \pi/2)$. Let $m = \tan\theta$. Then the equation of the line passing through (x_0, y_0) with slope m is

$$y = y_0 + m(x - x_0) \ ,$$

and the distance of this line from the center of the circle (i.e., the origin) is

$$d = \left|\frac{y_0 - mx_0}{\sqrt{m^2 + 1}}\right| \ .$$

We can use this distance formula to check whether the line intersects the circle (i.e., whether $d < 1$). If so, we consider the resulting chord a *random* chord.

This describes an experiment of dropping a long straw at random on a table on which a circle is drawn.

Write a program to simulate this experiment 10000 times and estimate the probability that the length of the chord is greater than $\sqrt{3}$. How does your estimate compare with the results of Example 2.6?

2.2 Continuous Density Functions

In the previous section we have seen how to simulate experiments with a whole continuum of possible outcomes and have gained some experience in thinking about such experiments. Now we turn to the general problem of assigning probabilities to the outcomes and events in such experiments. We shall restrict our attention here to those experiments whose sample space can be taken as a suitably chosen subset of the line, the plane, or some other Euclidean space. We begin with some simple examples.

Spinners

Example 2.7 The spinner experiment described in Example 2.1 has the interval $[0, 1)$ as the set of possible outcomes. We would like to construct a probability model in which each outcome is equally likely to occur. We saw that in such a model, it is necessary to assign the probability 0 to each outcome. This does not at all mean that the probability of *every* event must be zero. On the contrary, if we let the random variable X denote the outcome, then the probability

$$P(0 \leq X \leq 1)$$

that the head of the spinner comes to rest *somewhere* in the circle, should be equal to 1. Also, the probability that it comes to rest in the upper half of the circle should be the same as for the lower half, so that

$$P\left(0 \leq X < \frac{1}{2}\right) = P\left(\frac{1}{2} \leq X < 1\right) = \frac{1}{2} \ .$$

More generally, in our model, we would like the equation

$$P(c \leq X < d) = d - c$$

to be true for every choice of c and d.

If we let $E = [c, d]$, then we can write the above formula in the form

$$P(E) = \int_E f(x)\,dx \ ,$$

where $f(x)$ is the constant function with value 1. This should remind the reader of the corresponding formula in the discrete case for the probability of an event:

$$P(E) = \sum_{\omega \in E} m(\omega) \ .$$

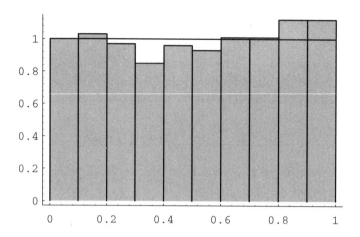

Figure 2.11: Spinner experiment.

The difference is that in the continuous case, the quantity being integrated, $f(x)$, is not the probability of the outcome x. (However, if one uses infinitesimals, one can consider $f(x)\,dx$ as the probability of the outcome x.)

In the continuous case, we will use the following convention. If the set of outcomes is a set of real numbers, then the individual outcomes will be referred to by small Roman letters such as x. If the set of outcomes is a subset of R^2, then the individual outcomes will be denoted by (x, y). In either case, it may be more convenient to refer to an individual outcome by using ω, as in Chapter 1.

Figure 2.11 shows the results of 1000 spins of the spinner. The function $f(x)$ is also shown in the figure. The reader will note that the area under $f(x)$ and above a given interval is approximately equal to the fraction of outcomes that fell in that interval. The function $f(x)$ is called the *density function* of the random variable X. The fact that the area under $f(x)$ and above an interval corresponds to a probability is the defining property of density functions. A precise definition of density functions will be given shortly. □

Darts

Example 2.8 A game of darts involves throwing a dart at a circular target of *unit radius*. Suppose we throw a dart once so that it hits the target, and we observe where it lands.

To describe the possible outcomes of this experiment, it is natural to take as our sample space the set Ω of all the points in the target. It is convenient to describe these points by their rectangular coordinates, relative to a coordinate system with origin at the center of the target, so that each pair (x, y) of coordinates with $x^2 + y^2 \leq 1$ describes a possible outcome of the experiment. Then $\Omega = \{(x, y) : x^2 + y^2 \leq 1\}$ is a subset of the Euclidean plane, and the event $E = \{(x, y) : y > 0\}$, for example, corresponds to the statement that the dart lands in the upper half of the target, and so forth. Unless there is reason to believe otherwise (and with experts at the

2.2. CONTINUOUS DENSITY FUNCTIONS

game there may well be!), it is natural to assume that the coordinates are chosen *at random*. (When doing this with a computer, each coordinate is chosen uniformly from the interval $[-1, 1]$. If the resulting point does not lie inside the unit circle, the point is not counted.) Then the arguments used in the preceding example show that the probability of any elementary event, consisting of a single outcome, must be zero, and suggest that the probability of the event that the dart lands in any subset E of the target should be determined by what fraction of the target area lies in E. Thus,

$$P(E) = \frac{\text{area of } E}{\text{area of target}} = \frac{\text{area of } E}{\pi} .$$

This can be written in the form

$$P(E) = \int_E f(x)\, dx ,$$

where $f(x)$ is the constant function with value $1/\pi$. In particular, if $E = \{\, (x, y) : x^2 + y^2 \le a^2 \,\}$ is the event that the dart lands within distance $a < 1$ of the center of the target, then

$$P(E) = \frac{\pi a^2}{\pi} = a^2 .$$

For example, the probability that the dart lies within a distance $1/2$ of the center is $1/4$. $\square$

Example 2.9 In the dart game considered above, suppose that, instead of observing where the dart lands, we observe how far it lands from the center of the target.

In this case, we take as our sample space the set Ω of all circles with centers at the center of the target. It is convenient to describe these circles by their radii, so that each circle is identified by its radius r, $0 \le r \le 1$. In this way, we may regard Ω as the subset $[0, 1]$ of the real line.

What probabilities should we assign to the events E of Ω? If

$$E = \{\, r : 0 \le r \le a \,\} ,$$

then E occurs if the dart lands within a distance a of the center, that is, within the circle of radius a, and we saw in the previous example that under our assumptions the probability of this event is given by

$$P([0, a]) = a^2 .$$

More generally, if

$$E = \{\, r : a \le r \le b \,\} ,$$

then by our basic assumptions,

$$\begin{aligned}
P(E) = P([a,b]) &= P([0,b]) - P([0,a]) \\
&= b^2 - a^2 \\
&= (b-a)(b+a) \\
&= 2(b-a)\frac{(b+a)}{2} .
\end{aligned}$$

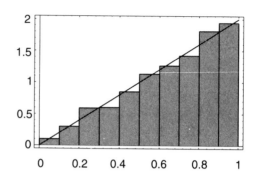

Figure 2.12: Distribution of dart distances in 400 throws.

Thus, $P(E) = 2$(length of E)(midpoint of E). Here we see that the probability assigned to the interval E depends not only on its length but also on its midpoint (i.e., not only on how long it is, but also on where it is). Roughly speaking, in this experiment, events of the form $E = [a, b]$ are more likely if they are near the rim of the target and less likely if they are near the center. (A common experience for beginners! The conclusion might well be different if the beginner is replaced by an expert.)

Again we can simulate this by computer. We divide the target area into ten concentric regions of equal thickness.

The computer program **Darts** throws n darts and records what fraction of the total falls in each of these concentric regions. The program **Areabargraph** then plots a bar graph with the *area* of the ith bar equal to the fraction of the total falling in the ith region. Running the program for 1000 darts resulted in the bar graph of Figure 2.12.

Note that here the heights of the bars are not all equal, but grow approximately linearly with r. In fact, the linear function $y = 2r$ appears to fit our bar graph quite well. This suggests that the probability that the dart falls within a distance a of the center should be given by the *area* under the graph of the function $y = 2r$ between 0 and a. This area is a^2, which agrees with the probability we have assigned above to this event. □

Sample Space Coordinates

These examples suggest that for continuous experiments of this sort we should assign probabilities for the outcomes to fall in a given interval by means of the area under a suitable function.

More generally, we suppose that suitable coordinates can be introduced into the sample space Ω, so that we can regard Ω as a subset of $\mathbf{R}^n$. We call such a sample space a *continuous sample space*. We let X be a random variable which represents the outcome of the experiment. Such a random variable is called a *continuous random variable*. We then define a density function for X as follows.

2.2. CONTINUOUS DENSITY FUNCTIONS

Density Functions of Continuous Random Variables

Definition 2.1 Let X be a continuous real-valued random variable. A *density function* for X is a real-valued function f which satisfies

$$P(a \leq X \leq b) = \int_a^b f(x)\,dx$$

for all $a, b \in \mathbf{R}$. □

We note that it is *not* the case that all continuous real-valued random variables possess density functions. However, in this book, we will only consider continuous random variables for which density functions exist.

In terms of the density $f(x)$, if E is a subset of $\mathbf{R}$, then

$$P(X \in E) = \int_E f(x)\,dx \ .$$

The notation here assumes that E is a subset of $\mathbf{R}$ for which $\int_E f(x)\,dx$ makes sense.

Example 2.10 (Example 2.7 continued) In the spinner experiment, we choose for our set of outcomes the interval $0 \leq x < 1$, and for our density function

$$f(x) = \begin{cases} 1, & \text{if } 0 \leq x < 1, \\ 0, & \text{otherwise.} \end{cases}$$

If E is the event that the head of the spinner falls in the upper half of the circle, then $E = \{\,x : 0 \leq x \leq 1/2\,\}$, and so

$$P(E) = \int_0^{1/2} 1\,dx = \frac{1}{2} \ .$$

More generally, if E is the event that the head falls in the interval $[a, b]$, then

$$P(E) = \int_a^b 1\,dx = b - a \ .$$

□

Example 2.11 (Example 2.8 continued) In the first dart game experiment, we choose for our sample space a disc of unit radius in the plane and for our density function the function

$$f(x,y) = \begin{cases} 1/\pi, & \text{if } x^2 + y^2 \leq 1, \\ 0, & \text{otherwise.} \end{cases}$$

The probability that the dart lands inside the subset E is then given by

$$\begin{aligned} P(E) &= \int\int_E \frac{1}{\pi}\,dx\,dy \\ &= \frac{1}{\pi} \cdot (\text{area of } E) \ . \end{aligned}$$

□

In these two examples, the density function is constant and does not depend on the particular outcome. It is often the case that experiments in which the coordinates are chosen *at random* can be described by *constant* density functions, and, as in Section 1.2, we call such density functions *uniform* or *equiprobable*. Not all experiments are of this type, however.

Example 2.12 (Example 2.9 continued) In the second dart game experiment, we choose for our sample space the unit interval on the real line and for our density the function

$$f(r) = \begin{cases} 2r, & \text{if } 0 < r < 1, \\ 0, & \text{otherwise}. \end{cases}$$

Then the probability that the dart lands at distance r, $a \leq r \leq b$, from the center of the target is given by

$$\begin{aligned} P([a,b]) &= \int_a^b 2r \, dr \\ &= b^2 - a^2 \ . \end{aligned}$$

Here again, since the density is small when r is near 0 and large when r is near 1, we see that in this experiment the dart is more likely to land near the rim of the target than near the center. In terms of the bar graph of Example 2.9, the heights of the bars approximate the density function, while the areas of the bars approximate the probabilities of the subintervals (see Figure 2.12). □

We see in this example that, unlike the case of discrete sample spaces, the value $f(x)$ of the density function for the outcome x is *not* the probability of x occurring (we have seen that this probability is always 0) and in general $f(x)$ is *not a probability at all*. In this example, if we take $\lambda = 2$ then $f(3/4) = 3/2$, which being bigger than 1, cannot be a probability.

Nevertheless, the density function f does contain all the probability information about the experiment, since the probabilities of all events can be derived from it. In particular, the probability that the outcome of the experiment falls in an interval $[a, b]$ is given by

$$P([a,b]) = \int_a^b f(x) \, dx \ ,$$

that is, by the *area* under the graph of the density function in the interval $[a, b]$. Thus, there is a close connection here between probabilities and areas. We have been guided by this close connection in making up our bar graphs; each bar is chosen so that its *area*, and not its height, represents the relative frequency of occurrence, and hence estimates the probability of the outcome falling in the associated interval.

In the language of the calculus, we can say that the probability of occurrence of an event of the form $[x, x + dx]$, where dx is small, is approximately given by

$$P([x, x + dx]) \approx f(x) dx \ ,$$

that is, by the area of the rectangle under the graph of f. Note that as $dx \to 0$, this probability $\to 0$, so that the probability $P(\{x\})$ of a single point is again 0, as in Example 2.7.

2.2. CONTINUOUS DENSITY FUNCTIONS

A glance at the graph of a density function tells us immediately which events of an experiment are more likely. Roughly speaking, we can say that where the density is large the events are more likely, and where it is small the events are less likely. In Example 2.4 the density function is largest at 1. Thus, given the two intervals $[0, a]$ and $[1, 1 + a]$, where a is a small positive real number, we see that X is more likely to take on a value in the second interval than in the first.

Cumulative Distribution Functions of Continuous Random Variables

We have seen that density functions are useful when considering continuous random variables. There is another kind of function, closely related to these density functions, which is also of great importance. These functions are called *cumulative distribution* functions.

Definition 2.2 Let X be a continuous real-valued random variable. Then the cumulative distribution function of X is defined by the equation

$$F_X(x) = P(X \leq x) \ .$$

$\square$

If X is a continuous real-valued random variable which possesses a density function, then it also has a cumulative distribution function, and the following theorem shows that the two functions are related in a very nice way.

Theorem 2.1 Let X be a continuous real-valued random variable with density function $f(x)$. Then the function defined by

$$F(x) = \int_{-\infty}^{x} f(t) \, dt$$

is the cumulative distribution function of X. Furthermore, we have

$$\frac{d}{dx} F(x) = f(x) \ .$$

Proof. By definition,
$$F(x) = P(X \leq x) \ .$$

Let $E = (-\infty, x]$. Then
$$P(X \leq x) = P(X \in E) \ ,$$

which equals
$$\int_{-\infty}^{x} f(t) \, dt \ .$$

Applying the Fundamental Theorem of Calculus to the first equation in the statement of the theorem yields the second statement. $\square$

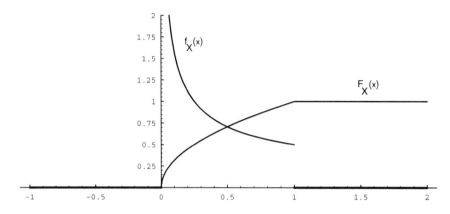

Figure 2.13: Distribution and density for $X = U^2$.

In many experiments, the density function of the relevant random variable is easy to write down. However, it is quite often the case that the cumulative distribution function is easier to obtain than the density function. (Of course, once we have the cumulative distribution function, the density function can easily be obtained by differentiation, as the above theorem shows.) We now give some examples which exhibit this phenomenon.

Example 2.13 A real number is chosen at random from $[0, 1]$ with uniform probability, and then this number is squared. Let X represent the result. What is the cumulative distribution function of X? What is the density of X?

We begin by letting U represent the chosen real number. Then $X = U^2$. If $0 \leq x \leq 1$, then we have

$$\begin{aligned} F_X(x) &= P(X \leq x) \\ &= P(U^2 \leq x) \\ &= P(U \leq \sqrt{x}) \\ &= \sqrt{x}\ . \end{aligned}$$

It is clear that X always takes on a value between 0 and 1, so the cumulative distribution function of X is given by

$$F_X(x) = \begin{cases} 0, & \text{if } x \leq 0, \\ \sqrt{x}, & \text{if } 0 \leq x \leq 1, \\ 1, & \text{if } x \geq 1. \end{cases}$$

From this we easily calculate that the density function of X is

$$f_X(x) = \begin{cases} 0, & \text{if } x \leq 0, \\ 1/(2\sqrt{x}), & \text{if } 0 \leq x \leq 1, \\ 0, & \text{if } x > 1. \end{cases}$$

Note that $F_X(x)$ is continuous, but $f_X(x)$ is not. (See Figure 2.13.) □

2.2. CONTINUOUS DENSITY FUNCTIONS

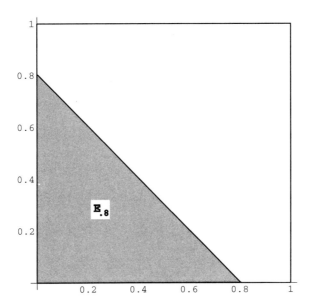

Figure 2.14: Calculation of distribution function for Example 2.14.

When referring to a continuous random variable X (say with a uniform density function), it is customary to say that "X is uniformly *distributed* on the interval $[a, b]$." It is also customary to refer to the cumulative distribution function of X as the distribution function of X. Thus, the word "distribution" is being used in several different ways in the subject of probability. (Recall that it also has a meaning when discussing discrete random variables.) When referring to the cumulative distribution function of a continuous random variable X, we will always use the word "cumulative" as a modifier, unless the use of another modifier, such as "normal" or "exponential," makes it clear. Since the phrase "uniformly densitied on the interval $[a, b]$" is not acceptable English, we will have to say "uniformly distributed" instead.

Example 2.14 In Example 2.4, we considered a random variable, defined to be the sum of two random real numbers chosen uniformly from $[0, 1]$. Let the random variables X and Y denote the two chosen real numbers. Define $Z = X + Y$. We will now derive expressions for the cumulative distribution function and the density function of Z.

Here we take for our sample space Ω the unit square in $\mathbf{R}^2$ with uniform density. A point $\omega \in \Omega$ then consists of a pair (x, y) of numbers chosen at random. Then $0 \leq Z \leq 2$. Let E_z denote the event that $Z \leq z$. In Figure 2.14, we show the set $E_{.8}$. The event E_z, for any z between 0 and 1, looks very similar to the shaded set in the figure. For $1 < z \leq 2$, the set E_z looks like the unit square with a triangle removed from the upper right-hand corner. We can now calculate the probability distribution F_Z of Z; it is given by

$$F_Z(z) = P(Z \leq z)$$
$$= \text{Area of } E_z$$

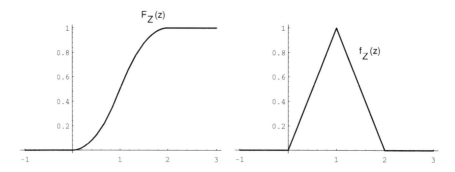

Figure 2.15: Distribution and density functions for Example 2.14.

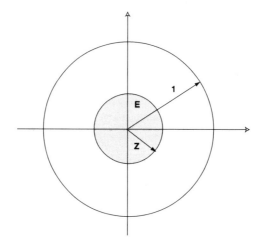

Figure 2.16: Calculation of F_z for Example 2.15.

$$= \begin{cases} 0, & \text{if } z < 0, \\ (1/2)z^2, & \text{if } 0 \leq z \leq 1, \\ 1 - (1/2)(2-z)^2, & \text{if } 1 \leq z \leq 2, \\ 1, & \text{if } 2 < z. \end{cases}$$

The density function is obtained by differentiating this function:

$$f_Z(z) = \begin{cases} 0, & \text{if } z < 0, \\ z, & \text{if } 0 \leq z \leq 1, \\ 2-z, & \text{if } 1 \leq z \leq 2, \\ 0, & \text{if } 2 < z. \end{cases}$$

The reader is referred to Figure 2.15 for the graphs of these functions. □

Example 2.15 In the dart game described in Example 2.8, what is the distribution of the distance of the dart from the center of the target? What is its density?

Here, as before, our sample space Ω is the unit disk in $\mathbf{R}^2$, with coordinates (X, Y). Let $Z = \sqrt{X^2 + Y^2}$ represent the distance from the center of the target. Let

2.2. CONTINUOUS DENSITY FUNCTIONS

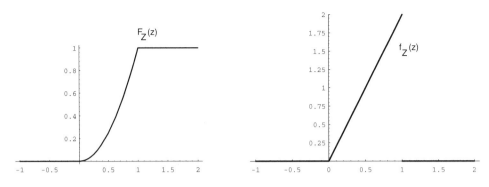

Figure 2.17: Distribution and density for $Z = \sqrt{X^2 + Y^2}$.

E be the event $\{Z \leq z\}$. Then the distribution function F_Z of Z (see Figure 2.16) is given by

$$\begin{aligned} F_Z(z) &= P(Z \leq z) \\ &= \frac{\text{Area of } E}{\text{Area of target}} . \end{aligned}$$

Thus, we easily compute that

$$F_Z(z) = \begin{cases} 0, & \text{if } z \leq 0, \\ z^2, & \text{if } 0 \leq z \leq 1, \\ 1, & \text{if } z > 1. \end{cases}$$

The density $f_Z(z)$ is given again by the derivative of $F_Z(z)$:

$$f_Z(z) = \begin{cases} 0, & \text{if } z \leq 0, \\ 2z, & \text{if } 0 \leq z \leq 1, \\ 0, & \text{if } z > 1. \end{cases}$$

The reader is referred to Figure 2.17 for the graphs of these functions.

We can verify this result by simulation, as follows: We choose values for X and Y at random from $[0, 1]$ with uniform distribution, calculate $Z = \sqrt{X^2 + Y^2}$, check whether $0 \leq Z \leq 1$, and present the results in a bar graph (see Figure 2.18). □

Example 2.16 Suppose Mr. and Mrs. Lockhorn agree to meet at the Hanover Inn between 5:00 and 6:00 P.M. on Tuesday. Suppose each arrives at a time between 5:00 and 6:00 chosen at random with uniform probability. What is the distribution function for the length of time that the first to arrive has to wait for the other? What is the density function?

Here again we can take the unit square to represent the sample space, and (X, Y) as the arrival times (after 5:00 P.M.) for the Lockhorns. Let $Z = |X - Y|$. Then we have $F_X(x) = x$ and $F_Y(y) = y$. Moreover (see Figure 2.19),

$$\begin{aligned} F_Z(z) &= P(Z \leq z) \\ &= P(|X - Y| \leq z) \\ &= \text{Area of } E . \end{aligned}$$

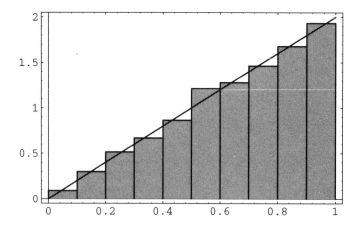

Figure 2.18: Simulation results for Example 2.15.

Thus, we have

$$F_Z(z) = \begin{cases} 0, & \text{if } z \leq 0, \\ 1-(1-z)^2, & \text{if } 0 \leq z \leq 1, \\ 1, & \text{if } z > 1. \end{cases}$$

The density $f_Z(z)$ is again obtained by differentiation:

$$f_Z(z) = \begin{cases} 0, & \text{if } z \leq 0, \\ 2(1-z), & \text{if } 0 \leq z \leq 1, \\ 0, & \text{if } z > 1. \end{cases}$$

$\square$

Example 2.17 There are many occasions where we observe a sequence of occurrences which occur at "random" times. For example, we might be observing emissions of a radioactive isotope, or cars passing a milepost on a highway, or light bulbs burning out. In such cases, we might define a random variable X to denote the time between successive occurrences. Clearly, X is a continuous random variable whose range consists of the non-negative real numbers. It is often the case that we can model X by using the *exponential density*. This density is given by the formula

$$f(t) = \begin{cases} \lambda e^{-\lambda t}, & \text{if } t \geq 0, \\ 0, & \text{if } t < 0. \end{cases}$$

The number λ is a non-negative real number, and represents the reciprocal of the average value of X. (This will be shown in Chapter 6.) Thus, if the average time between occurrences is 30 minutes, then $\lambda = 1/30$. A graph of this density function with $\lambda = 1/30$ is shown in Figure 2.20. One can see from the figure that even though the average value is 30, occasionally much larger values are taken on by X.

Suppose that we have bought a computer that contains a Warp 9 hard drive. The salesperson says that the average time between breakdowns of this type of hard drive is 30 months. It is often assumed that the length of time between breakdowns

2.2. CONTINUOUS DENSITY FUNCTIONS

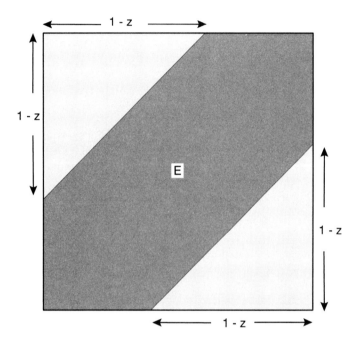

Figure 2.19: Calculation of F_Z.

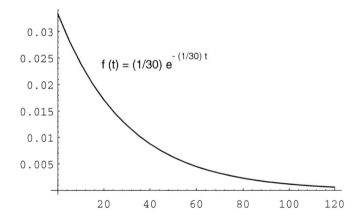

Figure 2.20: Exponential density with $\lambda = 1/30$.

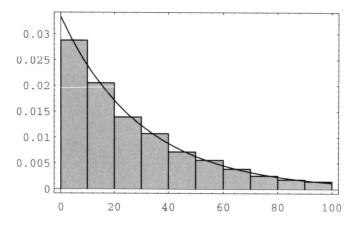

Figure 2.21: Residual lifespan of a hard drive.

is distributed according to the exponential density. We will assume that this model applies here, with $\lambda = 1/30$.

Now suppose that we have been operating our computer for 15 months. We assume that the original hard drive is still running. We ask how long we should expect the hard drive to continue to run. One could reasonably expect that the hard drive will run, on the average, another 15 months. (One might also guess that it will run more than 15 months, since the fact that it has already run for 15 months implies that we don't have a lemon.) The time which we have to wait is a new random variable, which we will call Y. Obviously, $Y = X - 15$. We can write a computer program to produce a sequence of simulated Y-values. To do this, we first produce a sequence of X's, and discard those values which are less than or equal to 15 (these values correspond to the cases where the hard drive has quit running before 15 months). To simulate a value of X, we compute the value of the expression

$$\left(-\frac{1}{\lambda}\right) \log(rnd) ,$$

where rnd represents a random real number between 0 and 1. (That this expression has the exponential density will be shown in Chapter 4.3.) Figure 2.21 shows an area bar graph of 10,000 simulated Y-values.

The average value of Y in this simulation is 29.74, which is closer to the original average life span of 30 months than to the value of 15 months which was guessed above. Also, the distribution of Y is seen to be close to the distribution of X. It is in fact the case that X and Y have the same distribution. This property is called the *memoryless property*, because the amount of time that we have to wait for an occurrence does not depend on how long we have already waited. The only continuous density function with this property is the exponential density. □

2.2. CONTINUOUS DENSITY FUNCTIONS

Assignment of Probabilities

A fundamental question in practice is: How shall we choose the probability density function in describing any given experiment? The answer depends to a great extent on the amount and kind of information available to us about the experiment. In some cases, we can see that the outcomes are equally likely. In some cases, we can see that the experiment resembles another already described by a known density. In some cases, we can run the experiment a large number of times and make a reasonable guess at the density on the basis of the observed distribution of outcomes, as we did in Chapter 1. In general, the problem of choosing the right density function for a given experiment is a central problem for the experimenter and is not always easy to solve (see Example 2.6). We shall not examine this question in detail here but instead shall assume that the right density is already known for each of the experiments under study.

The introduction of suitable coordinates to describe a continuous sample space, and a suitable density to describe its probabilities, is not always so obvious, as our final example shows.

Infinite Tree

Example 2.18 Consider an experiment in which a fair coin is tossed repeatedly, without stopping. We have seen in Example 1.6 that, for a coin tossed n times, the natural sample space is a binary tree with n stages. On this evidence we expect that for a coin tossed repeatedly, the natural sample space is a binary tree with an infinite number of stages, as indicated in Figure 2.22.

It is surprising to learn that, although the n-stage tree is obviously a finite sample space, the unlimited tree can be described as a continuous sample space. To see how this comes about, let us agree that a typical outcome of the unlimited coin tossing experiment can be described by a sequence of the form $\omega = \{H\ H\ T\ H\ T\ T\ H\ldots\}$. If we write 1 for H and 0 for T, then $\omega = \{1\ 1\ 0\ 1\ 0\ 0\ 1\ldots\}$. In this way, each outcome is described by a sequence of 0's and 1's.

Now suppose we think of this sequence of 0's and 1's as the binary expansion of some real number $x = .1101001\cdots$ lying between 0 and 1. (A *binary expansion* is like a decimal expansion but based on 2 instead of 10.) Then each outcome is described by a value of x, and in this way x becomes a coordinate for the sample space, taking on all real values between 0 and 1. (We note that it is possible for two different sequences to correspond to the same real number; for example, the sequences $\{T\ H\ H\ H\ H\ H\ldots\}$ and $\{H\ T\ T\ T\ T\ T\ldots\}$ both correspond to the real number 1/2. We will not concern ourselves with this apparent problem here.)

What probabilities should be assigned to the events of this sample space? Consider, for example, the event E consisting of all outcomes for which the first toss comes up heads and the second tails. Every such outcome has the form $.10****\cdots$, where $*$ can be either 0 or 1. Now if x is our real-valued coordinate, then the value of x for every such outcome must lie between $1/2 = .10000\cdots$ and $3/4 = .11000\cdots$, and moreover, every value of x between 1/2 and 3/4 has a binary expansion of the

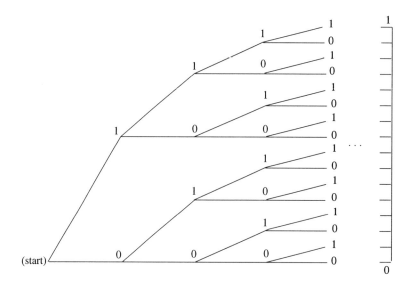

Figure 2.22: Tree for infinite number of tosses of a coin.

form $.10****\cdots$. This means that $\omega \in E$ if and only if $1/2 \leq x < 3/4$, and in this way we see that we can describe E by the interval $[1/2, 3/4)$. More generally, every event consisting of outcomes for which the results of the first n tosses are prescribed is described by a binary interval of the form $[k/2^n, (k+1)/2^n)$.

We have already seen in Section 1.2 that in the experiment involving n tosses, the probability of any one outcome must be exactly $1/2^n$. It follows that in the unlimited toss experiment, the probability of any event consisting of outcomes for which the results of the first n tosses are prescribed must also be $1/2^n$. But $1/2^n$ is exactly the length of the interval of x-values describing E! Thus we see that, just as with the spinner experiment, the probability of an event E is determined by what fraction of the unit interval lies in E.

Consider again the statement: The probability is $1/2$ that a fair coin will turn up heads when tossed. We have suggested that one interpretation of this statement is that if we toss the coin indefinitely the proportion of heads will approach $1/2$. That is, in our correspondence with binary sequences we expect to get a binary sequence with the proportion of 1's tending to $1/2$. The event E of binary sequences for which this is true is a proper subset of the set of all possible binary sequences. It does not contain, for example, the sequence $011011011\ldots$ (i.e., (011) repeated again and again). The event E is actually a very complicated subset of the binary sequences, but its probability can be determined as a limit of probabilities for events with a finite number of outcomes whose probabilities are given by finite tree measures. When the probability of E is computed in this way, its value is found to be 1. This remarkable result is known as the *Strong Law of Large Numbers* (or *Law of Averages*) and is one justification for our frequency concept of probability. We shall prove a weak form of this theorem in Chapter 8. □

Exercises

1. Suppose you choose *at random* a real number X from the interval $[2, 10]$.

    (a) Find the density function $f(x)$ and the probability of an event E for this experiment, where E is a subinterval $[a, b]$ of $[2, 10]$.

    (b) From (a), find the probability that $X > 5$, that $5 < X < 7$, and that $X^2 - 12X + 35 > 0$.

2. Suppose you choose a real number X from the interval $[2, 10]$ with a density function of the form
$$f(x) = Cx ,$$
where C is a constant.

    (a) Find C.

    (b) Find $P(E)$, where $E = [a, b]$ is a subinterval of $[2, 10]$.

    (c) Find $P(X > 5)$, $P(X < 7)$, and $P(X^2 - 12X + 35 > 0)$.

3. Same as Exercise 2, but suppose
$$f(x) = \frac{C}{x} .$$

4. Suppose you throw a dart at a circular target of radius 10 inches. Assuming that you hit the target and that the coordinates of the outcomes are chosen at random, find the probability that the dart falls

    (a) within 2 inches of the center.

    (b) within 2 inches of the rim.

    (c) within the first quadrant of the target.

    (d) within the first quadrant and within 2 inches of the rim.

5. Suppose you are watching a radioactive source that emits particles at a rate described by the exponential density
$$f(t) = \lambda e^{-\lambda t} ,$$
where $\lambda = 1$, so that the probability $P(0, T)$ that a particle will appear in the next T seconds is $P([0, T]) = \int_0^T \lambda e^{-\lambda t} \, dt$. Find the probability that a particle (not necessarily the first) will appear

    (a) within the next second.

    (b) within the next 3 seconds.

    (c) between 3 and 4 seconds from now.

    (d) after 4 seconds from now.

6 Assume that a new light bulb will burn out after t hours, where t is chosen from $[0, \infty)$ with an exponential density

$$f(t) = \lambda e^{-\lambda t} \ .$$

In this context, λ is often called the *failure rate* of the bulb.

(a) Assume that $\lambda = 0.01$, and find the probability that the bulb will *not* burn out before T hours. This probability is often called the *reliability* of the bulb.

(b) For what T is the reliability of the bulb $= 1/2$?

7 Choose a number B *at random* from the interval $[0, 1]$ with uniform density. Find the probability that

(a) $1/3 < B < 2/3$.

(b) $|B - 1/2| \leq 1/4$.

(c) $B < 1/4$ or $1 - B < 1/4$.

(d) $3B^2 < B$.

8 Choose independently two numbers B and C *at random* from the interval $[0, 1]$ with uniform density. Note that the point (B, C) is then chosen *at random* in the unit square. Find the probability that

(a) $B + C < 1/2$.

(b) $BC < 1/2$.

(c) $|B - C| < 1/2$.

(d) $\max\{B, C\} < 1/2$.

(e) $\min\{B, C\} < 1/2$.

(f) $B < 1/2$ and $1 - C < 1/2$.

(g) conditions (c) and (f) both hold.

(h) $B^2 + C^2 \leq 1/2$.

(i) $(B - 1/2)^2 + (C - 1/2)^2 < 1/4$.

9 Suppose that we have a sequence of occurrences. We assume that the time X between occurrences is exponentially distributed with $\lambda = 1/10$, so on the average, there is one occurrence every 10 minutes (see Example 2.17). You come upon this system at time 100, and wait until the next occurrence. Make a conjecture concerning how long, on the average, you will have to wait. Write a program to see if your conjecture is right.

10 As in Exercise 9, assume that we have a sequence of occurrences, but now assume that the time X between occurrences is uniformly distributed between 5 and 15. As before, you come upon this system at time 100, and wait until the next occurrence. Make a conjecture concerning how long, on the average, you will have to wait. Write a program to see if your conjecture is right.

2.2. CONTINUOUS DENSITY FUNCTIONS

11 For examples such as those in Exercises 9 and 10, it might seem that at least you should not have to wait on average *more* than 10 minutes if the average time between occurrences is 10 minutes. Alas, even this is not true. To see why, consider the following assumption about the times between occurrences. Assume that the time between occurrences is 3 minutes with probability .9 and 73 minutes with probability .1. Show by simulation that the average time between occurrences is 10 minutes, but that if you come upon this system at time 100, your average waiting time is more than 10 minutes.

12 Take a stick of unit length and break it into three pieces, choosing the break points at random. (The break points are assumed to be chosen simultaneously.) What is the probability that the three pieces can be used to form a triangle? *Hint*: The sum of the lengths of any two pieces must exceed the length of the third, so each piece must have length $< 1/2$. Now use Exercise 8(g).

13 Take a stick of unit length and break it into two pieces, choosing the break point at random. Now break the longer of the two pieces at a random point. What is the probability that the three pieces can be used to form a triangle?

14 Choose independently two numbers B and C *at random* from the interval $[-1, 1]$ with uniform distribution, and consider the quadratic equation

$$x^2 + Bx + C = 0 \ .$$

Find the probability that the roots of this equation

(a) are both real.

(b) are both positive.

Hints: (a) requires $0 \leq B^2 - 4C$, (b) requires $0 \leq B^2 - 4C$, $B \leq 0$, $0 \leq C$.

15 At the Tunbridge World's Fair, a coin toss game works as follows. Quarters are tossed onto a checkerboard. The management keeps all the quarters, but for each quarter landing entirely within one square of the checkerboard the management pays a dollar. Assume that the edge of each square is twice the diameter of a quarter, and that the outcomes are described by coordinates chosen *at random*. Is this a fair game?

16 Three points are chosen *at random* on a circle of *unit circumference*. What is the probability that the triangle defined by these points as vertices has three acute angles? *Hint*: One of the angles is obtuse if and only if all three points lie in the same semicircle. Take the circumference as the interval $[0, 1]$. Take one point at 0 and the others at B and C.

17 Write a program to choose a random number X in the interval $[2, 10]$ 1000 times and record what fraction of the outcomes satisfy $X > 5$, what fraction satisfy $5 < X < 7$, and what fraction satisfy $x^2 - 12x + 35 > 0$. How do these results compare with Exercise 1?

18 Write a program to choose a point (X, Y) at random in a square of side 20 inches, doing this 10,000 times, and recording what fraction of the outcomes fall within 19 inches of the center; of these, what fraction fall between 8 and 10 inches of the center; and, of these, what fraction fall within the first quadrant of the square. How do these results compare with those of Exercise 4?

19 Write a program to simulate the problem describe in Exercise 7 (see Exercise 17). How do the simulation results compare with the results of Exercise 7?

20 Write a program to simulate the problem described in Exercise 12.

21 Write a program to simulate the problem described in Exercise 16.

22 Write a program to carry out the following experiment. A coin is tossed 100 times and the number of heads that turn up is recorded. This experiment is then repeated 1000 times. Have your program plot a bar graph for the proportion of the 1000 experiments in which the number of heads is n, for each n in the interval $[35, 65]$. Does the bar graph look as though it can be fit with a normal curve?

23 Write a program that picks a random number between 0 and 1 and computes the negative of its logarithm. Repeat this process a large number of times and plot a bar graph to give the number of times that the outcome falls in each interval of length 0.1 in $[0, 10]$. On this bar graph plot a graph of the density $f(x) = e^{-x}$. How well does this density fit your graph?

Chapter 3

Combinatorics

3.1 Permutations

Many problems in probability theory require that we count the number of ways that a particular event can occur. For this, we study the topics of *permutations* and *combinations*. We consider permutations in this section and combinations in the next section.

Before discussing permutations, it is useful to introduce a general counting technique that will enable us to solve a variety of counting problems, including the problem of counting the number of possible permutations of n objects.

Counting Problems

Consider an experiment that takes place in several stages and is such that the number of outcomes m at the nth stage is independent of the outcomes of the previous stages. The number m may be different for different stages. We want to count the number of ways that the entire experiment can be carried out.

Example 3.1 You are eating at Émile's restaurant and the waiter informs you that you have (a) two choices for appetizers: soup or juice; (b) three for the main course: a meat, fish, or vegetable dish; and (c) two for dessert: ice cream or cake. How many possible choices do you have for your complete meal? We illustrate the possible meals by a tree diagram shown in Figure 3.1. Your menu is decided in three stages—at each stage the number of possible choices does not depend on what is chosen in the previous stages: two choices at the first stage, three at the second, and two at the third. From the tree diagram we see that the total number of choices is the product of the number of choices at each stage. In this examples we have $2 \cdot 3 \cdot 2 = 12$ possible menus. Our menu example is an example of the following general counting technique. □

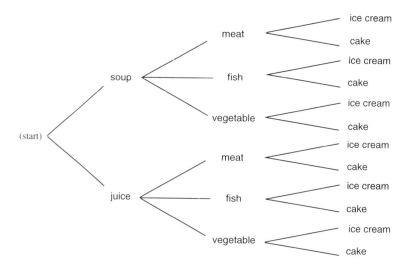

Figure 3.1: Tree for your menu.

A Counting Technique

A task is to be carried out in a sequence of r stages. There are n_1 ways to carry out the first stage; for each of these n_1 ways, there are n_2 ways to carry out the second stage; for each of these n_2 ways, there are n_3 ways to carry out the third stage, and so forth. Then the total number of ways in which the entire task can be accomplished is given by the product $N = n_1 \cdot n_2 \cdot \ldots \cdot n_r$.

Tree Diagrams

It will often be useful to use a tree diagram when studying probabilities of events relating to experiments that take place in stages and for which we are given the probabilities for the outcomes at each stage. For example, assume that the owner of Émile's restaurant has observed that 80 percent of his customers choose the soup for an appetizer and 20 percent choose juice. Of those who choose soup, 50 percent choose meat, 30 percent choose fish, and 20 percent choose the vegetable dish. Of those who choose juice for an appetizer, 30 percent choose meat, 40 percent choose fish, and 30 percent choose the vegetable dish. We can use this to estimate the probabilities at the first two stages as indicated on the tree diagram of Figure 3.2.

We choose for our sample space the set Ω of all possible paths $\omega = \omega_1, \omega_2, \ldots, \omega_6$ through the tree. How should we assign our probability distribution? For example, what probability should we assign to the customer choosing soup and then the meat? If $8/10$ of the customers choose soup and then $1/2$ of these choose meat, a proportion $8/10 \cdot 1/2 = 4/10$ of the customers choose soup and then meat. This suggests choosing our probability distribution for each path through the tree to be the *product* of the probabilities at each of the stages along the path. This results in the probability measure for the sample points ω indicated in Figure 3.2. (Note that $m(\omega_1) + \cdots + m(\omega_6) = 1$.) From this we see, for example, that the probability

3.1. PERMUTATIONS

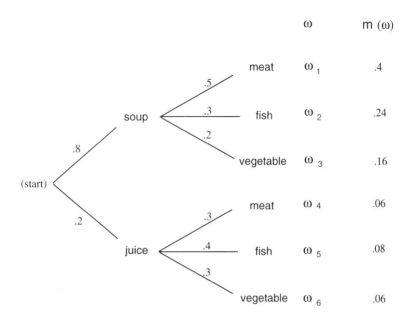

Figure 3.2: Two-stage probability assignment.

that a customer chooses meat is $m(\omega_1) + m(\omega_4) = .46$.

We shall say more about these tree measures when we discuss the concept of conditional probability in Chapter 4. We return now to more counting problems.

Example 3.2 We can show that there are at least two people in Columbus, Ohio, who have the same three initials. Assuming that each person has three initials, there are 26 possibilities for a person's first initial, 26 for the second, and 26 for the third. Therefore, there are $26^3 = 17{,}576$ possible sets of initials. This number is smaller than the number of people living in Columbus, Ohio; hence, there must be at least two people with the same three initials. □

We consider next the celebrated birthday problem—often used to show that naive intuition cannot always be trusted in probability.

Birthday Problem

Example 3.3 How many people do we need to have in a room to make it a favorable bet (probability of success greater than $1/2$) that two people in the room will have the same birthday?

Since there are 365 possible birthdays, it is tempting to guess that we would need about $1/2$ this number, or 183. You would surely win this bet. In fact, the number required for a favorable bet is only 23. To show this, we find the probability p_r that, in a room with r people, there is no duplication of birthdays; we will have a favorable bet if this probability is less than one half.

Number of people	Probability that all birthdays are different
20	.5885616
21	.5563117
22	.5243047
23	.4927028
24	.4616557
25	.4313003

Table 3.1: Birthday problem.

Assume that there are 365 possible birthdays for each person (we ignore leap years). Order the people from 1 to r. For a sample point ω, we choose a possible sequence of length r of birthdays each chosen as one of the 365 possible dates. There are 365 possibilities for the first element of the sequence, and for each of these choices there are 365 for the second, and so forth, making 365^r possible sequences of birthdays. We must find the number of these sequences that have no duplication of birthdays. For such a sequence, we can choose any of the 365 days for the first element, then any of the remaining 364 for the second, 363 for the third, and so forth, until we make r choices. For the rth choice, there will be $365 - r + 1$ possibilities. Hence, the total number of sequences with no duplications is

$$365 \cdot 364 \cdot 363 \cdot \ldots \cdot (365 - r + 1) \ .$$

Thus, assuming that each sequence is equally likely,

$$p_r = \frac{365 \cdot 364 \cdot \ldots \cdot (365 - r + 1)}{365^r} \ .$$

We denote the product

$$(n)(n-1)\cdots(n-r+1)$$

by $(n)_r$ (read "n down r," or "n lower r"). Thus,

$$p_r = \frac{(365)_r}{(365)^r} \ .$$

The program **Birthday** carries out this computation and prints the probabilities for $r = 20$ to 25. Running this program, we get the results shown in Table 3.1. As we asserted above, the probability for no duplication changes from greater than one half to less than one half as we move from 22 to 23 people. To see how unlikely it is that we would lose our bet for larger numbers of people, we have run the program again, printing out values from $r = 10$ to $r = 100$ in steps of 10. We see that in a room of 40 people the odds already heavily favor a duplication, and in a room of 100 the odds are overwhelmingly in favor of a duplication. We have assumed that birthdays are equally likely to fall on any particular day. Statistical evidence suggests that this is not true. However, it is intuitively clear (but not easy to prove) that this makes it even more likely to have a duplication with a group of 23 people. (See Exercise 19 to find out what happens on planets with more or fewer than 365 days per year.) □

Number of people	Probability that all birthdays are different
10	.8830518
20	.5885616
30	.2936838
40	.1087682
50	.0296264
60	.0058773
70	.0008404
80	.0000857
90	.0000062
100	.0000003

Table 3.2: Birthday problem.

We now turn to the topic of permutations.

Permutations

Definition 3.1 Let A be any finite set. A *permutation of A* is a one-to-one mapping of A onto itself. □

To specify a particular permutation we list the elements of A and, under them, show where each element is sent by the one-to-one mapping. For example, if $A = \{a, b, c\}$ a possible permutation σ would be

$$\sigma = \begin{pmatrix} a & b & c \\ b & c & a \end{pmatrix}.$$

By the permutation σ, a is sent to b, b is sent to c, and c is sent to a. The condition that the mapping be one-to-one means that no two elements of A are sent, by the mapping, into the same element of A.

We can put the elements of our set in some order and rename them 1, 2, ..., n. Then, a typical permutation of the set $A = \{a_1, a_2, a_3, a_4\}$ can be written in the form

$$\sigma = \begin{pmatrix} 1 & 2 & 3 & 4 \\ 2 & 1 & 4 & 3 \end{pmatrix},$$

indicating that a_1 went to a_2, a_2 to a_1, a_3 to a_4, and a_4 to a_3.

If we always choose the top row to be 1 2 3 4 then, to prescribe the permutation, we need only give the bottom row, with the understanding that this tells us where 1 goes, 2 goes, and so forth, under the mapping. When this is done, the permutation is often called a *rearrangement* of the n objects 1, 2, 3, ..., n. For example, all possible permutations, or rearrangements, of the numbers $A = \{1, 2, 3\}$ are:

$$123, \ 132, \ 213, \ 231, \ 312, \ 321 \ .$$

It is an easy matter to count the number of possible permutations of n objects. By our general counting principle, there are n ways to assign the first element, for

n	$n!$
0	1
1	1
2	2
3	6
4	24
5	120
6	720
7	5040
8	40320
9	362880
10	3628800

Table 3.3: Values of the factorial function.

each of these we have $n-1$ ways to assign the second object, $n-2$ for the third, and so forth. This proves the following theorem.

Theorem 3.1 The total number of permutations of a set A of n elements is given by $n \cdot (n-1) \cdot (n-2) \cdot \ldots \cdot 1$. □

It is sometimes helpful to consider orderings of subsets of a given set. This prompts the following definition.

Definition 3.2 Let A be an n-element set, and let k be an integer between 0 and n. Then a k-permutation of A is an ordered listing of a subset of A of size k. □

Using the same techniques as in the last theorem, the following result is easily proved.

Theorem 3.2 The total number of k-permutations of a set A of n elements is given by $n \cdot (n-1) \cdot (n-2) \cdot \ldots \cdot (n-k+1)$. □

Factorials

The number given in Theorem 3.1 is called *n factorial*, and is denoted by $n!$. The expression 0! is defined to be 1 to make certain formulas come out simpler. The first few values of this function are shown in Table 3.3. The reader will note that this function grows very rapidly.

The expression $n!$ will enter into many of our calculations, and we shall need to have some estimate of its magnitude when n is large. It is clearly not practical to make exact calculations in this case. We shall instead use a result called *Stirling's formula*. Before stating this formula we need a definition.

Definition 3.3 Let a_n and b_n be two sequences of numbers. We say that a_n is *asymptotically equal to* b_n, and write $a_n \sim b_n$, if

$$\lim_{n \to \infty} \frac{a_n}{b_n} = 1 \ .$$

3.1. PERMUTATIONS

n	$n!$	Approximation	Ratio
1	1	.922	1.084
2	2	1.919	1.042
3	6	5.836	1.028
4	24	23.506	1.021
5	120	118.019	1.016
6	720	710.078	1.013
7	5040	4980.396	1.011
8	40320	39902.395	1.010
9	362880	359536.873	1.009
10	3628800	3598696.619	1.008

Table 3.4: Stirling approximations to the factorial function.

$\square$

Example 3.4 If $a_n = n + \sqrt{n}$ and $b_n = n$ then, since $a_n/b_n = 1 + 1/\sqrt{n}$ and this ratio tends to 1 as n tends to infinity, we have $a_n \sim b_n$. $\square$

Theorem 3.3 (Stirling's Formula) The sequence $n!$ is asymptotically equal to

$$n^n e^{-n} \sqrt{2\pi n} \ .$$

$\square$

The proof of Stirling's formula may be found in most analysis texts. Let us verify this approximation by using the computer. The program **StirlingApproximations** prints $n!$, the Stirling approximation, and, finally, the ratio of these two numbers. Sample output of this program is shown in Table 3.4. Note that, while the ratio of the numbers is getting closer to 1, the difference between the exact value and the approximation is increasing, and indeed, this difference will tend to infinity as n tends to infinity, even though the ratio tends to 1. (This was also true in our Example 3.4 where $n + \sqrt{n} \sim n$, but the difference is $\sqrt{n}$.)

Generating Random Permutations

We now consider the question of generating a random permutation of the integers between 1 and n. Consider the following experiment. We start with a deck of n cards, labelled 1 through n. We choose a random card out of the deck, note its label, and put the card aside. We repeat this process until all n cards have been chosen. It is clear that each permutation of the integers from 1 to n can occur as a sequence of labels in this experiment, and that each sequence of labels is equally likely to occur. In our implementations of the computer algorithms, the above procedure is called **RandomPermutation**.

Number of fixed points	Fraction of permutations		
	n = 10	n = 20	n = 30
0	.362	.370	.358
1	.368	.396	.358
2	.202	.164	.192
3	.052	.060	.070
4	.012	.008	.020
5	.004	.002	.002
Average number of fixed points	.996	.948	1.042

Table 3.5: Fixed point distributions.

Fixed Points

There are many interesting problems that relate to properties of a permutation chosen at random from the set of all permutations of a given finite set. For example, since a permutation is a one-to-one mapping of the set onto itself, it is interesting to ask how many points are mapped onto themselves. We call such points *fixed points* of the mapping.

Let $p_k(n)$ be the probability that a random permutation of the set $\{1, 2, \ldots, n\}$ has exactly k fixed points. We will attempt to learn something about these probabilities using simulation. The program **FixedPoints** uses the procedure **RandomPermutation** to generate random permutations and count fixed points. The program prints the proportion of times that there are k fixed points as well as the average number of fixed points. The results of this program for 500 simulations for the cases $n = 10$, 20, and 30 are shown in Table 3.5. Notice the rather surprising fact that our estimates for the probabilities do not seem to depend very heavily on the number of elements in the permutation. For example, the probability that there are no fixed points, when $n = 10$, 20, or 30 is estimated to be between .35 and .37. We shall see later (see Example 3.12) that for $n \geq 10$ the exact probabilities $p_n(0)$ are, to six decimal place accuracy, equal to $1/e \approx .367879$. Thus, for all practical purposes, after $n = 10$ the probability that a random permutation of the set $\{1, 2, \ldots, n\}$ does not depend upon n. These simulations also suggest that the average number of fixed points is close to 1. It can be shown (see Example 6.8) that the average is exactly equal to 1 for all n.

More picturesque versions of the fixed-point problem are: You have arranged the books on your book shelf in alphabetical order by author and they get returned to your shelf at random; what is the probability that exactly k of the books end up in their correct position? (The library problem.) In a restaurant n hats are checked and they are hopelessly scrambled; what is the probability that no one gets his own hat back? (The hat check problem.) In the Historical Remarks at the end of this section, we give one method for solving the hat check problem exactly. Another method is given in Example 3.12.

3.1. PERMUTATIONS

Date	Snowfall in inches
1974	75
1975	88
1976	72
1977	110
1978	85
1979	30
1980	55
1981	86
1982	51
1983	64

Table 3.6: Snowfall in Hanover.

Year	1	2	3	4	5	6	7	8	9	10
Ranking	6	9	5	10	7	1	3	8	2	4

Table 3.7: Ranking of total snowfall.

Records

Here is another interesting probability problem that involves permutations. Estimates for the amount of measured snow in inches in Hanover, New Hampshire, in the ten years from 1974 to 1983 are shown in Table 3.6. Suppose we have started keeping records in 1974. Then our first year's snowfall could be considered a record snowfall starting from this year. A new record was established in 1975; the next record was established in 1977, and there were no new records established after this year. Thus, in this ten-year period, there were three records established: 1974, 1975, and 1977. The question that we ask is: How many records should we expect to be established in such a ten-year period? We can count the number of records in terms of a permutation as follows: We number the years from 1 to 10. The actual amounts of snowfall are not important but their relative sizes are. We can, therefore, change the numbers measuring snowfalls to numbers 1 to 10 by replacing the smallest number by 1, the next smallest by 2, and so forth. (We assume that there are no ties.) For our example, we obtain the data shown in Table 3.7.

This gives us a permutation of the numbers from 1 to 10 and, from this permutation, we can read off the records; they are in years 1, 2, and 4. Thus we can define records for a permutation as follows:

Definition 3.4 Let σ be a permutation of the set $\{1, 2, \ldots, n\}$. Then i is a *record* of σ if either $i = 1$ or $\sigma(j) < \sigma(i)$ for every $j = 1, \ldots, i - 1$. □

Now if we regard all rankings of snowfalls over an n-year period to be equally likely (and allow no ties), we can estimate the probability that there will be k records in n years as well as the average number of records by simulation.

We have written a program **Records** that counts the number of records in randomly chosen permutations. We have run this program for the cases $n = 10, 20, 30$.

For $n = 10$ the average number of records is 2.968, for 20 it is 3.656, and for 30 it is 3.960. We see now that the averages increase, but very slowly. We shall see later (see Example 6.11) that the average number is approximately $\log n$. Since $\log 10 = 2.3$, $\log 20 = 3$, and $\log 30 = 3.4$, this is consistent with the results of our simulations.

As remarked earlier, we shall be able to obtain formulas for exact results of certain problems of the above type. However, only minor changes in the problem make this impossible. The power of simulation is that minor changes in a problem do not make the simulation much more difficult. (See Exercise 20 for an interesting variation of the hat check problem.)

List of Permutations

Another method to solve problems that is not sensitive to small changes in the problem is to have the computer simply list all possible permutations and count the fraction that have the desired property. The program **AllPermutations** produces a list of all of the permutations of n. When we try running this program, we run into a limitation on the use of the computer. The number of permutations of n increases so rapidly that even to list all permutations of 20 objects is impractical.

Historical Remarks

Our basic counting principle stated that if you can do one thing in r ways and for each of these another thing in s ways, then you can do the pair in rs ways. This is such a self-evident result that you might expect that it occurred very early in mathematics. N. L. Biggs suggests that we might trace an example of this principle as follows: First, he relates a popular nursery rhyme dating back to at least 1730:

> As I was going to St. Ives,
> I met a man with seven wives,
> Each wife had seven sacks,
> Each sack had seven cats,
> Each cat had seven kits.
> Kits, cats, sacks and wives,
> How many were going to St. Ives?

(You need our principle only if you are not clever enough to realize that you are supposed to answer *one*, since only the narrator is going to St. Ives; the others are going in the other direction!)

He also gives a problem appearing on one of the oldest surviving mathematical manuscripts of about 1650 B.C., roughly translated as:

3.1. PERMUTATIONS

Houses	7
Cats	49
Mice	343
Wheat	2401
Hekat	16807
	19607

The following interpretation has been suggested: there are seven houses, each with seven cats; each cat kills seven mice; each mouse would have eaten seven heads of wheat, each of which would have produced seven hekat measures of grain. With this interpretation, the table answers the question of how many hekat measures were saved by the cats' actions. It is not clear why the writer of the table wanted to add the numbers together.[1]

One of the earliest uses of factorials occurred in Euclid's proof that there are infinitely many prime numbers. Euclid argued that there must be a prime number between n and $n!+1$ as follows: $n!$ and $n!+1$ cannot have common factors. Either $n!+1$ is prime or it has a proper factor. In the latter case, this factor cannot divide $n!$ and hence must be between n and $n!+1$. If this factor is not prime, then it has a factor that, by the same argument, must be bigger than n. In this way, we eventually reach a prime bigger than n, and this holds for all n.

The "$n!$" rule for the number of permutations seems to have occurred first in India. Examples have been found as early as 300 B.C., and by the eleventh century the general formula seems to have been well known in India and then in the Arab countries.

The *hat check problem* is found in an early probability book written by de Montmort and first printed in 1708.[2] It appears in the form of a game called *Treize*. In a simplified version of this game considered by de Montmort one turns over cards numbered 1 to 13, calling out 1, 2, ..., 13 as the cards are examined. De Montmort asked for the probability that no card that is turned up agrees with the number called out.

This probability is the same as the probability that a random permutation of 13 elements has no fixed point. De Montmort solved this problem by the use of a recursion relation as follows: let w_n be the number of permutations of n elements with no fixed point (such permutations are called *derangements*). Then $w_1 = 0$ and $w_2 = 1$.

Now assume that $n \geq 3$ and choose a derangement of the integers between 1 and n. Let k be the integer in the first position in this derangement. By the definition of derangement, we have $k \neq 1$. There are two possibilities of interest concerning the position of 1 in the derangement: either 1 is in the kth position or it is elsewhere. In the first case, the $n-2$ remaining integers can be positioned in w_{n-2} ways without resulting in any fixed points. In the second case, we consider the set of integers $\{1, 2, \ldots, k-1, k+1, \ldots, n\}$. The numbers in this set must occupy the positions $\{2, 3, \ldots, n\}$ so that none of the numbers other than 1 in this set are fixed, and

---

[1] N. L. Biggs, "The Roots of Combinatorics," *Historia Mathematica*, vol. 6 (1979), pp. 109–136.
[2] P. R. de Montmort, *Essay d'Analyse sur des Jeux de Hazard*, 2d ed. (Paris: Quillau, 1713).

also so that 1 is not in position k. The number of ways of achieving this kind of arrangement is just w_{n-1}. Since there are $n-1$ possible values of k, we see that

$$w_n = (n-1)w_{n-1} + (n-1)w_{n-2}$$

for $n \geq 3$. One might conjecture from this last equation that the sequence $\{w_n\}$ grows like the sequence $\{n!\}$.

In fact, it is easy to prove by induction that

$$w_n = nw_{n-1} + (-1)^n \ .$$

Then $p_i = w_i/i!$ satisfies

$$p_i - p_{i-1} = \frac{(-1)^i}{i!} \ .$$

If we sum from $i = 2$ to n, and use the fact that $p_1 = 0$, we obtain

$$p_n = \frac{1}{2!} - \frac{1}{3!} + \cdots + \frac{(-1)^n}{n!} \ .$$

This agrees with the first $n+1$ terms of the expansion for e^x for $x = -1$ and hence for large n is approximately $e^{-1} \approx .368$. David remarks that this was possibly the first use of the exponential function in probability.[3] We shall see another way to derive de Montmort's result in the next section, using a method known as the Inclusion-Exclusion method.

Recently, a related problem appeared in a column of Marilyn vos Savant.[4] Charles Price wrote to ask about his experience playing a certain form of solitaire, sometimes called "frustration solitaire." In this particular game, a deck of cards is shuffled, and then dealt out, one card at a time. As the cards are being dealt, the player counts from 1 to 13, and then starts again at 1. (Thus, each number is counted four times.) If a number that is being counted coincides with the rank of the card that is being turned up, then the player loses the game. Price found that he he rarely won and wondered how often he should win. Vos Savant remarked that the expected number of matches is 4 so it should be difficult to win the game.

Finding the chance of winning is a harder problem than the one that de Montmort solved because, when one goes through the entire deck, there are different patterns for the matches that might occur. For example matches may occur for two cards of the same rank, say two aces, or for two different ranks, say a two and a three.

A discussion of this problem can be found in Riordan.[5] In this book, it is shown that as $n \to \infty$, the probability of no matches tends to $1/e^4$.

The original game of Treize is more difficult to analyze than frustration solitaire. The game of Treize is played as follows. One person is chosen as dealer and the others are players. Each player, other than the dealer, puts up a stake. The dealer shuffles the cards and turns them up one at a time calling out, "Ace, two, three,...,

---

[3]F. N. David, *Games, Gods and Gambling* (London: Griffin, 1962), p. 146.
[4]M. vos Savant, Ask Marilyn, *Parade Magazine*, Boston Globe, 21 August 1994.
[5]J. Riordan, *An Introduction to Combinatorial Analysis*, (New York: John Wiley & Sons, 1958).

3.1. PERMUTATIONS

king," just as in frustration solitaire. If the dealer goes through the 13 cards without a match he pays the players an amount equal to their stake, and the deal passes to someone else. If there is a match the dealer collects the players' stakes; the players put up new stakes, and the dealer continues through the deck, calling out, "Ace, two, three, ..." If the dealer runs out of cards he reshuffles and continues the count where he left off. He continues until there is a run of 13 without a match and then a new dealer is chosen.

The question at this point is how much money can the dealer expect to win from each player. De Montmort found that if each player puts up a stake of 1, say, then the dealer will win approximately .801 from each player.

Peter Doyle calculated the exact amount that the dealer can expect to win. The answer is:

26516072156010218582227607912734182784642120482136091446715371962089931
52311343541724554334912870541440299239251607694113500080775917818512013
82176876653563173852874555859367254632009477403727395572807459384342747
87664965076063990538261189388143513547366316017004945507201764278828306
60117107953633142734382477922709835281753299035988581413688367655833113
24476153310720627474169719301806649152698704084383914217907906954976036
28528211590140316202120601549126920880824913325553882692055427830810368
57818861208758248800680978640438118582834877542560955550662878927123048
26997601700116233592793308297533642193505074540268925683193887821301442
70519791882/
33036929133582592220117220713156071114975101149831063364072138969878007
99647204708825303387525892236581323015628005621143427290625658974433971
65719454122908007086289841306087561302818991167357863623756067184986491
35353553622197448890223267101158801016285931351979294387223277033396967
79797069933475802423676949873661605184031477561560393380257070970711959
69641268242455013319879747054693517809383750593488858698672364846950539
88868628582609905586271001318150621134407056983214740221851567706672080
94586589378459432799868706334161812988630496327287254818458879353024498
00322425586446741048147720934108061350613503856973048971213063937040515
59533731591.

This is .803 to 3 decimal places. A description of the algorithm used to find this answer can be found on his Web page.[6] A discussion of this problem and other problems can be found in Doyle et al.[7]

The *birthday problem* does not seem to have a very old history. Problems of this type were first discussed by von Mises.[8] It was made popular in the 1950s by Feller's book.[9]

---
[6]P. Doyle, "Solution to Montmort's Probleme du Treize," http://math.ucsd.edu/~doyle/.
[7]P. Doyle, C. Grinstead, and J. Snell, "Frustration Solitaire," *UMAP Journal*, vol. 16, no. 2 (1995), pp. 137-145.
[8]R. von Mises, "Über Aufteilungs- und Besetzungs-Wahrscheinlichkeiten," *Revue de la Faculté des Sciences de l'Université d'Istanbul, N. S.* vol. 4 (1938-39), pp. 145-163.
[9]W. Feller, *Introduction to Probability Theory and Its Applications,* vol. 1, 3rd ed. (New York:

Stirling presented his formula

$$n! \sim \sqrt{2\pi n}\left(\frac{n}{e}\right)^n$$

in his work *Methodus Differentialis* published in 1730.[10] This approximation was used by de Moivre in establishing his celebrated central limit theorem that we will study in Chapter 9. De Moivre himself had independently established this approximation, but without identifying the constant π. Having established the approximation

$$\frac{2B}{\sqrt{n}}$$

for the central term of the binomial distribution, where the constant B was determined by an infinite series, de Moivre writes:

> ... my worthy and learned Friend, Mr. James Stirling, who had applied himself after me to that inquiry, found that the Quantity B did denote the Square-root of the Circumference of a Circle whose Radius is Unity, so that if that Circumference be called c the Ratio of the middle Term to the Sum of all Terms will be expressed by $2/\sqrt{nc}$....[11]

Exercises

1 Four people are to be arranged in a row to have their picture taken. In how many ways can this be done?

2 An automobile manufacturer has four colors available for automobile exteriors and three for interiors. How many different color combinations can he produce?

3 In a digital computer, a *bit* is one of the integers {0,1}, and a *word* is any string of 32 bits. How many different words are possible?

4 What is the probability that at least 2 of the presidents of the United States have died on the same day of the year? If you bet this has happened, would you win your bet?

5 There are three different routes connecting city A to city B. How many ways can a round trip be made from A to B and back? How many ways if it is desired to take a different route on the way back?

6 In arranging people around a circular table, we take into account their seats relative to each other, not the actual position of any one person. Show that n people can be arranged around a circular table in $(n-1)!$ ways.

John Wiley & Sons, 1968).
[10] J. Stirling, *Methodus Differentialis*, (London: Bowyer, 1730).
[11] A. de Moivre, *The Doctrine of Chances*, 3rd ed. (London: Millar, 1756).

3.1. PERMUTATIONS

7 Five people get on an elevator that stops at five floors. Assuming that each has an equal probability of going to any one floor, find the probability that they all get off at different floors.

8 A finite set Ω has n elements. Show that if we count the empty set and Ω as subsets, there are 2^n subsets of Ω.

9 A more refined inequality for approximating $n!$ is given by

$$\sqrt{2\pi n}\left(\frac{n}{e}\right)^n e^{1/(12n+1)} < n! < \sqrt{2\pi n}\left(\frac{n}{e}\right)^n e^{1/(12n)} \ .$$

Write a computer program to illustrate this inequality for $n = 1$ to 9.

10 A deck of ordinary cards is shuffled and 13 cards are dealt. What is the probability that the last card dealt is an ace?

11 There are n applicants for the director of computing. The applicants are interviewed independently by each member of the three-person search committee and ranked from 1 to n. A candidate will be hired if he or she is ranked first by at least two of the three interviewers. Find the probability that a candidate will be accepted if the members of the committee really have no ability at all to judge the candidates and just rank the candidates randomly. In particular, compare this probability for the case of three candidates and the case of ten candidates.

12 A symphony orchestra has in its repertoire 30 Haydn symphonies, 15 modern works, and 9 Beethoven symphonies. Its program always consists of a Haydn symphony followed by a modern work, and then a Beethoven symphony.

(a) How many different programs can it play?

(b) How many different programs are there if the three pieces can be played in any order?

(c) How many different three-piece programs are there if more than one piece from the same category can be played and they can be played in any order?

13 A certain state has license plates showing three numbers and three letters. How many different license plates are possible

(a) if the numbers must come before the letters?

(b) if there is no restriction on where the letters and numbers appear?

14 The door on the computer center has a lock which has five buttons numbered from 1 to 5. The combination of numbers that opens the lock is a sequence of five numbers and is reset every week.

(a) How many combinations are possible if every button must be used once?

(b) Assume that the lock can also have combinations that require you to push two buttons simultaneously and then the other three one at a time. How many more combinations does this permit?

15 A computing center has 3 processors that receive n jobs, with the jobs assigned to the processors purely at random so that all of the 3^n possible assignments are equally likely. Find the probability that exactly one processor has no jobs.

16 Prove that at least two people in Atlanta, Georgia, have the same initials, assuming no one has more than four initials.

17 Find a formula for the probability that among a set of n people, at least two have their birthdays in the same month of the year (assuming the months are equally likely for birthdays).

18 Consider the problem of finding the probability of more than one coincidence of birthdays in a group of n people. These include, for example, three people with the same birthday, or two pairs of people with the same birthday, or larger coincidences. Show how you could compute this probability, and write a computer program to carry out this computation. Use your program to find the smallest number of people for which it would be a favorable bet that there would be more than one coincidence of birthdays.

*19 Suppose that on planet Zorg a year has n days, and that the lifeforms there are equally likely to have hatched on any day of the year. We would like to estimate d, which is the minimum number of lifeforms needed so that the probability of at least two sharing a birthday exceeds $1/2$.

(a) In Example 3.3, it was shown that in a set of d lifeforms, the probability that no two life forms share a birthday is
$$\frac{(n)_d}{n^d},$$
where $(n)_d = (n)(n-1)\cdots(n-d+1)$. Thus, we would like to set this equal to $1/2$ and solve for d.

(b) Using Stirling's Formula, show that
$$\frac{(n)_d}{n^d} \sim \left(1 + \frac{d}{n-d}\right)^{n-d+1/2} e^{-d}.$$

(c) Now take the logarithm of the right-hand expression, and use the fact that for small values of x, we have
$$\log(1+x) \sim x - \frac{x^2}{2}.$$

(We are implicitly using the fact that d is of smaller order of magnitude than n. We will also use this fact in part (d).)

(d) Set the expression found in part (c) equal to $-\log(2)$, and solve for d as a function of n, thereby showing that

$$d \sim \sqrt{2(\log 2)\,n}\ .$$

Hint: If all three summands in the expression found in part (b) are used, one obtains a cubic equation in d. If the smallest of the three terms is thrown away, one obtains a quadratic equation in d.

(e) Use a computer to calculate the exact values of d for various values of n. Compare these values with the approximate values obtained by using the answer to part d).

20 At a mathematical conference, ten participants are randomly seated around a circular table for meals. Using simulation, estimate the probability that no two people sit next to each other at both lunch and dinner. Can you make an intelligent conjecture for the case of n participants when n is large?

21 Modify the program **AllPermutations** to count the number of permutations of n objects that have exactly j fixed points for $j = 0, 1, 2, \ldots, n$. Run your program for $n = 2$ to 6. Make a conjecture for the relation between the number that have 0 fixed points and the number that have exactly 1 fixed point. A proof of the correct conjecture can be found in Wilf.[12]

22 Mr. Wimply Dimple, one of London's most prestigious watch makers, has come to Sherlock Holmes in a panic, having discovered that someone has been producing and selling crude counterfeits of his best selling watch. The 16 counterfeits so far discovered bear stamped numbers, all of which fall between 1 and 56, and Dimple is anxious to know the extent of the forger's work. All present agree that it seems reasonable to assume that the counterfeits thus far produced bear consecutive numbers from 1 to whatever the total number is.

"Chin up, Dimple," opines Dr. Watson. "I shouldn't worry overly much if I were you; the Maximum Likelihood Principle, which estimates the total number as precisely that which gives the highest probability for the series of numbers found, suggests that we guess 56 itself as the total. Thus, your forgers are not a big operation, and we shall have them safely behind bars before your business suffers significantly."

"Stuff, nonsense, and bother your fancy principles, Watson," counters Holmes. "Anyone can see that, of course, there must be quite a few more than 56 watches—why the odds of our having discovered precisely the highest numbered watch made are laughably negligible. A much better guess would be *twice* 56."

(a) Show that Watson is correct that the Maximum Likelihood Principle gives 56.

---

[12]H. S. Wilf, "A Bijection in the Theory of Derangements," *Mathematics Magazine*, vol. 57, no. 1 (1984), pp. 37–40.

(b) Write a computer program to compare Holmes's and Watson's guessing strategies as follows: fix a total N and choose 16 integers randomly between 1 and N. Let m denote the largest of these. Then Watson's guess for N is m, while Holmes's is $2m$. See which of these is closer to N. Repeat this experiment (with N still fixed) a hundred or more times, and determine the proportion of times that each comes closer. Whose seems to be the better strategy?

23 Barbara Smith is interviewing candidates to be her secretary. As she interviews the candidates, she can determine the relative rank of the candidates but not the true rank. Thus, if there are six candidates and their true rank is 6, 1, 4, 2, 3, 5, (where 1 is best) then after she had interviewed the first three candidates she would rank them 3, 1, 2. As she interviews each candidate, she must either accept or reject the candidate. If she does not accept the candidate after the interview, the candidate is lost to her. She wants to decide on a strategy for deciding when to stop and accept a candidate that will maximize the probability of getting the best candidate. Assume that there are n candidates and they arrive in a random rank order.

(a) What is the probability that Barbara gets the best candidate if she interviews all of the candidates? What is it if she chooses the first candidate?

(b) Assume that Barbara decides to interview the first half of the candidates and then continue interviewing until getting a candidate better than any candidate seen so far. Show that she has a better than 25 percent chance of ending up with the best candidate.

24 For the task described in Exercise 23, it can be shown[13] that the best strategy is to pass over the first $k - 1$ candidates where k is the smallest integer for which
$$\frac{1}{k} + \frac{1}{k+1} + \cdots + \frac{1}{n-1} \leq 1 \ .$$
Using this strategy the probability of getting the best candidate is approximately $1/e = .368$. Write a program to simulate Barbara Smith's interviewing if she uses this optimal strategy, using $n = 10$, and see if you can verify that the probability of success is approximately $1/e$.

3.2 Combinations

Having mastered permutations, we now consider combinations. Let U be a set with n elements; we want to count the number of distinct subsets of the set U that have exactly j elements. The empty set and the set U are considered to be subsets of U. The empty set is usually denoted by ϕ.

---

[13] E. B. Dynkin and A. A. Yushkevich, *Markov Processes: Theorems and Problems*, trans. J. S. Wood (New York: Plenum, 1969).

3.2. COMBINATIONS

Example 3.5 Let $U = \{a, b, c\}$. The subsets of U are

$$\phi, \{a\}, \{b\}, \{c\}, \{a,b\}, \{a,c\}, \{b,c\}, \{a,b,c\} .$$

$\square$

Binomial Coefficients

The number of distinct subsets with j elements that can be chosen from a set with n elements is denoted by $\binom{n}{j}$, and is pronounced "n choose j." The number $\binom{n}{j}$ is called a *binomial coefficient*. This terminology comes from an application to algebra which will be discussed later in this section.

In the above example, there is one subset with no elements, three subsets with exactly 1 element, three subsets with exactly 2 elements, and one subset with exactly 3 elements. Thus, $\binom{3}{0} = 1$, $\binom{3}{1} = 3$, $\binom{3}{2} = 3$, and $\binom{3}{3} = 1$. Note that there are $2^3 = 8$ subsets in all. (We have already seen that a set with n elements has 2^n subsets; see Exercise 3.1.8.) It follows that

$$\binom{3}{0} + \binom{3}{1} + \binom{3}{2} + \binom{3}{3} = 2^3 = 8 ,$$

$$\binom{n}{0} = \binom{n}{n} = 1 .$$

Assume that $n > 0$. Then, since there is only one way to choose a set with no elements and only one way to choose a set with n elements, the remaining values of $\binom{n}{j}$ are determined by the following *recurrence relation*:

Theorem 3.4 For integers n and j, with $0 < j < n$, the binomial coefficients satisfy:

$$\binom{n}{j} = \binom{n-1}{j} + \binom{n-1}{j-1} . \tag{3.1}$$

Proof. We wish to choose a subset of j elements. Choose an element u of U. Assume first that we do not want u in the subset. Then we must choose the j elements from a set of $n-1$ elements; this can be done in $\binom{n-1}{j}$ ways. On the other hand, assume that we do want u in the subset. Then we must choose the other $j-1$ elements from the remaining $n-1$ elements of U; this can be done in $\binom{n-1}{j-1}$ ways. Since u is either in our subset or not, the number of ways that we can choose a subset of j elements is the sum of the number of subsets of j elements which have u as a member and the number which do not—this is what Equation 3.1 states. $\square$

The binomial coefficient $\binom{n}{j}$ is defined to be 0, if $j < 0$ or if $j > n$. With this definition, the restrictions on j in Theorem 3.4 are unnecessary.

	j = 0	1	2	3	4	5	6	7	8	9	10
n = 0	1										
1	1	1									
2	1	2	1								
3	1	3	3	1							
4	1	4	6	4	1						
5	1	5	10	10	5	1					
6	1	6	15	20	15	6	1				
7	1	7	21	35	35	21	7	1			
8	1	8	28	56	70	56	28	8	1		
9	1	9	36	84	126	126	84	36	9	1	
10	1	10	45	120	210	252	210	120	45	10	1

Figure 3.3: Pascal's triangle.

Pascal's Triangle

The relation 3.1, together with the knowledge that

$$\binom{n}{0} = \binom{n}{n} = 1 ,$$

determines completely the numbers $\binom{n}{j}$. We can use these relations to determine the famous *triangle of Pascal*, which exhibits all these numbers in matrix form (see Figure 3.3).

The nth row of this triangle has the entries $\binom{n}{0}$, $\binom{n}{1}$,..., $\binom{n}{n}$. We know that the first and last of these numbers are 1. The remaining numbers are determined by the recurrence relation Equation 3.1; that is, the entry $\binom{n}{j}$ for $0 < j < n$ in the nth row of Pascal's triangle is the *sum* of the entry immediately above and the one immediately to its left in the $(n-1)$st row. For example, $\binom{5}{2} = 6 + 4 = 10$.

This algorithm for constructing Pascal's triangle can be used to write a computer program to compute the binomial coefficients. You are asked to do this in Exercise 4.

While Pascal's triangle provides a way to construct recursively the binomial coefficients, it is also possible to give a formula for $\binom{n}{j}$.

Theorem 3.5 The binomial coefficients are given by the formula

$$\binom{n}{j} = \frac{(n)_j}{j!} . \qquad (3.2)$$

Proof. Each subset of size j of a set of size n can be ordered in $j!$ ways. Each of these orderings is a j-permutation of the set of size n. The number of j-permutations is $(n)_j$, so the number of subsets of size j is

$$\frac{(n)_j}{j!} .$$

This completes the proof. □

3.2. COMBINATIONS

The above formula can be rewritten in the form

$$\binom{n}{j} = \frac{n!}{j!(n-j)!} \ .$$

This immediately shows that

$$\binom{n}{j} = \binom{n}{n-j} \ .$$

When using Equation 3.2 in the calculation of $\binom{n}{j}$, if one alternates the multiplications and divisions, then all of the intermediate values in the calculation are integers. Furthermore, none of these intermediate values exceed the final value. (See Exercise 40.)

Another point that should be made concerning Equation 3.2 is that if it is used to *define* the binomial coefficients, then it is no longer necessary to require n to be a positive integer. The variable j must still be a non-negative integer under this definition. This idea is useful when extending the Binomial Theorem to general exponents. (The Binomial Theorem for non-negative integer exponents is given below as Theorem 3.7.)

Poker Hands

Example 3.6 Poker players sometimes wonder why a *four of a kind* beats a *full house*. A poker hand is a random subset of 5 elements from a deck of 52 cards. A hand has four of a kind if it has four cards with the same value—for example, four sixes or four kings. It is a full house if it has three of one value and two of a second—for example, three twos and two queens. Let us see which hand is more likely. How many hands have four of a kind? There are 13 ways that we can specify the value for the four cards. For each of these, there are 48 possibilities for the fifth card. Thus, the number of four-of-a-kind hands is $13 \cdot 48 = 624$. Since the total number of possible hands is $\binom{52}{5} = 2598960$, the probability of a hand with four of a kind is $624/2598960 = .00024$.

Now consider the case of a full house; how many such hands are there? There are 13 choices for the value which occurs three times; for each of these there are $\binom{4}{3} = 4$ choices for the particular three cards of this value that are in the hand. Having picked these three cards, there are 12 possibilities for the value which occurs twice; for each of these there are $\binom{4}{2} = 6$ possibilities for the particular pair of this value. Thus, the number of full houses is $13 \cdot 4 \cdot 12 \cdot 6 = 3744$, and the probability of obtaining a hand with a full house is $3744/2598960 = .0014$. Thus, while both types of hands are unlikely, you are six times more likely to obtain a full house than four of a kind. □

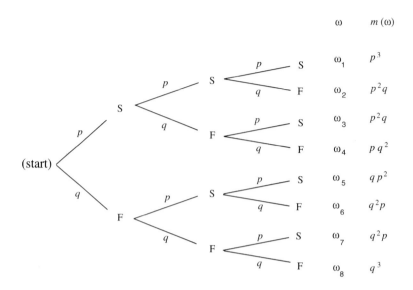

Figure 3.4: Tree diagram of three Bernoulli trials.

Bernoulli Trials

Our principal use of the binomial coefficients will occur in the study of one of the important chance processes called *Bernoulli trials*.

Definition 3.5 A *Bernoulli trials process* is a sequence of n chance experiments such that

1. Each experiment has two possible outcomes, which we may call *success* and *failure*.

2. The probability p of success on each experiment is the same for each experiment, and this probability is not affected by any knowledge of previous outcomes. The probability q of failure is given by $q = 1 - p$.

□

Example 3.7 The following are Bernoulli trials processes:

1. A coin is tossed ten times. The two possible outcomes are heads and tails. The probability of heads on any one toss is $1/2$.

2. An opinion poll is carried out by asking 1000 people, randomly chosen from the population, if they favor the Equal Rights Amendment—the two outcomes being yes and no. The probability p of a yes answer (i.e., a success) indicates the proportion of people in the entire population that favor this amendment.

3. A gambler makes a sequence of 1-dollar bets, betting each time on black at roulette at Las Vegas. Here a success is winning 1 dollar and a failure is losing

3.2. COMBINATIONS

1 dollar. Since in American roulette the gambler wins if the ball stops on one of 18 out of 38 positions and loses otherwise, the probability of winning is $p = 18/38 = .474$.

$\square$

To analyze a Bernoulli trials process, we choose as our sample space a binary tree and assign a probability measure to the paths in this tree. Suppose, for example, that we have three Bernoulli trials. The possible outcomes are indicated in the tree diagram shown in Figure 3.4. We define X to be the random variable which represents the outcome of the process, i.e., an ordered triple of S's and F's. The probabilities assigned to the branches of the tree represent the probability for each individual trial. Let the outcome of the ith trial be denoted by the random variable X_i, with distribution function m_i. Since we have assumed that outcomes on any one trial do not affect those on another, we assign the same probabilities at each level of the tree. An outcome ω for the entire experiment will be a path through the tree. For example, ω_3 represents the outcomes SFS. Our frequency interpretation of probability would lead us to expect a fraction p of successes on the first experiment; of these, a fraction q of failures on the second; and, of these, a fraction p of successes on the third experiment. This suggests assigning probability pqp to the outcome ω_3. More generally, we assign a distribution function $m(\omega)$ for paths ω by defining $m(\omega)$ to be the product of the branch probabilities along the path ω. Thus, the probability that the three events S on the first trial, F on the second trial, and S on the third trial occur is the product of the probabilities for the individual events. We shall see in the next chapter that this means that the events involved are *independent* in the sense that the knowledge of one event does not affect our prediction for the occurrences of the other events.

Binomial Probabilities

We shall be particularly interested in the probability that in n Bernoulli trials there are exactly j successes. We denote this probability by $b(n, p, j)$. Let us calculate the particular value $b(3, p, 2)$ from our tree measure. We see that there are three paths which have exactly two successes and one failure, namely ω_2, ω_3, and ω_5. Each of these paths has the same probability p^2q. Thus $b(3, p, 2) = 3p^2q$. Considering all possible numbers of successes we have

$$
\begin{aligned}
b(3, p, 0) &= q^3, \\
b(3, p, 1) &= 3pq^2, \\
b(3, p, 2) &= 3p^2q, \\
b(3, p, 3) &= p^3.
\end{aligned}
$$

We can, in the same manner, carry out a tree measure for n experiments and determine $b(n, p, j)$ for the general case of n Bernoulli trials.

Theorem 3.6 Given n Bernoulli trials with probability p of success on each experiment, the probability of exactly j successes is

$$b(n,p,j) = \binom{n}{j} p^j q^{n-j}$$

where $q = 1 - p$.

Proof. We construct a tree measure as described above. We want to find the sum of the probabilities for all paths which have exactly j successes and $n - j$ failures. Each such path is assigned a probability $p^j q^{n-j}$. How many such paths are there? To specify a path, we have to pick, from the n possible trials, a subset of j to be successes, with the remaining $n - j$ outcomes being failures. We can do this in $\binom{n}{j}$ ways. Thus the sum of the probabilities is

$$b(n,p,j) = \binom{n}{j} p^j q^{n-j} \ .$$

$\square$

Example 3.8 A fair coin is tossed six times. What is the probability that exactly three heads turn up? The answer is

$$b(6,.5,3) = \binom{6}{3} \left(\frac{1}{2}\right)^3 \left(\frac{1}{2}\right)^3 = 20 \cdot \frac{1}{64} = .3125 \ .$$

$\square$

Example 3.9 A die is rolled four times. What is the probability that we obtain exactly one 6? We treat this as Bernoulli trials with *success* = "rolling a 6" and *failure* = "rolling some number other than a 6." Then $p = 1/6$, and the probability of exactly one success in four trials is

$$b(4, 1/6, 1) = \binom{4}{1} \left(\frac{1}{6}\right)^1 \left(\frac{5}{6}\right)^3 = .386 \ .$$

$\square$

To compute binomial probabilities using the computer, multiply the function choose(n,k) by $p^k q^{n-k}$. The program **BinomialProbabilities** prints out the binomial probabilities $b(n,p,k)$ for k between $kmin$ and $kmax$, and the sum of these probabilities. We have run this program for $n = 100$, $p = 1/2$, $kmin = 45$, and $kmax = 55$; the output is shown in Table 3.8. Note that the individual probabilities are quite small. The probability of exactly 50 heads in 100 tosses of a coin is about .08. Our intuition tells us that this is the most likely outcome, which is correct; but, all the same, it is not a very likely outcome.

3.2. COMBINATIONS

k	$b(n,p,k)$
45	.0485
46	.0580
47	.0666
48	.0735
49	.0780
50	.0796
51	.0780
52	.0735
53	.0666
54	.0580
55	.0485

Table 3.8: Binomial probabilities for $n = 100$, $p = 1/2$.

Binomial Distributions

Definition 3.6 Let n be a positive integer, and let p be a real number between 0 and 1. Let B be the random variable which counts the number of successes in a Bernoulli trials process with parameters n and p. Then the distribution $b(n,p,k)$ of B is called the *binomial distribution*. □

We can get a better idea about the binomial distribution by graphing this distribution for different values of n and p (see Figure 3.5). The plots in this figure were generated using the program **BinomialPlot**.

We have run this program for $p = .5$ and $p = .3$. Note that even for $p = .3$ the graphs are quite symmetric. We shall have an explanation for this in Chapter 9. We also note that the highest probability occurs around the value np, but that these highest probabilities get smaller as n increases. We shall see in Chapter 6 that np is the *mean* or *expected* value of the binomial distribution $b(n,p,k)$.

The following example gives a nice way to see the binomial distribution, when $p = 1/2$.

Example 3.10 A *Galton board* is a board in which a large number of BB-shots are dropped from a chute at the top of the board and deflected off a number of pins on their way down to the bottom of the board. The final position of each slot is the result of a number of random deflections either to the left or the right. We have written a program **GaltonBoard** to simulate this experiment.

We have run the program for the case of 20 rows of pins and 10,000 shots being dropped. We show the result of this simulation in Figure 3.6.

Note that if we write 0 every time the shot is deflected to the left, and 1 every time it is deflected to the right, then the path of the shot can be described by a sequence of 0's and 1's of length n, just as for the n-fold coin toss.

The distribution shown in Figure 3.6 is an example of an empirical distribution, in the sense that it comes about by means of a sequence of experiments. As expected,

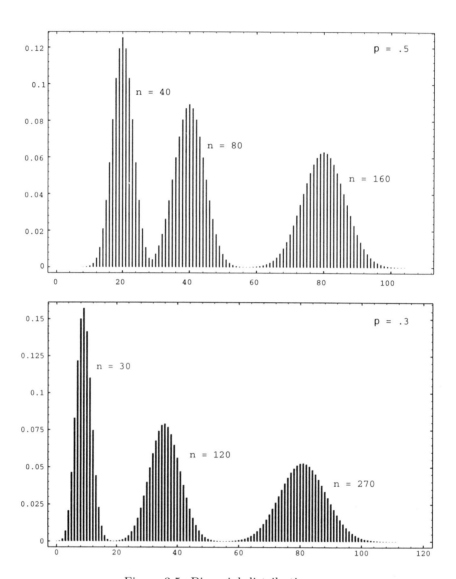

Figure 3.5: Binomial distributions.

3.2. COMBINATIONS

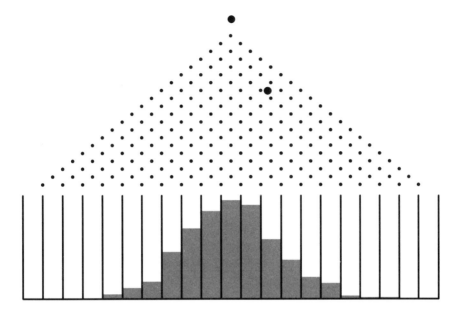

Figure 3.6: Simulation of the Galton board.

this empirical distribution resembles the corresponding binomial distribution with parameters $n = 20$ and $p = 1/2$. $\square$

Hypothesis Testing

Example 3.11 Suppose that ordinary aspirin has been found effective against headaches 60 percent of the time, and that a drug company claims that its new aspirin with a special headache additive is more effective. We can test this claim as follows: we call their claim the *alternate hypothesis*, and its negation, that the additive has no appreciable effect, the *null hypothesis*. Thus the null hypothesis is that $p = .6$, and the alternate hypothesis is that $p > .6$, where p is the probability that the new aspirin is effective.

We give the aspirin to n people to take when they have a headache. We want to find a number m, called the *critical value* for our experiment, such that we reject the null hypothesis if at least m people are cured, and otherwise we accept it. How should we determine this critical value?

First note that we can make two kinds of errors. The first, often called a *type 1 error* in statistics, is to reject the null hypothesis when in fact it is true. The second, called a *type 2 error*, is to accept the null hypothesis when it is false. To determine the probability of both these types of errors we introduce a function $\alpha(p)$, defined to be the probability that we reject the null hypothesis, where this probability is calculated under the assumption that the null hypothesis is true. In the present case, we have

$$\alpha(p) = \sum_{m \leq k \leq n} b(n, p, k) \ .$$

Note that $\alpha(.6)$ is the probability of a type 1 error, since this is the probability of a high number of successes for an ineffective additive. So for a given n we want to choose m so as to make $\alpha(.6)$ quite small, to reduce the likelihood of a type 1 error. But as m increases above the most probable value $np = .6n$, $\alpha(.6)$, being the upper tail of a binomial distribution, approaches 0. Thus *increasing* m makes a type 1 error less likely.

Now suppose that the additive really is effective, so that p is appreciably greater than .6; say $p = .8$. (This alternative value of p is chosen arbitrarily; the following calculations depend on this choice.) Then choosing m well below $np = .8n$ will increase $\alpha(.8)$, since now $\alpha(.8)$ is all but the lower tail of a binomial distribution. Indeed, if we put $\beta(.8) = 1 - \alpha(.8)$, then $\beta(.8)$ gives us the probability of a type 2 error, and so *decreasing* m makes a type 2 error less likely.

The manufacturer would like to guard against a type 2 error, since if such an error is made, then the test does not show that the new drug is better, when in fact it is. If the alternative value of p is chosen closer to the value of p given in the null hypothesis (in this case $p = .6$), then for a given test population, the value of β will increase. So, if the manufacturer's statistician chooses an alternative value for p which is close to the value in the null hypothesis, then it will be an expensive proposition (i.e., the test population will have to be large) to reject the null hypothesis with a small value of β.

What we hope to do then, for a given test population n, is to choose a value of m, if possible, which makes both these probabilities small. If we make a type 1 error we end up buying a lot of essentially ordinary aspirin at an inflated price; a type 2 error means we miss a bargain on a superior medication. Let us say that we want our critical number m to make each of these undesirable cases less than 5 percent probable.

We write a program **PowerCurve** to plot, for $n = 100$ and selected values of m, the function $\alpha(p)$, for p ranging from .4 to 1. The result is shown in Figure 3.7. We include in our graph a box (in dotted lines) from .6 to .8, with bottom and top at heights .05 and .95. Then a value for m satisfies our requirements if and only if the graph of α enters the box from the bottom, and leaves from the top (why?—which is the type 1 and which is the type 2 criterion?). As m increases, the graph of α moves to the right. A few experiments have shown us that $m = 69$ is the smallest value for m that thwarts a type 1 error, while $m = 73$ is the largest which thwarts a type 2. So we may choose our critical value between 69 and 73. If we're more intent on avoiding a type 1 error we favor 73, and similarly we favor 69 if we regard a type 2 error as worse. Of course, the drug company may not be happy with having as much as a 5 percent chance of an error. They might insist on having a 1 percent chance of an error. For this we would have to increase the number n of trials (see Exercise 28). □

Binomial Expansion

We next remind the reader of an application of the binomial coefficients to algebra. This is the *binomial expansion,* from which we get the term binomial coefficient.

3.2. COMBINATIONS

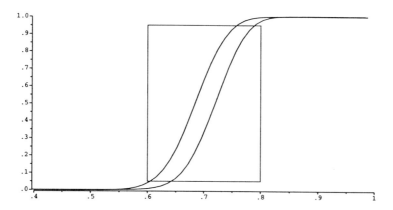

Figure 3.7: The power curve.

Theorem 3.7 (Binomial Theorem) The quantity $(a+b)^n$ can be expressed in the form
$$(a+b)^n = \sum_{j=0}^{n} \binom{n}{j} a^j b^{n-j} .$$

Proof. To see that this expansion is correct, write
$$(a+b)^n = (a+b)(a+b)\cdots(a+b) .$$

When we multiply this out we will have a sum of terms each of which results from a choice of an a or b for each of n factors. When we choose j a's and $(n-j)$ b's, we obtain a term of the form $a^j b^{n-j}$. To determine such a term, we have to specify j of the n terms in the product from which we choose the a. This can be done in $\binom{n}{j}$ ways. Thus, collecting these terms in the sum contributes a term $\binom{n}{j} a^j b^{n-j}$. □

For example, we have
$$\begin{aligned}
(a+b)^0 &= 1 \\
(a+b)^1 &= a+b \\
(a+b)^2 &= a^2 + 2ab + b^2 \\
(a+b)^3 &= a^3 + 3a^2b + 3ab^2 + b^3 .
\end{aligned}$$

We see here that the coefficients of successive powers do indeed yield Pascal's triangle.

Corollary 3.1 The sum of the elements in the nth row of Pascal's triangle is 2^n. If the elements in the nth row of Pascal's triangle are added with alternating signs, the sum is 0.

Proof. The first statement in the corollary follows from the fact that

$$2^n = (1+1)^n = \binom{n}{0} + \binom{n}{1} + \binom{n}{2} + \cdots + \binom{n}{n},$$

and the second from the fact that

$$0 = (1-1)^n = \binom{n}{0} - \binom{n}{1} + \binom{n}{2} - \cdots + (-1)^n \binom{n}{n}.$$

$\square$

The first statement of the corollary tells us that the number of subsets of a set of n elements is 2^n. We shall use the second statement in our next application of the binomial theorem.

We have seen that, when A and B are any two events (cf. Section 1.2),

$$P(A \cup B) = P(A) + P(B) - P(A \cap B).$$

We now extend this theorem to a more general version, which will enable us to find the probability that at least one of a number of events occurs.

Inclusion-Exclusion Principle

Theorem 3.8 Let P be a probability measure on a sample space Ω, and let $\{A_1, A_2, \ldots, A_n\}$ be a finite set of events. Then

$$P(A_1 \cup A_2 \cup \cdots \cup A_n) = \sum_{i=1}^{n} P(A_i) - \sum_{1 \leq i < j \leq n} P(A_i \cap A_j)$$

$$+ \sum_{1 \leq i < j < k \leq n} P(A_i \cap A_j \cap A_k) - \cdots . \quad (3.3)$$

That is, to find the probability that at least one of n events A_i occurs, first add the probability of each event, then subtract the probabilities of all possible two-way intersections, add the probability of all three-way intersections, and so forth.

Proof. If the outcome ω occurs in at least one of the events A_i, its probability is added exactly once by the left side of Equation 3.3. We must show that it is added exactly once by the right side of Equation 3.3. Assume that ω is in exactly k of the sets. Then its probability is added k times in the first term, subtracted $\binom{k}{2}$ times in the second, added $\binom{k}{3}$ times in the third term, and so forth. Thus, the total number of times that it is added is

$$\binom{k}{1} - \binom{k}{2} + \binom{k}{3} - \cdots (-1)^{k-1} \binom{k}{k}.$$

But

$$0 = (1-1)^k = \sum_{j=0}^{k} \binom{k}{j}(-1)^j = \binom{k}{0} - \sum_{j=1}^{k} \binom{k}{j}(-1)^{j-1}.$$

3.2. COMBINATIONS

Hence,
$$1 = \binom{k}{0} = \sum_{j=1}^{k} \binom{k}{j}(-1)^{j-1}.$$

If the outcome ω is not in any of the events A_i, then it is not counted on either side of the equation. $\square$

Hat Check Problem

Example 3.12 We return to the hat check problem discussed in Section 3.1, that is, the problem of finding the probability that a random permutation contains at least one fixed point. Recall that a permutation is a one-to-one map of a set $A = \{a_1, a_2, \ldots, a_n\}$ onto itself. Let A_i be the event that the ith element a_i remains fixed under this map. If we require that a_i is fixed, then the map of the remaining $n-1$ elements provides an arbitrary permutation of $(n-1)$ objects. Since there are $(n-1)!$ such permutations, $P(A_i) = (n-1)!/n! = 1/n$. Since there are n choices for a_i, the first term of Equation 3.3 is 1. In the same way, to have a particular pair (a_i, a_j) fixed, we can choose any permutation of the remaining $n-2$ elements; there are $(n-2)!$ such choices and thus

$$P(A_i \cap A_j) = \frac{(n-2)!}{n!} = \frac{1}{n(n-1)}.$$

The number of terms of this form in the right side of Equation 3.3 is

$$\binom{n}{2} = \frac{n(n-1)}{2!}.$$

Hence, the second term of Equation 3.3 is

$$-\frac{n(n-1)}{2!} \cdot \frac{1}{n(n-1)} = -\frac{1}{2!}.$$

Similarly, for any specific three events A_i, A_j, A_k,

$$P(A_i \cap A_j \cap A_k) = \frac{(n-3)!}{n!} = \frac{1}{n(n-1)(n-2)},$$

and the number of such terms is

$$\binom{n}{3} = \frac{n(n-1)(n-2)}{3!},$$

making the third term of Equation 3.3 equal to $1/3!$. Continuing in this way, we obtain

$$P(\text{at least one fixed point}) = 1 - \frac{1}{2!} + \frac{1}{3!} - \cdots (-1)^{n-1} \frac{1}{n!}$$

and

$$P(\text{no fixed point}) = \frac{1}{2!} - \frac{1}{3!} + \cdots (-1)^n \frac{1}{n!}.$$

n	Probability that no one gets his own hat back
3	.333333
4	.375
5	.366667
6	.368056
7	.367857
8	.367882
9	.367879
10	.367879

Table 3.9: Hat check problem.

From calculus we learn that

$$e^x = 1 + x + \frac{1}{2!}x^2 + \frac{1}{3!}x^3 + \cdots + \frac{1}{n!}x^n + \cdots.$$

Thus, if $x = -1$, we have

$$\begin{aligned} e^{-1} &= \frac{1}{2!} - \frac{1}{3!} + \cdots + \frac{(-1)^n}{n!} + \cdots \\ &= .3678794. \end{aligned}$$

Therefore, the probability that there is no fixed point, i.e., that none of the n people gets his own hat back, is equal to the sum of the first n terms in the expression for e^{-1}. This series converges very fast. Calculating the partial sums for $n = 3$ to 10 gives the data in Table 3.9.

After $n = 9$ the probabilities are essentially the same to six significant figures. Interestingly, the probability of no fixed point alternately increases and decreases as n increases. Finally, we note that our exact results are in good agreement with our simulations reported in the previous section. □

Choosing a Sample Space

We now have some of the tools needed to accurately describe sample spaces and to assign probability functions to those sample spaces. Nevertheless, in some cases, the description and assignment process is somewhat arbitrary. Of course, it is to be hoped that the description of the sample space and the subsequent assignment of a probability function will yield a model which accurately predicts what would happen if the experiment were actually carried out. As the following examples show, there are situations in which "reasonable" descriptions of the sample space do not produce a model which fits the data.

In Feller's book,[14] a pair of models is given which describe arrangements of certain kinds of elementary particles, such as photons and protons. It turns out that experiments have shown that certain types of elementary particles exhibit behavior

---

[14]W. Feller, *Introduction to Probability Theory and Its Applications* vol. 1, 3rd ed. (New York: John Wiley and Sons, 1968), p. 41

3.2. COMBINATIONS

which is accurately described by one model, called *"Bose-Einstein statistics,"* while other types of elementary particles can be modelled using *"Fermi-Dirac statistics."* Feller says:

> We have here an instructive example of the impossibility of selecting or justifying probability models by *a priori* arguments. In fact, no pure reasoning could tell that photons and protons would not obey the same probability laws.

We now give some examples of this description and assignment process.

Example 3.13 In the quantum mechanical model of the helium atom, various parameters can be used to classify the energy states of the atom. In the triplet spin state ($S = 1$) with orbital angular momentum 1 ($L = 1$), there are three possibilities, 0, 1, or 2, for the total angular momentum (J). (It is not assumed that the reader knows what any of this means; in fact, the example is more illustrative if the reader does *not* know anything about quantum mechanics.) We would like to assign probabilities to the three possibilities for J. The reader is undoubtedly resisting the idea of assigning the probability of 1/3 to each of these outcomes. She should now ask herself why she is resisting this assignment. The answer is probably because she does not have any "intuition" (i.e., experience) about the way in which helium atoms behave. In fact, in this example, the probabilities 1/9, 3/9, and 5/9 are assigned by the theory. The theory gives these assignments because these frequencies were observed *in experiments* and further parameters were developed in the theory to allow these frequencies to be predicted. □

Example 3.14 Suppose two pennies are flipped once each. There are several "reasonable" ways to describe the sample space. One way is to count the number of heads in the outcome; in this case, the sample space can be written $\{0, 1, 2\}$. Another description of the sample space is the set of all ordered pairs of H's and T's, i.e.,

$$\{(H, H), (H, T), (T, H), (T, T)\}.$$

Both of these descriptions are accurate ones, but it is easy to see that (at most) one of these, if assigned a constant probability function, can claim to accurately model reality. In this case, as opposed to the preceding example, the reader will probably say that the second description, with each outcome being assigned a probability of 1/4, is the "right" description. This conviction is due to experience; there is no proof that this is the way reality works. □

The reader is also referred to Exercise 26 for another example of this process.

Historical Remarks

The binomial coefficients have a long and colorful history leading up to Pascal's *Treatise on the Arithmetical Triangle*,[15] where Pascal developed many important

---
[15] B. Pascal, *Traité du Triangle Arithmétique* (Paris: Desprez, 1665).

1	1	1	1	1	1	1	1	1	1
1	2	3	4	5	6	7	8	9	
1	3	6	10	15	21	28	36		
1	4	10	20	35	56	84			
1	5	15	35	70	126				
1	6	21	56	126					
1	7	28	84						
1	8	36							
1	9								
1									

Table 3.10: Pascal's triangle.

natural numbers	1	2	3	4	5	6	7	8	9
triangular numbers	1	3	6	10	15	21	28	36	45
tetrahedral numbers	1	4	10	20	35	56	84	120	165

Table 3.11: Figurate numbers.

properties of these numbers. This history is set forth in the book *Pascal's Arithmetical Triangle* by A. W. F. Edwards.[16] Pascal wrote his triangle in the form shown in Table 3.10.

Edwards traces three different ways that the binomial coefficients arose. He refers to these as the *figurate numbers,* the *combinatorial numbers,* and the *binomial numbers.* They are all names for the same thing (which we have called binomial coefficients) but that they are all the same was not appreciated until the sixteenth century.

The *figurate numbers* date back to the Pythagorean interest in number patterns around 540 BC. The Pythagoreans considered, for example, triangular patterns shown in Figure 3.8. The sequence of numbers

$$1, 3, 6, 10, \ldots$$

obtained as the number of points in each triangle are called *triangular numbers.* From the triangles it is clear that the nth triangular number is simply the sum of the first n integers. The *tetrahedral numbers* are the sums of the triangular numbers and were obtained by the Greek mathematicians Theon and Nicomachus at the beginning of the second century BC. The tetrahedral number 10, for example, has the geometric representation shown in Figure 3.9. The first three types of figurate numbers can be represented in tabular form as shown in Table 3.11.

These numbers provide the first four rows of Pascal's triangle, but the table was not to be completed in the West until the sixteenth century.

In the East, Hindu mathematicians began to encounter the binomial coefficients in combinatorial problems. Bhaskara in his *Lilavati* of 1150 gave a rule to find the

[16] A. W. F. Edwards, *Pascal's Arithmetical Triangle* (London: Griffin, 1987).

3.2. COMBINATIONS 109

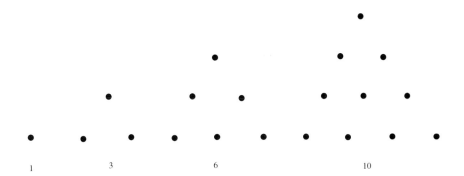

Figure 3.8: Pythagorean triangular patterns.

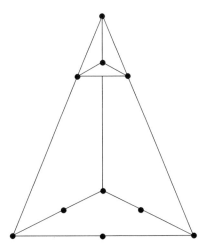

Figure 3.9: Geometric representation of the tetrahedral number 10.

```
11
12  22
13  23  33
14  24  34  44
15  25  35  45  55
16  26  36  46  56  66
```

Table 3.12: Outcomes for the roll of two dice.

number of medicinal preparations using 1, 2, 3, 4, 5, or 6 possible ingredients.[17] His rule is equivalent to our formula

$$\binom{n}{r} = \frac{(n)_r}{r!} .$$

The binomial numbers as coefficients of $(a+b)^n$ appeared in the works of mathematicians in China around 1100. There are references about this time to "the tabulation system for unlocking binomial coefficients." The triangle to provide the coefficients up to the eighth power is given by Chu Shih-chieh in a book written around 1303 (see Figure 3.10).[18] The original manuscript of Chu's book has been lost, but copies have survived. Edwards notes that there is an error in this copy of Chu's triangle. Can you find it? (*Hint*: Two numbers which should be equal are not.) Other copies do not show this error.

The first appearance of Pascal's triangle in the West seems to have come from calculations of Tartaglia in calculating the number of possible ways that n dice might turn up.[19] For one die the answer is clearly 6. For two dice the possibilities may be displayed as shown in Table 3.12.

Displaying them this way suggests the sixth triangular number $1 + 2 + 3 + 4 + 5 + 6 = 21$ for the throw of 2 dice. Tartaglia "on the first day of Lent, 1523, in Verona, having thought about the problem all night,"[20] realized that the extension of the figurate table gave the answers for n dice. The problem had suggested itself to Tartaglia from watching people casting their own horoscopes by means of a *Book of Fortune*, selecting verses by a process which included noting the numbers on the faces of three dice. The 56 ways that three dice can fall were set out on each page. The way the numbers were written in the book did not suggest the connection with figurate numbers, but a method of enumeration similar to the one we used for 2 dice does. Tartaglia's table was not published until 1556.

A table for the binomial coefficients was published in 1554 by the German mathematician Stifel.[21] Pascal's triangle appears also in Cardano's *Opus novum* of 1570.[22]

---

[17] ibid., p. 27.
[18] J. Needham, *Science and Civilization in China*, vol. 3 (New York: Cambridge University Press, 1959), p. 135.
[19] N. Tartaglia, *General Trattato di Numeri et Misure* (Vinegia, 1556).
[20] Quoted in Edwards, op. cit., p. 37.
[21] M. Stifel, *Arithmetica Integra* (Norimburgae, 1544).
[22] G. Cardano, *Opus Novum de Proportionibus Numerorum* (Basilea, 1570).

3.2. COMBINATIONS

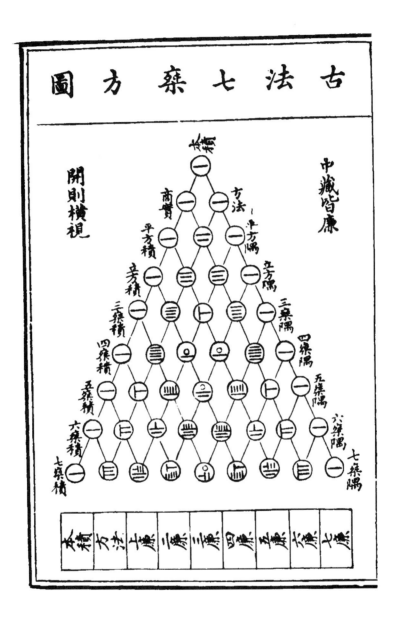

Figure 3.10: Chu Shih-chieh's triangle. [From J. Needham, *Science and Civilization in China*, vol. 3 (New York: Cambridge University Press, 1959), p. 135. Reprinted with permission.]

Cardano was interested in the problem of finding the number of ways to choose r objects out of n. Thus by the time of Pascal's work, his triangle had appeared as a result of looking at the figurate numbers, the combinatorial numbers, and the binomial numbers, and the fact that all three were the same was presumably pretty well understood.

Pascal's interest in the binomial numbers came from his letters with Fermat concerning a problem known as the problem of points. This problem, and the correspondence between Pascal and Fermat, were discussed in Chapter 1. The reader will recall that this problem can be described as follows: Two players A and B are playing a sequence of games and the first player to win n games wins the match. It is desired to find the probability that A wins the match at a time when A has won a games and B has won b games. (See Exercises 4.1.40-4.1.42.)

Pascal solved the problem by backward induction, much the way we would do today in writing a computer program for its solution. He referred to the combinatorial method of Fermat which proceeds as follows: If A needs c games and B needs d games to win, we require that the players continue to play until they have played $c + d - 1$ games. The winner in this extended series will be the same as the winner in the original series. The probability that A wins in the extended series and hence in the original series is

$$\sum_{r=c}^{c+d-1} \frac{1}{2^{c+d-1}} \binom{c+d-1}{r} .$$

Even at the time of the letters Pascal seemed to understand this formula.

Suppose that the first player to win n games wins the match, and suppose that each player has put up a stake of x. Pascal studied the value of winning a particular game. By this he meant the increase in the expected winnings of the winner of the particular game under consideration. He showed that the value of the first game is

$$\frac{1 \cdot 3 \cdot 5 \cdot \ldots \cdot (2n-1)}{2 \cdot 4 \cdot 6 \cdot \ldots \cdot (2n)} x .$$

His proof of this seems to use Fermat's formula and the fact that the above ratio of products of odd to products of even numbers is equal to the probability of exactly n heads in $2n$ tosses of a coin. (See Exercise 39.)

Pascal presented Fermat with the table shown in Table 3.13. He states:

> You will see as always, that the value of the first game is equal to that of the second which is easily shown by combinations. You will see, in the same way, that the numbers in the first line are always increasing; so also are those in the second; and those in the third. But those in the fourth line are decreasing, and those in the fifth, etc. This seems odd.[23]

The student can pursue this question further using the computer and Pascal's backward iteration method for computing the expected payoff at any point in the series.

---
[23] F. N. David, op. cit., p. 235.

3.2. COMBINATIONS

	if each one staken 256 in					
From my opponent's 256 positions I get, for the	6 games	5 games	4 games	3 games	2 games	1 games
1st game	63	70	80	96	128	256
2nd game	63	70	80	96	128	
3rd game	56	60	64	64		
4th game	42	40	32			
5th game	24	16				
6th game	8					

Table 3.13: Pascal's solution for the problem of points.

In his treatise, Pascal gave a formal proof of Fermat's combinatorial formula as well as proofs of many other basic properties of binomial numbers. Many of his proofs involved induction and represent some of the first proofs by this method. His book brought together all the different aspects of the numbers in the Pascal triangle as known in 1654, and, as Edwards states, "That the Arithmetical Triangle should bear Pascal's name cannot be disputed."[24]

The first serious study of the binomial distribution was undertaken by James Bernoulli in his *Ars Conjectandi* published in 1713.[25] We shall return to this work in the historical remarks in Chapter 8.

Exercises

1 Compute the following:

   (a) $\binom{6}{3}$

   (b) $b(5, .2, 4)$

   (c) $\binom{7}{2}$

   (d) $\binom{26}{26}$

   (e) $b(4, .2, 3)$

   (f) $\binom{6}{2}$

   (g) $\binom{10}{9}$

   (h) $b(8, .3, 5)$

2 In how many ways can we choose five people from a group of ten to form a committee?

3 How many seven-element subsets are there in a set of nine elements?

4 Using the relation Equation 3.1 write a program to compute Pascal's triangle, putting the results in a matrix. Have your program print the triangle for $n = 10$.

---

[24] A. W. F. Edwards, op. cit., p. ix.
[25] J. Bernoulli, *Ars Conjectandi* (Basil: Thurnisiorum, 1713).

5 Use the program **BinomialProbabilities** to find the probability that, in 100 tosses of a fair coin, the number of heads that turns up lies between 35 and 65, between 40 and 60, and between 45 and 55.

6 Charles claims that he can distinguish between beer and ale 75 percent of the time. Ruth bets that he cannot and, in fact, just guesses. To settle this, a bet is made: Charles is to be given ten small glasses, each having been filled with beer or ale, chosen by tossing a fair coin. He wins the bet if he gets seven or more correct. Find the probability that Charles wins if he has the ability that he claims. Find the probability that Ruth wins if Charles is guessing.

7 Show that
$$b(n,p,j) = \frac{p}{q}\left(\frac{n-j+1}{j}\right)b(n,p,j-1),$$
for $j \geq 1$. Use this fact to determine the value or values of j which give $b(n,p,j)$ its greatest value. *Hint*: Consider the successive ratios as j increases.

8 A die is rolled 30 times. What is the probability that a 6 turns up exactly 5 times? What is the most probable number of times that a 6 will turn up?

9 Find integers n and r such that the following equation is true:
$$\binom{13}{5} + 2\binom{13}{6} + \binom{13}{7} = \binom{n}{r}.$$

10 In a ten-question true-false exam, find the probability that a student gets a grade of 70 percent or better by guessing. Answer the same question if the test has 30 questions, and if the test has 50 questions.

11 A restaurant offers apple and blueberry pies and stocks an equal number of each kind of pie. Each day ten customers request pie. They choose, with equal probabilities, one of the two kinds of pie. How many pieces of each kind of pie should the owner provide so that the probability is about .95 that each customer gets the pie of his or her own choice?

12 A poker hand is a set of 5 cards randomly chosen from a deck of 52 cards. Find the probability of a

(a) royal flush (ten, jack, queen, king, ace in a single suit).

(b) straight flush (five in a sequence in a single suit, but not a royal flush).

(c) four of a kind (four cards of the same face value).

(d) full house (one pair and one triple, each of the same face value).

(e) flush (five cards in a single suit but not a straight or royal flush).

(f) straight (five cards in a sequence, not all the same suit). (Note that in straights, an ace counts high or low.)

13 If a set has $2n$ elements, show that it has more subsets with n elements than with any other number of elements.

3.2. COMBINATIONS

14 Let $b(2n, .5, n)$ be the probability that in $2n$ tosses of a fair coin exactly n heads turn up. Using Stirling's formula (Theorem 3.3), show that $b(2n, .5, n) \sim 1/\sqrt{\pi n}$. Use the program **BinomialProbabilities** to compare this with the exact value for $n = 10$ to 25.

15 A baseball player, Smith, has a batting average of .300 and in a typical game comes to bat three times. Assume that Smith's hits in a game can be considered to be a Bernoulli trials process with probability .3 for *success*. Find the probability that Smith gets 0, 1, 2, and 3 hits.

16 The Siwash University football team plays eight games in a season, winning three, losing three, and ending two in a tie. Show that the number of ways that this can happen is
$$\binom{8}{3}\binom{5}{3} = \frac{8!}{3!\,3!\,2!}\ .$$

17 Using the technique of Exercise 16, show that the number of ways that one can put n different objects into three boxes with a in the first, b in the second, and c in the third is $n!/(a!\,b!\,c!)$.

18 Baumgartner, Prosser, and Crowell are grading a calculus exam. There is a true-false question with ten parts. Baumgartner notices that one student has only two out of the ten correct and remarks, "The student was not even bright enough to have flipped a coin to determine his answers." "Not so clear," says Prosser. "With 340 students I bet that if they all flipped coins to determine their answers there would be at least one exam with two or fewer answers correct." Crowell says, "I'm with Prosser. In fact, I bet that we should expect at least one exam in which no answer is correct if everyone is just guessing." Who is right in all of this?

19 A gin hand consists of 10 cards from a deck of 52 cards. Find the probability that a gin hand has

(a) all 10 cards of the same suit.

(b) exactly 4 cards in one suit and 3 in two other suits.

(c) a 4, 3, 2, 1, distribution of suits.

20 A six-card hand is dealt from an ordinary deck of cards. Find the probability that:

(a) All six cards are hearts.

(b) There are three aces, two kings, and one queen.

(c) There are three cards of one suit and three of another suit.

21 A lady wishes to color her fingernails on one hand using at most two of the colors red, yellow, and blue. How many ways can she do this?

22 How many ways can six indistinguishable letters be put in three mail boxes? *Hint*: One representation of this is given by a sequence |LL|L|LLL| where the |'s represent the partitions for the boxes and the L's the letters. Any possible way can be so described. Note that we need two bars at the ends and the remaining two bars and the six L's can be put in any order.

23 Using the method for the hint in Exercise 22, show that r indistinguishable objects can be put in n boxes in

$$\binom{n+r-1}{n-1} = \binom{n+r-1}{r}$$

different ways.

24 A travel bureau estimates that when 20 tourists go to a resort with ten hotels they distribute themselves as if the bureau were putting 20 indistinguishable objects into ten distinguishable boxes. Assuming this model is correct, find the probability that no hotel is left vacant when the first group of 20 tourists arrives.

25 An elevator takes on six passengers and stops at ten floors. We can assign two different equiprobable measures for the ways that the passengers are discharged: (a) we consider the passengers to be distinguishable or (b) we consider them to be indistinguishable (see Exercise 23 for this case). For each case, calculate the probability that all the passengers get off at different floors.

26 You are playing *heads or tails* with Prosser but you suspect that his coin is unfair. Von Neumann suggested that you proceed as follows: Toss Prosser's coin twice. If the outcome is HT call the result *win*. if it is TH call the result *lose*. If it is TT or HH ignore the outcome and toss Prosser's coin twice again. Keep going until you get either an HT or a TH and call the result win or lose in a single play. Repeat this procedure for each play. Assume that Prosser's coin turns up heads with probability p.

(a) Find the probability of HT, TH, HH, TT with two tosses of Prosser's coin.

(b) Using part (a), show that the probability of a win on any one play is 1/2, no matter what p is.

27 John claims that he has extrasensory powers and can tell which of two symbols is on a card turned face down (see Example 3.11). To test his ability he is asked to do this for a sequence of trials. Let the null hypothesis be that he is just guessing, so that the probability is 1/2 of his getting it right each time, and let the alternative hypothesis be that he can name the symbol correctly more than half the time. Devise a test with the property that the probability of a type 1 error is less than .05 and the probability of a type 2 error is less than .05 if John can name the symbol correctly 75 percent of the time.

3.2. COMBINATIONS

28 In Example 3.11 assume the alternative hypothesis is that $p = .8$ and that it is desired to have the probability of each type of error less than .01. Use the program **PowerCurve** to determine values of n and m that will achieve this. Choose n as small as possible.

29 A drug is assumed to be effective with an unknown probability p. To estimate p the drug is given to n patients. It is found to be effective for m patients. The *method of maximum likelihood* for estimating p states that we should choose the value for p that gives the highest probability of getting what we got on the experiment. Assuming that the experiment can be considered as a Bernoulli trials process with probability p for success, show that the maximum likelihood estimate for p is the proportion m/n of successes.

30 Recall that in the World Series the first team to win four games wins the series. The series can go at most seven games. Assume that the Red Sox and the Mets are playing the series. Assume that the Mets win each game with probability p. Fermat observed that even though the series might not go seven games, the probability that the Mets win the series is the same as the probability that they win four or more game in a series that was forced to go seven games no matter who wins the individual games.

(a) Using the program **PowerCurve** of Example 3.11 find the probability that the Mets win the series for the cases $p = .5$, $p = .6$, $p = .7$.

(b) Assume that the Mets have probability .6 of winning each game. Use the program **PowerCurve** to find a value of n so that, if the series goes to the first team to win more than half the games, the Mets will have a 95 percent chance of winning the series. Choose n as small as possible.

31 Each of the four engines on an airplane functions correctly on a given flight with probability .99, and the engines function independently of each other. Assume that the plane can make a safe landing if at least two of its engines are functioning correctly. What is the probability that the engines will allow for a safe landing?

32 A small boy is lost coming down Mount Washington. The leader of the search team estimates that there is a probability p that he came down on the east side and a probability $1 - p$ that he came down on the west side. He has n people in his search team who will search independently and, if the boy is on the side being searched, each member will find the boy with probability u. Determine how he should divide the n people into two groups to search the two sides of the mountain so that he will have the highest probability of finding the boy. How does this depend on u?

*33 $2n$ balls are chosen at random from a total of $2n$ red balls and $2n$ blue balls. Find a combinatorial expression for the probability that the chosen balls are equally divided in color. Use Stirling's formula to estimate this probability.

Using **BinomialProbabilities**, compare the exact value with Stirling's approximation for $n = 20$.

34 Assume that every time you buy a box of Wheaties, you receive one of the pictures of the n players on the New York Yankees. Over a period of time, you buy $m \geq n$ boxes of Wheaties.

   (a) Use Theorem 3.8 to show that the probability that you get all n pictures is
   $$1 - \binom{n}{1}\left(\frac{n-1}{n}\right)^m + \binom{n}{2}\left(\frac{n-2}{n}\right)^m - \cdots$$
   $$+ (-1)^{n-1}\binom{n}{n-1}\left(\frac{1}{n}\right)^m.$$

   Hint: Let E_k be the event that you do not get the kth player's picture.

   (b) Write a computer program to compute this probability. Use this program to find, for given n, the smallest value of m which will give probability $\geq .5$ of getting all n pictures. Consider $n = 50$, 100, and 150 and show that $m = n \log n + n \log 2$ is a good estimate for the number of boxes needed. (For a derivation of this estimate, see Feller.[26])

*35 Prove the following *binomial identity*
$$\binom{2n}{n} = \sum_{j=0}^{n} \binom{n}{j}^2.$$

Hint: Consider an urn with n red balls and n blue balls inside. Show that each side of the equation equals the number of ways to choose n balls from the urn.

36 Let j and n be positive integers, with $j \leq n$. An experiment consists of choosing, at random, a j-tuple of *positive* integers whose sum is at most n.

   (a) Find the size of the sample space. *Hint*: Consider n indistinguishable balls placed in a row. Place j markers between consecutive pairs of balls, with no two markers between the same pair of balls. (We also allow one of the n markers to be placed at the end of the row of balls.) Show that there is a 1-1 correspondence between the set of possible positions for the markers and the set of j-tuples whose size we are trying to count.

   (b) Find the probability that the j-tuple selected contains at least one 1.

37 Let $n \pmod{m}$ denote the remainder when the integer n is divided by the integer m. Write a computer program to compute the numbers $\binom{n}{j} \pmod{m}$ where $\binom{n}{j}$ is a binomial coefficient and m is an integer. You can do this by using the recursion relations for generating binomial coefficients, doing all the

---

[26] W. Feller, *Introduction to Probability Theory and its Applications*, vol. I, 3rd ed. (New York: John Wiley & Sons, 1968), p. 106.

arithmetic using the basic function $\text{mod}(n, m)$. Try to write your program to make as large a table as possible. Run your program for the cases $m = 2$ to 7. Do you see any patterns? In particular, for the case $m = 2$ and n a power of 2, verify that all the entries in the $(n-1)$st row are 1. (The corresponding binomial numbers are odd.) Use your pictures to explain why this is true.

38 Lucas[27] proved the following general result relating to Exercise 37. If p is any prime number, then $\binom{n}{j}$ (mod p) can be found as follows: Expand n and j in base p as $n = s_0 + s_1 p + s_2 p^2 + \cdots + s_k p^k$ and $j = r_0 + r_1 p + r_2 p^2 + \cdots + r_k p^k$, respectively. (Here k is chosen large enough to represent all numbers from 0 to n in base p using k digits.) Let $s = (s_0, s_1, s_2, \ldots, s_k)$ and $r = (r_0, r_1, r_2, \ldots, r_k)$. Then

$$\binom{n}{j} \pmod{p} = \prod_{i=0}^{k} \binom{s_i}{r_i} \pmod{p}.$$

For example, if $p = 7$, $n = 12$, and $j = 9$, then

$$\begin{aligned} 12 &= 5 \cdot 7^0 + 1 \cdot 7^1, \\ 9 &= 2 \cdot 7^0 + 1 \cdot 7^1, \end{aligned}$$

so that

$$\begin{aligned} s &= (5, 1), \\ r &= (2, 1), \end{aligned}$$

and this result states that

$$\binom{12}{9} \pmod{p} = \binom{5}{2}\binom{1}{1} \pmod{7}.$$

Since $\binom{12}{9} = 220 = 3 \pmod{7}$, and $\binom{5}{2} = 10 = 3 \pmod{7}$, we see that the result is correct for this example.

Show that this result implies that, for $p = 2$, the $(p^k - 1)$st row of your triangle in Exercise 37 has no zeros.

39 Prove that the probability of exactly n heads in $2n$ tosses of a fair coin is given by the product of the odd numbers up to $2n - 1$ divided by the product of the even numbers up to $2n$.

40 Let n be a positive integer, and assume that j is a positive integer not exceeding $n/2$. Show that in Theorem 3.5, if one alternates the multiplications and divisions, then all of the intermediate values in the calculation are integers. Show also that none of these intermediate values exceed the final value.

---

[27] E. Lucas, "Théorie des Functions Numériques Simplement Periodiques," *American J. Math.*, vol. 1 (1878), pp. 184-240, 289-321.

3.3 Card Shuffling

Much of this section is based upon an article by Brad Mann,[28] which is an exposition of an article by David Bayer and Persi Diaconis.[29]

Riffle Shuffles

Given a deck of n cards, how many times must we shuffle it to make it "random"? Of course, the answer depends upon the method of shuffling which is used and what we mean by "random." We shall begin the study of this question by considering a standard model for the riffle shuffle.

We begin with a deck of n cards, which we will assume are labelled in increasing order with the integers from 1 to n. A riffle shuffle consists of a cut of the deck into two stacks and an interleaving of the two stacks. For example, if $n = 6$, the initial ordering is $(1, 2, 3, 4, 5, 6)$, and a cut might occur between cards 2 and 3. This gives rise to two stacks, namely $(1, 2)$ and $(3, 4, 5, 6)$. These are interleaved to form a new ordering of the deck. For example, these two stacks might form the ordering $(1, 3, 4, 2, 5, 6)$. In order to discuss such shuffles, we need to assign a probability measure to the set of all possible shuffles. There are several reasonable ways in which this can be done. We will give several different assignment strategies, and show that they are equivalent. (This does not mean that this assignment is the only reasonable one.) First, we assign the binomial probability $b(n, 1/2, k)$ to the event that the cut occurs after the kth card. Next, we assume that all possible interleavings, given a cut, are equally likely. Thus, to complete the assignment of probabilities, we need to determine the number of possible interleavings of two stacks of cards, with k and $n - k$ cards, respectively.

We begin by writing the second stack in a line, with spaces in between each pair of consecutive cards, and with spaces at the beginning and end (so there are $n - k + 1$ spaces). We choose, with replacement, k of these spaces, and place the cards from the first stack in the chosen spaces. This can be done in

$$\binom{n}{k}$$

ways. Thus, the probability of a given interleaving should be

$$\frac{1}{\binom{n}{k}}.$$

Next, we note that if the new ordering is not the identity ordering, it is the result of a unique cut-interleaving pair. If the new ordering is the identity, it is the result of any one of $n + 1$ cut-interleaving pairs.

We define a *rising sequence* in an ordering to be a maximal subsequence of consecutive integers in increasing order. For example, in the ordering

$$(2, 3, 5, 1, 4, 7, 6),$$

---

[28] B. Mann, "How Many Times Should You Shuffle a Deck of Cards?", *UMAP Journal*, vol. 15, no. 4 (1994), pp. 303–331.
[29] D. Bayer and P. Diaconis, "Trailing the Dovetail Shuffle to its Lair," *Annals of Applied Probability*, vol. 2, no. 2 (1992), pp. 294–313.

3.3. CARD SHUFFLING

there are 4 rising sequences; they are (1), (2, 3, 4), (5, 6), and (7). It is easy to see that an ordering is the result of a riffle shuffle applied to the identity ordering if and only if it has no more than two rising sequences. (If the ordering has two rising sequences, then these rising sequences correspond to the two stacks induced by the cut, and if the ordering has one rising sequence, then it is the identity ordering.) Thus, the sample space of orderings obtained by applying a riffle shuffle to the identity ordering is naturally described as the set of all orderings with at most two rising sequences.

It is now easy to assign a probability measure to this sample space. Each ordering with two rising sequences is assigned the value

$$\frac{b(n, 1/2, k)}{\binom{n}{k}} = \frac{1}{2^n} ,$$

and the identity ordering is assigned the value

$$\frac{n+1}{2^n} .$$

There is another way to view a riffle shuffle. We can imagine starting with a deck cut into two stacks as before, with the same probabilities assignment as before i.e., the binomial distribution. Once we have the two stacks, we take cards, one by one, off of the bottom of the two stacks, and place them onto one stack. If there are k_1 and k_2 cards, respectively, in the two stacks at some point in this process, then we make the assumption that the probabilities that the next card to be taken comes from a given stack is proportional to the current stack size. This implies that the probability that we take the next card from the first stack equals

$$\frac{k_1}{k_1 + k_2} ,$$

and the corresponding probability for the second stack is

$$\frac{k_2}{k_1 + k_2} .$$

We shall now show that this process assigns the uniform probability to each of the possible interleavings of the two stacks.

Suppose, for example, that an interleaving came about as the result of choosing cards from the two stacks in some order. The probability that this result occurred is the product of the probabilities at each point in the process, since the choice of card at each point is assumed to be independent of the previous choices. Each factor of this product is of the form

$$\frac{k_i}{k_1 + k_2} ,$$

where $i = 1$ or 2, and the denominator of each factor equals the number of cards left to be chosen. Thus, the denominator of the probability is just $n!$. At the moment when a card is chosen from a stack that has i cards in it, the numerator of the corresponding factor in the probability is i, and the number of cards in this stack decreases by 1. Thus, the numerator is seen to be $k!(n-k)!$, since all cards in both stacks are eventually chosen. Therefore, this process assigns the probability

$$\frac{1}{\binom{n}{k}}$$

to each possible interleaving.

We now turn to the question of what happens when we riffle shuffle s times. It should be clear that if we start with the identity ordering, we obtain an ordering with at most 2^s rising sequences, since a riffle shuffle creates at most two rising sequences from every rising sequence in the starting ordering. In fact, it is not hard to see that each such ordering is the result of s riffle shuffles. The question becomes, then, in how many ways can an ordering with r rising sequences come about by applying s riffle shuffles to the identity ordering? In order to answer this question, we turn to the idea of an a-shuffle.

a-Shuffles

There are several ways to visualize an a-shuffle. One way is to imagine a creature with a hands who is given a deck of cards to riffle shuffle. The creature naturally cuts the deck into a stacks, and then riffles them together. (Imagine that!) Thus, the ordinary riffle shuffle is a 2-shuffle. As in the case of the ordinary 2-shuffle, we allow some of the stacks to have 0 cards. Another way to visualize an a-shuffle is to think about its inverse, called an a-unshuffle. This idea is described in the proof of the next theorem.

We will now show that an a-shuffle followed by a b-shuffle is equivalent to an ab-shuffle. This means, in particular, that s riffle shuffles in succession are equivalent to one 2^s-shuffle. This equivalence is made precise by the following theorem.

Theorem 3.9 Let a and b be two positive integers. Let $S_{a,b}$ be the set of all ordered pairs in which the first entry is an a-shuffle and the second entry is a b-shuffle. Let S_{ab} be the set of all ab-shuffles. Then there is a 1-1 correspondence between $S_{a,b}$ and S_{ab} with the following property. Suppose that (T_1, T_2) corresponds to T_3. If T_1 is applied to the identity ordering, and T_2 is applied to the resulting ordering, then the final ordering is the same as the ordering that is obtained by applying T_3 to the identity ordering.

Proof. The easiest way to describe the required correspondence is through the idea of an unshuffle. An a-unshuffle begins with a deck of n cards. One by one, cards are taken from the top of the deck and placed, with equal probability, on the bottom of any one of a stacks, where the stacks are labelled from 0 to $a-1$. After all of the cards have been distributed, we combine the stacks to form one stack by placing stack i on top of stack $i+1$, for $0 \le i \le a-1$. It is easy to see that if one starts with a deck, there is exactly one way to cut the deck to obtain the a stacks generated by the a-unshuffle, and with these a stacks, there is exactly one way to interleave them

3.3. CARD SHUFFLING

to obtain the deck in the order that it was in before the unshuffle was performed. Thus, this a-unshuffle corresponds to a unique a-shuffle, and this a-shuffle is the inverse of the original a-unshuffle.

If we apply an ab-unshuffle U_3 to a deck, we obtain a set of ab stacks, which are then combined, in order, to form one stack. We label these stacks with ordered pairs of integers, where the first coordinate is between 0 and $a-1$, and the second coordinate is between 0 and $b-1$. Then we label each card with the label of its stack. The number of possible labels is ab, as required. Using this labelling, we can describe how to find a b-unshuffle and an a-unshuffle, such that if these two unshuffles are applied in this order to the deck, we obtain the same set of ab stacks as were obtained by the ab-unshuffle.

To obtain the b-unshuffle U_2, we sort the deck into b stacks, with the ith stack containing all of the cards with second coordinate i, for $0 \leq i \leq b-1$. Then these stacks are combined to form one stack. The a-unshuffle U_1 proceeds in the same manner, except that the first coordinates of the labels are used. The resulting a stacks are then combined to form one stack.

The above description shows that the cards ending up on top are all those labelled $(0,0)$. These are followed by those labelled $(0,1)$, $(0,2)$, ..., $(0, b-1)$, $(1,0)$, $(1,1)$, ..., $(a-1, b-1)$. Furthermore, the relative order of any pair of cards with the same labels is never altered. But this is exactly the same as an ab-unshuffle, if, at the beginning of such an unshuffle, we label each of the cards with one of the labels $(0,0)$, $(0,1)$, ..., $(0, b-1)$, $(1,0)$, $(1,1)$, ..., $(a-1, b-1)$. This completes the proof. $\square$

In Figure 3.11, we show the labels for a 2-unshuffle of a deck with 10 cards. There are 4 cards with the label 0 and 6 cards with the label 1, so if the 2-unshuffle is performed, the first stack will have 4 cards and the second stack will have 6 cards. When this unshuffle is performed, the deck ends up in the identity ordering.

In Figure 3.12, we show the labels for a 4-unshuffle of the same deck (because there are four labels being used). This figure can also be regarded as an example of a pair of 2-unshuffles, as described in the proof above. The first 2-unshuffle will use the second coordinate of the labels to determine the stacks. In this case, the two stacks contain the cards whose values are

$$\{5, 1, 6, 2, 7\} \text{ and } \{8, 9, 3, 4, 10\} .$$

After this 2-unshuffle has been performed, the deck is in the order shown in Figure 3.11, as the reader should check. If we wish to perform a 4-unshuffle on the deck, using the labels shown, we sort the cards lexicographically, obtaining the four stacks

$$\{1, 2\}, \{3, 4\}, \{5, 6, 7\}, \text{ and } \{8, 9, 10\} .$$

When these stacks are combined, we once again obtain the identity ordering of the deck. The point of the above theorem is that both sorting procedures always lead to the same initial ordering.

Theorem 3.10 If D is any ordering that is the result of applying an a-shuffle and then a b-shuffle to the identity ordering, then the probability assigned to D by this

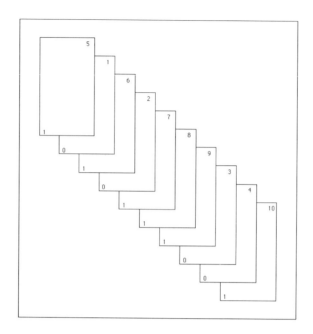

Figure 3.11: Before a 2-unshuffle.

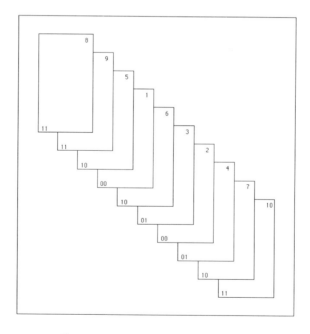

Figure 3.12: Before a 4-unshuffle.

3.3. CARD SHUFFLING

pair of operations is the same as the probability assigned to D by the process of applying an ab-shuffle to the identity ordering.

Proof. Call the sample space of a-shuffles S_a. If we label the stacks by the integers from 0 to $a-1$, then each cut-interleaving pair, i.e., shuffle, corresponds to exactly one n-digit base a integer, where the ith digit in the integer is the stack of which the ith card is a member. Thus, the number of cut-interleaving pairs is equal to the number of n-digit base a integers, which is a^n. Of course, not all of these pairs leads to different orderings. The number of pairs leading to a given ordering will be discussed later. For our purposes it is enough to point out that it is the cut-interleaving pairs that determine the probability assignment.

The previous theorem shows that there is a 1-1 correspondence between $S_{a,b}$ and S_{ab}. Furthermore, corresponding elements give the same ordering when applied to the identity ordering. Given any ordering D, let m_1 be the number of elements of $S_{a,b}$ which, when applied to the identity ordering, result in D. Let m_2 be the number of elements of S_{ab} which, when applied to the identity ordering, result in D. The previous theorem implies that $m_1 = m_2$. Thus, both sets assign the probability

$$\frac{m_1}{(ab)^n}$$

to D. This completes the proof. $\square$

Connection with the Birthday Problem

There is another point that can be made concerning the labels given to the cards by the successive unshuffles. Suppose that we 2-unshuffle an n-card deck until the labels on the cards are all different. It is easy to see that this process produces each permutation with the same probability, i.e., this is a random process. To see this, note that if the labels become distinct on the sth 2-unshuffle, then one can think of this sequence of 2-unshuffles as one 2^s-unshuffle, in which all of the stacks determined by the unshuffle have at most one card in them (remember, the stacks correspond to the labels). If each stack has at most one card in it, then given any two cards in the deck, it is equally likely that the first card has a lower or a higher label than the second card. Thus, each possible ordering is equally likely to result from this 2^s-unshuffle.

Let T be the random variable that counts the number of 2-unshuffles until all labels are distinct. One can think of T as giving a measure of how long it takes in the unshuffling process until randomness is reached. Since shuffling and unshuffling are inverse processes, T also measures the number of shuffles necessary to achieve randomness. Suppose that we have an n-card deck, and we ask for $P(T \leq s)$. This equals $1 - P(T > s)$. But $T > s$ if and only if it is the case that not all of the labels after s 2-unshuffles are distinct. This is just the birthday problem; we are asking for the probability that at least two people have the same birthday, given that we have n people and there are 2^s possible birthdays. Using our formula from

Example 3.3, we find that

$$P(T > s) = 1 - \binom{2^s}{n}\frac{n!}{2^{sn}}. \qquad (3.4)$$

In Chapter 6, we will define the average value of a random variable. Using this idea, and the above equation, one can calculate the average value of the random variable T (see Exercise 6.1.41). For example, if $n = 52$, then the average value of T is about 11.7. This means that, on the average, about 12 riffle shuffles are needed for the process to be considered random.

Cut-Interleaving Pairs and Orderings

As was noted in the proof of Theorem 3.10, not all of the cut-interleaving pairs lead to different orderings. However, there is an easy formula which gives the number of such pairs that lead to a given ordering.

Theorem 3.11 If an ordering of length n has r rising sequences, then the number of cut-interleaving pairs under an a-shuffle of the identity ordering which lead to the ordering is

$$\binom{n+a-r}{n}.$$

Proof. To see why this is true, we need to count the number of ways in which the cut in an a-shuffle can be performed which will lead to a given ordering with r rising sequences. We can disregard the interleavings, since once a cut has been made, at most one interleaving will lead to a given ordering. Since the given ordering has r rising sequences, $r - 1$ of the division points in the cut are determined. The remaining $a - 1 - (r - 1) = a - r$ division points can be placed anywhere. The number of places to put these remaining division points is $n + 1$ (which is the number of spaces between the consecutive pairs of cards, including the positions at the beginning and the end of the deck). These places are chosen with repetition allowed, so the number of ways to make these choices is

$$\binom{n+a-r}{a-r} = \binom{n+a-r}{n}.$$

In particular, this means that if D is an ordering that is the result of applying an a-shuffle to the identity ordering, and if D has r rising sequences, then the probability assigned to D by this process is

$$\frac{\binom{n+a-r}{n}}{a^n}.$$

This completes the proof. $\square$

The above theorem shows that the essential information about the probability assigned to an ordering under an a-shuffle is just the number of rising sequences in

3.3. CARD SHUFFLING

the ordering. Thus, if we determine the number of orderings which contain exactly r rising sequences, for each r between 1 and n, then we will have determined the distribution function of the random variable which consists of applying a random a-shuffle to the identity ordering.

The number of orderings of $\{1, 2, \ldots, n\}$ with r rising sequences is denoted by $A(n, r)$, and is called an Eulerian number. There are many ways to calculate the values of these numbers; the following theorem gives one recursive method which follows immediately from what we already know about a-shuffles.

Theorem 3.12 Let a and n be positive integers. Then

$$a^n = \sum_{r=1}^{a} \binom{n+a-r}{n} A(n,r) . \tag{3.5}$$

Thus,

$$A(n,a) = a^n - \sum_{r=1}^{a-1} \binom{n+a-r}{n} A(n,r) .$$

In addition,

$$A(n,1) = 1 .$$

Proof. The second equation can be used to calculate the values of the Eulerian numbers, and follows immediately from the Equation 3.5. The last equation is a consequence of the fact that the only ordering of $\{1, 2, \ldots, n\}$ with one rising sequence is the identity ordering. Thus, it remains to prove Equation 3.5. We will count the set of a-shuffles of a deck with n cards in two ways. First, we know that there are a^n such shuffles (this was noted in the proof of Theorem 3.10). But there are $A(n,r)$ orderings of $\{1, 2, \ldots, n\}$ with r rising sequences, and Theorem 3.11 states that for each such ordering, there are exactly

$$\binom{n+a-r}{n}$$

cut-interleaving pairs that lead to the ordering. Therefore, the right-hand side of Equation 3.5 counts the set of a-shuffles of an n-card deck. This completes the proof. □

Random Orderings and Random Processes

We now turn to the second question that was asked at the beginning of this section: What do we mean by a "random" ordering? It is somewhat misleading to think about a given ordering as being random or not random. If we want to choose a random ordering from the set of all orderings of $\{1, 2, \ldots, n\}$, we mean that we want every ordering to be chosen with the same probability, i.e., any ordering is as "random" as any other.

The word "random" should really be used to describe a process. We will say that a process that produces an object from a (finite) set of objects is a random process if each object in the set is produced with the same probability by the process. In the present situation, the objects are the orderings, and the process which produces these objects is the shuffling process. It is easy to see that no a-shuffle is really a random process, since if T_1 and T_2 are two orderings with a different number of rising sequences, then they are produced by an a-shuffle, applied to the identity ordering, with different probabilities.

Variation Distance

Instead of requiring that a sequence of shuffles yield a process which is random, we will define a measure that describes how far away a given process is from a random process. Let X be any process which produces an ordering of $\{1, 2, \ldots, n\}$. Define $f_X(\pi)$ be the probability that X produces the ordering π. (Thus, X can be thought of as a random variable with distribution function f.) Let Ω_n be the set of all orderings of $\{1, 2, \ldots, n\}$. Finally, let $u(\pi) = 1/|\Omega_n|$ for all $\pi \in \Omega_n$. The function u is the distribution function of a process which produces orderings and which is random. For each ordering $\pi \in \Omega_n$, the quantity

$$|f_X(\pi) - u(\pi)|$$

is the difference between the actual and desired probabilities that X produces π. If we sum this over all orderings π and call this sum S, we see that $S = 0$ if and only if X is random, and otherwise S is positive. It is easy to show that the maximum value of S is 2, so we will multiply the sum by $1/2$ so that the value falls in the interval $[0, 1]$. Thus, we obtain the following sum as the formula for the *variation distance* between the two processes:

$$\| f_X - u \| = \frac{1}{2} \sum_{\pi \in \Omega_n} |f_X(\pi) - u(\pi)| \ .$$

Now we apply this idea to the case of shuffling. We let X be the process of s successive riffle shuffles applied to the identity ordering. We know that it is also possible to think of X as one 2^s-shuffle. We also know that f_X is constant on the set of all orderings with r rising sequences, where r is any positive integer. Finally, we know the value of f_X on an ordering with r rising sequences, and we know how many such orderings there are. Thus, in this specific case, we have

$$\| f_X - u \| = \frac{1}{2} \sum_{r=1}^{n} A(n,r) \left| \binom{2^s + n - r}{n} / 2^{ns} - \frac{1}{n!} \right| \ .$$

Since this sum has only n summands, it is easy to compute this for moderate sized values of n. For $n = 52$, we obtain the list of values given in Table 3.14.

To help in understanding these data, they are shown in graphical form in Figure 3.13. The program **VariationList** produces the data shown in both Table 3.14 and Figure 3.13. One sees that until 5 shuffles have occurred, the output of X is very far from random. After 5 shuffles, the distance from the random process is essentially halved each time a shuffle occurs.

3.3. CARD SHUFFLING

Number of Riffle Shuffles	Variation Distance
1	1
2	1
3	1
4	0.9999995334
5	0.9237329294
6	0.6135495966
7	0.3340609995
8	0.1671586419
9	0.0854201934
10	0.0429455489
11	0.0215023760
12	0.0107548935
13	0.0053779101
14	0.0026890130

Table 3.14: Distance to the random process.

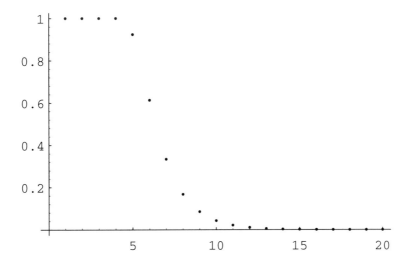

Figure 3.13: Distance to the random process.

Given the distribution functions $f_X(\pi)$ and $u(\pi)$ as above, there is another way to view the variation distance $\| f_X - u \|$. Given any event T (which is a subset of S_n), we can calculate its probability under the process X and under the uniform process. For example, we can imagine that T represents the set of all permutations in which the first player in a 7-player poker game is dealt a straight flush (five consecutive cards in the same suit). It is interesting to consider how much the probability of this event after a certain number of shuffles differs from the probability of this event if all permutations are equally likely. This difference can be thought of as describing how close the process X is to the random process with respect to the event T.

Now consider the event T such that the absolute value of the difference between these two probabilities is as large as possible. It can be shown that this absolute value is the variation distance between the process X and the uniform process. (The reader is asked to prove this fact in Exercise 4.)

We have just seen that, for a deck of 52 cards, the variation distance between the 7-riffle shuffle process and the random process is about .334. It is of interest to find an event T such that the difference between the probabilities that the two processes produce T is close to .334. An event with this property can be described in terms of the game called New-Age Solitaire.

New-Age Solitaire

This game was invented by Peter Doyle. It is played with a standard 52-card deck. We deal the cards face up, one at a time, onto a discard pile. If an ace is encountered, say the ace of Hearts, we use it to start a Heart pile. Each suit pile must be built up in order, from ace to king, using only subsequently dealt cards. Once we have dealt all of the cards, we pick up the discard pile and continue. We define the Yin suits to be Hearts and Clubs, and the Yang suits to be Diamonds and Spades. The game ends when either both Yin suit piles have been completed, or both Yang suit piles have been completed. It is clear that if the ordering of the deck is produced by the random process, then the probability that the Yin suit piles are completed first is exactly 1/2.

Now suppose that we buy a new deck of cards, break the seal on the package, and riffle shuffle the deck 7 times. If one tries this, one finds that the Yin suits win about 75% of the time. This is 25% more than we would get if the deck were in truly random order. This deviation is reasonably close to the theoretical maximum of 33.4% obtained above.

Why do the Yin suits win so often? In a brand new deck of cards, the suits are in the following order, from top to bottom: ace through king of Hearts, ace through king of Clubs, king through ace of Diamonds, and king through ace of Spades. Note that if the cards were not shuffled at all, then the Yin suit piles would be completed on the first pass, before any Yang suit cards are even seen. If we were to continue playing the game until the Yang suit piles are completed, it would take 13 passes through the deck to do this. Thus, one can see that in a new deck, the Yin suits are in the most advantageous order and the Yang suits are in the least advantageous

3.3. CARD SHUFFLING

order. Under 7 riffle shuffles, the relative advantage of the Yin suits over the Yang suits is preserved to a certain extent.

Exercises

1. Given any ordering σ of $\{1, 2, \ldots, n\}$, we can define σ^{-1}, the inverse ordering of σ, to be the ordering in which the ith element is the position occupied by i in σ. For example, if $\sigma = (1, 3, 5, 2, 4, 7, 6)$, then $\sigma^{-1} = (1, 4, 2, 5, 3, 7, 6)$. (If one thinks of these orderings as permutations, then σ^{-1} is the inverse of σ.)

   A *fall* occurs between two positions in an ordering if the left position is occupied by a larger number than the right position. It will be convenient to say that every ordering has a fall after the last position. In the above example, σ^{-1} has four falls. They occur after the second, fourth, sixth, and seventh positions. Prove that the number of rising sequences in an ordering σ equals the number of falls in σ^{-1}.

2. Show that if we start with the identity ordering of $\{1, 2, \ldots, n\}$, then the probability that an a-shuffle leads to an ordering with exactly r rising sequences equals
$$\frac{\binom{n+a-r}{n}}{a^n} A(n, r) ,$$
for $1 \leq r \leq a$.

3. Let D be a deck of n cards. We have seen that there are a^n a-shuffles of D. A coding of the set of a-unshuffles was given in the proof of Theorem 3.9. We will now give a coding of the a-shuffles which corresponds to the coding of the a-unshuffles. Let S be the set of all n-tuples of integers, each between 0 and $a - 1$. Let $M = (m_1, m_2, \ldots, m_n)$ be any element of S. Let n_i be the number of i's in M, for $0 \leq i \leq a - 1$. Suppose that we start with the deck in increasing order (i.e., the cards are numbered from 1 to n). We label the first n_0 cards with a 0, the next n_1 cards with a 1, etc. Then the a-shuffle corresponding to M is the shuffle which results in the ordering in which the cards labelled i are placed in the positions in M containing the label i. The cards with the same label are placed in these positions in increasing order of their numbers. For example, if $n = 6$ and $a = 3$, let $M = (1, 0, 2, 2, 0, 2)$. Then $n_0 = 2$, $n_1 = 1$, and $n_2 = 3$. So we label cards 1 and 2 with a 0, card 3 with a 1, and cards 4, 5, and 6 with a 2. Then cards 1 and 2 are placed in positions 2 and 5, card 3 is placed in position 1, and cards 4, 5, and 6 are placed in positions 3, 4, and 6, resulting in the ordering $(3, 1, 4, 5, 2, 6)$.

   (a) Using this coding, show that the probability that in an a-shuffle, the first card (i.e., card number 1) moves to the ith position, is given by the following expression:
   $$\frac{(a-1)^{i-1}a^{n-i} + (a-2)^{i-1}(a-1)^{n-i} + \cdots + 1^{i-1}2^{n-i}}{a^n} .$$

(b) Give an accurate estimate for the probability that in three riffle shuffles of a 52-card deck, the first card ends up in one of the first 26 positions. Using a computer, accurately estimate the probability of the same event after seven riffle shuffles.

4 Let X denote a particular process that produces elements of S_n, and let U denote the uniform process. Let the distribution functions of these processes be denoted by f_X and u, respectively. Show that the variation distance $\| f_X - u \|$ is equal to

$$\max_{T \subset S_n} \sum_{\pi \in T} \Bigl(f_X(\pi) - u(\pi) \Bigr) \ .$$

Hint: Write the permutations in S_n in decreasing order of the difference $f_X(\pi) - u(\pi)$.

5 Consider the process described in the text in which an n-card deck is repeatedly labelled and 2-unshuffled, in the manner described in the proof of Theorem 3.9. (See Figures 3.10 and 3.13.) The process continues until the labels are all different. Show that the process never terminates until at least $\lceil \log_2(n) \rceil$ unshuffles have been done.

Chapter 4

Conditional Probability

4.1 Discrete Conditional Probability

Conditional Probability

In this section we ask and answer the following question. Suppose we assign a distribution function to a sample space and then learn that an event E has occurred. How should we change the probabilities of the remaining events? We shall call the new probability for an event F the *conditional probability of F given E* and denote it by $P(F|E)$.

Example 4.1 An experiment consists of rolling a die once. Let X be the outcome. Let F be the event $\{X = 6\}$, and let E be the event $\{X > 4\}$. We assign the distribution function $m(\omega) = 1/6$ for $\omega = 1, 2, \ldots, 6$. Thus, $P(F) = 1/6$. Now suppose that the die is rolled and we are told that the event E has occurred. This leaves only two possible outcomes: 5 and 6. In the absence of any other information, we would still regard these outcomes to be equally likely, so the probability of F becomes $1/2$, making $P(F|E) = 1/2$. □

Example 4.2 In the Life Table (see Appendix C), one finds that in a population of 100,000 females, 89.835% can expect to live to age 60, while 57.062% can expect to live to age 80. Given that a woman is 60, what is the probability that she lives to age 80?

This is an example of a conditional probability. In this case, the original sample space can be thought of as a set of 100,000 females. The events E and F are the subsets of the sample space consisting of all women who live at least 60 years, and at least 80 years, respectively. We consider E to be the new sample space, and note that F is a subset of E. Thus, the size of E is 89,835, and the size of F is 57,062. So, the probability in question equals $57{,}062/89{,}835 = .6352$. Thus, a woman who is 60 has a 63.52% chance of living to age 80. □

Example 4.3 Consider our voting example from Section 1.2: three candidates A, B, and C are running for office. We decided that A and B have an equal chance of winning and C is only 1/2 as likely to win as A. Let A be the event "A wins," B that "B wins," and C that "C wins." Hence, we assigned probabilities $P(A) = 2/5$, $P(B) = 2/5$, and $P(C) = 1/5$.

Suppose that before the election is held, A drops out of the race. As in Example 4.1, it would be natural to assign new probabilities to the events B and C which are proportional to the original probabilities. Thus, we would have $P(B|\,A) = 2/3$, and $P(C|\,A) = 1/3$. It is important to note that any time we assign probabilities to real-life events, the resulting distribution is only useful if we take into account all relevant information. In this example, we may have knowledge that most voters who favor A will vote for C if A is no longer in the race. This will clearly make the probability that C wins greater than the value of 1/3 that was assigned above. $\square$

In these examples we assigned a distribution function and then were given new information that determined a new sample space, consisting of the outcomes that are still possible, and caused us to assign a new distribution function to this space.

We want to make formal the procedure carried out in these examples. Let $\Omega = \{\omega_1, \omega_2, \ldots, \omega_r\}$ be the original sample space with distribution function $m(\omega_j)$ assigned. Suppose we learn that the event E has occurred. We want to assign a new distribution function $m(\omega_j|E)$ to Ω to reflect this fact. Clearly, if a sample point ω_j is not in E, we want $m(\omega_j|E) = 0$. Moreover, in the absence of information to the contrary, it is reasonable to assume that the probabilities for ω_k in E should have the same relative magnitudes that they had before we learned that E had occurred. For this we require that

$$m(\omega_k|E) = cm(\omega_k)$$

for all ω_k in E, with c some positive constant. But we must also have

$$\sum_E m(\omega_k|E) = c \sum_E m(\omega_k) = 1 \ .$$

Thus,

$$c = \frac{1}{\sum_E m(\omega_k)} = \frac{1}{P(E)} \ .$$

(Note that this requires us to assume that $P(E) > 0$.) Thus, we will define

$$m(\omega_k|E) = \frac{m(\omega_k)}{P(E)}$$

for ω_k in E. We will call this new distribution the *conditional distribution* given E. For a general event F, this gives

$$P(F|E) = \sum_{F \cap E} m(\omega_k|E) = \sum_{F \cap E} \frac{m(\omega_k)}{P(E)} = \frac{P(F \cap E)}{P(E)} \ .$$

We call $P(F|E)$ the *conditional probability of F occurring given that E occurs*, and compute it using the formula

$$P(F|E) = \frac{P(F \cap E)}{P(E)} \ .$$

4.1. DISCRETE CONDITIONAL PROBABILITY

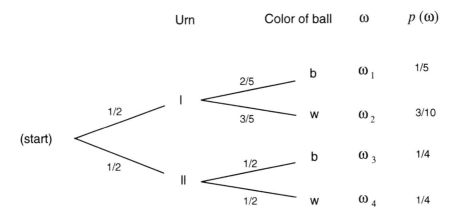

Figure 4.1: Tree diagram.

Example 4.4 (Example 4.1 continued) Let us return to the example of rolling a die. Recall that F is the event $X = 6$, and E is the event $X > 4$. Note that $E \cap F$ is the event F. So, the above formula gives

$$\begin{aligned} P(F|E) &= \frac{P(F \cap E)}{P(E)} \\ &= \frac{1/6}{1/3} \\ &= \frac{1}{2}, \end{aligned}$$

in agreement with the calculations performed earlier. □

Example 4.5 We have two urns, I and II. Urn I contains 2 black balls and 3 white balls. Urn II contains 1 black ball and 1 white ball. An urn is drawn at random and a ball is chosen at random from it. We can represent the sample space of this experiment as the paths through a tree as shown in Figure 4.1. The probabilities assigned to the paths are also shown.

Let B be the event "a black ball is drawn," and I the event "urn I is chosen." Then the branch weight 2/5, which is shown on one branch in the figure, can now be interpreted as the conditional probability $P(B|I)$.

Suppose we wish to calculate $P(I|B)$. Using the formula, we obtain

$$\begin{aligned} P(I|B) &= \frac{P(I \cap B)}{P(B)} \\ &= \frac{P(I \cap B)}{P(B \cap I) + P(B \cap II)} \\ &= \frac{1/5}{1/5 + 1/4} = \frac{4}{9}. \end{aligned}$$

□

CHAPTER 4. CONDITIONAL PROBABILITY

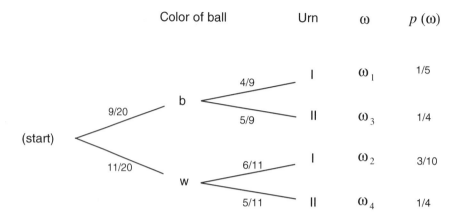

Figure 4.2: Reverse tree diagram.

Bayes Probabilities

Our original tree measure gave us the probabilities for drawing a ball of a given color, given the urn chosen. We have just calculated the *inverse probability* that a particular urn was chosen, given the color of the ball. Such an inverse probability is called a *Bayes probability* and may be obtained by a formula that we shall develop later. Bayes probabilities can also be obtained by simply constructing the tree measure for the two-stage experiment carried out in reverse order. We show this tree in Figure 4.2.

The paths through the reverse tree are in one-to-one correspondence with those in the forward tree, since they correspond to individual outcomes of the experiment, and so they are assigned the same probabilities. From the forward tree, we find that the probability of a black ball is

$$\frac{1}{2} \cdot \frac{2}{5} + \frac{1}{2} \cdot \frac{1}{2} = \frac{9}{20} .$$

The probabilities for the branches at the second level are found by simple division. For example, if x is the probability to be assigned to the top branch at the second level, we must have

$$\frac{9}{20} \cdot x = \frac{1}{5}$$

or $x = 4/9$. Thus, $P(I|B) = 4/9$, in agreement with our previous calculations. The reverse tree then displays all of the inverse, or Bayes, probabilities.

Example 4.6 We consider now a problem called the *Monty Hall* problem. This has long been a favorite problem but was revived by a letter from Craig Whitaker to Marilyn vos Savant for consideration in her column in *Parade Magazine*.[1] Craig wrote:

---

[1] Marilyn vos Savant, Ask Marilyn, *Parade Magazine*, 9 September; 2 December; 17 February 1990, reprinted in Marilyn vos Savant, *Ask Marilyn*, St. Martins, New York, 1992.

4.1. DISCRETE CONDITIONAL PROBABILITY

> Suppose you're on Monty Hall's *Let's Make a Deal!* You are given the choice of three doors, behind one door is a car, the others, goats. You pick a door, say 1, Monty opens another door, say 3, which has a goat. Monty says to you "Do you want to pick door 2?" Is it to your advantage to switch your choice of doors?

Marilyn gave a solution concluding that you should switch, and if you do, your probability of winning is 2/3. Several irate readers, some of whom identified themselves as having a PhD in mathematics, said that this is absurd since after Monty has ruled out one door there are only two possible doors and they should still each have the same probability 1/2 so there is no advantage to switching. Marilyn stuck to her solution and encouraged her readers to simulate the game and draw their own conclusions from this. We also encourage the reader to do this (see Exercise 11).

Other readers complained that Marilyn had not described the problem completely. In particular, the way in which certain decisions were made during a play of the game were not specified. This aspect of the problem will be discussed in Section 4.3. We will assume that the car was put behind a door by rolling a three-sided die which made all three choices equally likely. Monty knows where the car is, and always opens a door with a goat behind it. Finally, we assume that if Monty has a choice of doors (i.e., the contestant has picked the door with the car behind it), he chooses each door with probability 1/2. Marilyn clearly expected her readers to assume that the game was played in this manner.

As is the case with most apparent paradoxes, this one can be resolved through careful analysis. We begin by describing a simpler, related question. We say that a contestant is using the "stay" strategy if he picks a door, and, if offered a chance to switch to another door, declines to do so (i.e., he stays with his original choice). Similarly, we say that the contestant is using the "switch" strategy if he picks a door, and, if offered a chance to switch to another door, takes the offer. Now suppose that a contestant decides in advance to play the "stay" strategy. His only action in this case is to pick a door (and decline an invitation to switch, if one is offered). What is the probability that he wins a car? The same question can be asked about the "switch" strategy.

Using the "stay" strategy, a contestant will win the car with probability 1/3, since 1/3 of the time the door he picks will have the car behind it. On the other hand, if a contestant plays the "switch" strategy, then he will win whenever the door he originally picked does not have the car behind it, which happens 2/3 of the time.

This very simple analysis, though correct, does not quite solve the problem that Craig posed. Craig asked for the conditional probability that you win if you switch, given that you have chosen door 1 and that Monty has chosen door 3. To solve this problem, we set up the problem before getting this information and then compute the conditional probability given this information. This is a process that takes place in several stages; the car is put behind a door, the contestant picks a door, and finally Monty opens a door. Thus it is natural to analyze this using a tree measure. Here we make an additional assumption that if Monty has a choice

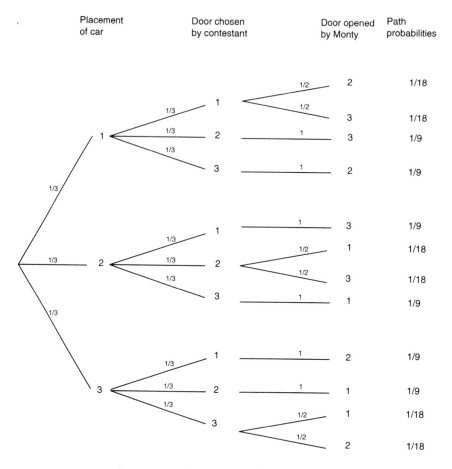

Figure 4.3: The Monty Hall problem.

of doors (i.e., the contestant has picked the door with the car behind it) then he picks each door with probability 1/2. The assumptions we have made determine the branch probabilities and these in turn determine the tree measure. The resulting tree and tree measure are shown in Figure 4.3. It is tempting to reduce the tree's size by making certain assumptions such as: "Without loss of generality, we will assume that the contestant always picks door 1." We have chosen not to make any such assumptions, in the interest of clarity.

Now the given information, namely that the contestant chose door 1 and Monty chose door 3, means only two paths through the tree are possible (see Figure 4.4). For one of these paths, the car is behind door 1 and for the other it is behind door 2. The path with the car behind door 2 is twice as likely as the one with the car behind door 1. Thus the conditional probability is 2/3 that the car is behind door 2 and 1/3 that it is behind door 1, so if you switch you have a 2/3 chance of winning the car, as Marilyn claimed.

At this point, the reader may think that the two problems above are the same, since they have the same answers. Recall that we assumed in the original problem

4.1. DISCRETE CONDITIONAL PROBABILITY

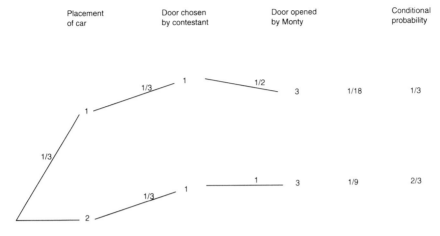

Figure 4.4: Conditional probabilities for the Monty Hall problem.

if the contestant chooses the door with the car, so that Monty has a choice of two doors, he chooses each of them with probability 1/2. Now suppose instead that in the case that he has a choice, he chooses the door with the larger number with probability 3/4. In the "switch" vs. "stay" problem, the probability of winning with the "switch" strategy is still 2/3. However, in the original problem, if the contestant switches, he wins with probability 4/7. The reader can check this by noting that the same two paths as before are the only two possible paths in the tree. The path leading to a win, if the contestant switches, has probability 1/3, while the path which leads to a loss, if the contestant switches, has probability 1/4.

□

Independent Events

It often happens that the knowledge that a certain event E has occurred has no effect on the probability that some other event F has occurred, that is, that $P(F|E) = P(F)$. One would expect that in this case, the equation $P(E|F) = P(E)$ would also be true. In fact (see Exercise 1), each equation implies the other. If these equations are true, we might say the F is *independent* of E. For example, you would not expect the knowledge of the outcome of the first toss of a coin to change the probability that you would assign to the possible outcomes of the second toss, that is, you would not expect that the second toss depends on the first. This idea is formalized in the following definition of independent events.

Definition 4.1 Two events E and F are *independent* if both E and F have positive probability and if
$$P(E|F) = P(E) ,$$
and
$$P(F|E) = P(F) .$$

□

As noted above, if both $P(E)$ and $P(F)$ are positive, then each of the above equations imply the other, so that to see whether two events are independent, only one of these equations must be checked (see Exercise 1).

The following theorem provides another way to check for independence.

Theorem 4.1 If $P(E) > 0$ and $P(F) > 0$, then E and F are independent if and only if
$$P(E \cap F) = P(E)P(F) .$$

Proof. Assume first that E and F are independent. Then $P(E|F) = P(E)$, and so
$$\begin{aligned} P(E \cap F) &= P(E|F)P(F) \\ &= P(E)P(F) . \end{aligned}$$

Assume next that $P(E \cap F) = P(E)P(F)$. Then
$$P(E|F) = \frac{P(E \cap F)}{P(F)} = P(E) .$$

Also,
$$P(F|E) = \frac{P(F \cap E)}{P(E)} = P(F) .$$

Therefore, E and F are independent. $\square$

Example 4.7 Suppose that we have a coin which comes up heads with probability p, and tails with probability q. Now suppose that this coin is tossed twice. Using a frequency interpretation of probability, it is reasonable to assign to the outcome (H, H) the probability p^2, to the outcome (H, T) the probability pq, and so on. Let E be the event that heads turns up on the first toss and F the event that tails turns up on the second toss. We will now check that with the above probability assignments, these two events are independent, as expected. We have $P(E) = p^2 + pq = p$, $P(F) = pq + q^2 = q$. Finally $P(E \cap F) = pq$, so $P(E \cap F) = P(E)P(F)$. $\square$

Example 4.8 It is often, but not always, intuitively clear when two events are independent. In Example 4.7, let A be the event "the first toss is a head" and B the event "the two outcomes are the same." Then
$$P(B|A) = \frac{P(B \cap A)}{P(A)} = \frac{P\{HH\}}{P\{HH,HT\}} = \frac{1/4}{1/2} = \frac{1}{2} = P(B).$$

Therefore, A and B are independent, but the result was not so obvious. $\square$

4.1. DISCRETE CONDITIONAL PROBABILITY

Example 4.9 Finally, let us give an example of two events that are not independent. In Example 4.7, let I be the event "heads on the first toss" and J the event "two heads turn up." Then $P(I) = 1/2$ and $P(J) = 1/4$. The event $I \cap J$ is the event "heads on both tosses" and has probability $1/4$. Thus, I and J are not independent since $P(I)P(J) = 1/8 \neq P(I \cap J)$. $\square$

We can extend the concept of independence to any finite set of events A_1, A_2, ..., A_n.

Definition 4.2 A set of events $\{A_1, A_2, \ldots, A_n\}$ is said to be *mutually independent* if for any subset $\{A_i, A_j, \ldots, A_m\}$ of these events we have

$$P(A_i \cap A_j \cap \cdots \cap A_m) = P(A_i)P(A_j)\cdots P(A_m),$$

or equivalently, if for any sequence $\bar{A}_1, \bar{A}_2, \ldots, \bar{A}_n$ with $\bar{A}_j = A_j$ or $\tilde{A}_j$,

$$P(\bar{A}_1 \cap \bar{A}_2 \cap \cdots \cap \bar{A}_n) = P(\bar{A}_1)P(\bar{A}_2)\cdots P(\bar{A}_n).$$

(For a proof of the equivalence in the case $n = 3$, see Exercise 33.) $\square$

Using this terminology, it is a fact that any sequence $(S, S, F, F, S, \ldots, S)$ of possible outcomes of a Bernoulli trials process forms a sequence of mutually independent events.

It is natural to ask: If all pairs of a set of events are independent, is the whole set mutually independent? The answer is *not necessarily,* and an example is given in Exercise 7.

It is important to note that the statement

$$P(A_1 \cap A_2 \cap \cdots \cap A_n) = P(A_1)P(A_2)\cdots P(A_n)$$

does not imply that the events A_1, A_2, ..., A_n are mutually independent (see Exercise 8).

Joint Distribution Functions and Independence of Random Variables

It is frequently the case that when an experiment is performed, several different quantities concerning the outcomes are investigated.

Example 4.10 Suppose we toss a coin three times. The basic random variable $\bar{X}$ corresponding to this experiment has eight possible outcomes, which are the ordered triples consisting of H's and T's. We can also define the random variable X_i, for $i = 1, 2, 3$, to be the outcome of the ith toss. If the coin is fair, then we should assign the probability $1/8$ to each of the eight possible outcomes. Thus, the distribution functions of X_1, X_2, and X_3 are identical; in each case they are defined by $m(H) = m(T) = 1/2$. $\square$

If we have several random variables $X_1, X_2, \ldots, X_n$ which correspond to a given experiment, then we can consider the joint random variable $\bar{X} = (X_1, X_2, \ldots, X_n)$ defined by taking an outcome ω of the experiment, and writing, as an n-tuple, the corresponding n outcomes for the random variables $X_1, X_2, \ldots, X_n$. Thus, if the random variable X_i has, as its set of possible outcomes the set R_i, then the set of possible outcomes of the joint random variable $\bar{X}$ is the Cartesian product of the R_i's, i.e., the set of all n-tuples of possible outcomes of the X_i's.

Example 4.11 (Example 4.10 continued) In the coin-tossing example above, let X_i denote the outcome of the ith toss. Then the joint random variable $\bar{X} = (X_1, X_2, X_3)$ has eight possible outcomes.

Suppose that we now define Y_i, for $i = 1, 2, 3$, as the number of heads which occur in the first i tosses. Then Y_i has $\{0, 1, \ldots, i\}$ as possible outcomes, so at first glance, the set of possible outcomes of the joint random variable $\bar{Y} = (Y_1, Y_2, Y_3)$ should be the set

$$\{(a_1, a_2, a_3) \ : \ 0 \le a_1 \le 1, 0 \le a_2 \le 2, 0 \le a_3 \le 3\} \ .$$

However, the outcome $(1, 0, 1)$ cannot occur, since we must have $a_1 \le a_2 \le a_3$. The solution to this problem is to define the probability of the outcome $(1, 0, 1)$ to be 0.

We now illustrate the assignment of probabilities to the various outcomes for the joint random variables $\bar{X}$ and $\bar{Y}$. In the first case, each of the eight outcomes should be assigned the probability $1/8$, since we are assuming that we have a fair coin. In the second case, since Y_i has $i + 1$ possible outcomes, the set of possible outcomes has size 24. Only eight of these 24 outcomes can actually occur, namely the ones satisfying $a_1 \le a_2 \le a_3$. Each of these outcomes corresponds to exactly one of the outcomes of the random variable $\bar{X}$, so it is natural to assign probability $1/8$ to each of these. We assign probability 0 to the other 16 outcomes. In each case, the probability function is called a joint distribution function. □

We collect the above ideas in a definition.

Definition 4.3 Let $X_1, X_2, \ldots, X_n$ be random variables associated with an experiment. Suppose that the sample space (i.e., the set of possible outcomes) of X_i is the set R_i. Then the joint random variable $\bar{X} = (X_1, X_2, \ldots, X_n)$ is defined to be the random variable whose outcomes consist of ordered n-tuples of outcomes, with the ith coordinate lying in the set R_i. The sample space Ω of $\bar{X}$ is the Cartesian product of the R_i's:
$$\Omega = R_1 \times R_1 \times \cdots \times R_n \ .$$
The joint distribution function of $\bar{X}$ is the function which gives the probability of each of the outcomes of $\bar{X}$. □

Example 4.12 (Example 4.10 continued) We now consider the assignment of probabilities in the above example. In the case of the random variable $\bar{X}$, the probability of any outcome (a_1, a_2, a_3) is just the product of the probabilities $P(X_i = a_i)$,

4.1. DISCRETE CONDITIONAL PROBABILITY

	Not smoke	Smoke	Total
Not cancer	40	10	50
Cancer	7	3	10
Totals	47	13	60

Table 4.1: Smoking and cancer.

		S	
		0	1
C	0	40/60	10/60
	1	7/60	3/60

Table 4.2: Joint distribution.

for $i = 1, 2, 3$. However, in the case of $\bar{Y}$, the probability assigned to the outcome $(1, 1, 0)$ is not the product of the probabilities $P(Y_1 = 1)$, $P(Y_2 = 1)$, and $P(Y_3 = 0)$. The difference between these two situations is that the value of X_i does not affect the value of X_j, if $i \neq j$, while the values of Y_i and Y_j affect one another. For example, if $Y_1 = 1$, then Y_2 cannot equal 0. This prompts the next definition. $\square$

Definition 4.4 The random variables X_1, X_2, ..., X_n are *mutually independent* if

$$P(X_1 = r_1, X_2 = r_2, \ldots, X_n = r_n)$$
$$= P(X_1 = r_1)P(X_2 = r_2) \cdots P(X_n = r_n)$$

for any choice of $r_1, r_2, \ldots, r_n$. Thus, if $X_1, X_2, \ldots, X_n$ are mutually independent, then the joint distribution function of the random variable

$$\bar{X} = (X_1, X_2, \ldots, X_n)$$

is just the product of the individual distribution functions. When two random variables are mutually independent, we shall say more briefly that they are *independent*. $\square$

Example 4.13 In a group of 60 people, the numbers who do or do not smoke and do or do not have cancer are reported as shown in Table 4.1. Let Ω be the sample space consisting of these 60 people. A person is chosen at random from the group. Let $C(\omega) = 1$ if this person has cancer and 0 if not, and $S(\omega) = 1$ if this person smokes and 0 if not. Then the joint distribution of $\{C, S\}$ is given in Table 4.2. For example $P(C = 0, S = 0) = 40/60$, $P(C = 0, S = 1) = 10/60$, and so forth. The distributions of the individual random variables are called *marginal distributions*. The marginal distributions of C and S are:

$$p_C = \begin{pmatrix} 0 & 1 \\ 50/60 & 10/60 \end{pmatrix},$$

$$p_S = \begin{pmatrix} 0 & 1 \\ 47/60 & 13/60 \end{pmatrix}.$$

The random variables S and C are not independent, since

$$P(C=1, S=1) = \frac{3}{60} = .05,$$
$$P(C=1)P(S=1) = \frac{10}{60} \cdot \frac{13}{60} = .036.$$

Note that we would also see this from the fact that

$$P(C=1|S=1) = \frac{3}{13} = .23,$$
$$P(C=1) = \frac{1}{6} = .167.$$

□

Independent Trials Processes

The study of random variables proceeds by considering special classes of random variables. One such class that we shall study is the class of *independent trials*.

Definition 4.5 A sequence of random variables X_1, X_2, ..., X_n that are mutually independent and that have the same distribution is called a sequence of independent trials or an *independent trials process*.

Independent trials processes arise naturally in the following way. We have a single experiment with sample space $R = \{r_1, r_2, \ldots, r_s\}$ and a distribution function

$$m_X = \begin{pmatrix} r_1 & r_2 & \cdots & r_s \\ p_1 & p_2 & \cdots & p_s \end{pmatrix}.$$

We repeat this experiment n times. To describe this total experiment, we choose as sample space the space

$$\Omega = R \times R \times \cdots \times R,$$

consisting of all possible sequences $\omega = (\omega_1, \omega_2, \ldots, \omega_n)$ where the value of each ω_j is chosen from R. We assign a distribution function to be the *product distribution*

$$m(\omega) = m(\omega_1) \cdot \ldots \cdot m(\omega_n),$$

with $m(\omega_j) = p_k$ when $\omega_j = r_k$. Then we let X_j denote the jth coordinate of the outcome $(r_1, r_2, \ldots, r_n)$. The random variables X_1, ..., X_n form an independent trials process. □

Example 4.14 An experiment consists of rolling a die three times. Let X_i represent the outcome of the ith roll, for $i = 1, 2, 3$. The common distribution function is

$$m_i = \begin{pmatrix} 1 & 2 & 3 & 4 & 5 & 6 \\ 1/6 & 1/6 & 1/6 & 1/6 & 1/6 & 1/6 \end{pmatrix}.$$

4.1. DISCRETE CONDITIONAL PROBABILITY

The sample space is $R^3 = R \times R \times R$ with $R = \{1, 2, 3, 4, 5, 6\}$. If $\omega = (1, 3, 6)$, then $X_1(\omega) = 1$, $X_2(\omega) = 3$, and $X_3(\omega) = 6$ indicating that the first roll was a 1, the second was a 3, and the third was a 6. The probability assigned to any sample point is

$$m(\omega) = \frac{1}{6} \cdot \frac{1}{6} \cdot \frac{1}{6} = \frac{1}{216} \; .$$

$\Box$

Example 4.15 Consider next a Bernoulli trials process with probability p for success on each experiment. Let $X_j(\omega) = 1$ if the jth outcome is success and $X_j(\omega) = 0$ if it is a failure. Then X_1, X_2, ..., X_n is an independent trials process. Each X_j has the same distribution function

$$m_j = \begin{pmatrix} 0 & 1 \\ q & p \end{pmatrix},$$

where $q = 1 - p$.

If $S_n = X_1 + X_2 + \cdots + X_n$, then

$$P(S_n = j) = \binom{n}{j} p^j q^{n-j} \; ,$$

and S_n has, as distribution, the binomial distribution $b(n, p, j)$. $\Box$

Bayes' Formula

In our examples, we have considered conditional probabilities of the following form: Given the outcome of the second stage of a two-stage experiment, find the probability for an outcome at the first stage. We have remarked that these probabilities are called *Bayes probabilities*.

We return now to the calculation of more general Bayes probabilities. Suppose we have a set of events H_1, H_2, ..., H_m that are pairwise disjoint and such that

$$\Omega = H_1 \cup H_2 \cup \cdots \cup H_m \; .$$

We call these events *hypotheses*. We also have an event E that gives us some information about which hypothesis is correct. We call this event *evidence*.

Before we receive the evidence, then, we have a set of *prior probabilities* $P(H_1)$, $P(H_2)$, ..., $P(H_m)$ for the hypotheses. If we know the correct hypothesis, we know the probability for the evidence. That is, we know $P(E|H_i)$ for all i. We want to find the probabilities for the hypotheses given the evidence. That is, we want to find the conditional probabilities $P(H_i|E)$. These probabilities are called the *posterior probabilities*.

To find these probabilities, we write them in the form

$$P(H_i|E) = \frac{P(H_i \cap E)}{P(E)} \; . \tag{4.1}$$

	Number having		The results		
Disease	this disease	+ +	+ −	− +	− −
d_1	3215	2110	301	704	100
d_2	2125	396	132	1187	410
d_3	4660	510	3568	73	509
Total	10000				

Table 4.3: Diseases data.

We can calculate the numerator from our given information by

$$P(H_i \cap E) = P(H_i)P(E|H_i) . \quad (4.2)$$

Since one and only one of the events H_1, H_2, ..., H_m can occur, we can write the probability of E as

$$P(E) = P(H_1 \cap E) + P(H_2 \cap E) + \cdots + P(H_m \cap E) .$$

Using Equation 4.2, the above expression can be seen to equal

$$P(H_1)P(E|H_1) + P(H_2)P(E|H_2) + \cdots + P(H_m)P(E|H_m) . \quad (4.3)$$

Using (4.1), (4.2), and (4.3) yields *Bayes' formula*:

$$P(H_i|E) = \frac{P(H_i)P(E|H_i)}{\sum_{k=1}^{m} P(H_k)P(E|H_k)} .$$

Although this is a very famous formula, we will rarely use it. If the number of hypotheses is small, a simple tree measure calculation is easily carried out, as we have done in our examples. If the number of hypotheses is large, then we should use a computer.

Bayes probabilities are particularly appropriate for medical diagnosis. A doctor is anxious to know which of several diseases a patient might have. She collects evidence in the form of the outcomes of certain tests. From statistical studies the doctor can find the prior probabilities of the various diseases before the tests, and the probabilities for specific test outcomes, given a particular disease. What the doctor wants to know is the posterior probability for the particular disease, given the outcomes of the tests.

Example 4.16 A doctor is trying to decide if a patient has one of three diseases d_1, d_2, or d_3. Two tests are to be carried out, each of which results in a positive (+) or a negative (−) outcome. There are four possible test patterns ++, +−, −+, and −−. National records have indicated that, for 10,000 people having one of these three diseases, the distribution of diseases and test results are as in Table 4.3.

From this data, we can estimate the prior probabilities for each of the diseases and, given a particular disease, the probability of a particular test outcome. For example, the prior probability of disease d_1 may be estimated to be $3215/10{,}000 = .3215$. The probability of the test result +−, given disease d_1, may be estimated to be $301/3125 = .094$.

4.1. DISCRETE CONDITIONAL PROBABILITY

		d_1	d_2	d_3
+	+	.700	.132	.168
+	−	.076	.033	.891
−	+	.357	.605	.038
−	−	.098	.405	.497

Table 4.4: Posterior probabilities.

We can now use Bayes' formula to compute various posterior probabilities. The computer program **Bayes** computes these posterior probabilities. The results for this example are shown in Table 4.4.

We note from the outcomes that, when the test result is ++, the disease d_1 has a significantly higher probability than the other two. When the outcome is +−, this is true for disease d_3. When the outcome is −+, this is true for disease d_2. Note that these statements might have been guessed by looking at the data. If the outcome is −−, the most probable cause is d_3, but the probability that a patient has d_2 is only slightly smaller. If one looks at the data in this case, one can see that it might be hard to guess which of the two diseases d_2 and d_3 is more likely. □

Our final example shows that one has to be careful when the prior probabilities are small.

Example 4.17 A doctor gives a patient a test for a particular cancer. Before the results of the test, the only evidence the doctor has to go on is that 1 woman in 1000 has this cancer. Experience has shown that, in 99 percent of the cases in which cancer is present, the test is positive; and in 95 percent of the cases in which it is not present, it is negative. If the test turns out to be positive, what probability should the doctor assign to the event that cancer is present? An alternative form of this question is to ask for the relative frequencies of false positives and cancers.

We are given that prior(cancer) = .001 and prior(not cancer) = .999. We know also that $P(+|\text{cancer}) = .99$, $P(-|\text{cancer}) = .01$, $P(+|\text{not cancer}) = .05$, and $P(-|\text{not cancer}) = .95$. Using this data gives the result shown in Figure 4.5.

We see now that the probability of cancer given a positive test has only increased from .001 to .019. While this is nearly a twenty-fold increase, the probability that the patient has the cancer is still small. Stated in another way, among the positive results, 98.1 percent are false positives, and 1.9 percent are cancers. When a group of second-year medical students was asked this question, over half of the students incorrectly guessed the probability to be greater than .5. □

Historical Remarks

Conditional probability was used long before it was formally defined. Pascal and Fermat considered the *problem of points*: given that team A has won m games and team B has won n games, what is the probability that A will win the series? (See Exercises 40–42.) This is clearly a conditional probability problem.

In his book, Huygens gave a number of problems, one of which was:

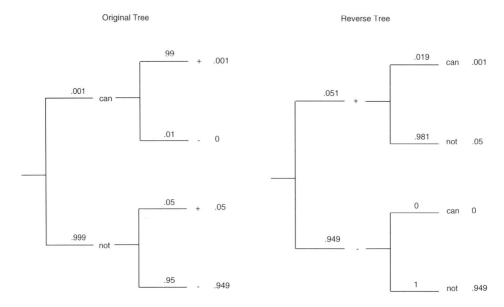

Figure 4.5: Forward and reverse tree diagrams.

> Three gamblers, A, B and C, take 12 balls of which 4 are white and 8 black. They play with the rules that the drawer is blindfolded, A is to draw first, then B and then C, the winner to be the one who first draws a white ball. What is the ratio of their chances?[2]

From his answer it is clear that Huygens meant that each ball is replaced after drawing. However, John Hudde, the mayor of Amsterdam, assumed that he meant to sample without replacement and corresponded with Huygens about the difference in their answers. Hacking remarks that "Neither party can understand what the other is doing."[3]

By the time of de Moivre's book, *The Doctrine of Chances,* these distinctions were well understood. De Moivre defined independence and dependence as follows:

> Two Events are independent, when they have no connexion one with the other, and that the happening of one neither forwards nor obstructs the happening of the other.
>
> Two Events are dependent, when they are so connected together as that the Probability of either's happening is altered by the happening of the other.[4]

De Moivre used sampling with and without replacement to illustrate that the probability that two independent events both happen is the product of their probabilities, and for dependent events that:

---

[2]Quoted in F. N. David, *Games, Gods and Gambling* (London: Griffin, 1962), p. 119.
[3]I. Hacking, *The Emergence of Probability* (Cambridge: Cambridge University Press, 1975), p. 99.
[4]A. de Moivre, *The Doctrine of Chances,* 3rd ed. (New York: Chelsea, 1967), p. 6.

4.1. DISCRETE CONDITIONAL PROBABILITY

The Probability of the happening of two Events dependent, is the product of the Probability of the happening of one of them, by the Probability which the other will have of happening, when the first is considered as having happened; and the same Rule will extend to the happening of as many Events as may be assigned.[5]

The formula that we call Bayes' formula, and the idea of computing the probability of a hypothesis given evidence, originated in a famous essay of Thomas Bayes. Bayes was an ordained minister in Tunbridge Wells near London. His mathematical interests led him to be elected to the Royal Society in 1742, but none of his results were published within his lifetime. The work upon which his fame rests, "An Essay Toward Solving a Problem in the Doctrine of Chances," was published in 1763, three years after his death.[6] Bayes reviewed some of the basic concepts of probability and then considered a new kind of inverse probability problem requiring the use of conditional probability.

Bernoulli, in his study of processes that we now call Bernoulli trials, had proven his famous law of large numbers which we will study in Chapter 8. This theorem assured the experimenter that if he knew the probability p for success, he could predict that the proportion of successes would approach this value as he increased the number of experiments. Bernoulli himself realized that in most interesting cases you do not know the value of p and saw his theorem as an important step in showing that you could determine p by experimentation.

To study this problem further, Bayes started by assuming that the probability p for success is itself determined by a random experiment. He assumed in fact that this experiment was such that this value for p is equally likely to be any value between 0 and 1. Without knowing this value we carry out n experiments and observe m successes. Bayes proposed the problem of finding the conditional probability that the unknown probability p lies between a and b. He obtained the answer:

$$P(a \leq p < b | m \text{ successes in } n \text{ trials}) = \frac{\int_a^b x^m (1-x)^{n-m} \, dx}{\int_0^1 x^m (1-x)^{n-m} \, dx} \ .$$

We shall see in the next section how this result is obtained. Bayes clearly wanted to show that the conditional distribution function, given the outcomes of more and more experiments, becomes concentrated around the true value of p. Thus, Bayes was trying to solve an *inverse problem*. The computation of the integrals was too difficult for exact solution except for small values of j and n, and so Bayes tried approximate methods. His methods were not very satisfactory and it has been suggested that this discouraged him from publishing his results.

However, his paper was the first in a series of important studies carried out by Laplace, Gauss, and other great mathematicians to solve inverse problems. They studied this problem in terms of errors in measurements in astronomy. If an astronomer were to know the true value of a distance and the nature of the random

---

[5]ibid, p. 7.
[6]T. Bayes, "An Essay Toward Solving a Problem in the Doctrine of Chances," *Phil. Trans. Royal Soc. London*, vol. 53 (1763), pp. 370–418.

errors caused by his measuring device he could predict the probabilistic nature of his measurements. In fact, however, he is presented with the inverse problem of knowing the nature of the random errors, and the values of the measurements, and wanting to make inferences about the unknown true value.

As Maistrov remarks, the formula that we have called Bayes' formula does not appear in his essay. Laplace gave it this name when he studied these inverse problems.[7] The computation of inverse probabilities is fundamental to statistics and has led to an important branch of statistics called Bayesian analysis, assuring Bayes eternal fame for his brief essay.

Exercises

1. Assume that E and F are two events with positive probabilities. Show that if $P(E|F) = P(E)$, then $P(F|E) = P(F)$.

2. A coin is tossed three times. What is the probability that exactly two heads occur, given that

    (a) the first outcome was a head?

    (b) the first outcome was a tail?

    (c) the first two outcomes were heads?

    (d) the first two outcomes were tails?

    (e) the first outcome was a head and the third outcome was a head?

3. A die is rolled twice. What is the probability that the sum of the faces is greater than 7, given that

    (a) the first outcome was a 4?

    (b) the first outcome was greater than 3?

    (c) the first outcome was a 1?

    (d) the first outcome was less than 5?

4. A card is drawn at random from a deck of cards. What is the probability that

    (a) it is a heart, given that it is red?

    (b) it is higher than a 10, given that it is a heart? (Interpret J, Q, K, A as 11, 12, 13, 14.)

    (c) it is a jack, given that it is red?

5. A coin is tossed three times. Consider the following events
    A: Heads on the first toss.
    B: Tails on the second.
    C: Heads on the third toss.
    D: All three outcomes the same (HHH or TTT).
    E: Exactly one head turns up.

---

[7]L. E. Maistrov, *Probability Theory: A Historical Sketch*, trans. and ed. Samual Kotz (New York: Academic Press, 1974), p. 100.

(a) Which of the following pairs of these events are independent?
   (1) A, B
   (2) A, D
   (3) A, E
   (4) D, E

(b) Which of the following triples of these events are independent?
   (1) A, B, C
   (2) A, B, D
   (3) C, D, E

6 From a deck of five cards numbered 2, 4, 6, 8, and 10, respectively, a card is drawn at random and replaced. This is done three times. What is the probability that the card numbered 2 was drawn exactly two times, given that the sum of the numbers on the three draws is 12?

7 A coin is tossed twice. Consider the following events.
A: Heads on the first toss.
B: Heads on the second toss.
C: The two tosses come out the same.

(a) Show that A, B, C are pairwise independent but not independent.

(b) Show that C is independent of A and B but not of $A \cap B$.

8 Let $\Omega = \{a, b, c, d, e, f\}$. Assume that $m(a) = m(b) = 1/8$ and $m(c) = m(d) = m(e) = m(f) = 3/16$. Let A, B, and C be the events $A = \{d, e, a\}$, $B = \{c, e, a\}$, $C = \{c, d, a\}$. Show that $P(A \cap B \cap C) = P(A)P(B)P(C)$ but no two of these events are independent.

9 What is the probability that a family of two children has

(a) two boys given that it has at least one boy?

(b) two boys given that the first child is a boy?

10 In Example 4.2, we used the Life Table (see Appendix C) to compute a conditional probability. The number 93,753 in the table, corresponding to 40-year-old males, means that of all the males born in the United States in 1950, 93.753% were alive in 1990. Is it reasonable to use this as an estimate for the probability of a male, born this year, surviving to age 40?

11 Simulate the Monty Hall problem. Carefully state any assumptions that you have made when writing the program. Which version of the problem do you think that you are simulating?

12 In Example 4.17, how large must the prior probability of cancer be to give a posterior probability of .5 for cancer given a positive test?

13 Two cards are drawn from a bridge deck. What is the probability that the second card drawn is red?

14 If $P(\tilde{B}) = 1/4$ and $P(A|B) = 1/2$, what is $P(A \cap B)$?

15 (a) What is the probability that your bridge partner has exactly two aces, given that she has at least one ace?

(b) What is the probability that your bridge partner has exactly two aces, given that she has the ace of spades?

16 Prove that for any three events A, B, C, each having positive probability,

$$P(A \cap B \cap C) = P(A)P(B|A)P(C|A \cap B) .$$

17 Prove that if A and B are independent so are

(a) A and $\tilde{B}$.

(b) $\tilde{A}$ and $\tilde{B}$.

18 A doctor assumes that a patient has one of three diseases d_1, d_2, or d_3. Before any test, he assumes an equal probability for each disease. He carries out a test that will be positive with probability .8 if the patient has d_1, .6 if he has disease d_2, and .4 if he has disease d_3. Given that the outcome of the test was positive, what probabilities should the doctor now assign to the three possible diseases?

19 In a poker hand, John has a very strong hand and bets 5 dollars. The probability that Mary has a better hand is .04. If Mary had a better hand she would raise with probability .9, but with a poorer hand she would only raise with probability .1. If Mary raises, what is the probability that she has a better hand than John does?

20 The Polya urn model for contagion is as follows: We start with an urn which contains one white ball and one black ball. At each second we choose a ball at random from the urn and replace this ball and add one more of the color chosen. Write a program to simulate this model, and see if you can make any predictions about the proportion of white balls in the urn after a large number of draws. Is there a tendency to have a large fraction of balls of the same color in the long run?

21 It is desired to find the probability that in a bridge deal each player receives an ace. A student argues as follows. It does not matter where the first ace goes. The second ace must go to one of the other three players and this occurs with probability 3/4. Then the next must go to one of two, an event of probability 1/2, and finally the last ace must go to the player who does not have an ace. This occurs with probability 1/4. The probability that all these events occur is the product $(3/4)(1/2)(1/4) = 3/32$. Is this argument correct?

22 One coin in a collection of 65 has two heads. The rest are fair. If a coin, chosen at random from the lot and then tossed, turns up heads 6 times in a row, what is the probability that it is the two-headed coin?

4.1. DISCRETE CONDITIONAL PROBABILITY

23 You are given two urns and fifty balls. Half of the balls are white and half are black. You are asked to distribute the balls in the urns with no restriction placed on the number of either type in an urn. How should you distribute the balls in the urns to maximize the probability of obtaining a white ball if an urn is chosen at random and a ball drawn out at random? Justify your answer.

24 A fair coin is thrown n times. Show that the conditional probability of a head on any specified trial, given a total of k heads over the n trials, is k/n ($k > 0$).

25 (Johnsonbough[8]) A coin with probability p for heads is tossed n times. Let E be the event "a head is obtained on the first toss' and F_k the event 'exactly k heads are obtained." For which pairs (n, k) are E and F_k independent?

26 Suppose that A and B are events such that $P(A|B) = P(B|A)$ and $P(A \cup B) = 1$ and $P(A \cap B) > 0$. Prove that $P(A) > 1/2$.

27 (Chung[9]) In London, half of the days have some rain. The weather forecaster is correct 2/3 of the time, i.e., the probability that it rains, given that she has predicted rain, and the probability that it does not rain, given that she has predicted that it won't rain, are both equal to 2/3. When rain is forecast, Mr. Pickwick takes his umbrella. When rain is not forecast, he takes it with probability 1/3. Find

(a) the probability that Pickwick has no umbrella, given that it rains.

(b) the probability that it doesn't rain, given that he brings his umbrella.

28 Probability theory was used in a famous court case: *People v. Collins*.[10] In this case a purse was snatched from an elderly person in a Los Angeles suburb. A couple seen running from the scene were described as a black man with a beard and a mustache and a blond girl with hair in a ponytail. Witnesses said they drove off in a partly yellow car. Malcolm and Janet Collins were arrested. He was black and though clean shaven when arrested had evidence of recently having had a beard and a mustache. She was blond and usually wore her hair in a ponytail. They drove a partly yellow Lincoln. The prosecution called a professor of mathematics as a witness who suggested that a conservative set of probabilities for the characteristics noted by the witnesses would be as shown in Table 4.5.

The prosecution then argued that the probability that all of these characteristics are met by a randomly chosen couple is the product of the probabilities or 1/12,000,000, which is very small. He claimed this was proof beyond a reasonable doubt that the defendants were guilty. The jury agreed and handed down a verdict of guilty of second-degree robbery.

---
[8] R. Johnsonbough, "Problem #103," *Two Year College Math Journal*, vol. 8 (1977), p. 292.
[9] K. L. Chung, *Elementary Probability Theory With Stochastic Processes*, 3rd ed. (New York: Springer-Verlag, 1979), p. 152.
[10] M. W. Gray, "Statistics and the Law," *Mathematics Magazine*, vol. 56 (1983), pp. 67–81.

man with mustache	1/4
girl with blond hair	1/3
girl with ponytail	1/10
black man with beard	1/10
interracial couple in a car	1/1000
partly yellow car	1/10

Table 4.5: Collins case probabilities.

If you were the lawyer for the Collins couple how would you have countered the above argument? (The appeal of this case is discussed in Exercise 5.1.34.)

29 A student is applying to Harvard and Dartmouth. He estimates that he has a probability of .5 of being accepted at Dartmouth and .3 of being accepted at Harvard. He further estimates the probability that he will be accepted by both is .2. What is the probability that he is accepted by Dartmouth if he is accepted by Harvard? Is the event "accepted at Harvard" independent of the event "accepted at Dartmouth"?

30 Luxco, a wholesale lightbulb manufacturer, has two factories. Factory A sells bulbs in lots that consists of 1000 regular and 2000 *softglow* bulbs each. Random sampling has shown that on the average there tend to be about 2 bad regular bulbs and 11 bad softglow bulbs per lot. At factory B the lot size is reversed—there are 2000 regular and 1000 softglow per lot—and there tend to be 5 bad regular and 6 bad softglow bulbs per lot.

The manager of factory A asserts, "We're obviously the better producer; our bad bulb rates are .2 percent and .55 percent compared to B's .25 percent and .6 percent. We're better at both regular and softglow bulbs by half of a tenth of a percent each."

"Au contraire," counters the manager of B, "each of our 3000 bulb lots contains only 11 bad bulbs, while A's 3000 bulb lots contain 13. So our .37 percent bad bulb rate beats their .43 percent."

Who is right?

31 Using the Life Table for 1981 given in Appendix C, find the probability that a male of age 60 in 1981 lives to age 80. Find the same probability for a female.

32 (a) There has been a blizzard and Helen is trying to drive from Woodstock to Tunbridge, which are connected like the top graph in Figure 4.6. Here p and q are the probabilities that the two roads are passable. What is the probability that Helen can get from Woodstock to Tunbridge?

(b) Now suppose that Woodstock and Tunbridge are connected like the middle graph in Figure 4.6. What now is the probability that she can get from W to T? Note that if we think of the roads as being components of a system, then in (a) and (b) we have computed the *reliability* of a system whose components are (a) *in series* and (b) *in parallel*.

4.1. DISCRETE CONDITIONAL PROBABILITY

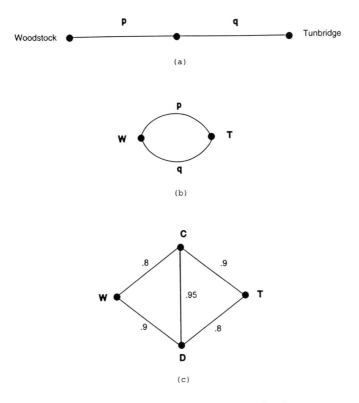

Figure 4.6: From Woodstock to Tunbridge.

(c) Now suppose W and T are connected like the bottom graph in Figure 4.6. Find the probability of Helen's getting from W to T. *Hint*: If the road from C to D is impassable, it might as well not be there at all; if it is passable, then figure out how to use part (b) twice.

33 Let A_1, A_2, and A_3 be events, and let B_i represent either A_i or its complement $\tilde{A}_i$. Then there are eight possible choices for the triple (B_1, B_2, B_3). Prove that the events A_1, A_2, A_3 are independent if and only if

$$P(B_1 \cap B_2 \cap B_3) = P(B_1)P(B_2)P(B_3) ,$$

for all eight of the possible choices for the triple (B_1, B_2, B_3).

34 Four women, A, B, C, and D, check their hats, and the hats are returned in a random manner. Let Ω be the set of all possible permutations of A, B, C, D. Let $X_j = 1$ if the jth woman gets her own hat back and 0 otherwise. What is the distribution of X_j? Are the X_i's mutually independent?

35 A box has numbers from 1 to 10. A number is drawn at random. Let X_1 be the number drawn. This number is replaced, and the ten numbers mixed. A second number X_2 is drawn. Find the distributions of X_1 and X_2. Are X_1 and X_2 independent? Answer the same questions if the first number is not replaced before the second is drawn.

	Y			
	-1	0	1	2
X -1	0	1/36	1/6	1/12
0	1/18	0	1/18	0
1	0	1/36	1/6	1/12
2	1/12	0	1/12	1/6

Table 4.6: Joint distribution.

36 A die is thrown twice. Let X_1 and X_2 denote the outcomes. Define $X = \min(X_1, X_2)$. Find the distribution of X.

*37 Given that $P(X = a) = r$, $P(\max(X, Y) = a) = s$, and $P(\min(X, Y) = a) = t$, show that you can determine $u = P(Y = a)$ in terms of r, s, and t.

38 A fair coin is tossed three times. Let X be the number of heads that turn up on the first two tosses and Y the number of heads that turn up on the third toss. Give the distribution of

(a) the random variables X and Y.

(b) the random variable $Z = X + Y$.

(c) the random variable $W = X - Y$.

39 Assume that the random variables X and Y have the joint distribution given in Table 4.6.

(a) What is $P(X \geq 1 \text{ and } Y \leq 0)$?

(b) What is the conditional probability that $Y \leq 0$ given that $X = 2$?

(c) Are X and Y independent?

(d) What is the distribution of $Z = XY$?

40 In the *problem of points*, discussed in the historical remarks in Section 3.2, two players, A and B, play a series of points in a game with player A winning each point with probability p and player B winning each point with probability $q = 1 - p$. The first player to win N points wins the game. Assume that $N = 3$. Let X be a random variable that has the value 1 if player A wins the series and 0 otherwise. Let Y be a random variable with value the number of points played in a game. Find the distribution of X and Y when $p = 1/2$. Are X and Y independent in this case? Answer the same questions for the case $p = 2/3$.

41 The letters between Pascal and Fermat, which are often credited with having started probability theory, dealt mostly with the *problem of points* described in Exercise 40. Pascal and Fermat considered the problem of finding a fair division of stakes if the game must be called off when the first player has won r games and the second player has won s games, with $r < N$ and $s < N$. Let $P(r, s)$ be the probability that player A wins the game if he has already won r points and player B has won s points. Then

(a) $P(r, N) = 0$ if $r < N$,

(b) $P(N, s) = 1$ if $s < N$,

(c) $P(r, s) = pP(r + 1, s) + qP(r, s + 1)$ if $r < N$ and $s < N$;

and (1), (2), and (3) determine $P(r, s)$ for $r \leq N$ and $s \leq N$. Pascal used these facts to find $P(r, s)$ by working backward: He first obtained $P(N-1, j)$ for $j = N-1, N-2, \ldots, 0$; then, from these values, he obtained $P(N-2, j)$ for $j = N-1, N-2, \ldots, 0$ and, continuing backward, obtained all the values $P(r, s)$. Write a program to compute $P(r, s)$ for given N, a, b, and p. *Warning*: Follow Pascal and you will be able to run $N = 100$; use recursion and you will not be able to run $N = 20$.

42 Fermat solved the *problem of points* (see Exercise 40) as follows: He realized that the problem was difficult because the possible ways the play might go are not equally likely. For example, when the first player needs two more games and the second needs three to win, two possible ways the series might go for the first player are WLW and LWLW. These sequences are not equally likely. To avoid this difficulty, Fermat extended the play, adding fictitious plays so that the series went the maximum number of games needed (four in this case). He obtained equally likely outcomes and used, in effect, the Pascal triangle to calculate $P(r, s)$. Show that this leads to a *formula* for $P(r, s)$ even for the case $p \neq 1/2$.

43 The Yankees are playing the Dodgers in a world series. The Yankees win each game with probability .6. What is the probability that the Yankees win the series? (The series is won by the first team to win four games.)

44 C. L. Anderson[11] has used Fermat's argument for the *problem of points* to prove the following result due to J. G. Kingston. You are playing the *game of points* (see Exercise 40) but, at each point, when you serve you win with probability p, and when your opponent serves you win with probability $\bar{p}$. You will serve first, but you can choose one of the following two conventions for serving: for the first convention you alternate service (tennis), and for the second the person serving continues to serve until he loses a point and then the other player serves (racquetball). The first player to win N points wins the game. The problem is to show that the probability of winning the game is the same under either convention.

(a) Show that, under either convention, you will serve at most N points and your opponent at most $N - 1$ points.

(b) Extend the number of points to $2N - 1$ so that you serve N points and your opponent serves $N - 1$. For example, you serve any additional points necessary to make N serves and then your opponent serves any additional points necessary to make him serve $N - 1$ points. The winner

---
[11]C. L. Anderson, "Note on the Advantage of First Serve," *Journal of Combinatorial Theory*, Series A, vol. 23 (1977), p. 363.

is now the person, in the extended game, who wins the most points. Show that playing these additional points has not changed the winner.

(c) Show that (a) and (b) prove that you have the same probability of winning the game under either convention.

45 In the previous problem, assume that $p = 1 - \bar{p}$.

(a) Show that under either service convention, the first player will win more often than the second player if and only if $p > .5$.

(b) In volleyball, a team can only win a point while it is serving. Thus, any individual "play" either ends with a point being awarded to the serving team or with the service changing to the other team. The first team to win N points wins the game. (We ignore here the additional restriction that the winning team must be ahead by at least two points at the end of the game.) Assume that each team has the same probability of winning the play when it is serving, i.e., that $p = 1 - \bar{p}$. Show that in this case, the team that serves first will win more than half the time, as long as $p > 0$. (If $p = 0$, then the game never ends.) *Hint*: Define p' to be the probability that a team wins the next point, given that it is serving. If we write $q = 1 - p$, then one can show that

$$p' = \frac{p}{1 - q^2} .$$

If one now considers this game in a slightly different way, one can see that the second service convention in the preceding problem can be used, with p replaced by p'.

46 A poker hand consists of 5 cards dealt from a deck of 52 cards. Let X and Y be, respectively, the number of aces and kings in a poker hand. Find the joint distribution of X and Y.

47 Let X_1 and X_2 be independent random variables and let $Y_1 = \phi_1(X_1)$ and $Y_2 = \phi_2(X_2)$.

(a) Show that

$$P(Y_1 = r, Y_2 = s) = \sum_{\substack{\phi_1(a)=r \\ \phi_2(b)=s}} P(X_1 = a, X_2 = b) .$$

(b) Using (a), show that $P(Y_1 = r, Y_2 = s) = P(Y_1 = r)P(Y_2 = s)$ so that Y_1 and Y_2 are independent.

48 Let Ω be the sample space of an experiment. Let E be an event with $P(E) > 0$ and define $m_E(\omega)$ by $m_E(\omega) = m(\omega|E)$. Prove that $m_E(\omega)$ is a distribution function on E, that is, that $m_E(\omega) \geq 0$ and that $\sum_{\omega \in \Omega} m_E(\omega) = 1$. The function m_E is called the *conditional distribution given* E.

4.1. DISCRETE CONDITIONAL PROBABILITY

49 You are given two urns each containing two biased coins. The coins in urn I come up heads with probability p_1, and the coins in urn II come up heads with probability $p_2 \neq p_1$. You are given a choice of (a) choosing an urn at random and tossing the two coins in this urn or (b) choosing one coin from each urn and tossing these two coins. You win a prize if both coins turn up heads. Show that you are better off selecting choice (a).

50 Prove that, if A_1, A_2, ..., A_n are independent events defined on a sample space Ω and if $0 < P(A_j) < 1$ for all j, then Ω must have at least 2^n points.

51 Prove that if

$$P(A|C) \geq P(B|C) \text{ and } P(A|\tilde{C}) \geq P(B|\tilde{C}) \,,$$

then $P(A) \geq P(B)$.

52 A coin is in one of n boxes. The probability that it is in the ith box is p_i. If you search in the ith box and it is there, you find it with probability a_i. Show that the probability p that the coin is in the jth box, given that you have looked in the ith box and not found it, is

$$p = \begin{cases} p_j/(1 - a_i p_i), & \text{if } j \neq i, \\ (1 - a_i) p_i/(1 - a_i p_i), & \text{if } j = i. \end{cases}$$

53 George Wolford has suggested the following variation on the Linda problem (see Exercise 1.2.25). The registrar is carrying John and Mary's registration cards and drops them in a puddle. When he pickes them up he cannot read the names but on the first card he picked up he can make out Mathematics 23 and Government 35, and on the second card he can make out only Mathematics 23. He asks you if you can help him decide which card belongs to Mary. You know that Mary likes government but does not like mathematics. You know nothing about John and assume that he is just a typical Dartmouth student. From this you estimate:

$$\begin{aligned} P(\text{Mary takes Government 35}) &= .5 \,, \\ P(\text{Mary takes Mathematics 23}) &= .1 \,, \\ P(\text{John takes Government 35}) &= .3 \,, \\ P(\text{John takes Mathematics 23}) &= .2 \,. \end{aligned}$$

Assume that their choices for courses are independent events. Show that the card with Mathematics 23 and Government 35 showing is more likely to be Mary's than John's. The conjunction fallacy referred to in the Linda problem would be to assume that the event "Mary takes Mathematics 23 and Government 35" is more likely than the event "Mary takes Mathematics 23." Why are we not making this fallacy here?

54 (Suggested by Eisenberg and Ghosh[12]) A deck of playing cards can be described as a Cartesian product

$$\text{Deck} = \text{Suit} \times \text{Rank} ,$$

where Suit = $\{\clubsuit, \diamondsuit, \heartsuit, \spadesuit\}$ and Rank = $\{2, 3, \ldots, 10, J, Q, K, A\}$. This just means that every card may be thought of as an ordered pair like $(\diamondsuit, 2)$. By a *suit event* we mean any event A contained in Deck which is described in terms of Suit alone. For instance, if A is "the suit is red," then

$$A = \{\diamondsuit, \heartsuit\} \times \text{Rank} ,$$

so that A consists of all cards of the form $(\diamondsuit, r)$ or $(\heartsuit, r)$ where r is any rank. Similarly, a *rank event* is any event described in terms of rank alone.

(a) Show that if A is any suit event and B any rank event, then A and B are *independent*. (We can express this briefly by saying that suit and rank are independent.)

(b) Throw away the ace of spades. Show that now no nontrivial (i.e., neither empty nor the whole space) suit event A is independent of any nontrivial rank event B. *Hint*: Here independence comes down to

$$c/51 = (a/51) \cdot (b/51) ,$$

where a, b, c are the respective sizes of A, B and $A \cap B$. It follows that 51 must divide ab, hence that 3 must divide one of a and b, and 17 the other. But the possible sizes for suit and rank events preclude this.

(c) Show that the deck in (b) nevertheless does have pairs A, B of nontrivial independent events. *Hint*: Find 2 events A and B of sizes 3 and 17, respectively, which intersect in a single point.

(d) Add a joker to a full deck. Show that now there is no pair A, B of nontrivial independent events. *Hint*: See the hint in (b); 53 is prime.

The following problems are suggested by Stanley Gudder in his article "Do Good Hands Attract?"[13] He says that event A *attracts* event B if $P(B|A) > P(B)$ and *repels* B if $P(B|A) < P(B)$.

55 Let R_i be the event that the ith player in a poker game has a royal flush. Show that a royal flush (A,K,Q,J,10 of one suit) attracts another royal flush, that is $P(R_2|R_1) > P(R_2)$. Show that a royal flush repels full houses.

56 Prove that A attracts B if and only if B attracts A. Hence we can say that A and B are *mutually attractive* if A attracts B.

---

[12]B. Eisenberg and B. K. Ghosh, "Independent Events in a Discrete Uniform Probability Space," *The American Statistician*, vol. 41, no. 1 (1987), pp. 52–56.

[13]S. Gudder, "Do Good Hands Attract?" *Mathematics Magazine*, vol. 54, no. 1 (1981), pp. 13–16.

4.1. DISCRETE CONDITIONAL PROBABILITY

57 Prove that A neither attracts nor repels B if and only if A and B are independent.

58 Prove that A and B are mutually attractive if and only if $P(B|A) > P(B|\tilde{A})$.

59 Prove that if A attracts B, then A repels $\tilde{B}$.

60 Prove that if A attracts both B and C, and A repels $B \cap C$, then A attracts $B \cup C$. Is there any example in which A attracts both B and C and repels $B \cup C$?

61 Prove that if $B_1, B_2, \ldots, B_n$ are mutually disjoint and collectively exhaustive, and if A attracts some B_i, then A must repel some B_j.

62 (a) Suppose that you are looking in your desk for a letter from some time ago. Your desk has eight drawers, and you assess the probability that it is in any particular drawer is 10% (so there is a 20% chance that it is not in the desk at all). Suppose now that you start searching systematically through your desk, one drawer at a time. In addition, suppose that you have not found the letter in the first i drawers, where $0 \le i \le 7$. Let p_i denote the probability that the letter will be found in the next drawer, and let q_i denote the probability that the letter will be found in some subsequent drawer (both p_i and q_i are conditional probabilities, since they are based upon the assumption that the letter is not in the first i drawers). Show that the p_i's increase and the q_i's decrease. (This problem is from Falk et al.[14])

(b) The following data appeared in an article in the Wall Street Journal.[15] For the ages 20, 30, 40, 50, and 60, the probability of a woman in the U.S. developing cancer in the next ten years is 0.5%, 1.2%, 3.2%, 6.4%, and 10.8%, respectively. At the same set of ages, the probability of a woman in the U.S. eventually developing cancer is 39.6%, 39.5%, 39.1%, 37.5%, and 34.2%, respectively. Do you think that the problem in part (a) gives an explanation for these data?

63 Here are two variations of the Monty Hall problem that are discussed by Granberg.[16]

(a) Suppose that everything is the same except that Monty forgot to find out in advance which door has the car behind it. In the spirit of "the show must go on," he makes a guess at which of the two doors to open and gets lucky, opening a door behind which stands a goat. Now should the contestant switch?

---

[14] R. Falk, A. Lipson, and C. Konold, "The ups and downs of the hope function in a fruitless search," in *Subjective Probability*, G. Wright and P. Ayton, (eds.) (Chichester: Wiley, 1994), pgs. 353-377.

[15] C. Crossen, "Fright by the numbers: Alarming disease data are frequently flawed," *Wall Street Journal*, 11 April 1996, p. B1.

[16] D. Granberg, "To switch or not to switch," in *The power of logical thinking*, M. vos Savant, (New York: St. Martin's 1996).

(b) You have observed the show for a long time and found that the car is put behind door A 45% of the time, behind door B 40% of the time and behind door C 15% of the time. Assume that everything else about the show is the same. Again you pick door A. Monty opens a door with a goat and offers to let you switch. Should you? Suppose you knew in advance that Monty was going to give you a chance to switch. Should you have initially chosen door A?

4.2 Continuous Conditional Probability

In situations where the sample space is continuous we will follow the same procedure as in the previous section. Thus, for example, if X is a continuous random variable with density function $f(x)$, and if E is an event with positive probability, we define a conditional density function by the formula

$$f(x|E) = \begin{cases} f(x)/P(E), & \text{if } x \in E, \\ 0, & \text{if } x \notin E. \end{cases}$$

Then for any event F, we have

$$P(F|E) = \int_F f(x|E)\, dx \ .$$

The expression $P(F|E)$ is called the conditional probability of F given E. As in the previous section, it is easy to obtain an alternative expression for this probability:

$$P(F|E) = \int_F f(x|E)\, dx = \int_{E \cap F} \frac{f(x)}{P(E)}\, dx = \frac{P(E \cap F)}{P(E)} \ .$$

We can think of the conditional density function as being 0 except on E, and normalized to have integral 1 over E. Note that if the original density is a uniform density corresponding to an experiment in which all events of equal size are *equally likely*, then the same will be true for the conditional density.

Example 4.18 In the spinner experiment (cf. Example 2.1), suppose we know that the spinner has stopped with head in the upper half of the circle, $0 \le x \le 1/2$. What is the probability that $1/6 \le x \le 1/3$?

Here $E = [0, 1/2]$, $F = [1/6, 1/3]$, and $F \cap E = F$. Hence

$$\begin{aligned} P(F|E) &= \frac{P(F \cap E)}{P(E)} \\ &= \frac{1/6}{1/2} \\ &= \frac{1}{3}\ , \end{aligned}$$

which is reasonable, since F is 1/3 the size of E. The conditional density function here is given by

4.2. CONTINUOUS CONDITIONAL PROBABILITY

$$f(x|E) = \begin{cases} 2, & \text{if } 0 \leq x < 1/2, \\ 0, & \text{if } 1/2 \leq x < 1. \end{cases}$$

Thus the conditional density function is nonzero only on $[0, 1/2]$, and is uniform there. □

Example 4.19 In the dart game (cf. Example 2.8), suppose we know that the dart lands in the upper half of the target. What is the probability that its distance from the center is less than $1/2$?

Here $E = \{(x, y) : y \geq 0\}$, and $F = \{(x, y) : x^2 + y^2 < (1/2)^2\}$. Hence,

$$\begin{aligned} P(F|E) &= \frac{P(F \cap E)}{P(E)} = \frac{(1/\pi)[(1/2)(\pi/4)]}{(1/\pi)(\pi/2)} \\ &= 1/4 \ . \end{aligned}$$

Here again, the size of $F \cap E$ is $1/4$ the size of E. The conditional density function is

$$f((x,y)|E) = \begin{cases} f(x,y)/P(E) = 2/\pi, & \text{if } (x, y) \in E, \\ 0, & \text{if } (x, y) \notin E. \end{cases}$$

□

Example 4.20 We return to the exponential density (cf. Example 2.17). We suppose that we are observing a lump of plutonium-239. Our experiment consists of waiting for an emission, then starting a clock, and recording the length of time X that passes until the next emission. Experience has shown that X has an exponential density with some parameter λ, which depends upon the size of the lump. Suppose that when we perform this experiment, we notice that the clock reads r seconds, and is still running. What is the probability that there is no emission in a further s seconds?

Let $G(t)$ be the probability that the next particle is emitted after time t. Then

$$\begin{aligned} G(t) &= \int_t^\infty \lambda e^{-\lambda x} \, dx \\ &= -e^{-\lambda x}\Big|_t^\infty = e^{-\lambda t} \ . \end{aligned}$$

Let E be the event "the next particle is emitted after time r" and F the event "the next particle is emitted after time $r + s$." Then

$$\begin{aligned} P(F|E) &= \frac{P(F \cap E)}{P(E)} \\ &= \frac{G(r+s)}{G(r)} \\ &= \frac{e^{-\lambda(r+s)}}{e^{-\lambda r}} \\ &= e^{-\lambda s} \ . \end{aligned}$$

This tells us the rather surprising fact that the probability that we have to wait s seconds more for an emission, given that there has been no emission in r seconds, is *independent* of the time r. This property (called the *memoryless* property) was introduced in Example 2.17. When trying to model various phenomena, this property is helpful in deciding whether the exponential density is appropriate.

The fact that the exponential density is memoryless means that it is reasonable to assume if one comes upon a lump of a radioactive isotope at some random time, then the amount of time until the next emission has an exponential density with the same parameter as the time between emissions. A well-known example, known as the "bus paradox," replaces the emissions by buses. The apparent paradox arises from the following two facts: 1) If you know that, on the average, the buses come by every 30 minutes, then if you come to the bus stop at a random time, you should only have to wait, on the average, for 15 minutes for a bus, and 2) Since the buses arrival times are being modelled by the exponential density, then no matter when you arrive, you will have to wait, on the average, for 30 minutes for a bus.

The reader can now see that in Exercises 2.2.9, 2.2.10, and 2.2.11, we were asking for simulations of conditional probabilities, under various assumptions on the distribution of the interarrival times. If one makes a reasonable assumption about this distribution, such as the one in Exercise 2.2.10, then the average waiting time is more nearly one-half the average interarrival time. $\square$

Independent Events

If E and F are two events with positive probability in a continuous sample space, then, as in the case of discrete sample spaces, we define E and F to be *independent* if $P(E|F) = P(E)$ and $P(F|E) = P(F)$. As before, each of the above equations imply the other, so that to see whether two events are independent, only one of these equations must be checked. It is also the case that, if E and F are independent, then $P(E \cap F) = P(E)P(F)$.

Example 4.21 (Example 4.18 continued) In the dart game (see Example 4.18), let E be the event that the dart lands in the *upper* half of the target ($y \geq 0$) and F the event that the dart lands in the *right* half of the target ($x \geq 0$). Then $P(E \cap F)$ is the probability that the dart lies in the first quadrant of the target, and

$$
\begin{aligned}
P(E \cap F) &= \frac{1}{\pi} \int_{E \cap F} 1 \, dx dy \\
&= \text{Area}\,(E \cap F) \\
&= \text{Area}\,(E) \, \text{Area}\,(F) \\
&= \left(\frac{1}{\pi} \int_E 1 \, dx dy \right) \left(\frac{1}{\pi} \int_F 1 \, dx dy \right) \\
&= P(E)P(F)
\end{aligned}
$$

so that E and F are independent. What makes this work is that the events E and F are described by restricting different coordinates. This idea is made more precise below. $\square$

Joint Density and Cumulative Distribution Functions

In a manner analogous with discrete random variables, we can define joint density functions and cumulative distribution functions for multi-dimensional continuous random variables.

Definition 4.6 Let $X_1, X_2, \ldots, X_n$ be continuous random variables associated with an experiment, and let $\bar{X} = (X_1, X_2, \ldots, X_n)$. Then the joint cumulative distribution function of $\bar{X}$ is defined by

$$F(x_1, x_2, \ldots, x_n) = P(X_1 \leq x_1, X_2 \leq x_2, \ldots, X_n \leq x_n) \ .$$

The joint density function of $\bar{X}$ satisfies the following equation:

$$F(x_1, x_2, \ldots, x_n) = \int_{-\infty}^{x_1} \int_{-\infty}^{x_2} \cdots \int_{-\infty}^{x_n} f(t_1, t_2, \ldots t_n) \, dt_n dt_{n-1} \ldots dt_1 \ .$$

□

It is straightforward to show that, in the above notation,

$$f(x_1, x_2, \ldots, x_n) = \frac{\partial^n F(x_1, x_2, \ldots, x_n)}{\partial x_1 \partial x_2 \cdots \partial x_n} \ . \tag{4.4}$$

Independent Random Variables

As with discrete random variables, we can define mutual independence of continuous random variables.

Definition 4.7 Let $X_1, X_2, \ldots, X_n$ be continuous random variables with cumulative distribution functions $F_1(x), F_2(x), \ldots, F_n(x)$. Then these random variables are *mutually independent* if

$$F(x_1, x_2, \ldots, x_n) = F_1(x_1) F_2(x_2) \cdots F_n(x_n)$$

for any choice of $x_1, x_2, \ldots, x_n$. Thus, if $X_1, X_2, \ldots, X_n$ are mutually independent, then the joint cumulative distribution function of the random variable $\bar{X} = (X_1, X_2, \ldots, X_n)$ is just the product of the individual cumulative distribution functions. When two random variables are mutually independent, we shall say more briefly that they are *independent*. □

Using Equation 4.4, the following theorem can easily be shown to hold for mutually independent continuous random variables.

Theorem 4.2 Let $X_1, X_2, \ldots, X_n$ be continuous random variables with density functions $f_1(x), f_2(x), \ldots, f_n(x)$. Then these random variables are *mutually independent* if and only if

$$f(x_1, x_2, \ldots, x_n) = f_1(x_1) f_2(x_2) \cdots f_n(x_n)$$

for any choice of $x_1, x_2, \ldots, x_n$. □

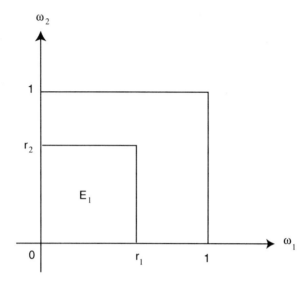

Figure 4.7: X_1 and X_2 are independent.

Let's look at some examples.

Example 4.22 In this example, we define three random variables, X_1, X_2, and X_3. We will show that X_1 and X_2 are independent, and that X_1 and X_3 are not independent. Choose a point $\omega = (\omega_1, \omega_2)$ at random from the unit square. Set $X_1 = \omega_1^2$, $X_2 = \omega_2^2$, and $X_3 = \omega_1 + \omega_2$. Find the joint distributions $F_{12}(r_1, r_2)$ and $F_{23}(r_2, r_3)$.

We have already seen (see Example 2.13) that
$$\begin{aligned} F_1(r_1) &= P(-\infty < X_1 \leq r_1) \\ &= \sqrt{r_1}, \quad \text{if } 0 \leq r_1 \leq 1 , \end{aligned}$$

and similarly,
$$F_2(r_2) = \sqrt{r_2} ,$$
if $0 \leq r_2 \leq 1$. Now we have (see Figure 4.7)
$$\begin{aligned} F_{12}(r_1, r_2) &= P(X_1 \leq r_1 \text{ and } X_2 \leq r_2) \\ &= P(\omega_1 \leq \sqrt{r_1} \text{ and } \omega_2 \leq \sqrt{r_2}) \\ &= \text{Area}(E_1) \\ &= \sqrt{r_1}\sqrt{r_2} \\ &= F_1(r_1) F_2(r_2) . \end{aligned}$$

In this case $F_{12}(r_1, r_2) = F_1(r_1) F_2(r_2)$ so that X_1 and X_2 are independent. On the other hand, if $r_1 = 1/4$ and $r_3 = 1$, then (see Figure 4.8)
$$F_{13}(1/4, 1) = P(X_1 \leq 1/4, X_3 \leq 1)$$

4.2. CONTINUOUS CONDITIONAL PROBABILITY

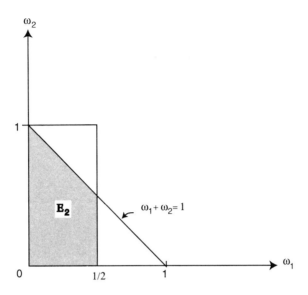

Figure 4.8: X_1 and X_3 are not independent.

$$
\begin{aligned}
&= P(\omega_1 \leq 1/2,\ \omega_1 + \omega_2 \leq 1) \\
&= \text{Area}\,(E_2) \\
&= \frac{1}{2} - \frac{1}{8} = \frac{3}{8}\ .
\end{aligned}
$$

Now recalling that

$$F_3(r_3) = \begin{cases} 0, & \text{if } r_3 < 0, \\ (1/2)r_3^2, & \text{if } 0 \leq r_3 \leq 1, \\ 1 - (1/2)(2 - r_3)^2, & \text{if } 1 \leq r_3 \leq 2, \\ 1, & \text{if } 2 < r_3, \end{cases}$$

(see Example 2.14), we have $F_1(1/4)F_3(1) = (1/2)(1/2) = 1/4$. Hence, X_1 and X_3 are not independent random variables. A similar calculation shows that X_2 and X_3 are not independent either. □

Although we shall not prove it here, the following theorem is a useful one. The statement also holds for mutually independent discrete random variables. A proof may be found in Rényi.[17]

Theorem 4.3 Let $X_1, X_2, \ldots, X_n$ be mutually independent continuous random variables and let $\phi_1(x), \phi_2(x), \ldots, \phi_n(x)$ be continuous functions. Then $\phi_1(X_1)$, $\phi_2(X_2), \ldots, \phi_n(X_n)$ are mutually independent. □

Independent Trials

Using the notion of independence, we can now formulate for continuous sample spaces the notion of independent trials (see Definition 4.5).

---

[17] A. Rényi, *Probability Theory* (Budapest: Akadémiai Kiadó, 1970), p. 183.

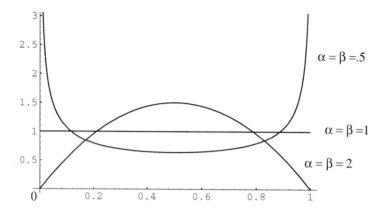

Figure 4.9: Beta density for $\alpha = \beta = .5, 1, 2$.

Definition 4.8 A sequence X_1, X_2, ..., X_n of random variables X_i that are mutually independent and have the same density is called an *independent trials process*. □

As in the case of discrete random variables, these independent trials processes arise naturally in situations where an experiment described by a single random variable is repeated n times.

Beta Density

We consider next an example which involves a sample space with both discrete and continuous coordinates. For this example we shall need a new density function called the *beta density*. This density has two parameters α, β and is defined by

$$B(\alpha, \beta, x) = \begin{cases} (1/B(\alpha, \beta))x^{\alpha-1}(1-x)^{\beta-1}, & \text{if } 0 \leq x \leq 1, \\ 0, & \text{otherwise.} \end{cases}$$

Here α and β are any positive numbers, and the beta function $B(\alpha, \beta)$ is given by the area under the graph of $x^{\alpha-1}(1-x)^{\beta-1}$ between 0 and 1:

$$B(\alpha, \beta) = \int_0^1 x^{\alpha-1}(1-x)^{\beta-1}\,dx\ .$$

Note that when $\alpha = \beta = 1$ the beta density if the uniform density. When α and β are greater than 1 the density is bell-shaped, but when they are less than 1 it is U-shaped as suggested by the examples in Figure 4.9.

We shall need the values of the beta function only for integer values of α and β, and in this case

$$B(\alpha, \beta) = \frac{(\alpha-1)!\,(\beta-1)!}{(\alpha+\beta-1)!}\ .$$

Example 4.23 In medical problems it is often assumed that a drug is effective with a probability x each time it is used and the various trials are independent, so that

4.2. CONTINUOUS CONDITIONAL PROBABILITY

one is, in effect, tossing a biased coin with probability x for heads. Before further experimentation, you do not know the value x but past experience might give some information about its possible values. It is natural to represent this information by sketching a density function to determine a distribution for x. Thus, we are considering x to be a continuous random variable, which takes on values between 0 and 1. If you have no knowledge at all, you would sketch the uniform density. If past experience suggests that x is very likely to be near 2/3 you would sketch a density with maximum at 2/3 and a spread reflecting your uncertainly in the estimate of 2/3. You would then want to find a density function that reasonably fits your sketch. The beta densities provide a class of densities that can be fit to most sketches you might make. For example, for $\alpha > 1$ and $\beta > 1$ it is bell-shaped with the parameters α and β determining its peak and its spread.

Assume that the experimenter has chosen a beta density to describe the state of his knowledge about x before the experiment. Then he gives the drug to n subjects and records the number i of successes. The number i is a discrete random variable, so we may conveniently describe the set of possible outcomes of this experiment by referring to the ordered pair (x, i).

We let $m(i|x)$ denote the probability that we observe i successes given the value of x. By our assumptions, $m(i|x)$ is the binomial distribution with probability x for success:

$$m(i|x) = b(n, x, i) = \binom{n}{i} x^i (1-x)^j ,$$

where $j = n - i$.

If x is chosen at random from $[0, 1]$ with a beta density $B(\alpha, \beta, x)$, then the density function for the outcome of the pair (x, i) is

$$\begin{aligned} f(x, i) &= m(i|x) B(\alpha, \beta, x) \\ &= \binom{n}{i} x^i (1-x)^j \frac{1}{B(\alpha, \beta)} x^{\alpha-1} (1-x)^{\beta-1} \\ &= \binom{n}{i} \frac{1}{B(\alpha, \beta)} x^{\alpha+i-1} (1-x)^{\beta+j-1} . \end{aligned}$$

Now let $m(i)$ be the probability that we observe i successes *not* knowing the value of x. Then

$$\begin{aligned} m(i) &= \int_0^1 m(i|x) B(\alpha, \beta, x) \, dx \\ &= \binom{n}{i} \frac{1}{B(\alpha, \beta)} \int_0^1 x^{\alpha+i-1} (1-x)^{\beta+j-1} \, dx \\ &= \binom{n}{i} \frac{B(\alpha+i, \beta+j)}{B(\alpha, \beta)} . \end{aligned}$$

Hence, the probability density $f(x|i)$ for x, given that i successes were observed, is

$$f(x|i) = \frac{f(x, i)}{m(i)}$$

$$= \frac{x^{\alpha+i-1}(1-x)^{\beta+j-1}}{B(\alpha+i, \beta+j)}, \quad (4.5)$$

that is, $f(x|i)$ is another beta density. This says that if we observe i successes and j failures in n subjects, then the new density for the probability that the drug is effective is again a beta density but with parameters $\alpha + i$, $\beta + j$.

Now we assume that before the experiment we choose a beta density with parameters α and β, and that in the experiment we obtain i successes in n trials. We have just seen that in this case, the new density for x is a beta density with parameters $\alpha + i$ and $\beta + j$.

Now we wish to calculate the probability that the drug is effective on the next subject. For any particular real number t between 0 and 1, the probability that x has the value t is given by the expression in Equation 4.5. Given that x has the value t, the probability that the drug is effective on the next subject is just t. Thus, to obtain the probability that the drug is effective on the next subject, we integrate the product of the expression in Equation 4.5 and t over all possible values of t. We obtain:

$$\frac{1}{B(\alpha+i, \beta+j)} \int_0^1 t \cdot t^{\alpha+i-1}(1-t)^{\beta+j-1}\, dt$$
$$= \frac{B(\alpha+i+1, \beta+j)}{B(\alpha+i, \beta+j)}$$
$$= \frac{(\alpha+i)!\,(\beta+j-1)!}{(\alpha+\beta+i+j)!} \cdot \frac{(\alpha+\beta+i+j-1)!}{(\alpha+i-1)!\,(\beta+j-1)!}$$
$$= \frac{\alpha+i}{\alpha+\beta+n}\ .$$

If n is large, then our estimate for the probability of success after the experiment is approximately the proportion of successes observed in the experiment, which is certainly a reasonable conclusion. $\square$

The next example is another in which the true probabilities are unknown and must be estimated based upon experimental data.

Example 4.24 (Two-armed bandit problem) You are in a casino and confronted by two slot machines. Each machine pays off either 1 dollar or nothing. The probability that the first machine pays off a dollar is x and that the second machine pays off a dollar is y. We assume that x and y are random numbers chosen independently from the interval $[0, 1]$ and unknown to you. You are permitted to make a series of ten plays, each time choosing one machine or the other. How should you choose to maximize the number of times that you win?

One strategy that sounds reasonable is to calculate, at every stage, the probability that each machine will pay off and choose the machine with the higher probability. Let win(i), for $i = 1$ or 2, be the number of times that you have won on the ith machine. Similarly, let lose(i) be the number of times you have lost on the ith machine. Then, from Example 4.23, the probability $p(i)$ that you win if you

4.2. CONTINUOUS CONDITIONAL PROBABILITY

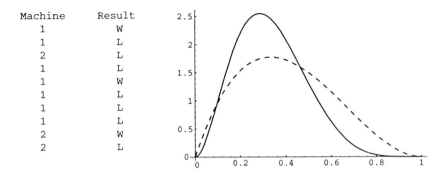

Figure 4.10: Play the best machine.

choose the ith machine is

$$p(i) = \frac{\text{win}(i) + 1}{\text{win}(i) + \text{lose}(i) + 2} \ .$$

Thus, if $p(1) > p(2)$ you would play machine 1 and otherwise you would play machine 2. We have written a program **TwoArm** to simulate this experiment. In the program, the user specifies the initial values for x and y (but these are unknown to the experimenter). The program calculates at each stage the two conditional densities for x and y, given the outcomes of the previous trials, and then computes $p(i)$, for $i = 1, 2$. It then chooses the machine with the highest value for the probability of winning for the next play. The program prints the machine chosen on each play and the outcome of this play. It also plots the new densities for x (solid line) and y (dotted line), showing only the current densities. We have run the program for ten plays for the case $x = .6$ and $y = .7$. The result is shown in Figure 4.10.

The run of the program shows the weakness of this strategy. Our initial probability for winning on the better of the two machines is .7. We start with the poorer machine and our outcomes are such that we always have a probability greater than .6 of winning and so we just keep playing this machine even though the other machine is better. If we had lost on the first play we would have switched machines. Our final density for y is the same as our initial density, namely, the uniform density. Our final density for x is different and reflects a much more accurate knowledge about x. The computer did pretty well with this strategy, winning seven out of the ten trials, but ten trials are not enough to judge whether this is a good strategy in the long run.

Another popular strategy is the *play-the-winner strategy*. As the name suggests, for this strategy we choose the same machine when we win and switch machines when we lose. The program **TwoArm** will simulate this strategy as well. In Figure 4.11, we show the results of running this program with the play-the-winner strategy and the same true probabilities of .6 and .7 for the two machines. After ten plays our densities for the unknown probabilities of winning suggest to us that the second machine is indeed the better of the two. We again won seven out of the ten trials.

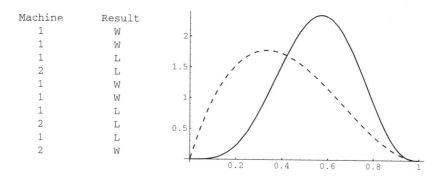

Machine	Result
1	W
1	W
1	L
2	L
1	W
1	W
1	L
2	L
1	L
2	W

Figure 4.11: Play the winner.

Neither of the strategies that we simulated is the best one in terms of maximizing our average winnings. This best strategy is very complicated but is reasonably approximated by the play-the-winner strategy. Variations on this example have played an important role in the problem of clinical tests of drugs where experimenters face a similar situation. □

Exercises

1 Pick a point x at random (with uniform density) in the interval $[0,1]$. Find the probability that $x > 1/2$, given that

(a) $x > 1/4$.

(b) $x < 3/4$.

(c) $|x - 1/2| < 1/4$.

(d) $x^2 - x + 2/9 < 0$.

2 A radioactive material emits α-particles at a rate described by the density function
$$f(t) = .1 e^{-.1t} \ .$$
Find the probability that a particle is emitted in the first 10 seconds, given that

(a) no particle is emitted in the first second.

(b) no particle is emitted in the first 5 seconds.

(c) a particle is emitted in the first 3 seconds.

(d) a particle is emitted in the first 20 seconds.

3 The Acme Super light bulb is known to have a useful life described by the density function
$$f(t) = .01 e^{-.01t} \ ,$$
where time t is measured in hours.

(a) Find the *failure rate* of this bulb (see Exercise 2.2.6).

(b) Find the *reliability* of this bulb after 20 hours.

(c) Given that it lasts 20 hours, find the probability that the bulb lasts another 20 hours.

(d) Find the probability that the bulb burns out in the forty-first hour, given that it lasts 40 hours.

4 Suppose you toss a dart at a circular target of radius 10 inches. Given that the dart lands in the upper half of the target, find the probability that

(a) it lands in the right half of the target.

(b) its distance from the center is less than 5 inches.

(c) its distance from the center is greater than 5 inches.

(d) it lands within 5 inches of the point $(0, 5)$.

5 Suppose you choose two numbers x and y, independently at random from the interval $[0, 1]$. Given that their sum lies in the interval $[0, 1]$, find the probability that

(a) $|x - y| < 1$.

(b) $xy < 1/2$.

(c) $\max\{x, y\} < 1/2$.

(d) $x^2 + y^2 < 1/4$.

(e) $x > y$.

6 Find the conditional density functions for the following experiments.

(a) A number x is chosen at random in the interval $[0, 1]$, given that $x > 1/4$.

(b) A number t is chosen at random in the interval $[0, \infty)$ with exponential density e^{-t}, given that $1 < t < 10$.

(c) A dart is thrown at a circular target of radius 10 inches, given that it falls in the upper half of the target.

(d) Two numbers x and y are chosen at random in the interval $[0, 1]$, given that $x > y$.

7 Let x and y be chosen at random from the interval $[0, 1]$. Show that the events $x > 1/3$ and $y > 2/3$ are independent events.

8 Let x and y be chosen at random from the interval $[0, 1]$. Which pairs of the following events are independent?

(a) $x > 1/3$.

(b) $y > 2/3$.

(c) $x > y$.

(d) $x + y < 1$.

9 Suppose that X and Y are continuous random variables with density functions $f_X(x)$ and $f_Y(y)$, respectively. Let $f(x, y)$ denote the joint density function of (X, Y). Show that

$$\int_{-\infty}^{\infty} f(x, y)\, dy = f_X(x)\ ,$$

and

$$\int_{-\infty}^{\infty} f(x, y)\, dx = f_Y(y)\ .$$

*10 In Exercise 2.2.12 you proved the following: If you take a stick of unit length and break it into three pieces, choosing the breaks at random (i.e., choosing two real numbers independently and uniformly from [0, 1]), then the probability that the three pieces form a triangle is 1/4. Consider now a similar experiment: First break the stick at random, then break the longer piece at random. Show that the two experiments are actually quite different, as follows:

(a) Write a program which simulates both cases for a run of 1000 trials, prints out the proportion of successes for each run, and repeats this process ten times. (Call a trial a success if the three pieces do form a triangle.) Have your program pick (x, y) at random in the unit square, and in each case use x and y to find the two breaks. For each experiment, have it plot (x, y) if (x, y) gives a success.

(b) Show that in the second experiment the theoretical probability of success is actually $2 \log 2 - 1$.

11 A coin has an unknown bias p that is assumed to be uniformly distributed between 0 and 1. The coin is tossed n times and heads turns up j times and tails turns up k times. We have seen that the probability that heads turns up next time is

$$\frac{j+1}{n+2}\ .$$

Show that this is the same as the probability that the next ball is black for the Polya urn model of Exercise 4.1.20. Use this result to explain why, in the Polya urn model, the proportion of black balls does not tend to 0 or 1 as one might expect but rather to a uniform distribution on the interval [0, 1].

12 Previous experience with a drug suggests that the probability p that the drug is effective is a random quantity having a beta density with parameters $\alpha = 2$ and $\beta = 3$. The drug is used on ten subjects and found to be successful in four out of the ten patients. What density should we now assign to the probability p? What is the probability that the drug will be successful the next time it is used?

4.3. PARADOXES

13 Write a program to allow you to compare the strategies play-the-winner and play-the-best-machine for the two-armed bandit problem of Example 4.24. Have your program determine the initial payoff probabilities for each machine by choosing a pair of random numbers between 0 and 1. Have your program carry out 20 plays and keep track of the number of wins for each of the two strategies. Finally, have your program make 1000 repetitions of the 20 plays and compute the average winning per 20 plays. Which strategy seems to be the best? Repeat these simulations with 20 replaced by 100. Does your answer to the above question change?

14 Consider the two-armed bandit problem of Example 4.24. Bruce Barnes proposed the following strategy, which is a variation on the play-the-best-machine strategy. The machine with the greatest probability of winning is played *unless* the following two conditions hold: (a) the difference in the probabilities for winning is less than .08, and (b) the ratio of the number of times played on the more often played machine to the number of times played on the less often played machine is greater than 1.4. If the above two conditions hold, then the machine with the smaller probability of winning is played. Write a program to simulate this strategy. Have your program choose the initial payoff probabilities at random from the unit interval $[0, 1]$, make 20 plays, and keep track of the number of wins. Repeat this experiment 1000 times and obtain the average number of wins per 20 plays. Implement a second strategy—for example, play-the-best-machine or one of your own choice, and see how this second strategy compares with Bruce's on average wins.

4.3 Paradoxes

Much of this section is based on an article by Snell and Vanderbei.[18]

One must be very careful in dealing with problems involving conditional probability. The reader will recall that in the Monty Hall problem (Example 4.6), if the contestant chooses the door with the car behind it, then Monty has a choice of doors to open. We made an assumption that in this case, he will choose each door with probability $1/2$. We then noted that if this assumption is changed, the answer to the original question changes. In this section, we will study other examples of the same phenomenon.

Example 4.25 Consider a family with two children. Given that one of the children is a boy, what is the probability that both children are boys?

One way to approach this problem is to say that the other child is equally likely to be a boy or a girl, so the probability that both children are boys is $1/2$. The "textbook" solution would be to draw the tree diagram and then form the conditional tree by deleting paths to leave only those paths that are consistent with the given

---

[18] J. L. Snell and R. Vanderbei, "Three Bewitching Paradoxes," in *Topics in Contemporary Probability and Its Applications*, CRC Press, Boca Raton, 1995.

176   CHAPTER 4. CONDITIONAL PROBABILITY

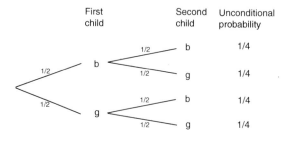

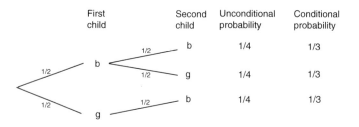

Figure 4.12: Tree for Example 4.25.

information. The result is shown in Figure 4.12. We see that the probability of two boys given a boy in the family is not 1/2 but rather 1/3.   □

This problem and others like it are discussed in Bar-Hillel and Falk.[19] These authors stress that the answer to conditional probabilities of this kind can change depending upon how the information given was actually obtained. For example, they show that 1/2 is the correct answer for the following scenario.

Example 4.26 Mr. Smith is the father of two. We meet him walking along the street with a young boy whom he proudly introduces as his son. What is the probability that Mr. Smith's other child is also a boy?

As usual we have to make some additional assumptions. For example, we will assume that if Mr. Smith has a boy and a girl, he is equally likely to choose either one to accompany him on his walk. In Figure 4.13 we show the tree analysis of this problem and we see that 1/2 is, indeed, the correct answer.   □

Example 4.27 It is not so easy to think of reasonable scenarios that would lead to the classical 1/3 answer. An attempt was made by Stephen Geller in proposing this problem to Marilyn vos Savant.[20] Geller's problem is as follows: A shopkeeper says she has two new baby beagles to show you, but she doesn't know whether they're both male, both female, or one of each sex. You tell her that you want only a male, and she telephones the fellow who's giving them a bath. "Is at least one a male?"

---

[19]M. Bar-Hillel and R. Falk, "Some teasers concerning conditional probabilities," *Cognition*, vol. 11 (1982), pgs. 109-122.

[20]M. vos Savant, "Ask Marilyn," *Parade Magazine*, 9 September; 2 December; 17 February 1990, reprinted in Marilyn vos Savant, *Ask Marilyn*, St. Martins, New York, 1992.

4.3. PARADOXES

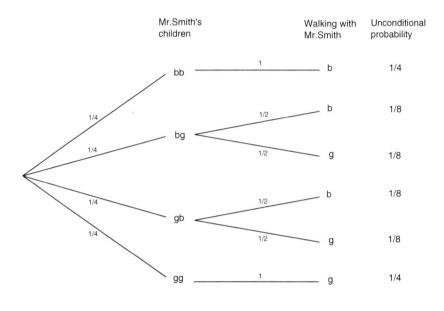

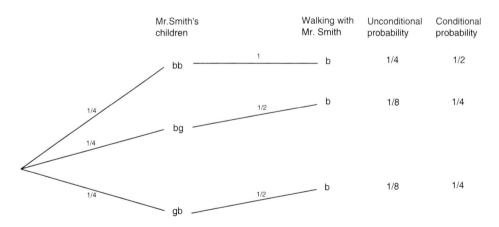

Figure 4.13: Tree for Example 4.26.

she asks. "Yes," she informs you with a smile. What is the probability that the *other* one is male?

The reader is asked to decide whether the model which gives an answer of 1/3 is a reasonable one to use in this case. □

In the preceding examples, the apparent paradoxes could easily be resolved by clearly stating the model that is being used and the assumptions that are being made. We now turn to some examples in which the paradoxes are not so easily resolved.

Example 4.28 Two envelopes each contain a certain amount of money. One envelope is given to Ali and the other to Baba and they are told that one envelope contains twice as much money as the other. However, neither knows who has the larger prize. Before anyone has opened their envelope, Ali is asked if she would like to trade her envelope with Baba. She reasons as follows: Assume that the amount in my envelope is x. If I switch, I will end up with $x/2$ with probability 1/2, and $2x$ with probability 1/2. If I were given the opportunity to play this game many times, and if I were to switch each time, I would, on average, get

$$\frac{1}{2}\frac{x}{2} + \frac{1}{2}2x = \frac{5}{4}x \ .$$

This is greater than my average winnings if I didn't switch.

Of course, Baba is presented with the same opportunity and reasons in the same way to conclude that he too would like to switch. So they switch and each thinks that his/her net worth just went up by 25%.

Since neither has yet opened any envelope, this process can be repeated and so again they switch. Now they are back with their original envelopes and yet they think that their fortune has increased 25% twice. By this reasoning, they could convince themselves that by repeatedly switching the envelopes, they could become arbitrarily wealthy. Clearly, something is wrong with the above reasoning, but where is the mistake?

One of the tricks of making paradoxes is to make them slightly more difficult than is necessary to further befuddle us. As John Finn has suggested, in this paradox we could just have well started with a simpler problem. Suppose Ali and Baba know that I am going to give then either an envelope with $5 or one with $10 and I am going to toss a coin to decide which to give to Ali, and then give the other to Baba. Then Ali can argue that Baba has $2x$ with probability 1/2 and $x/2$ with probability 1/2. This leads Ali to the same conclusion as before. But now it is clear that this is nonsense, since if Ali has the envelope containing $5, Baba cannot possibly have half of this, namely $2.50, since that was not even one of the choices. Similarly, if Ali has $10, Baba cannot have twice as much, namely $20. In fact, in this simpler problem the possibly outcomes are given by the tree diagram in Figure 4.14. From the diagram, it is clear that neither is made better off by switching. □

In the above example, Ali's reasoning is incorrect because he infers that if the amount in his envelope is x, then the probability that his envelope contains the

4.3. PARADOXES 179

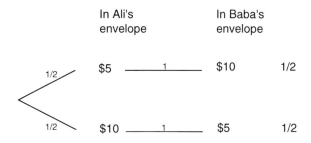

Figure 4.14: John Finn's version of Example 4.28.

smaller amount is 1/2, and the probability that her envelope contains the larger amount is also 1/2. In fact, these conditional probabilities depend upon the distribution of the amounts that are placed in the envelopes.

For definiteness, let X denote the positive integer-valued random variable which represents the smaller of the two amounts in the envelopes. Suppose, in addition, that we are given the distribution of X, i.e., for each positive integer x, we are given the value of

$$p_x = P(X = x) .$$

(In Finn's example, $p_5 = 1$, and $p_n = 0$ for all other values of n.) Then it is easy to calculate the conditional probability that an envelope contains the smaller amount, given that it contains x dollars. The two possible sample points are $(x, x/2)$ and $(x, 2x)$. If x is odd, then the first sample point has probability 0, since $x/2$ is not an integer, so the desired conditional probability is 1 that x is the smaller amount. If x is even, then the two sample points have probabilities $p_{x/2}$ and p_x, respectively, so the conditional probability that x is the smaller amount is

$$\frac{p_x}{p_{x/2} + p_x} ,$$

which is not necessarily equal to 1/2.

Steven Brams and D. Marc Kilgour[21] study the problem, for different distributions, of whether or not one should switch envelopes, if one's objective is to maximize the long-term average winnings. Let x be the amount in your envelope. They show that for any distribution of X, there is at least one value of x such that you should switch. They give an example of a distribution for which there is exactly one value of x such that you should switch (see Exercise 5). Perhaps the most interesting case is a distribution in which you should always switch. We now give this example.

Example 4.29 Suppose that we have two envelopes in front of us, and that one envelope contains twice the amount of money as the other (both amounts are positive integers). We are given one of the envelopes, and asked if we would like to switch.

---

[21]S. J. Brams and D. M. Kilgour, "The Box Problem: To Switch or Not to Switch," *Mathematics Magazine*, vol. 68, no. 1 (1995), p. 29.

As above, we let X denote the smaller of the two amounts in the envelopes, and let
$$p_x = P(X = x) \ .$$
We are now in a position where we can calculate the long-term average winnings, if we switch. (This long-term average is an example of a probabilistic concept known as expectation, and will be discussed in Chapter 6.) Given that one of the two sample points has occurred, the probability that it is the point $(x, x/2)$ is
$$\frac{p_{x/2}}{p_{x/2} + p_x} \ ,$$
and the probability that it is the point $(x, 2x)$ is
$$\frac{p_x}{p_{x/2} + p_x} \ .$$
Thus, if we switch, our long-term average winnings are
$$\frac{p_{x/2}}{p_{x/2} + p_x} \frac{x}{2} + \frac{p_x}{p_{x/2} + p_x} 2x \ .$$
If this is greater than x, then it pays in the long run for us to switch. Some routine algebra shows that the above expression is greater than x if and only if
$$\frac{p_{x/2}}{p_{x/2} + p_x} < \frac{2}{3} \ . \tag{4.6}$$

It is interesting to consider whether there is a distribution on the positive integers such that the inequality 4.6 is true for all even values of x. Brams and Kilgour[22] give the following example.

We define p_x as follows:
$$p_x = \begin{cases} \frac{1}{3}\left(\frac{2}{3}\right)^{k-1}, & \text{if } x = 2^k, \\ 0, & \text{otherwise.} \end{cases}$$
It is easy to calculate (see Exercise 4) that for all relevant values of x, we have
$$\frac{p_{x/2}}{p_{x/2} + p_x} = \frac{3}{5} \ ,$$
which means that the inequality 4.6 is always true. □

So far, we have been able to resolve paradoxes by clearly stating the assumptions being made and by precisely stating the models being used. We end this section by describing a paradox which we cannot resolve.

Example 4.30 Suppose that we have two envelopes in front of us, and we are told that the envelopes contain X and Y dollars, respectively, where X and Y are different positive integers. We randomly choose one of the envelopes, and we open

---
[22]ibid.

4.3. PARADOXES

it, revealing X, say. Is it possible to determine, with probability greater than $1/2$, whether X is the smaller of the two dollar amounts?

Even if we have no knowledge of the joint distribution of X and Y, the surprising answer is yes! Here's how to do it. Toss a fair coin until the first time that heads turns up. Let Z denote the number of tosses required plus $1/2$. If $Z > X$, then we say that X is the smaller of the two amounts, and if $Z < X$, then we say that X is the larger of the two amounts.

First, if Z lies between X and Y, then we are sure to be correct. Since X and Y are unequal, Z lies between them with positive probability. Second, if Z is not between X and Y, then Z is either greater than both X and Y, or is less than both X and Y. In either case, X is the smaller of the two amounts with probability $1/2$, by symmetry considerations (remember, we chose the envelope at random). Thus, the probability that we are correct is greater than $1/2$. $\square$

Exercises

1. One of the first conditional probability paradoxes was provided by Bertrand.[23] It is called the *Box Paradox*. A cabinet has three drawers. In the first drawer there are two gold balls, in the second drawer there are two silver balls, and in the third drawer there is one silver and one gold ball. A drawer is picked at random and a ball chosen at random from the two balls in the drawer. Given that a gold ball was drawn, what is the probability that the drawer with the two gold balls was chosen?

   In the next two problems, the reader is asked to assume that the deck has only four cards: the ace of hearts, the ace of spades, the king of hearts, and the king of spades. (The reader is also invited to solve these problems using a standard 52-card deck.)

2. The following problem is called the *two aces problem*. This problem, dating back to 1936, has been attributed to the English mathematician J. H. C. Whitehead (see Gridgeman[24]). This problem was also submitted to Marilyn vos Savant by the master of mathematical puzzles Martin Gardner, who remarks that it is one of his favorites.

   A bridge hand has been dealt. Are the following two conditional probabilities equal? If a given hand has an ace, what is the probability that the given hand has a second ace? Given that the hand has the ace of hearts, what is the probability that the hand has a second ace?

3. In the preceding exercise, it is natural to ask "How do we get the information that the given hand has an ace?" Gridgeman considers two different ways that we might get this information.

---

[23] J. Bertrand, *Calcul des Probabilités*, Gauthier-Uillars, 1888.
[24] N. T. Gridgeman, Letter, *American Statistician*, 21 (1967), pgs. 38-39.

(a) Assume that the person holding the hand is asked to "Name an ace in your hand" and answers "The ace of hearts." What is the probability that he has a second ace?

(b) Suppose the person holding the hand is asked the more direct question "Do you have the ace of hearts?" and the answer is yes. What is the probability that he has a second ace?

4 Using the notation introduced in Example 4.29, show that in the example of Brams and Kilgour, if x is a positive power of 2, then

$$\frac{p_{x/2}}{p_{x/2} + p_x} = \frac{3}{5}.$$

5 Using the notation introduced in Example 4.29, let

$$p_x = \begin{cases} \frac{2}{3}\left(\frac{1}{3}\right)^k, & \text{if } x = 2^k, \\ 0, & \text{otherwise.} \end{cases}$$

Show that there is exactly one value of x such that if your envelope contains x, then you should switch.

*6 (For bridge players only. From Sutherland.[25]) Suppose that we are the declarer in a hand of bridge, and we have the king, 9, 8, 7, and 2 of a certain suit, while the dummy has the ace, 10, 5, and 4 of the same suit. Suppose that we want to play this suit in such a way as to maximize the probability of having no losers in the suit. We begin by leading the 2 to the ace, and we note that the queen drops on our left. We then lead the 10 from the dummy, and our right-hand opponent plays the six (after playing the three on the first round). Should we finesse or play for the drop?

---

[25] E. Sutherland, "Restricted Choice — Fact or Fiction?", *Canadian Master Point*, November 1, 1993.

Chapter 5

Important Distributions and Densities

5.1 Important Distributions

In this chapter, we describe the discrete probability distributions and the continuous probability densities that occur most often in the analysis of experiments. We will also show how one simulates these distributions and densities on a computer.

Discrete Uniform Distribution

In Chapter 1, we saw that in many cases, we assume that all outcomes of an experiment are equally likely. If X is a random variable which represents the outcome of an experiment of this type, then we say that X is uniformly distributed. If the sample space S is of size n, where $0 < n < \infty$, then the distribution function $m(\omega)$ is defined to be $1/n$ for all $\omega \in S$. As is the case with all of the discrete probability distributions discussed in this chapter, this experiment can be simulated on a computer using the program **GeneralSimulation**. However, in this case, a faster algorithm can be used instead. (This algorithm was described in Chapter 1; we repeat the description here for completeness.) The expression

$$1 + \lfloor n\,(rnd) \rfloor$$

takes on as a value each integer between 1 and n with probability $1/n$ (the notation $\lfloor x \rfloor$ denotes the greatest integer not exceeding x). Thus, if the possible outcomes of the experiment are labelled $\omega_1\, \omega_2,\, \ldots,\, \omega_n$, then we use the above expression to represent the subscript of the output of the experiment.

If the sample space is a countably infinite set, such as the set of positive integers, then it is not possible to have an experiment which is uniform on this set (see Exercise 3). If the sample space is an uncountable set, with positive, finite length, such as the interval $[0, 1]$, then we use continuous density functions (see Section 5.2).

Binomial Distribution

The binomial distribution with parameters n, p, and k was defined in Chapter 3. It is the distribution of the random variable which counts the number of heads which occur when a coin is tossed n times, assuming that on any one toss, the probability that a head occurs is p. The distribution function is given by the formula

$$b(n,p,k) = \binom{n}{k} p^k q^{n-k} ,$$

where $q = 1 - p$.

One straightforward way to simulate a binomial random variable X is to compute the sum of n independent $0-1$ random variables, each of which take on the value 1 with probability p. This method requires n calls to a random number generator to obtain one value of the random variable. When n is relatively large (say at least 30), the Central Limit Theorem (see Chapter 9) implies that the binomial distribution is well-approximated by the corresponding normal density function (which is defined in Section 5.2) with parameters $\mu = np$ and $\sigma = \sqrt{npq}$. Thus, in this case we can compute a value Y of a normal random variable with these parameters, and if $-1/2 \leq Y < n + 1/2$, we can use the value

$$\lfloor Y + 1/2 \rfloor$$

to represent the random variable X. If $Y < -1/2$ or $Y > n + 1/2$, we reject Y and compute another value. We will see in the next section how we can quickly simulate normal random variables.

Geometric Distribution

Consider a Bernoulli trials process continued for an infinite number of trials; for example, a coin tossed an infinite sequence of times. We showed in Section 2.2 how to assign a probability measure to the infinite tree. Thus, we can determine the distribution for any random variable X relating to the experiment provided $P(X = a)$ can be computed in terms of a finite number of trials. For example, let T be the number of trials up to and including the first success. Then

$$\begin{aligned} P(T=1) &= p , \\ P(T=2) &= qp , \\ P(T=3) &= q^2 p , \end{aligned}$$

and in general,

$$P(T = n) = q^{n-1} p .$$

To show that this is a distribution, we must show that

$$p + qp + q^2 p + \cdots = 1 .$$

5.1. IMPORTANT DISTRIBUTIONS

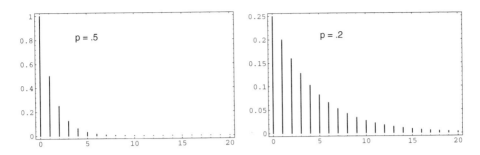

Figure 5.1: Geometric distributions.

The left-hand expression is just a geometric series with first term p and common ratio q, so its sum is
$$\frac{p}{1-q}$$
which equals 1.

In Figure 5.1 we have plotted this distribution using the program **GeometricPlot** for the cases $p = .5$ and $p = .2$. We see that as p decreases we are more likely to get large values for T, as would be expected. In both cases, the most probable value for T is 1. This will always be true since
$$\frac{P(T=j+1)}{P(T=j)} = q < 1 \ .$$

In general, if $0 < p < 1$, and $q = 1 - p$, then we say that the random variable T has a *geometric distribution* if
$$P(T = j) = q^{j-1}p \ ,$$
for $j = 1, 2, 3, \ldots$.

To simulate the geometric distribution with parameter p, we can simply compute a sequence of random numbers in $[0, 1)$, stopping when an entry does not exceed p. However, for small values of p, this is time-consuming (taking, on the average, $1/p$ steps). We now describe a method whose running time does not depend upon the size of p. Let X be a geometrically distributed random variable with parameter p, where $0 < p < 1$. Now, define Y to be the smallest integer satisfying the inequality
$$1 - q^Y \geq rnd \ . \tag{5.1}$$
Then we have
$$\begin{aligned} P(Y = j) &= P\Big(1 - q^j \geq rnd > 1 - q^{j-1}\Big) \\ &= q^{j-1} - q^j \\ &= q^{j-1}(1 - q) \\ &= q^{j-1}p \ . \end{aligned}$$

Thus, Y is geometrically distributed with parameter p. To generate Y, all we have to do is solve Equation 5.1 for Y. We obtain

$$Y = \left\lfloor \frac{\log(1 - rnd)}{\log q} \right\rfloor .$$

Since $\log(1-rnd)$ and $\log(rnd)$ are identically distributed, Y can also be generated using the equation

$$Y = \left\lfloor \frac{\log \; rnd}{\log q} \right\rfloor .$$

Example 5.1 The geometric distribution plays an important role in the theory of queues, or waiting lines. For example, suppose a line of customers waits for service at a counter. It is often assumed that, in each small time unit, either 0 or 1 new customers arrive at the counter. The probability that a customer arrives is p and that no customer arrives is $q = 1 - p$. Then the time T until the next arrival has a geometric distribution. It is natural to ask for the probability that no customer arrives in the next k time units, that is, for $P(T > k)$. This is given by

$$P(T > k) = \sum_{j=k+1}^{\infty} q^{j-1} p \;\; = \;\; q^k (p + qp + q^2 p + \cdots)$$
$$= \;\; q^k .$$

This probability can also be found by noting that we are asking for no successes (i.e., arrivals) in a sequence of k consecutive time units, where the probability of a success in any one time unit is p. Thus, the probability is just q^k, since arrivals in any two time units are independent events.

It is often assumed that the length of time required to service a customer also has a geometric distribution but with a different value for p. This implies a rather special property of the service time. To see this, let us compute the conditional probability

$$P(T > r + s \,|\, T > r) = \frac{P(T > r + s)}{P(T > r)} = \frac{q^{r+s}}{q^r} = q^s .$$

Thus, the probability that the customer's service takes s more time units is independent of the length of time r that the customer has already been served. Because of this interpretation, this property is called the "memoryless" property, and is also obeyed by the exponential distribution. (Fortunately, not too many service stations have this property.) □

Negative Binomial Distribution

Suppose we are given a coin which has probability p of coming up heads when it is tossed. We fix a positive integer k, and toss the coin until the kth head appears. We let X represent the number of tosses. When $k = 1$, X is geometrically distributed.

5.1. IMPORTANT DISTRIBUTIONS

For a general k, we say that X has a negative binomial distribution. We now calculate the probability distribution of X. If $X = x$, then it must be true that there were exactly $k - 1$ heads thrown in the first $x - 1$ tosses, and a head must have been thrown on the xth toss. There are

$$\binom{x-1}{k-1}$$

sequences of length x with these properties, and each of them is assigned the same probability, namely

$$p^{k-1} q^{x-k} \ .$$

Therefore, if we define

$$u(x, k, p) = P(X = x) \ ,$$

then

$$u(x, k, p) = \binom{x-1}{k-1} p^k q^{x-k} \ .$$

One can simulate this on a computer by simulating the tossing of a coin. The following algorithm is, in general, much faster. We note that X can be understood as the sum of k outcomes of a geometrically distributed experiment with parameter p. Thus, we can use the following sum as a means of generating X:

$$\sum_{j=1}^{k} \left\lfloor \frac{\log \ rnd_j}{\log \ q} \right\rfloor \ .$$

Example 5.2 A fair coin is tossed until the second time a head turns up. The distribution for the number of tosses is $u(x, 2, p)$. Thus the probability that x tosses are needed to obtain two heads is found by letting $k = 2$ in the above formula. We obtain

$$u(x, 2, 1/2) = \binom{x-1}{1} \frac{1}{2^x} \ ,$$

for $x = 2, 3, \ldots$.

In Figure 5.2 we give a graph of the distribution for $k = 2$ and $p = .25$. Note that the distribution is quite asymmetric, with a long tail reflecting the fact that large values of x are possible. □

Poisson Distribution

The Poisson distribution arises in many situations. It is safe to say that it is one of the three most important discrete probability distributions (the other two being the uniform and the binomial distributions). The Poisson distribution can be viewed as arising from the binomial distribution or from the exponential density. We shall now explain its connection with the former; its connection with the latter will be explained in the next section.

Suppose that we have a situation in which a certain kind of occurrence happens at random over a period of time. For example, the occurrences that we are interested

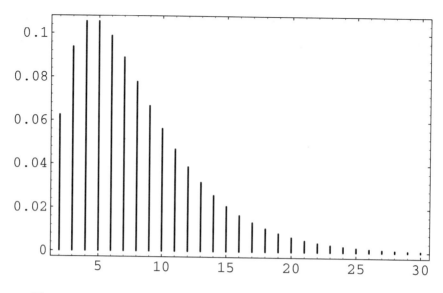

Figure 5.2: Negative binomial distribution with $k = 2$ and $p = .25$.

in might be incoming telephone calls to a police station in a large city. We want to model this situation so that we can consider the probabilities of events such as more than 10 phone calls occurring in a 5-minute time interval. Presumably, in our example, there would be more incoming calls between 6:00 and 7:00 P.M. than between 4:00 and 5:00 A.M., and this fact would certainly affect the above probability. Thus, to have a hope of computing such probabilities, we must assume that the average rate, i.e., the average number of occurrences per minute, is a constant. This rate we will denote by λ. (Thus, in a given 5-minute time interval, we would expect about 5λ occurrences.) This means that if we were to apply our model to the two time periods given above, we would simply use different rates for the two time periods, thereby obtaining two different probabilities for the given event.

Our next assumption is that the number of occurrences in two non-overlapping time intervals are independent. In our example, this means that the events that there are j calls between 5:00 and 5:15 P.M. and k calls between 6:00 and 6:15 P.M. on the same day are independent.

We can use the binomial distribution to model this situation. We imagine that a given time interval is broken up into n subintervals of equal length. If the subintervals are sufficiently short, we can assume that two or more occurrences happen in one subinterval with a probability which is negligible in comparison with the probability of at most one occurrence. Thus, in each subinterval, we are assuming that there is either 0 or 1 occurrence. This means that the sequence of subintervals can be thought of as a sequence of Bernoulli trials, with a success corresponding to an occurrence in the subinterval.

To decide upon the proper value of p, the probability of an occurrence in a given subinterval, we reason as follows. On the average, there are λt occurrences in a

5.1. IMPORTANT DISTRIBUTIONS

time interval of length t. If this time interval is divided into n subintervals, then we would expect, using the Bernoulli trials interpretation, that there should be np occurrences. Thus, we want

$$\lambda t = np \;,$$

so

$$p = \frac{\lambda t}{n} \;.$$

We now wish to consider the random variable X, which counts the number of occurrences in a given time interval. We want to calculate the distribution of X. For ease of calculation, we will assume that the time interval is of length 1; for time intervals of arbitrary length t, see Exercise 11. We know that

$$P(X=0) = b(n,p,0) = (1-p)^n = \left(1 - \frac{\lambda}{n}\right)^n \;.$$

For large n, this is approximately $e^{-\lambda}$. It is easy to calculate that for any fixed k, we have

$$\frac{b(n,p,k)}{b(n,p,k-1)} = \frac{\lambda - (k-1)p}{kq}$$

which, for large n (and therefore small p) is approximately λ/k. Thus, we have

$$P(X=1) \approx \lambda e^{-\lambda} \;,$$

and in general,

$$P(X=k) \approx \frac{\lambda^k}{k!} e^{-\lambda} \;. \tag{5.2}$$

The above distribution is the Poisson distribution. We note that it must be checked that the distribution given in Equation 5.2 really *is* a distribution, i.e., that its values are non-negative and sum to 1. (See Exercise 12.)

The Poisson distribution is used as an approximation to the binomial distribution when the parameters n and p are large and small, respectively (see Examples 5.3 and 5.4). However, the Poisson distribution also arises in situations where it may not be easy to interpret or measure the parameters n and p (see Example 5.5).

Example 5.3 A typesetter makes, on the average, one mistake per 1000 words. Assume that he is setting a book with 100 words to a page. Let S_{100} be the number of mistakes that he makes on a single page. Then the exact probability distribution for S_{100} would be obtained by considering S_{100} as a result of 100 Bernoulli trials with $p = 1/1000$. The expected value of S_{100} is $\lambda = 100(1/1000) = .1$. The exact probability that $S_{100} = j$ is $b(100, 1/1000, j)$, and the Poisson approximation is

$$\frac{e^{-.1}(.1)^j}{j!}.$$

In Table 5.1 we give, for various values of n and p, the exact values computed by the binomial distribution and the Poisson approximation. □

	Poisson	Binomial	Poisson	Binomial	Poisson	Binomial
		$n = 100$		$n = 100$		$n = 1000$
j	$\lambda = .1$	$p = .001$	$\lambda = 1$	$p = .01$	$\lambda = 10$	$p = .01$
0	.9048	.9048	.3679	.3660	.0000	.0000
1	.0905	.0905	.3679	.3697	.0005	.0004
2	.0045	.0045	.1839	.1849	.0023	.0022
3	.0002	.0002	.0613	.0610	.0076	.0074
4	.0000	.0000	.0153	.0149	.0189	.0186
5			.0031	.0029	.0378	.0374
6			.0005	.0005	.0631	.0627
7			.0001	.0001	.0901	.0900
8			.0000	.0000	.1126	.1128
9					.1251	.1256
10					.1251	.1257
11					.1137	.1143
12					.0948	.0952
13					.0729	.0731
14					.0521	.0520
15					.0347	.0345
16					.0217	.0215
17					.0128	.0126
18					.0071	.0069
19					.0037	.0036
20					.0019	.0018
21					.0009	.0009
22					.0004	.0004
23					.0002	.0002
24					.0001	.0001
25					.0000	.0000

Table 5.1: Poisson approximation to the binomial distribution.

5.1. IMPORTANT DISTRIBUTIONS

Example 5.4 In his book,[1] Feller discusses the statistics of flying bomb hits in the south of London during the Second World War.

Assume that you live in a district of size 10 blocks by 10 blocks so that the total district is divided into 100 small squares. How likely is it that the square in which you live will receive no hits if the total area is hit by 400 bombs?

We assume that a particular bomb will hit your square with probability $1/100$. Since there are 400 bombs, we can regard the number of hits that your square receives as the number of *successes* in a Bernoulli trials process with $n = 400$ and $p = 1/100$. Thus we can use the Poisson distribution with $\lambda = 400 \cdot 1/100 = 4$ to approximate the probability that your square will receive j hits. This probability is $p(j) = e^{-4}4^j/j!$. The expected number of squares that receive exactly j hits is then $100 \cdot p(j)$. It is easy to write a program **LondonBombs** to simulate this situation and compare the expected number of squares with j hits with the observed number. In Exercise 26 you are asked to compare the actual observed data with that predicted by the Poisson distribution.

In Figure 5.3, we have shown the simulated hits, together with a spike graph showing both the observed and predicted frequencies. The observed frequencies are shown as squares, and the predicted frequencies are shown as dots. □

If the reader would rather not consider flying bombs, he is invited to instead consider an analogous situation involving cookies and raisins. We assume that we have made enough cookie dough for 500 cookies. We put 600 raisins in the dough, and mix it thoroughly. One way to look at this situation is that we have 500 cookies, and after placing the cookies in a grid on the table, we throw 600 raisins at the cookies. (See Exercise 22.)

Example 5.5 Suppose that in a certain fixed amount A of blood, the average human has 40 white blood cells. Let X be the random variable which gives the number of white blood cells in a random sample of size A from a random individual. We can think of X as binomially distributed with each white blood cell in the body representing a trial. If a given white blood cell turns up in the sample, then the trial corresponding to that blood cell was a success. Then p should be taken as the ratio of A to the total amount of blood in the individual, and n will be the number of white blood cells in the individual. Of course, in practice, neither of these parameters is very easy to measure accurately, but presumably the number 40 is easy to measure. But for the average human, we then have $40 = np$, so we can think of X as being Poisson distributed, with parameter $\lambda = 40$. In this case, it is easier to model the situation using the Poisson distribution than the binomial distribution. □

To simulate a Poisson random variable on a computer, a good way is to take advantage of the relationship between the Poisson distribution and the exponential density. This relationship and the resulting simulation algorithm will be described in the next section.

---

[1] ibid., p. 161.

192   CHAPTER 5. DISTRIBUTIONS AND DENSITIES

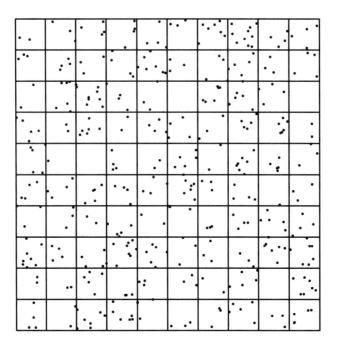

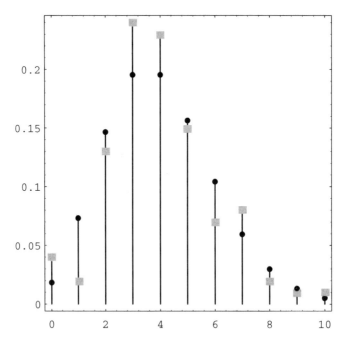

Figure 5.3: Flying bomb hits.

5.1. IMPORTANT DISTRIBUTIONS

Hypergeometric Distribution

Suppose that we have a set of N balls, of which k are red and $N-k$ are blue. We choose n of these balls, without replacement, and define X to be the number of red balls in our sample. The distribution of X is called the hypergeometric distribution. We note that this distribution depends upon three parameters, namely N, k, and n. There does not seem to be a standard notation for this distribution; we will use the notation $h(N, k, n, x)$ to denote $P(X = x)$. This probability can be found by noting that there are

$$\binom{N}{n}$$

different samples of size n, and the number of such samples with exactly x red balls is obtained by multiplying the number of ways of choosing x red balls from the set of k red balls and the number of ways of choosing $n - x$ blue balls from the set of $N - k$ blue balls. Hence, we have

$$h(N, k, n, x) = \frac{\binom{k}{x}\binom{N-k}{n-x}}{\binom{N}{n}}.$$

This distribution can be generalized to the case where there are more than two types of objects. (See Exercise 40.)

If we let N and k tend to ∞, in such a way that the ratio k/N remains fixed, then the hypergeometric distribution tends to the binomial distribution with parameters n and $p = k/N$. This is reasonable because if N and k are much larger than n, then whether we choose our sample with or without replacement should not affect the probabilities very much, and the experiment consisting of choosing with replacement yields a binomially distributed random variable (see Exercise 44).

An example of how this distribution might be used is given in Exercises 36 and 37. We now give another example involving the hypergeometric distribution. It illustrates a statistical test called Fisher's Exact Test.

Example 5.6 It is often of interest to consider two traits, such as eye color and hair color, and to ask whether there is an association between the two traits. Two traits are associated if knowing the value of one of the traits for a given person allows us to predict the value of the other trait for that person. The stronger the association, the more accurate the predictions become. If there is no association between the traits, then we say that the traits are independent. In this example, we will use the traits of gender and political party, and we will assume that there are only two possible genders, female and male, and only two possible political parties, Democratic and Republican.

Suppose that we have collected data concerning these traits. To test whether there is an association between the traits, we first assume that there is no association between the two traits. This gives rise to an "expected" data set, in which knowledge of the value of one trait is of no help in predicting the value of the other trait. Our collected data set usually differs from this expected data set. If it differs by quite a bit, then we would tend to reject the assumption of independence of the traits. To

	Democrat	Republican	
Female	24	4	28
Male	8	14	22
	32	18	50

Table 5.2: Observed data.

	Democrat	Republican	
Female	s_{11}	s_{12}	t_{11}
Male	s_{21}	s_{22}	t_{12}
	t_{21}	t_{22}	n

Table 5.3: General data table.

nail down what is meant by "quite a bit," we decide which possible data sets differ from the expected data set by at least as much as ours does, and then we compute the probability that any of these data sets would occur under the assumption of independence of traits. If this probability is small, then it is unlikely that the difference between our collected data set and the expected data set is due entirely to chance.

Suppose that we have collected the data shown in Table 5.2. The row and column sums are called marginal totals, or marginals. In what follows, we will denote the row sums by t_{11} and t_{12}, and the column sums by t_{21} and t_{22}. The ijth entry in the table will be denoted by s_{ij}. Finally, the size of the data set will be denoted by n. Thus, a general data table will look as shown in Table 5.3. We now explain the model which will be used to construct the "expected" data set. In the model, we assume that the two traits are independent. We then put t_{21} yellow balls and t_{22} green balls, corresponding to the Democratic and Republican marginals, into an urn. We draw t_{11} balls, without replacement, from the urn, and call these balls females. The t_{12} balls remaining in the urn are called males. In the specific case under consideration, the probability of getting the actual data under this model is given by the expression

$$\frac{\binom{32}{24}\binom{18}{4}}{\binom{50}{28}},$$

i.e., a value of the hypergeometric distribution.

We are now ready to construct the expected data set. If we choose 28 balls out of 50, we should expect to see, on the average, the same percentage of yellow balls in our sample as in the urn. Thus, we should expect to see, on the average, $28(32/50) = 17.92 \approx 18$ yellow balls in our sample. (See Exercise 36.) The other expected values are computed in exactly the same way. Thus, the expected data set is shown in Table 5.4. We note that the value of s_{11} determines the other three values in the table, since the marginals are all fixed. Thus, in considering the possible data sets that could appear in this model, it is enough to consider the various possible values of s_{11}. In the specific case at hand, what is the probability

5.1. IMPORTANT DISTRIBUTIONS

	Democrat	Republican	
Female	18	10	28
Male	14	8	22
	32	18	50

Table 5.4: Expected data.

of drawing exactly a yellow balls, i.e., what is the probability that $s_{11} = a$? It is

$$\frac{\binom{32}{a}\binom{18}{28-a}}{\binom{50}{28}}. \tag{5.3}$$

We are now ready to decide whether our actual data differs from the expected data set by an amount which is greater than could be reasonably attributed to chance alone. We note that the expected number of female Democrats is 18, but the actual number in our data is 24. The other data sets which differ from the expected data set by more than ours correspond to those where the number of female Democrats equals 25, 26, 27, or 28. Thus, to obtain the required probability, we sum the expression in (5.3) from $a = 24$ to $a = 28$. We obtain a value of .000395. Thus, we should reject the hypothesis that the two traits are independent. □

Finally, we turn to the question of how to simulate a hypergeometric random variable X. Let us assume that the parameters for X are N, k, and n. We imagine that we have a set of N balls, labelled from 1 to N. We decree that the first k of these balls are red, and the rest are blue. Suppose that we have chosen m balls, and that j of them are red. Then there are $k - j$ red balls left, and $N - m$ balls left. Thus, our next choice will be red with probability

$$\frac{k-j}{N-m}.$$

So at this stage, we choose a random number in $[0, 1]$, and report that a red ball has been chosen if and only if the random number does not exceed the above expression. Then we update the values of m and j, and continue until n balls have been chosen.

Benford Distribution

Our next example of a distribution comes from the study of leading digits in data sets. It turns out that many data sets that occur "in real life" have the property that the first digits of the data are not uniformly distributed over the set $\{1, 2, \ldots, 9\}$. Rather, it appears that the digit 1 is most likely to occur, and that the distribution is monotonically decreasing on the set of possible digits. The Benford distribution appears, in many cases, to fit such data. Many explanations have been given for the occurrence of this distribution. Possibly the most convincing explanation is that this distribution is the only one that is invariant under a change of scale. If one thinks of certain data sets as somehow "naturally occurring," then the distribution should be unaffected by which units are chosen in which to represent the data, i.e., the distribution should be invariant under change of scale.

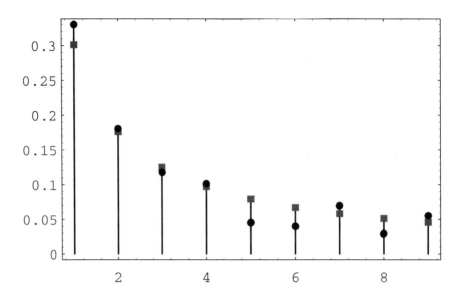

Figure 5.4: Leading digits in President Clinton's tax returns.

Theodore Hill[2] gives a general description of the Benford distribution, when one considers the first d digits of integers in a data set. We will restrict our attention to the first digit. In this case, the Benford distribution has distribution function

$$f(k) = \log_{10}(k+1) - \log_{10}(k) ,$$

for $1 \leq k \leq 9$.

Mark Nigrini[3] has advocated the use of the Benford distribution as a means of testing suspicious financial records such as bookkeeping entries, checks, and tax returns. His idea is that if someone were to "make up" numbers in these cases, the person would probably produce numbers that are fairly uniformly distributed, while if one were to use the actual numbers, the leading digits would roughly follow the Benford distribution. As an example, Negrini analyzed President Clinton's tax returns for a 13-year period. In Figure 5.4, the Benford distribution values are shown as squares, and the President's tax return data are shown as circles. One sees that in this example, the Benford distribution fits the data very well.

This distribution was discovered by the astronomer Simon Newcomb who stated the following in his paper on the subject: "That the ten digits do not occur with equal frequency must be evident to anyone making use of logarithm tables, and noticing how much faster the first pages wear out than the last ones. The first significant figure is oftener 1 than any other digit, and the frequency diminishes up to 9."[4]

---

[2]T. P. Hill, "The Significant Digit Phenomenon," *American Mathematical Monthly*, vol. 102, no. 4 (April 1995), pgs. 322-327.

[3]M. Nigrini, "Detecting Biases and Irregularities in Tabulated Data," working paper

[4]S. Newcomb, "Note on the frequency of use of the different digits in natural numbers," *American Journal of Mathematics*, vol. 4 (1881), pgs. 39-40.

5.1. IMPORTANT DISTRIBUTIONS

Exercises

1. For which of the following random variables would it be appropriate to assign a uniform distribution?

   (a) Let X represent the roll of one die.

   (b) Let X represent the number of heads obtained in three tosses of a coin.

   (c) A roulette wheel has 38 possible outcomes: 0, 00, and 1 through 36. Let X represent the outcome when a roulette wheel is spun.

   (d) Let X represent the birthday of a randomly chosen person.

   (e) Let X represent the number of tosses of a coin necessary to achieve a head for the first time.

2. Let n be a positive integer. Let S be the set of integers between 1 and n. Consider the following process: We remove a number from S and write it down. We repeat this until S is empty. The result is a permutation of the integers from 1 to n. Let X denote this permutation. Is X uniformly distributed?

3. Let X be a random variable which can take on countably many values. Show that X cannot be uniformly distributed.

4. Suppose we are attending a college which has 3000 students. We wish to choose a subset of size 100 from the student body. Let X represent the subset, chosen using the following possible strategies. For which strategies would it be appropriate to assign the uniform distribution to X? If it is appropriate, what probability should we assign to each outcome?

   (a) Take the first 100 students who enter the cafeteria to eat lunch.

   (b) Ask the Registrar to sort the students by their Social Security number, and then take the first 100 in the resulting list.

   (c) Ask the Registrar for a set of cards, with each card containing the name of exactly one student, and with each student appearing on exactly one card. Throw the cards out of a third-story window, then walk outside and pick up the first 100 cards that you find.

5. Under the same conditions as in the preceding exercise, can you describe a procedure which, if used, would produce each possible outcome with the same probability? Can you describe such a procedure that does not rely on a computer or a calculator?

6. Let X_1, X_2, ..., X_n be n mutually independent random variables, each of which is uniformly distributed on the integers from 1 to k. Let Y denote the minimum of the X_i's. Find the distribution of Y.

7. A die is rolled until the first time T that a six turns up.

   (a) What is the probability distribution for T?

(b) Find $P(T > 3)$.

(c) Find $P(T > 6 | T > 3)$.

8 If a coin is tossed a sequence of times, what is the probability that the first head will occur after the fifth toss, given that it has not occurred in the first two tosses?

9 A worker for the Department of Fish and Game is assigned the job of estimating the number of trout in a certain lake of modest size. She proceeds as follows: She catches 100 trout, tags each of them, and puts them back in the lake. One month later, she catches 100 more trout, and notes that 10 of them have tags.

(a) Without doing any fancy calculations, give a rough estimate of the number of trout in the lake.

(b) Let N be the number of trout in the lake. Find an expression, in terms of N, for the probability that the worker would catch 10 tagged trout out of the 100 trout that she caught the second time.

(c) Find the value of N which maximizes the expression in part (b). This value is called the *maximum likelihood estimate* for the unknown quantity N. *Hint*: Consider the ratio of the expressions for successive values of N.

10 A census in the United States is an attempt to count everyone in the country. It is inevitable that many people are not counted. The U. S. Census Bureau proposed a way to estimate the number of people who were not counted by the latest census. Their proposal was as follows: In a given locality, let N denote the actual number of people who live there. Assume that the census counted n_1 people living in this area. Now, another census was taken in the locality, and n_2 people were counted. In addition, n_{12} people were counted both times.

(a) Given N, n_1, and n_2, let X denote the number of people counted both times. Find the probability that $X = k$, where k is a fixed positive integer between 0 and n_2.

(b) Now assume that $X = n_{12}$. Find the value of N which maximizes the expression in part (a). *Hint*: Consider the ratio of the expressions for successive values of N.

11 Suppose that X is a random variable which represents the number of calls coming in to a police station in a one-minute interval. In the text, we showed that X could be modelled using a Poisson distribution with parameter λ, where this parameter represents the average number of incoming calls per minute. Now suppose that Y is a random variable which represents the number of incoming calls in an interval of length t. Show that the distribution of Y is given by

$$P(Y = k) = e^{-\lambda t} \frac{(\lambda t)^k}{k!} ,$$

i.e., Y is Poisson with parameter λt. *Hint*: Suppose a Martian were to observe the police station. Let us also assume that the basic time interval used on Mars is exactly t Earth minutes. Finally, we will assume that the Martian understands the derivation of the Poisson distribution in the text. What would she write down for the distribution of Y?

12 Show that the values of the Poisson distribution given in Equation 5.2 sum to 1.

13 The Poisson distribution with parameter $\lambda = .3$ has been assigned for the outcome of an experiment. Let X be the outcome function. Find $P(X = 0)$, $P(X = 1)$, and $P(X > 1)$.

14 On the average, only 1 person in 1000 has a particular rare blood type.

(a) Find the probability that, in a city of 10,000 people, no one has this blood type.

(b) How many people would have to be tested to give a probability greater than 1/2 of finding at least one person with this blood type?

15 Write a program for the user to input n, p, j and have the program print out the exact value of $b(n, p, k)$ and the Poisson approximation to this value.

16 Assume that, during each second, a Dartmouth switchboard receives one call with probability .01 and no calls with probability .99. Use the Poisson approximation to estimate the probability that the operator will miss at most one call if she takes a 5-minute coffee break.

17 The probability of a royal flush in a poker hand is $p = 1/649{,}740$. How large must n be to render the probability of having no royal flush in n hands smaller than $1/e$?

18 A baker blends 600 raisins and 400 chocolate chips into a dough mix and, from this, makes 500 cookies.

(a) Find the probability that a randomly picked cookie will have no raisins.

(b) Find the probability that a randomly picked cookie will have exactly two chocolate chips.

(c) Find the probability that a randomly chosen cookie will have at least two bits (raisins or chips) in it.

19 The probability that, in a bridge deal, one of the four hands has all hearts is approximately 6.3×10^{-12}. In a city with about 50,000 bridge players the resident probability expert is called on the average once a year (usually late at night) and told that the caller has just been dealt a hand of all hearts. Should she suspect that some of these callers are the victims of practical jokes?

20 An advertiser drops 10,000 leaflets on a city which has 2000 blocks. Assume that each leaflet has an equal chance of landing on each block. What is the probability that a particular block will receive no leaflets?

21 In a class of 80 students, the professor calls on 1 student chosen at random for a recitation in each class period. There are 32 class periods in a term.

(a) Write a formula for the exact probability that a given student is called upon j times during the term.

(b) Write a formula for the Poisson approximation for this probability. Using your formula estimate the probability that a given student is called upon more than twice.

22 Assume that we are making raisin cookies. We put a box of 600 raisins into our dough mix, mix up the dough, then make from the dough 500 cookies. We then ask for the probability that a randomly chosen cookie will have 0, 1, 2, ... raisins. Consider the cookies as trials in an experiment, and let X be the random variable which gives the number of raisins in a given cookie. Then we can regard the number of raisins in a cookie as the result of $n = 600$ independent trials with probability $p = 1/500$ for success on each trial. Since n is large and p is small, we can use the Poisson approximation with $\lambda = 600(1/500) = 1.2$. Determine the probability that a given cookie will have at least five raisins.

23 For a certain experiment, the Poisson distribution with parameter $\lambda = m$ has been assigned. Show that a most probable outcome for the experiment is the integer value k such that $m - 1 \leq k \leq m$. Under what conditions will there be two most probable values? *Hint*: Consider the ratio of successive probabilities.

24 When John Kemeny was chair of the Mathematics Department at Dartmouth College, he received an average of ten letters each day. On a certain weekday he received no mail and wondered if it was a holiday. To decide this he computed the probability that, in ten years, he would have at least 1 day without any mail. He assumed that the number of letters he received on a given day has a Poisson distribution. What probability did he find? *Hint*: Apply the Poisson distribution twice. First, to find the probability that, in 3000 days, he will have at least 1 day without mail, assuming each year has about 300 days on which mail is delivered.

25 Reese Prosser never puts money in a 10-cent parking meter in Hanover. He assumes that there is a probability of .05 that he will be caught. The first offense costs nothing, the second costs 2 dollars, and subsequent offenses cost 5 dollars each. Under his assumptions, how does the expected cost of parking 100 times without paying the meter compare with the cost of paying the meter each time?

Number of deaths	Number of corps with x deaths in a given year
0	144
1	91
2	32
3	11
4	2

Table 5.5: Mule kicks.

26 Feller[5] discusses the statistics of flying bomb hits in an area in the south of London during the Second World War. The area in question was divided into $24 \times 24 = 576$ small areas. The total number of hits was 537. There were 229 squares with 0 hits, 211 with 1 hit, 93 with 2 hits, 35 with 3 hits, 7 with 4 hits, and 1 with 5 or more. Assuming the hits were purely random, use the Poisson approximation to find the probability that a particular square would have exactly k hits. Compute the expected number of squares that would have 0, 1, 2, 3, 4, and 5 or more hits and compare this with the observed results.

27 Assume that the probability that there is a significant accident in a nuclear power plant during one year's time is .001. If a country has 100 nuclear plants, estimate the probability that there is at least one such accident during a given year.

28 An airline finds that 4 percent of the passengers that make reservations on a particular flight will not show up. Consequently, their policy is to sell 100 reserved seats on a plane that has only 98 seats. Find the probability that every person who shows up for the flight will find a seat available.

29 The king's coinmaster boxes his coins 500 to a box and puts 1 counterfeit coin in each box. The king is suspicious, but, instead of testing all the coins in 1 box, he tests 1 coin chosen at random out of each of 500 boxes. What is the probability that he finds at least one fake? What is it if the king tests 2 coins from each of 250 boxes?

30 (From Kemeny[6]) Show that, if you make 100 bets on the number 17 at roulette at Monte Carlo (see Example 6.13), you will have a probability greater than 1/2 of coming out ahead. What is your expected winning?

31 In one of the first studies of the Poisson distribution, von Bortkiewicz[7] considered the frequency of deaths from kicks in the Prussian army corps. From the study of 14 corps over a 20-year period, he obtained the data shown in Table 5.5. Fit a Poisson distribution to this data and see if you think that the Poisson distribution is appropriate.

---

[5] ibid., p. 161.
[6] Private communication.
[7] L. von Bortkiewicz, *Das Gesetz der Kleinen Zahlen* (Leipzig: Teubner, 1898), p. 24.

32 It is often assumed that the auto traffic that arrives at the intersection during a unit time period has a Poisson distribution with expected value m. Assume that the number of cars X that arrive at an intersection from the north in unit time has a Poisson distribution with parameter $\lambda = m$ and the number Y that arrive from the west in unit time has a Poisson distribution with parameter $\lambda = \bar{m}$. If X and Y are independent, show that the total number $X + Y$ that arrive at the intersection in unit time has a Poisson distribution with parameter $\lambda = m + \bar{m}$.

33 Cars coming along Magnolia Street come to a fork in the road and have to choose either Willow Street or Main Street to continue. Assume that the number of cars that arrive at the fork in unit time has a Poisson distribution with parameter $\lambda = 4$. A car arriving at the fork chooses Main Street with probability 3/4 and Willow Street with probability 1/4. Let X be the random variable which counts the number of cars that, in a given unit of time, pass by Joe's Barber Shop on Main Street. What is the distribution of X?

34 In the appeal of the *People v. Collins* case (see Exercise 4.1.28), the counsel for the defense argued as follows: Suppose, for example, there are 5,000,000 couples in the Los Angeles area and the probability that a randomly chosen couple fits the witnesses' description is 1/12,000,000. Then the probability that there are two such couples given that there is at least one is not at all small. Find this probability. (The California Supreme Court overturned the initial guilty verdict.)

35 A manufactured lot of brass turnbuckles has S items of which D are defective. A sample of s items is drawn without replacement. Let X be a random variable that gives the number of defective items in the sample. Let $p(d) = P(X = d)$.

(a) Show that
$$p(d) = \frac{\binom{D}{d}\binom{S-D}{s-d}}{\binom{S}{s}}.$$

Thus, X is hypergeometric.

(b) Prove the following identity, known as *Euler's formula*:
$$\sum_{d=0}^{\min(D,s)} \binom{D}{d}\binom{S-D}{s-d} = \binom{S}{s}.$$

36 A bin of 1000 turnbuckles has an unknown number D of defectives. A sample of 100 turnbuckles has 2 defectives. The *maximum likelihood estimate* for D is the number of defectives which gives the highest probability for obtaining the number of defectives observed in the sample. Guess this number D and then write a computer program to verify your guess.

37 There are an unknown number of moose on Isle Royale (a National Park in Lake Superior). To estimate the number of moose, 50 moose are captured and

5.1. IMPORTANT DISTRIBUTIONS

tagged. Six months later 200 moose are captured and it is found that 8 of these were tagged. Estimate the number of moose on Isle Royale from these data, and then verify your guess by computer program (see Exercise 36).

38 A manufactured lot of buggy whips has 20 items, of which 5 are defective. A random sample of 5 items is chosen to be inspected. Find the probability that the sample contains exactly one defective item

  (a) if the sampling is done with replacement.

  (b) if the sampling is done without replacement.

39 Suppose that N and k tend to ∞ in such a way that k/N remains fixed. Show that
$$h(N, k, n, x) \to b(n, k/N, x) .$$

40 A bridge deck has 52 cards with 13 cards in each of four suits: spades, hearts, diamonds, and clubs. A hand of 13 cards is dealt from a shuffled deck. Find the probability that the hand has

  (a) a distribution of suits 4, 4, 3, 2 (for example, four spades, four hearts, three diamonds, two clubs).

  (b) a distribution of suits 5, 3, 3, 2.

41 Write a computer algorithm that simulates a hypergeometric random variable with parameters N, k, and n.

42 You are presented with four different dice. The first one has two sides marked 0 and four sides marked 4. The second one has a 3 on every side. The third one has a 2 on four sides and a 6 on two sides, and the fourth one has a 1 on three sides and a 5 on three sides. You allow your friend to pick any of the four dice he wishes. Then you pick one of the remaining three and you each roll your die. The person with the largest number showing wins a dollar. Show that you can choose your die so that you have probability 2/3 of winning no matter which die your friend picks. (See Tenney and Foster.[8])

43 The students in a certain class were classified by hair color and eye color. The conventions used were: Brown and black hair were considered dark, and red and blonde hair were considered light; black and brown eyes were considered dark, and blue and green eyes were considered light. They collected the data shown in Table 5.6. Are these traits independent? (See Example 5.6.)

44 Suppose that in the hypergeometric distribution, we let N and k tend to ∞ in such a way that the ratio k/N approaches a real number p between 0 and 1. Show that the hypergeometric distribution tends to the binomial distribution with parameters n and p.

---

[8]R. L. Tenney and C. C. Foster, *Non transitive Dominance*, Math. Mag. 49 (1976) no. 3, pgs. 115-120.

	Dark Eyes	Light Eyes	
Dark Hair	28	15	43
Light Hair	9	23	32
	37	38	75

Table 5.6: Observed data.

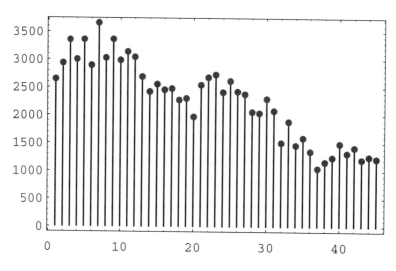

Figure 5.5: Distribution of choices in the Powerball lottery.

45 (a) Compute the leading digits of the first 100 powers of 2, and see how well these data fit the Benford distribution.

(b) Multiply each number in the data set of part (a) by 3, and compare the distribution of the leading digits with the Benford distribution.

46 In the Powerball lottery, contestants pick 5 different integers between 1 and 45, and in addition, pick a bonus integer from the same range (the bonus integer can equal one of the first five integers chosen). Some contestants choose the numbers themselves, and others let the computer choose the numbers. The data shown in Table 5.7 are the contestant-chosen numbers in a certain state on May 3, 1996. A spike graph of the data is shown in Figure 5.5.

The goal of this problem is to check the hypothesis that the chosen numbers are uniformly distributed. To do this, compute the value v of the random variable χ^2 given in Example 5.10. In the present case, this random variable has 44 degrees of freedom. One can find, in a χ^2 table, the value $v_0 = 59.43$, which represents a number with the property that a χ^2-distributed random variable takes on values that exceed v_0 only 5% of the time. Does your computed value of v exceed v_0? If so, you should reject the hypothesis that the contestants' choices are uniformly distributed.

5.2. IMPORTANT DENSITIES

Integer	Times Chosen	Integer	Times Chosen	Integer	Times Chosen
1	2646	2	2934	3	3352
4	3000	5	3357	6	2892
7	3657	8	3025	9	3362
10	2985	11	3138	12	3043
13	2690	14	2423	15	2556
16	2456	17	2479	18	2276
19	2304	20	1971	21	2543
22	2678	23	2729	24	2414
25	2616	26	2426	27	2381
28	2059	29	2039	30	2298
31	2081	32	1508	33	1887
34	1463	35	1594	36	1354
37	1049	38	1165	39	1248
40	1493	41	1322	42	1423
43	1207	44	1259	45	1224

Table 5.7: Numbers chosen by contestants in the Powerball lottery.

5.2 Important Densities

In this section, we will introduce some important probability density functions and give some examples of their use. We will also consider the question of how one simulates a given density using a computer.

Continuous Uniform Density

The simplest density function corresponds to the random variable U whose value represents the outcome of the experiment consisting of choosing a real number at random from the interval $[a, b]$.

$$f(\omega) = \begin{cases} 1/(b-a), & \text{if } a \leq \omega \leq b, \\ 0, & \text{otherwise.} \end{cases}$$

It is easy to simulate this density on a computer. We simply calculate the expression

$$(b-a)rnd + a \ .$$

Exponential and Gamma Densities

The exponential density function is defined by

$$f(x) = \begin{cases} \lambda e^{-\lambda x}, & \text{if } 0 \leq x < \infty, \\ 0, & \text{otherwise.} \end{cases}$$

Here λ is any positive constant, depending on the experiment. The reader has seen this density in Example 2.17. In Figure 5.6 we show graphs of several exponential densities for different choices of λ. The exponential density is often used to

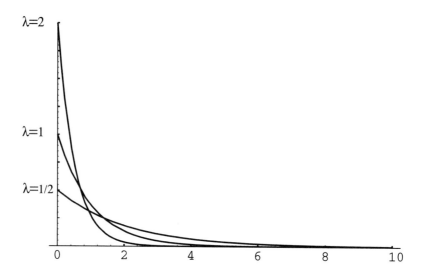

Figure 5.6: Exponential densities.

describe experiments involving a question of the form: How long until something happens? For example, the exponential density is often used to study the time between emissions of particles from a radioactive source.

The cumulative distribution function of the exponential density is easy to compute. Let T be an exponentially distributed random variable with parameter λ. If $x \geq 0$, then we have

$$\begin{aligned} F(x) &= P(T \leq x) \\ &= \int_0^x \lambda e^{-\lambda t}\, dt \\ &= 1 - e^{-\lambda x}\ . \end{aligned}$$

Both the exponential density and the geometric distribution share a property known as the "memoryless" property. This property was introduced in Example 5.1; it says that

$$P(T > r + s \,|\, T > r) = P(T > s)\ .$$

This can be demonstrated to hold for the exponential density by computing both sides of this equation. The right-hand side is just

$$1 - F(s) = e^{-\lambda s}\ ,$$

while the left-hand side is

$$\frac{P(T > r + s)}{P(T > r)} = \frac{1 - F(r + s)}{1 - F(s)}$$

5.2. IMPORTANT DENSITIES

$$= \frac{e^{-\lambda(r+s)}}{e^{-\lambda r}}$$
$$= e^{-\lambda s} .$$

There is a very important relationship between the exponential density and the Poisson distribution. We begin by defining X_1, X_2, ... to be a sequence of independent exponentially distributed random variables with parameter λ. We might think of X_i as denoting the amount of time between the ith and $(i+1)$st emissions of a particle by a radioactive source. (As we shall see in Chapter 6, we can think of the parameter λ as representing the reciprocal of the average length of time between emissions. This parameter is a quantity that might be measured in an actual experiment of this type.)

We now consider a time interval of length t, and we let Y denote the random variable which counts the number of emissions that occur in the time interval. We would like to calculate the distribution function of Y (clearly, Y is a discrete random variable). If we let S_n denote the sum $X_1 + X_2 + \cdots + X_n$, then it is easy to see that
$$P(Y = n) = P(S_n \leq t \text{ and } S_{n+1} > t) .$$
Since the event $S_{n+1} \leq t$ is a subset of the event $S_n \leq t$, the above probability is seen to be equal to
$$P(S_n \leq t) - P(S_{n+1} \leq t) . \tag{5.4}$$
We will show in Chapter 7 that the density of S_n is given by the following formula:
$$g_n(x) = \begin{cases} \lambda \frac{(\lambda x)^{n-1}}{(n-1)!} e^{-\lambda x}, & \text{if } x > 0, \\ 0, & \text{otherwise.} \end{cases}$$
This density is an example of a gamma density with parameters λ and n. The general gamma density allows n to be any positive real number. We shall not discuss this general density.

It is easy to show by induction on n that the cumulative distribution function of S_n is given by:
$$G_n(x) = \begin{cases} 1 - e^{-\lambda x}\left(1 + \frac{\lambda x}{1!} + \cdots + \frac{(\lambda x)^{n-1}}{(n-1)!}\right), & \text{if } x > 0, \\ 0, & \text{otherwise.} \end{cases}$$
Using this expression, the quantity in (5.4) is easy to compute; we obtain
$$e^{-\lambda t} \frac{(\lambda t)^n}{n!} ,$$
which the reader will recognize as the probability that a Poisson-distributed random variable, with parameter λt, takes on the value n.

The above relationship will allow us to simulate a Poisson distribution, once we have found a way to simulate an exponential density. The following random variable does the job:
$$Y = -\frac{1}{\lambda} \log(rnd) . \tag{5.5}$$

Using Corollary 5.2 (below), one can derive the above expression (see Exercise 3). We content ourselves for now with a short calculation that should convince the reader that the random variable Y has the required property. We have

$$\begin{aligned} P(Y \leq y) &= P\left(-\frac{1}{\lambda}\log(rnd) \leq y\right) \\ &= P(\log(rnd) \geq -\lambda y) \\ &= P(rnd \geq e^{-\lambda y}) \\ &= 1 - e^{-\lambda y} \ . \end{aligned}$$

This last expression is seen to be the cumulative distribution function of an exponentially distributed random variable with parameter λ.

To simulate a Poisson random variable W with parameter λ, we simply generate a sequence of values of an exponentially distributed random variable with the same parameter, and keep track of the subtotals S_k of these values. We stop generating the sequence when the subtotal first exceeds λ. Assume that we find that

$$S_n \leq \lambda < S_{n+1} \ .$$

Then the value n is returned as a simulated value for W.

Example 5.7 (Queues) Suppose that customers arrive at random times at a service station with one server, and suppose that each customer is served immediately if no one is ahead of him, but must wait his turn in line otherwise. How long should each customer expect to wait? (We define the waiting time of a customer to be the length of time between the time that he arrives and the time that he begins to be served.)

Let us assume that the interarrival times between successive customers are given by random variables $X_1, X_2, \ldots, X_n$ that are mutually independent and identically distributed with an exponential cumulative distribution function given by

$$F_X(t) = 1 - e^{-\lambda t}.$$

Let us assume, too, that the service times for successive customers are given by random variables $Y_1, Y_2, \ldots, Y_n$ that again are mutually independent and identically distributed with another exponential cumulative distribution function given by

$$F_Y(t) = 1 - e^{-\mu t}.$$

The parameters λ and μ represent, respectively, the reciprocals of the average time between arrivals of customers and the average service time of the customers. Thus, for example, the larger the value of λ, the smaller the average time between arrivals of customers. We can guess that the length of time a customer will spend in the queue depends on the relative sizes of the average interarrival time and the average service time.

It is easy to verify this conjecture by simulation. The program **Queue** simulates this queueing process. Let $N(t)$ be the number of customers in the queue at time t.

5.2. IMPORTANT DENSITIES

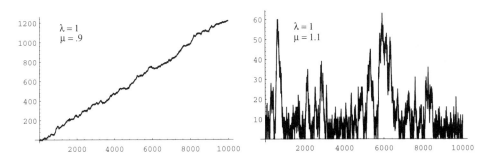

Figure 5.7: Queue sizes.

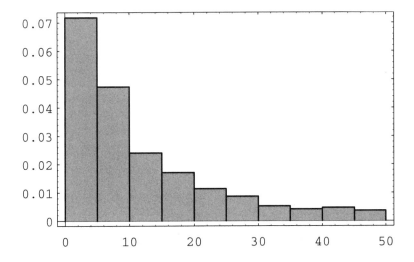

Figure 5.8: Waiting times.

Then we plot $N(t)$ as a function of t for different choices of the parameters λ and μ (see Figure 5.7).

We note that when $\lambda < \mu$, then $1/\lambda > 1/\mu$, so the average interarrival time is greater than the average service time, i.e., customers are served more quickly, on average, than new ones arrive. Thus, in this case, it is reasonable to expect that $N(t)$ remains small. However, if $\lambda > \mu$ then customers arrive more quickly than they are served, and, as expected, $N(t)$ appears to grow without limit.

We can now ask: How long will a customer have to wait in the queue for service? To examine this question, we let W_i be the length of time that the ith customer has to remain in the system (waiting in line and being served). Then we can present these data in a bar graph, using the program **Queue**, to give some idea of how the W_i are distributed (see Figure 5.8). (Here $\lambda = 1$ and $\mu = 1.1$.)

We see that these waiting times appear to be distributed exponentially. This is always the case when $\lambda < \mu$. The proof of this fact is too complicated to give here, but we can verify it by simulation for different choices of λ and μ, as above. □

Functions of a Random Variable

Before continuing our list of important densities, we pause to consider random variables which are functions of other random variables. We will prove a general theorem that will allow us to derive expressions such as Equation 5.5.

Theorem 5.1 Let X be a continuous random variable, and suppose that $\phi(x)$ is a strictly increasing function on the range of X. Define $Y = \phi(X)$. Suppose that X and Y have cumulative distribution functions F_X and F_Y respectively. Then these functions are related by
$$F_Y(y) = F_X(\phi^{-1}(y)).$$
If $\phi(x)$ is strictly decreasing on the range of X, then
$$F_Y(y) = 1 - F_X(\phi^{-1}(y)) \ .$$

Proof. Since ϕ is a strictly increasing function on the range of X, the events $(X \leq \phi^{-1}(y))$ and $(\phi(X) \leq y)$ are equal. Thus, we have
$$\begin{aligned} F_Y(y) &= P(Y \leq y) \\ &= P(\phi(X) \leq y) \\ &= P(X \leq \phi^{-1}(y)) \\ &= F_X(\phi^{-1}(y)) \ . \end{aligned}$$

If $\phi(x)$ is strictly decreasing on the range of X, then we have
$$\begin{aligned} F_Y(y) &= P(Y \leq y) \\ &= P(\phi(X) \leq y) \\ &= P(X \geq \phi^{-1}(y)) \\ &= 1 - P(X < \phi^{-1}(y)) \\ &= 1 - F_X(\phi^{-1}(y)) \ . \end{aligned}$$

This completes the proof. $\square$

Corollary 5.1 Let X be a continuous random variable, and suppose that $\phi(x)$ is a strictly increasing function on the range of X. Define $Y = \phi(X)$. Suppose that the density functions of X and Y are f_X and f_Y, respectively. Then these functions are related by
$$f_Y(y) = f_X(\phi^{-1}(y)) \frac{d}{dy}\phi^{-1}(y) \ .$$
If $\phi(x)$ is strictly decreasing on the range of X, then
$$f_Y(y) = -f_X(\phi^{-1}(y)) \frac{d}{dy}\phi^{-1}(y) \ .$$

5.2. IMPORTANT DENSITIES

Proof. This result follows from Theorem 5.1 by using the Chain Rule. $\square$

If the function ϕ is neither strictly increasing nor strictly decreasing, then the situation is somewhat more complicated but can be treated by the same methods. For example, suppose that $Y = X^2$, then $\phi(x) = x^2$, and

$$\begin{aligned} F_Y(y) &= P(Y \leq y) \\ &= P(-\sqrt{y} \leq X \leq +\sqrt{y}) \\ &= P(X \leq +\sqrt{y}) - P(X \leq -\sqrt{y}) \\ &= F_X(\sqrt{y}) - F_X(-\sqrt{y}) \ . \end{aligned}$$

Moreover,

$$\begin{aligned} f_Y(y) &= \frac{d}{dy} F_Y(y) \\ &= \frac{d}{dy}(F_X(\sqrt{y}) - F_X(-\sqrt{y})) \\ &= \left(f_X(\sqrt{y}) + f_X(-\sqrt{y})\right)\frac{1}{2\sqrt{y}} \ . \end{aligned}$$

We see that in order to express F_Y in terms of F_X when $Y = \phi(X)$, we have to express $P(Y \leq y)$ in terms of $P(X \leq x)$, and this process will depend in general upon the structure of ϕ.

Simulation

Theorem 5.1 tells us, among other things, how to simulate on the computer a random variable Y with a prescribed cumulative distribution function F. We assume that $F(y)$ is strictly increasing for those values of y where $0 < F(y) < 1$. For this purpose, let U be a random variable which is uniformly distributed on $[0, 1]$. Then U has cumulative distribution function $F_U(u) = u$. Now, if F is the prescribed cumulative distribution function for Y, then to write Y in terms of U we first solve the equation

$$F(y) = u$$

for y in terms of u. We obtain $y = F^{-1}(u)$. Note that since F is an increasing function this equation always has a unique solution (see Figure 5.9). Then we set $Z = F^{-1}(U)$ and obtain, by Theorem 5.1,

$$F_Z(y) = F_U(F(y)) = F(y) \ ,$$

since $F_U(u) = u$. Therefore, Z and Y have the same cumulative distribution function. Summarizing, we have the following.

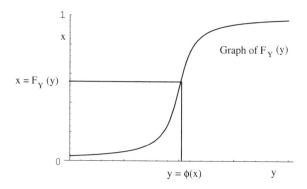

figure 5.9: Converting a uniform distribution F_U into a prescribed distribution F_Y.

Corollary 5.2 If $F(y)$ is a given cumulative distribution function that is strictly increasing when $0 < F(y) < 1$ and if U is a random variable with uniform distribution on $[0, 1]$, then
$$Y = F^{-1}(U)$$
has the cumulative distribution $F(y)$. □

Thus, to simulate a random variable with a given cumulative distribution F we need only set $Y = F^{-1}(\text{rnd})$.

Normal Density

We now come to the most important density function, the normal density function. We have seen in Chapter 3 that the binomial distribution functions are bell-shaped, even for moderate size values of n. We recall that a binomially-distributed random variable with parameters n and p can be considered to be the sum of n mutually independent 0-1 random variables. A very important theorem in probability theory, called the Central Limit Theorem, states that under very general conditions, if we sum a large number of mutually independent random variables, then the distribution of the sum can be closely approximated by a certain specific continuous density, called the normal density. This theorem will be discussed in Chapter 9.

The normal density function with parameters μ and σ is defined as follows:
$$f_X(x) = \frac{1}{\sqrt{2\pi}\sigma} e^{-(x-\mu)^2/2\sigma^2} .$$

The parameter μ represents the "center" of the density (and in Chapter 6, we will show that it is the average, or expected, value of the density). The parameter σ is a measure of the "spread" of the density, and thus it is assumed to be positive. (In Chapter 6, we will show that σ is the standard deviation of the density.) We note that it is not at all obvious that the above function is a density, i.e., that its

5.2. IMPORTANT DENSITIES

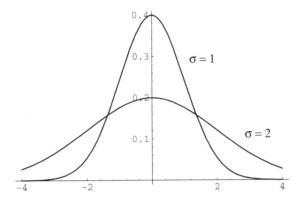

Figure 5.10: Normal density for two sets of parameter values.

integral over the real line equals 1. The cumulative distribution function is given by the formula

$$F_X(x) = \int_{-\infty}^{x} \frac{1}{\sqrt{2\pi}\sigma} e^{-(u-\mu)^2/2\sigma^2} \, du \ .$$

In Figure 5.10 we have included for comparison a plot of the normal density for the cases $\mu = 0$ and $\sigma = 1$, and $\mu = 0$ and $\sigma = 2$.

One cannot write F_X in terms of simple functions. This leads to several problems. First of all, values of F_X must be computed using numerical integration. Extensive tables exist containing values of this function (see Appendix A). Secondly, we cannot write F_X^{-1} in closed form, so we cannot use Corollary 5.2 to help us simulate a normal random variable. For this reason, special methods have been developed for simulating a normal distribution. One such method relies on the fact that if U and V are independent random variables with uniform densities on $[0, 1]$, then the random variables

$$X = \sqrt{-2 \log U} \cos 2\pi V$$

and

$$Y = \sqrt{-2 \log U} \sin 2\pi V$$

are independent, and have normal density functions with parameters $\mu = 0$ and $\sigma = 1$. (This is not obvious, nor shall we prove it here. See Box and Muller.[9])

Let Z be a normal random variable with parameters $\mu = 0$ and $\sigma = 1$. A normal random variable with these parameters is said to be a *standard* normal random variable. It is an important and useful fact that if we write

$$X = \sigma Z + \mu \ ,$$

then X is a normal random variable with parameters μ and σ. To show this, we will use Theorem 5.1. We have $\phi(z) = \sigma z + \mu$, $\phi^{-1}(x) = (x - \mu)/\sigma$, and

$$F_X(x) \;=\; F_Z\left(\frac{x-\mu}{\sigma}\right),$$

---

[9] G. E. P. Box and M. E. Muller, *A Note on the Generation of Random Normal Deviates*, Ann. of Math. Stat. 29 (1958), pgs. 610-611.

$$f_X(x) = f_Z\left(\frac{x-\mu}{\sigma}\right) \cdot \frac{1}{\sigma}$$
$$= \frac{1}{\sqrt{2\pi}\sigma} e^{-(x-\mu)^2/2\sigma^2} .$$

The reader will note that this last expression is the density function with parameters μ and σ, as claimed.

We have seen above that it is possible to simulate a standard normal random variable Z. If we wish to simulate a normal random variable X with parameters μ and σ, then we need only transform the simulated values for Z using the equation $X = \sigma Z + \mu$.

Suppose that we wish to calculate the value of a cumulative distribution function for the normal random variable X, with parameters μ and σ. We can reduce this calculation to one concerning the standard normal random variable Z as follows:

$$F_X(x) = P(X \leq x)$$
$$= P\left(Z \leq \frac{x-\mu}{\sigma}\right)$$
$$= F_Z\left(\frac{x-\mu}{\sigma}\right) .$$

This last expression can be found in a table of values of the cumulative distribution function for a standard normal random variable. Thus, we see that it is unnecessary to make tables of normal distribution functions with arbitrary μ and σ.

The process of changing a normal random variable to a standard normal random variable is known as standardization. If X has a normal distribution with parameters μ and σ and if

$$Z = \frac{X-\mu}{\sigma},$$

then Z is said to be the standardized version of X.

The following example shows how we use the standardized version of a normal random variable X to compute specific probabilities relating to X.

Example 5.8 Suppose that X is a normally distributed random variable with parameters $\mu = 10$ and $\sigma = 3$. Find the probability that X is between 4 and 16.

To solve this problem, we note that $Z = (X - 10)/3$ is the standardized version of X. So, we have

$$P(4 \leq X \leq 16) = P(X \leq 16) - P(X \leq 4)$$
$$= F_X(16) - F_X(4)$$
$$= F_Z\left(\frac{16-10}{3}\right) - F_Z\left(\frac{4-10}{3}\right)$$
$$= F_Z(2) - F_Z(-2) .$$

5.2. IMPORTANT DENSITIES

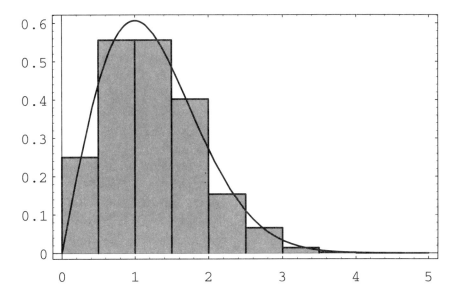

Figure 5.11: Distribution of dart distances in 1000 drops.

This last expression can be evaluated by using tabulated values of the standard normal distribution function (see 12.3); when we use this table, we find that $F_Z(2) = .9772$ and $F_Z(-2) = .0228$. Thus, the answer is .9544.

In Chapter 6, we will see that the parameter μ is the mean, or average value, of the random variable X. The parameter σ is a measure of the spread of the random variable, and is called the standard deviation. Thus, the question asked in this example is of a typical type, namely, what is the probability that a random variable has a value within two standard deviations of its average value. □

Maxwell and Rayleigh Densities

Example 5.9 Suppose that we drop a dart on a large table top, which we consider as the xy-plane, and suppose that the x and y coordinates of the dart point are independent and have a normal distribution with parameters $\mu = 0$ and $\sigma = 1$. How is the distance of the point from the origin distributed?

This problem arises in physics when it is assumed that a moving particle in R^n has components of the velocity that are mutually independent and normally distributed and it is desired to find the density of the speed of the particle. The density in the case $n = 3$ is called the Maxwell density.

The density in the case $n = 2$ (i.e. the dart board experiment described above) is called the Rayleigh density. We can simulate this case by picking independently a pair of coordinates (x, y), each from a normal distribution with $\mu = 0$ and $\sigma = 1$ on $(-\infty, \infty)$, calculating the distance $r = \sqrt{x^2 + y^2}$ of the point (x, y) from the origin, repeating this process a large number of times, and then presenting the results in a bar graph. The results are shown in Figure 5.11.

	Female	Male	
A	37	56	93
B	63	60	123
C	47	43	90
Below C	5	8	13
	152	167	319

Table 5.8: Calculus class data.

	Female	Male	
A	44.3	48.7	93
B	58.6	64.4	123
C	42.9	47.1	90
Below C	6.2	6.8	13
	152	167	319

Table 5.9: Expected data.

We have also plotted the theoretical density

$$f(r) = re^{-r^2/2} \ .$$

This will be derived in Chapter 7; see Example 7.7. □

Chi-Squared Density

We return to the problem of independence of traits discussed in Example 5.6. It is frequently the case that we have two traits, each of which have several different values. As was seen in the example, quite a lot of calculation was needed even in the case of two values for each trait. We now give another method for testing independence of traits, which involves much less calculation.

Example 5.10 Suppose that we have the data shown in Table 5.8 concerning grades and gender of students in a Calculus class. We can use the same sort of model in this situation as was used in Example 5.6. We imagine that we have an urn with 319 balls of two colors, say blue and red, corresponding to females and males, respectively. We now draw 93 balls, without replacement, from the urn. These balls correspond to the grade of A. We continue by drawing 123 balls, which correspond to the grade of B. When we finish, we have four sets of balls, with each ball belonging to exactly one set. (We could have stipulated that the balls were of four colors, corresponding to the four possible grades. In this case, we would draw a subset of size 152, which would correspond to the females. The balls remaining in the urn would correspond to the males. The choice does not affect the final determination of whether we should reject the hypothesis of independence of traits.)

The expected data set can be determined in exactly the same way as in Example 5.6. If we do this, we obtain the expected values shown in Table 5.9. Even if

5.2. IMPORTANT DENSITIES

the traits are independent, we would still expect to see some differences between the numbers in corresponding boxes in the two tables. However, if the differences are large, then we might suspect that the two traits are not independent. In Example 5.6, we used the probability distribution of the various possible data sets to compute the probability of finding a data set that differs from the expected data set by at least as much as the actual data set does. We could do the same in this case, but the amount of computation is enormous.

Instead, we will describe a single number which does a good job of measuring how far a given data set is from the expected one. To quantify how far apart the two sets of numbers are, we could sum the squares of the differences of the corresponding numbers. (We could also sum the absolute values of the differences, but we would not want to sum the differences.) Suppose that we have data in which we expect to see 10 objects of a certain type, but instead we see 18, while in another case we expect to see 50 objects of a certain type, but instead we see 58. Even though the two differences are about the same, the first difference is more surprising than the second, since the expected number of outcomes in the second case is quite a bit larger than the expected number in the first case. One way to correct for this is to divide the individual squares of the differences by the expected number for that box. Thus, if we label the values in the eight boxes in the first table by O_i (for observed values) and the values in the eight boxes in the second table by E_i (for expected values), then the following expression might be a reasonable one to use to measure how far the observed data is from what is expected:

$$\sum_{i=1}^{8} \frac{(O_i - E_i)^2}{E_i} .$$

This expression is a random variable, which is usually denoted by the symbol χ^2, pronounced "ki-squared." It is called this because, under the assumption of independence of the two traits, the density of this random variable can be computed and is approximately equal to a density called the chi-squared density. We choose not to give the explicit expression for this density, since it involves the gamma function, which we have not discussed. The chi-squared density is, in fact, a special case of the general gamma density.

In applying the chi-squared density, tables of values of this density are used, as in the case of the normal density. The chi-squared density has one parameter n, which is called the number of degrees of freedom. The number n is usually easy to determine from the problem at hand. For example, if we are checking two traits for independence, and the two traits have a and b values, respectively, then the number of degrees of freedom of the random variable χ^2 is $(a-1)(b-1)$. So, in the example at hand, the number of degrees of freedom is 3.

We recall that in this example, we are trying to test for independence of the two traits of gender and grades. If we assume these traits are independent, then the ball-and-urn model given above gives us a way to simulate the experiment. Using a computer, we have performed 1000 experiments, and for each one, we have calculated a value of the random variable χ^2. The results are shown in Figure 5.12, together with the chi-squared density function with three degrees of freedom.

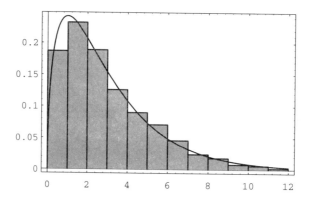

Figure 5.12: Chi-squared density with three degrees of freedom.

As we stated above, if the value of the random variable χ^2 is large, then we would tend not to believe that the two traits are independent. But how large is large? The actual value of this random variable for the data above is 4.13. In Figure 5.12, we have shown the chi-squared density with 3 degrees of freedom. It can be seen that the value 4.13 is larger than most of the values taken on by this random variable.

Typically, a statistician will compute the value v of the random variable χ^2, just as we have done. Then, by looking in a table of values of the chi-squared density, a value v_0 is determined which is only exceeded 5% of the time. If $v \geq v_0$, the statistician rejects the hypothesis that the two traits are independent. In the present case, $v_0 = 7.815$, so we would not reject the hypothesis that the two traits are independent. □

Cauchy Density

The following example is from Feller.[10]

Example 5.11 Suppose that a mirror is mounted on a vertical axis, and is free to revolve about that axis. The axis of the mirror is 1 foot from a straight wall of infinite length. A pulse of light is shown onto the mirror, and the reflected ray hits the wall. Let ϕ be the angle between the reflected ray and the line that is perpendicular to the wall and that runs through the axis of the mirror. We assume that ϕ is uniformly distributed between $-\pi/2$ and $\pi/2$. Let X represent the distance between the point on the wall that is hit by the reflected ray and the point on the wall that is closest to the axis of the mirror. We now determine the density of X.

Let B be a fixed positive quantity. Then $X \geq B$ if and only if $\tan(\phi) \geq B$, which happens if and only if $\phi \geq \arctan(B)$. This happens with probability

$$\frac{\pi/2 - \arctan(B)}{\pi}.$$

---

[10]W. Feller, *An Introduction to Probability Theory and Its Applications,*, vol. 2, (New York: Wiley, 1966)

5.2. IMPORTANT DENSITIES

Thus, for positive B, the cumulative distribution function of X is

$$F(B) = 1 - \frac{\pi/2 - \arctan(B)}{\pi} .$$

Therefore, the density function for positive B is

$$f(B) = \frac{1}{\pi(1+B^2)} .$$

Since the physical situation is symmetric with respect to $\phi = 0$, it is easy to see that the above expression for the density is correct for negative values of B as well.

The Law of Large Numbers, which we will discuss in Chapter 8, states that in many cases, if we take the average of independent values of a random variable, then the average approaches a specific number as the number of values increases. It turns out that if one does this with a Cauchy-distributed random variable, the average does not approach any specific number. $\square$

Exercises

1 Choose a number U from the unit interval $[0,1]$ with uniform distribution. Find the cumulative distribution and density for the random variables

   (a) $Y = U + 2$.

   (b) $Y = U^3$.

2 Choose a number U from the interval $[0,1]$ with uniform distribution. Find the cumulative distribution and density for the random variables

   (a) $Y = 1/(U+1)$.

   (b) $Y = \log(U+1)$.

3 Use Corollary 5.2 to derive the expression for the random variable given in Equation 5.5. *Hint*: The random variables $1 - rnd$ and rnd are identically distributed.

4 Suppose we know a random variable Y as a function of the uniform random variable U: $Y = \phi(U)$, and suppose we have calculated the cumulative distribution function $F_Y(y)$ and thence the density $f_Y(y)$. How can we check whether our answer is correct? An easy simulation provides the answer: Make a bar graph of $Y = \phi(rnd)$ and compare the result with the graph of $f_Y(y)$. These graphs should look similar. Check your answers to Exercises 1 and 2 by this method.

5 Choose a number U from the interval $[0,1]$ with uniform distribution. Find the cumulative distribution and density for the random variables

   (a) $Y = |U - 1/2|$.

   (b) $Y = (U - 1/2)^2$.

6 Check your results for Exercise 5 by simulation as described in Exercise 4.

7 Explain how you can generate a random variable whose cumulative distribution function is
$$F(x) = \begin{cases} 0, & \text{if } x < 0, \\ x^2, & \text{if } 0 \leq x \leq 1, \\ 1, & \text{if } x > 1. \end{cases}$$

8 Write a program to generate a sample of 1000 random outcomes each of which is chosen from the distribution given in Exercise 7. Plot a bar graph of your results and compare this empirical density with the density for the cumulative distribution given in Exercise 7.

9 Let U, V be random numbers chosen independently from the interval $[0, 1]$ with uniform distribution. Find the cumulative distribution and density of each of the variables

(a) $Y = U + V$.

(b) $Y = |U - V|$.

10 Let U, V be random numbers chosen independently from the interval $[0, 1]$. Find the cumulative distribution and density for the random variables

(a) $Y = \max(U, V)$.

(b) $Y = \min(U, V)$.

11 Write a program to simulate the random variables of Exercises 9 and 10 and plot a bar graph of the results. Compare the resulting empirical density with the density found in Exercises 9 and 10.

12 A number U is chosen at random in the interval $[0, 1]$. Find the probability that

(a) $R = U^2 < 1/4$.

(b) $S = U(1 - U) < 1/4$.

(c) $T = U/(1 - U) < 1/4$.

13 Find the cumulative distribution function F and the density function f for each of the random variables R, S, and T in Exercise 12.

14 A point P in the unit square has coordinates X and Y chosen at random in the interval $[0, 1]$. Let D be the distance from P to the nearest edge of the square, and E the distance to the nearest corner. What is the probability that

(a) $D < 1/4$?

(b) $E < 1/4$?

15 In Exercise 14 find the cumulative distribution F and density f for the random variable D.

5.2. IMPORTANT DENSITIES

16 Let X be a random variable with density function

$$f_X(x) = \begin{cases} cx(1-x), & \text{if } 0 < x < 1, \\ 0, & \text{otherwise.} \end{cases}$$

(a) What is the value of c?

(b) What is the cumulative distribution function F_X for X?

(c) What is the probability that $X < 1/4$?

17 Let X be a random variable with cumulative distribution function

$$F(x) = \begin{cases} 0, & \text{if } x < 0, \\ \sin^2(\pi x/2), & \text{if } 0 \leq x \leq 1, \\ 1, & \text{if } 1 < x. \end{cases}$$

(a) What is the density function f_X for X?

(b) What is the probability that $X < 1/4$?

18 Let X be a random variable with cumulative distribution function F_X, and let $Y = X + b$, $Z = aX$, and $W = aX + b$, where a and b are any constants. Find the cumulative distribution functions F_Y, F_Z, and F_W. *Hint*: The cases $a > 0$, $a = 0$, and $a < 0$ require different arguments.

19 Let X be a random variable with density function f_X, and let $Y = X + b$, $Z = aX$, and $W = aX + b$, where $a \neq 0$. Find the density functions f_Y, f_Z, and f_W. (See Exercise 18.)

20 Let X be a random variable uniformly distributed over $[c, d]$, and let $Y = aX + b$. For what choice of a and b is Y uniformly distributed over $[0, 1]$?

21 Let X be a random variable with cumulative distribution function F strictly increasing on the range of X. Let $Y = F(X)$. Show that Y is uniformly distributed in the interval $[0, 1]$. (The formula $X = F^{-1}(Y)$ then tells us how to construct X from a uniform random variable Y.)

22 Let X be a random variable with cumulative distribution function F. The *median* of X is the value m for which $F(m) = 1/2$. Then $X < m$ with probability $1/2$ and $X > m$ with probability $1/2$. Find m if X is

(a) uniformly distributed over the interval $[a, b]$.

(b) normally distributed with parameters μ and σ.

(c) exponentially distributed with parameter λ.

23 Let X be a random variable with density function f_X. The *mean* of X is the value $\mu = \int x f_x(x)\, dx$. Then μ gives an average value for X (see Section 6.3). Find μ if X is distributed uniformly, normally, or exponentially, as in Exercise 22.

Test Score	Letter grade
$\mu + \sigma < x$	A
$\mu < x < \mu + \sigma$	B
$\mu - \sigma < x < \mu$	C
$\mu - 2\sigma < x < \mu - \sigma$	D
$x < \mu - 2\sigma$	F

Table 5.10: Grading on the curve.

24 Let X be a random variable with density function f_X. The *mode* of X is the value M for which $f(M)$ is maximum. Then values of X near M are most likely to occur. Find M if X is distributed normally or exponentially, as in Exercise 22. What happens if X is distributed uniformly?

25 Let X be a random variable normally distributed with parameters $\mu = 70$, $\sigma = 10$. Estimate

(a) $P(X > 50)$.

(b) $P(X < 60)$.

(c) $P(X > 90)$.

(d) $P(60 < X < 80)$.

26 Bridies' Bearing Works manufactures bearing shafts whose diameters are normally distributed with parameters $\mu = 1$, $\sigma = .002$. The buyer's specifications require these diameters to be $1.000 \pm .003$ cm. What fraction of the manufacturer's shafts are likely to be rejected? If the manufacturer improves her quality control, she can reduce the value of σ. What value of σ will ensure that no more than 1 percent of her shafts are likely to be rejected?

27 A final examination at Podunk University is constructed so that the test scores are approximately normally distributed, with parameters μ and σ. The instructor assigns letter grades to the test scores as shown in Table 5.10 (this is the process of "grading on the curve").

What fraction of the class gets A, B, C, D, F?

28 (Ross[11]) An expert witness in a paternity suit testifies that the length (in days) of a pregnancy, from conception to delivery, is approximately normally distributed, with parameters $\mu = 270$, $\sigma = 10$. The defendant in the suit is able to prove that he was out of the country during the period from 290 to 240 days before the birth of the child. What is the probability that the defendant was in the country when the child was conceived?

29 Suppose that the time (in hours) required to repair a car is an exponentially distributed random variable with parameter $\lambda = 1/2$. What is the probability that the repair time exceeds 4 hours? If it exceeds 4 hours what is the probability that it exceeds 8 hours?

---

[11] S. Ross, *A First Course in Probability Theory*, 2d ed. (New York: Macmillan, 1984).

5.2. IMPORTANT DENSITIES

30 Suppose that the number of years a car will run is exponentially distributed with parameter $\mu = 1/4$. If Prosser buys a used car today, what is the probability that it will still run after 4 years?

31 Let U be a uniformly distributed random variable on $[0,1]$. What is the probability that the equation

$$x^2 + 4Ux + 1 = 0$$

has two distinct real roots x_1 and x_2?

32 Write a program to simulate the random variables whose densities are given by the following, making a suitable bar graph of each and comparing the exact density with the bar graph.

(a) $f_X(x) = e^{-x}$ on $[0, \infty)$ (but just do it on $[0,10]$).

(b) $f_X(x) = 2x$ on $[0,1]$.

(c) $f_X(x) = 3x^2$ on $[0,1]$.

(d) $f_X(x) = 4|x - 1/2|$ on $[0,1]$.

33 Suppose we are observing a process such that the time between occurrences is exponentially distributed with $\lambda = 1/30$ (i.e., the average time between occurrences is 30 minutes). Suppose that the process starts at a certain time and we start observing the process 3 hours later. Write a program to simulate this process. Let T denote the length of time that we have to wait, after we start our observation, for an occurrence. Have your program keep track of T. What is an estimate for the average value of T?

34 Jones puts in two new lightbulbs: a 60 watt bulb and a 100 watt bulb. It is claimed that the lifetime of the 60 watt bulb has an exponential density with average lifetime 200 hours ($\lambda = 1/200$). The 100 watt bulb also has an exponential density but with average lifetime of only 100 hours ($\lambda = 1/100$). Jones wonders what is the probability that the 100 watt bulb will outlast the 60 watt bulb.

If X and Y are two independent random variables with exponential densities $f(x) = \lambda e^{-\lambda x}$ and $g(x) = \mu e^{-\mu x}$, respectively, then the probability that X is less than Y is given by

$$P(X < Y) = \int_0^\infty f(x)(1 - G(x))\,dx,$$

where $G(x)$ is the cumulative distribution function for $g(x)$. Explain why this is the case. Use this to show that

$$P(X < Y) = \frac{\lambda}{\lambda + \mu}$$

and to answer Jones's question.

35 Consider the simple queueing process of Example 5.7. Suppose that you watch the size of the queue. If there are j people in the queue the next time the queue size changes it will either decrease to $j-1$ or increase to $j+1$. Use the result of Exercise 34 to show that the probability that the queue size decreases to $j-1$ is $\mu/(\mu+\lambda)$ and the probability that it increases to $j+1$ is $\lambda/(\mu+\lambda)$. When the queue size is 0 it can only increase to 1. Write a program to simulate the queue size. Use this simulation to help formulate a conjecture containing conditions on μ and λ that will ensure that the queue will have times when it is empty.

36 Let X be a random variable having an exponential density with parameter λ. Find the density for the random variable $Y = rX$, where r is a positive real number.

37 Let X be a random variable having a normal density and consider the random variable $Y = e^X$. Then Y has a *log normal* density. Find this density of Y.

38 Let X_1 and X_2 be independent random variables and for $i = 1, 2$, let $Y_i = \phi_i(X_i)$, where ϕ_i is strictly increasing on the range of X_i. Show that Y_1 and Y_2 are independent. Note that the same result is true without the assumption that the ϕ_i's are strictly increasing, but the proof is more difficult.

Chapter 6

Expected Value and Variance

6.1 Expected Value of Discrete Random Variables

When a large collection of numbers is assembled, as in a census, we are usually interested not in the individual numbers, but rather in certain descriptive quantities such as the average or the median. In general, the same is true for the probability distribution of a numerically-valued random variable. In this and in the next section, we shall discuss two such descriptive quantities: the *expected value* and the *variance*. Both of these quantities apply only to numerically-valued random variables, and so we assume, in these sections, that all random variables have numerical values. To give some intuitive justification for our definition, we consider the following game.

Average Value

A die is rolled. If an odd number turns up, we win an amount equal to this number; if an even number turns up, we lose an amount equal to this number. For example, if a two turns up we lose 2, and if a three comes up we win 3. We want to decide if this is a reasonable game to play. We first try simulation. The program **Die** carries out this simulation.

The program prints the frequency and the relative frequency with which each outcome occurs. It also calculates the average winnings. We have run the program twice. The results are shown in Table 6.1.

In the first run we have played the game 100 times. In this run our average gain is $-.57$. It looks as if the game is unfavorable, and we wonder how unfavorable it really is. To get a better idea, we have played the game 10,000 times. In this case our average gain is $-.4949$.

We note that the relative frequency of each of the six possible outcomes is quite close to the probability $1/6$ for this outcome. This corresponds to our frequency interpretation of probability. It also suggests that for very large numbers of plays, our average gain should be

$$\mu \;=\; 1\left(\frac{1}{6}\right) - 2\left(\frac{1}{6}\right) + 3\left(\frac{1}{6}\right) - 4\left(\frac{1}{6}\right) + 5\left(\frac{1}{6}\right) - 6\left(\frac{1}{6}\right)$$

| | n = 100 | | n = 10000 | |
Winning	Frequency	Relative Frequency	Frequency	Relative Frequency
1	17	.17	1681	.1681
-2	17	.17	1678	.1678
3	16	.16	1626	.1626
-4	18	.18	1696	.1696
5	16	.16	1686	.1686
-6	16	.16	1633	.1633

Table 6.1: Frequencies for dice game.

$$= \frac{9}{6} - \frac{12}{6} = -\frac{3}{6} = -.5 \ .$$

This agrees quite well with our average gain for 10,000 plays.

We note that the value we have chosen for the average gain is obtained by taking the possible outcomes, multiplying by the probability, and adding the results. This suggests the following definition for the expected outcome of an experiment.

Expected Value

Definition 6.1 Let X be a numerically-valued discrete random variable with sample space Ω and distribution function $m(x)$. The *expected value* $E(X)$ is defined by
$$E(X) = \sum_{x \in \Omega} x m(x) \ ,$$
provided this sum converges absolutely. We often refer to the expected value as the *mean*, and denote $E(X)$ by μ for short. If the above sum does not converge absolutely, then we say that X does not have an expected value. □

Example 6.1 Let an experiment consist of tossing a fair coin three times. Let X denote the number of heads which appear. Then the possible values of X are $0, 1, 2$ and 3. The corresponding probabilities are $1/8, 3/8, 3/8$, and $1/8$. Thus, the expected value of X equals
$$0\left(\frac{1}{8}\right) + 1\left(\frac{3}{8}\right) + 2\left(\frac{3}{8}\right) + 3\left(\frac{1}{8}\right) = \frac{3}{2} \ .$$
Later in this section we shall see a quicker way to compute this expected value, based on the fact that X can be written as a sum of simpler random variables. □

Example 6.2 Suppose that we toss a fair coin until a head first comes up, and let X represent the number of tosses which were made. Then the possible values of X are $1, 2, \ldots$, and the distribution function of X is defined by
$$m(i) = \frac{1}{2^i} \ .$$

6.1. EXPECTED VALUE

(This is just the geometric distribution with parameter 1/2.) Thus, we have

$$\begin{aligned} E(X) &= \sum_{i=1}^{\infty} i \frac{1}{2^i} \\ &= \sum_{i=1}^{\infty} \frac{1}{2^i} + \sum_{i=2}^{\infty} \frac{1}{2^i} + \cdots \\ &= 1 + \frac{1}{2} + \frac{1}{2^2} + \cdots \\ &= 2 \ . \end{aligned}$$

$\square$

Example 6.3 (Example 6.2 continued) Suppose that we flip a coin until a head first appears, and if the number of tosses equals n, then we are paid 2^n dollars. What is the expected value of the payment?

We let Y represent the payment. Then,

$$P(Y = 2^n) = \frac{1}{2^n} \ ,$$

for $n \geq 1$. Thus,

$$E(Y) = \sum_{n=1}^{\infty} 2^n \frac{1}{2^n} \ ,$$

which is a divergent sum. Thus, Y has no expectation. This example is called the *St. Petersburg Paradox*. The fact that the above sum is infinite suggests that a player should be willing to pay any fixed amount per game for the privilege of playing this game. The reader is asked to consider how much he or she would be willing to pay for this privilege. It is unlikely that the reader's answer is more than 10 dollars; therein lies the paradox.

In the early history of probability, various mathematicians gave ways to resolve this paradox. One idea (due to G. Cramer) consists of assuming that the amount of money in the world is finite. He thus assumes that there is some fixed value of n such that if the number of tosses equals or exceeds n, the payment is 2^n dollars. The reader is asked to show in Exercise 20 that the expected value of the payment is now finite.

Daniel Bernoulli and Cramer also considered another way to assign value to the payment. Their idea was that the value of a payment is some function of the payment; such a function is now called a utility function. Examples of reasonable utility functions might include the square-root function or the logarithm function. In both cases, the value of $2n$ dollars is less than twice the value of n dollars. It can easily be shown that in both cases, the expected utility of the payment is finite (see Exercise 20). $\square$

Example 6.4 Let T be the time for the first success in a Bernoulli trials process. Then we take as sample space Ω the integers 1, 2, ... and assign the geometric distribution
$$m(j) = P(T = j) = q^{j-1}p \ .$$
Thus,
$$\begin{aligned} E(T) &= 1 \cdot p + 2qp + 3q^2 p + \cdots \\ &= p(1 + 2q + 3q^2 + \cdots) \ . \end{aligned}$$
Now if $|x| < 1$, then
$$1 + x + x^2 + x^3 + \cdots = \frac{1}{1-x} \ .$$
Differentiating this formula, we get
$$1 + 2x + 3x^2 + \cdots = \frac{1}{(1-x)^2} \ ,$$
so
$$E(T) = \frac{p}{(1-q)^2} = \frac{p}{p^2} = \frac{1}{p} \ .$$
In particular, we see that if we toss a fair coin a sequence of times, the expected time until the first heads is $1/(1/2) = 2$. If we roll a die a sequence of times, the expected number of rolls until the first six is $1/(1/6) = 6$. $\square$

Interpretation of Expected Value

In statistics, one is frequently concerned with the average value of a set of data. The following example shows that the ideas of average value and expected value are very closely related.

Example 6.5 The heights, in inches, of the women on the Swarthmore basketball team are 5' 9", 5' 9", 5' 6", 5' 8", 5' 11", 5' 5", 5' 7", 5' 6", 5' 6", 5' 7", 5' 10", and 6' 0".

A statistician would compute the average height (in inches) as follows:
$$\frac{69 + 69 + 66 + 68 + 71 + 65 + 67 + 66 + 66 + 67 + 70 + 72}{12} = 67.9 \ .$$
One can also interpret this number as the expected value of a random variable. To see this, let an experiment consist of choosing one of the women at random, and let X denote her height. Then the expected value of X equals 67.9. $\square$

Of course, just as with the frequency interpretation of probability, to interpret expected value as an average outcome requires further justification. We know that for any finite experiment the average of the outcomes is not predictable. However, we shall eventually prove that the average will usually be close to $E(X)$ if we repeat the experiment a large number of times. We first need to develop some properties of the expected value. Using these properties, and those of the concept of the variance

6.1. EXPECTED VALUE

X	Y
HHH	1
HHT	2
HTH	3
HTT	2
THH	2
THT	3
TTH	2
TTT	1

Table 6.2: Tossing a coin three times.

to be introduced in the next section, we shall be able to prove the *Law of Large Numbers*. This theorem will justify mathematically both our frequency concept of probability and the interpretation of expected value as the average value to be expected in a large number of experiments.

Expectation of a Function of a Random Variable

Suppose that X is a discrete random variable with sample space Ω, and $\phi(x)$ is a real-valued function with domain Ω. Then $\phi(X)$ is a real-valued random variable. One way to determine the expected value of $\phi(X)$ is to first determine the distribution function of this random variable, and then use the definition of expectation. However, there is a better way to compute the expected value of $\phi(X)$, as demonstrated in the next example.

Example 6.6 Suppose a coin is tossed 9 times, with the result

$$HHHTTTTHT.$$

The first set of three heads is called a *run*. There are three more runs in this sequence, namely the next four tails, the next head, and the next tail. We do not consider the first two tosses to constitute a run, since the third toss has the same value as the first two.

Now suppose an experiment consists of tossing a fair coin three times. Find the expected number of runs. It will be helpful to think of two random variables, X and Y, associated with this experiment. We let X denote the sequence of heads and tails that results when the experiment is performed, and Y denote the number of runs in the outcome X. The possible outcomes of X and the corresponding values of Y are shown in Table 6.2.

To calculate $E(Y)$ using the definition of expectation, we first must find the distribution function $m(y)$ of Y i.e., we group together those values of X with a common value of Y and add their probabilities. In this case, we calculate that the distribution function of Y is: $m(1) = 1/4$, $m(2) = 1/2$, and $m(3) = 1/4$. One easily finds that $E(Y) = 2$.

Now suppose we didn't group the values of X with a common Y-value, but instead, for each X-value x, we multiply the probability of x and the corresponding value of Y, and add the results. We obtain

$$1\left(\frac{1}{8}\right) + 2\left(\frac{1}{8}\right) + 3\left(\frac{1}{8}\right) + 2\left(\frac{1}{8}\right) + 2\left(\frac{1}{8}\right) + 3\left(\frac{1}{8}\right) + 2\left(\frac{1}{8}\right) + 1\left(\frac{1}{8}\right),$$

which equals 2.

This illustrates the following general principle. If X and Y are two random variables, and Y can be written as a function of X, then one can compute the expected value of Y using the distribution function of X. $\square$

Theorem 6.1 If X is a discrete random variable with sample space Ω and distribution function $m(x)$, and if $\phi : \Omega \to \mathbf{R}$ is a function, then

$$E(\phi(X)) = \sum_{x \in \Omega} \phi(x) m(x) ,$$

provided the series converges absolutely. $\square$

The proof of this theorem is straightforward, involving nothing more than grouping values of X with a common Y-value, as in Example 6.6.

The Sum of Two Random Variables

Many important results in probability theory concern sums of random variables. We first consider what it means to add two random variables.

Example 6.7 We flip a coin and let X have the value 1 if the coin comes up heads and 0 if the coin comes up tails. Then, we roll a die and let Y denote the face that comes up. What does $X + Y$ mean, and what is its distribution? This question is easily answered in this case, by considering, as we did in Chapter 4, the joint random variable $Z = (X, Y)$, whose outcomes are ordered pairs of the form (x, y), where $0 \le x \le 1$ and $1 \le y \le 6$. The description of the experiment makes it reasonable to assume that X and Y are independent, so the distribution function of Z is uniform, with $1/12$ assigned to each outcome. Now it is an easy matter to find the set of outcomes of $X + Y$, and its distribution function. $\square$

In Example 6.1, the random variable X denoted the number of heads which occur when a fair coin is tossed three times. It is natural to think of X as the sum of the random variables X_1, X_2, X_3, where X_i is defined to be 1 if the ith toss comes up heads, and 0 if the ith toss comes up tails. The expected values of the X_i's are extremely easy to compute. It turns out that the expected value of X can be obtained by simply adding the expected values of the X_i's. This fact is stated in the following theorem.

6.1. EXPECTED VALUE

Theorem 6.2 Let X and Y be random variables with finite expected values. Then
$$E(X+Y) = E(X) + E(Y),$$
and if c is any constant, then
$$E(cX) = cE(X).$$

Proof. Let the sample spaces of X and Y be denoted by Ω_X and Ω_Y, and suppose that
$$\Omega_X = \{x_1, x_2, \ldots\}$$
and
$$\Omega_Y = \{y_1, y_2, \ldots\}.$$
Then we can consider the random variable $X+Y$ to be the result of applying the function $\phi(x,y) = x+y$ to the joint random variable (X,Y). Then, by Theorem 6.1, we have

$$\begin{aligned}
E(X+Y) &= \sum_j \sum_k (x_j + y_k) P(X = x_j, Y = y_k) \\
&= \sum_j \sum_k x_j P(X = x_j, Y = y_k) + \sum_j \sum_k y_k P(X = x_j, Y = y_k) \\
&= \sum_j x_j P(X = x_j) + \sum_k y_k P(Y = y_k).
\end{aligned}$$

The last equality follows from the fact that
$$\sum_k P(X = x_j, Y = y_k) = P(X = x_j)$$
and
$$\sum_j P(X = x_j, Y = y_k) = P(Y = y_k).$$

Thus,
$$E(X+Y) = E(X) + E(Y).$$

If c is any constant,
$$\begin{aligned}
E(cX) &= \sum_j c x_j P(X = x_j) \\
&= c \sum_j x_j P(X = x_j) \\
&= cE(X).
\end{aligned}$$

□

X			Y
a	b	c	3
a	c	b	1
b	a	c	1
b	c	a	0
c	a	b	0
c	b	a	1

Table 6.3: Number of fixed points.

It is easy to prove by mathematical induction that *the expected value of the sum of any finite number of random variables is the sum of the expected values of the individual random variables.*

It is important to note that mutual independence of the summands was not needed as a hypothesis in the Theorem 6.2 and its generalization. The fact that expectations add, whether or not the summands are mutually independent, is sometimes referred to as the First Fundamental Mystery of Probability.

Example 6.8 Let Y be the number of fixed points in a random permutation of the set $\{a, b, c\}$. To find the expected value of Y, it is helpful to consider the basic random variable associated with this experiment, namely the random variable X which represents the random permutation. There are six possible outcomes of X, and we assign to each of them the probability 1/6 see Table 6.3. Then we can calculate $E(Y)$ using Theorem 6.1, as

$$3\left(\frac{1}{6}\right) + 1\left(\frac{1}{6}\right) + 1\left(\frac{1}{6}\right) + 0\left(\frac{1}{6}\right) + 0\left(\frac{1}{6}\right) + 1\left(\frac{1}{6}\right) = 1 \ .$$

We now give a very quick way to calculate the average number of fixed points in a random permutation of the set $\{1, 2, 3, \ldots, n\}$. Let Z denote the random permutation. For each i, $1 \leq i \leq n$, let X_i equal 1 if Z fixes i, and 0 otherwise. So if we let F denote the number of fixed points in Z, then

$$F = X_1 + X_2 + \cdots + X_n \ .$$

Therefore, Theorem 6.2 implies that

$$E(F) = E(X_1) + E(X_2) + \cdots + E(X_n) \ .$$

But it is easy to see that for each i,

$$E(X_i) = \frac{1}{n} \ ,$$

so

$$E(F) = 1 \ .$$

This method of calculation of the expected value is frequently very useful. It applies whenever the random variable in question can be written as a sum of simpler random variables. We emphasize again that it is not necessary that the summands be mutually independent. □

6.1. EXPECTED VALUE

Bernoulli Trials

Theorem 6.3 Let S_n be the number of successes in n Bernoulli trials with probability p for success on each trial. Then the expected number of successes is np. That is,
$$E(S_n) = np .$$

Proof. Let X_j be a random variable which has the value 1 if the jth outcome is a success and 0 if it is a failure. Then, for each X_j,
$$E(X_j) = 0 \cdot (1-p) + 1 \cdot p = p .$$
Since
$$S_n = X_1 + X_2 + \cdots + X_n ,$$
and the expected value of the sum is the sum of the expected values, we have
$$\begin{aligned} E(S_n) &= E(X_1) + E(X_2) + \cdots + E(X_n) \\ &= np . \end{aligned}$$
$\square$

Poisson Distribution

Recall that the Poisson distribution with parameter λ was obtained as a limit of binomial distributions with parameters n and p, where it was assumed that $np = \lambda$, and $n \to \infty$. Since for each n, the corresponding binomial distribution has expected value λ, it is reasonable to guess that the expected value of a Poisson distribution with parameter λ also has expectation equal to λ. This is in fact the case, and the reader is invited to show this (see Exercise 21).

Independence

If X and Y are two random variables, it is not true in general that $E(X \cdot Y) = E(X)E(Y)$. However, this is true if X and Y are *independent*.

Theorem 6.4 If X and Y are independent random variables, then
$$E(X \cdot Y) = E(X)E(Y) .$$

Proof. Suppose that
$$\Omega_X = \{x_1, x_2, \ldots\}$$
and
$$\Omega_Y = \{y_1, y_2, \ldots\}$$

are the sample spaces of X and Y, respectively. Using Theorem 6.1, we have

$$E(X \cdot Y) = \sum_j \sum_k x_j y_k P(X = x_j,\ Y = y_k)\ .$$

But if X and Y are independent,

$$P(X = x_j, Y = y_k) = P(X = x_j)P(Y = y_k)\ .$$

Thus,

$$\begin{aligned} E(X \cdot Y) &= \sum_j \sum_k x_j y_k P(X = x_j) P(Y = y_k) \\ &= \left(\sum_j x_j P(X = x_j)\right)\left(\sum_k y_k P(Y = y_k)\right) \\ &= E(X)E(Y)\ . \end{aligned}$$

$\square$

Example 6.9 A coin is tossed twice. $X_i = 1$ if the ith toss is heads and 0 otherwise. We know that X_1 and X_2 are independent. They each have expected value $1/2$. Thus $E(X_1 \cdot X_2) = E(X_1)E(X_2) = (1/2)(1/2) = 1/4$. $\square$

We next give a simple example to show that the expected values need not multiply if the random variables are not independent.

Example 6.10 Consider a single toss of a coin. We define the random variable X to be 1 if heads turns up and 0 if tails turns up, and we set $Y = 1 - X$. Then $E(X) = E(Y) = 1/2$. But $X \cdot Y = 0$ for either outcome. Hence, $E(X \cdot Y) = 0 \neq E(X)E(Y)$. $\square$

We return to our records example of Section 3.1 for another application of the result that the expected value of the sum of random variables is the sum of the expected values of the individual random variables.

Records

Example 6.11 We start keeping snowfall records this year and want to find the expected number of records that will occur in the next n years. The first year is necessarily a record. The second year will be a record if the snowfall in the second year is greater than that in the first year. By symmetry, this probability is $1/2$. More generally, let X_j be 1 if the jth year is a record and 0 otherwise. To find $E(X_j)$, we need only find the probability that the jth year is a record. But the record snowfall for the first j years is equally likely to fall in any one of these years,

6.1. EXPECTED VALUE

so $E(X_j) = 1/j$. Therefore, if S_n is the total number of records observed in the first n years,
$$E(S_n) = 1 + \frac{1}{2} + \frac{1}{3} + \cdots + \frac{1}{n} .$$
This is the famous *divergent harmonic series*. It is easy to show that
$$E(S_n) \sim \log n$$
as $n \to \infty$.

Therefore, in ten years the expected number of records is approximately $\log 10 = 2.3$; the exact value is the sum of the first ten terms of the harmonic series which is 2.9. We see that, even for such a small value as $n = 10$, $\log n$ is not a bad approximation. $\square$

Craps

Example 6.12 In the game of craps, the player makes a bet and rolls a pair of dice. If the sum of the numbers is 7 or 11 the player wins, if it is 2, 3, or 12 the player loses. If any other number results, say r, then r becomes the player's point and he continues to roll until either r or 7 occurs. If r comes up first he wins, and if 7 comes up first he loses. The program **Craps** simulates playing this game a number of times.

We have run the program for 1000 plays in which the player bets 1 dollar each time. The player's average winnings were $-.006$. The game of craps would seem to be only slightly unfavorable. Let us calculate the expected winnings on a single play and see if this is the case. We construct a two-stage tree measure as shown in Figure 6.1.

The first stage represents the possible sums for his first roll. The second stage represents the possible outcomes for the game if it has not ended on the first roll. In this stage we are representing the possible outcomes of a sequence of rolls required to determine the final outcome. The branch probabilities for the first stage are computed in the usual way assuming all 36 possibilites for outcomes for the pair of dice are equally likely. For the second stage we assume that the game will eventually end, and we compute the conditional probabilities for obtaining either the point or a 7. For example, assume that the player's point is 6. Then the game will end when one of the eleven pairs, $(1,5)$, $(2,4)$, $(3,3)$, $(4,2)$, $(5,1)$, $(1,6)$, $(2,5)$, $(3,4)$, $(4,3)$, $(5,2)$, $(6,1)$, occurs. We assume that each of these possible pairs has the same probability. Then the player wins in the first five cases and loses in the last six. Thus the probability of winning is 5/11 and the probability of losing is 6/11. From the path probabilities, we can find the probability that the player wins 1 dollar; it is 244/495. The probability of losing is then 251/495. Thus if X is his winning for a dollar bet,
$$\begin{aligned} E(X) &= 1\left(\frac{244}{495}\right) + (-1)\left(\frac{251}{495}\right) \\ &= -\frac{7}{495} \approx -.0141 . \end{aligned}$$

CHAPTER 6. EXPECTED VALUE AND VARIANCE

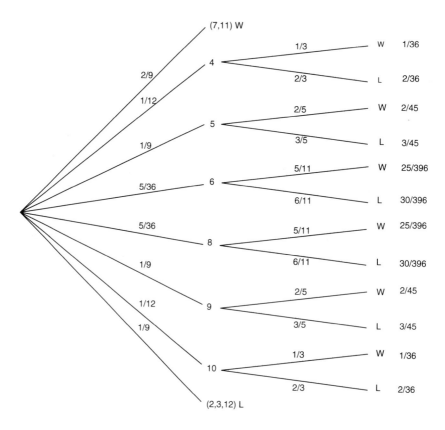

Figure 6.1: Tree measure for craps.

6.1. EXPECTED VALUE

The game is unfavorable, but only slightly. The player's expected gain in n plays is $-n(.0141)$. If n is not large, this is a small expected loss for the player. The casino makes a large number of plays and so can afford a small average gain per play and still expect a large profit. □

Roulette

Example 6.13 In Las Vegas, a roulette wheel has 38 slots numbered 0, 00, 1, 2, ..., 36. The 0 and 00 slots are green, and half of the remaining 36 slots are red and half are black. A croupier spins the wheel and throws an ivory ball. If you bet 1 dollar on red, you win 1 dollar if the ball stops in a red slot, and otherwise you lose a dollar. We wish to calculate the expected value of your winnings, if you bet 1 dollar on red.

Let X be the random variable which denotes your winnings in a 1 dollar bet on red in Las Vegas roulette. Then the distribution of X is given by

$$m_X = \begin{pmatrix} -1 & 1 \\ 20/38 & 18/38 \end{pmatrix},$$

and one can easily calculate (see Exercise 5) that

$$E(X) \approx -.0526 .$$

We now consider the roulette game in Monte Carlo, and follow the treatment of Sagan.[1] In the roulette game in Monte Carlo there is only one 0. If you bet 1 franc on red and a 0 turns up, then, depending upon the casino, one or more of the following options may be offered:
(a) You get 1/2 of your bet back, and the casino gets the other half of your bet.
(b) Your bet is put "in prison," which we will denote by P_1. If red comes up on the next turn, you get your bet back (but you don't win any money). If black or 0 comes up, you lose your bet.
(c) Your bet is put in prison P_1, as before. If red comes up on the next turn, you get your bet back, and if black comes up on the next turn, then you lose your bet. If a 0 comes up on the next turn, then your bet is put into double prison, which we will denote by P_2. If your bet is in double prison, and if red comes up on the next turn, then your bet is moved back to prison P_1 and the game proceeds as before. If your bet is in double prison, and if black or 0 come up on the next turn, then you lose your bet. We refer the reader to Figure 6.2, where a tree for this option is shown. In this figure, S is the starting position, W means that you win your bet, L means that you lose your bet, and E means that you break even.

It is interesting to compare the expected winnings of a 1 franc bet on red, under each of these three options. We leave the first two calculations as an exercise (see Exercise 37). Suppose that you choose to play alternative (c). The calculation for this case illustrates the way that the early French probabilists worked problems like this.

---
[1] H. Sagan, *Markov Chains in Monte Carlo*, Math. Mag., vol. 54, no. 1 (1981), pp. 3-10.

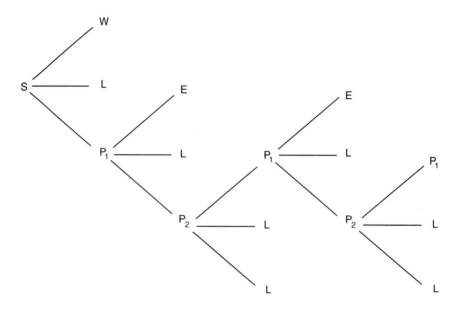

Figure 6.2: Tree for 2-prison Monte Carlo roulette.

Suppose you bet on red, you choose alternative (c), and a 0 comes up. Your possible future outcomes are shown in the tree diagram in Figure 6.3. Assume that your money is in the first prison and let x be the probability that you lose your franc. From the tree diagram we see that

$$x = \frac{18}{37} + \frac{1}{37} P(\text{you lose your franc} \mid \text{your franc is in } P_2) .$$

Also,

$$P(\text{you lose your franc} \mid \text{your franc is in } P_2) = \frac{19}{37} + \frac{18}{37} x .$$

So, we have

$$x = \frac{18}{37} + \frac{1}{37}\left(\frac{19}{37} + \frac{18}{37} x\right) .$$

Solving for x, we obtain $x = 685/1351$. Thus, starting at S, the probability that you lose your bet equals

$$\frac{18}{37} + \frac{1}{37} x = \frac{25003}{49987} .$$

To find the probability that you win when you bet on red, note that you can only win if red comes up on the first turn, and this happens with probability 18/37. Thus your expected winnings are

$$1 \cdot \frac{18}{37} - 1 \cdot \frac{25003}{49987} = -\frac{687}{49987} \approx -.0137 .$$

It is interesting to note that the more romantic option (c) is less favorable than option (a) (see Exercise 37).

6.1. EXPECTED VALUE

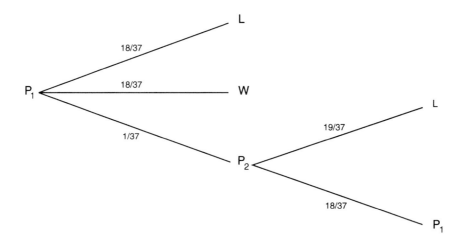

Figure 6.3: Your money is put in prison.

If you bet 1 dollar on the number 17, then the distribution function for your winnings X is

$$P_X = \begin{pmatrix} -1 & 35 \\ 36/37 & 1/37 \end{pmatrix},$$

and the expected winnings are

$$-1 \cdot \frac{36}{37} + 35 \cdot \frac{1}{37} = -\frac{1}{37} \approx -.027\ .$$

Thus, at Monte Carlo different bets have different expected values. In Las Vegas almost all bets have the same expected value of $-2/38 = -.0526$ (see Exercises 4 and 5). $\square$

Conditional Expectation

Definition 6.2 If F is any event and X is a random variable with sample space $\Omega = \{x_1, x_2, \ldots\}$, then the *conditional expectation given F* is defined by

$$E(X|F) = \sum_j x_j P(X = x_j | F)\ .$$

Conditional expectation is used most often in the form provided by the following theorem. $\square$

Theorem 6.5 Let X be a random variable with sample space Ω. If $F_1, F_2, \ldots, F_r$ are events such that $F_i \cap F_j = \emptyset$ for $i \neq j$ and $\Omega = \cup_j F_j$, then

$$E(X) = \sum_j E(X|F_j) P(F_j)\ .$$

Proof. We have

$$\sum_j E(X|F_j)P(F_j) = \sum_j \sum_k x_k P(X = x_k|F_j)P(F_j)$$
$$= \sum_j \sum_k x_k P(X = x_k \text{ and } F_j \text{ occurs})$$
$$= \sum_k \sum_j x_k P(X = x_k \text{ and } F_j \text{ occurs})$$
$$= \sum_k x_k P(X = x_k)$$
$$= E(X) \ .$$

$\square$

Example 6.14 (Example 6.12 continued) Let T be the number of rolls in a single play of craps. We can think of a single play as a two-stage process. The first stage consists of a single roll of a pair of dice. The play is over if this roll is a 2, 3, 7, 11, or 12. Otherwise, the player's point is established, and the second stage begins. This second stage consists of a sequence of rolls which ends when either the player's point or a 7 is rolled. We record the outcomes of this two-stage experiment using the random variables X and S, where X denotes the first roll, and S denotes the number of rolls in the second stage of the experiment (of course, S is sometimes equal to 0). Note that $T = S + 1$. Then by Theorem 6.5

$$E(T) = \sum_{j=2}^{12} E(T|X=j) P(X=j) \ .$$

If $j = 7$, 11 or 2, 3, 12, then $E(T|X = j) = 1$. If $j = 4, 5, 6, 8, 9$, or 10, we can use Example 6.4 to calculate the expected value of S. In each of these cases, we continue rolling until we get either a j or a 7. Thus, S is geometrically distributed with parameter p, which depends upon j. If $j = 4$, for example, the value of p is $3/36 + 6/36 = 1/4$. Thus, in this case, the expected number of additional rolls is $1/p = 4$, so $E(T|X = 4) = 1 + 4 = 5$. Carrying out the corresponding calculations for the other possible values of j and using Theorem 6.5 gives

$$E(T) = 1\left(\frac{12}{36}\right) + \left(1 + \frac{36}{3+6}\right)\left(\frac{3}{36}\right) + \left(1 + \frac{36}{4+6}\right)\left(\frac{4}{36}\right)$$
$$+ \left(1 + \frac{36}{5+6}\right)\left(\frac{5}{36}\right) + \left(1 + \frac{36}{5+6}\right)\left(\frac{5}{36}\right)$$
$$+ \left(1 + \frac{36}{4+6}\right)\left(\frac{4}{36}\right) + \left(1 + \frac{36}{3+6}\right)\left(\frac{3}{36}\right)$$
$$= \frac{557}{165}$$
$$\approx 3.375\ldots \ .$$

$\square$

Martingales

We can extend the notion of fairness to a player playing a sequence of games by using the concept of conditional expectation.

Example 6.15 Let $S_1, S_2, \ldots, S_n$ be Peter's accumulated fortune in playing heads or tails (see Example 1.4). Then

$$E(S_n | S_{n-1} = a, \ldots, S_1 = r) = \frac{1}{2}(a+1) + \frac{1}{2}(a-1) = a \ .$$

We note that Peter's expected fortune after the next play is equal to his present fortune. When this occurs, we say the game is *fair*. A fair game is also called a *martingale*. If the coin is biased and comes up heads with probability p and tails with probability $q = 1 - p$, then

$$E(S_n | S_{n-1} = a, \ldots, S_1 = r) = p(a+1) + q(a-1) = a + p - q \ .$$

Thus, if $p < q$, this game is unfavorable, and if $p > q$, it is favorable. □

If you are in a casino, you will see players adopting elaborate *systems* of play to try to make unfavorable games favorable. Two such systems, the martingale doubling system and the more conservative Labouchere system, were described in Exercises 1.1.9 and 1.1.10. Unfortunately, such systems cannot change even a fair game into a favorable game.

Even so, it is a favorite pastime of many people to develop systems of play for gambling games and for other games such as the stock market. We close this section with a simple illustration of such a system.

Stock Prices

Example 6.16 Let us assume that a stock increases or decreases in value each day by 1 dollar, each with probability 1/2. Then we can identify this simplified model with our familiar game of heads or tails. We assume that a buyer, Mr. Ace, adopts the following strategy. He buys the stock on the first day at its price V. He then waits until the price of the stock increases by one to $V + 1$ and sells. He then continues to watch the stock until its price falls back to V. He buys again and waits until it goes up to $V + 1$ and sells. Thus he holds the stock in intervals during which it increases by 1 dollar. In each such interval, he makes a profit of 1 dollar. However, we assume that he can do this only for a finite number of trading days. Thus he can lose if, in the last interval that he holds the stock, it does not get back up to $V + 1$; and this is the only we he can lose. In Figure 6.4 we illustrate a typical history if Mr. Ace must stop in twenty days. Mr. Ace holds the stock under his system during the days indicated by broken lines. We note that for the history shown in Figure 6.4, his system nets him a gain of 4 dollars.

We have written a program **StockSystem** to simulate the fortune of Mr. Ace if he uses his sytem over an n-day period. If one runs this program a large number

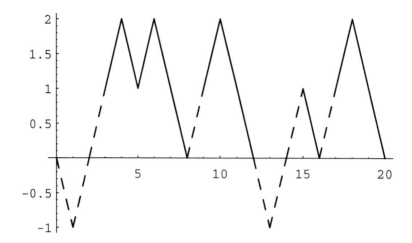

Figure 6.4: Mr. Ace's system.

of times, for $n = 20$, say, one finds that his expected winnings are very close to 0, but the probability that he is ahead after 20 days is significantly greater than 1/2. For small values of n, the exact distribution of winnings can be calculated. The distribution for the case $n = 20$ is shown in Figure 6.5. Using this distribution, it is easy to calculate that the expected value of his winnings is exactly 0. This is another instance of the fact that a fair game (a martingale) remains fair under quite general systems of play.

Although the expected value of his winnings is 0, the probability that Mr. Ace is ahead after 20 days is about .610. Thus, he would be able to tell his friends that his system gives him a better chance of being ahead than that of someone who simply buys the stock and holds it, if our simple random model is correct. There have been a number of studies to determine how random the stock market is. □

Historical Remarks

With the Law of Large Numbers to bolster the frequency interpretation of probability, we find it natural to justify the definition of expected value in terms of the average outcome over a large number of repetitions of the experiment. The concept of expected value was used before it was formally defined; and when it was used, it was considered not as an average value but rather as the appropriate value for a gamble. For example, recall Pascal's way of finding the value of a three-game series that had to be called off before it is finished.

Pascal first observed that if each player has only one game to win, then the stake of 64 pistoles should be divided evenly. Then he considered the case where one player has won two games and the other one.

> Then consider, Sir, if the first man wins, he gets 64 pistoles, if he loses he gets 32. Thus if they do not wish to risk this last game, but wish to separate without playing it, the first man must say: "I am certain

6.1. EXPECTED VALUE

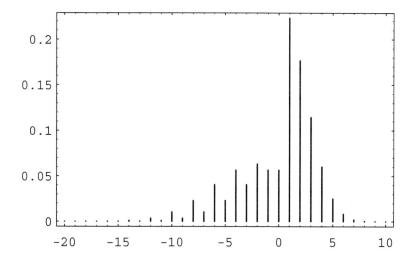

Figure 6.5: Winnings distribution for $n = 20$.

to get 32 pistoles, even if I lose I still get them; but as for the other 32 pistoles, perhaps I will get them, perhaps you will get them, the chances are equal. Let us then divide these 32 pistoles in half and give one half to me as well as my 32 which are mine for sure." He will then have 48 pistoles and the other 16.[2]

Note that Pascal reduced the problem to a symmetric bet in which each player gets the same amount and takes it as obvious that in this case the stakes should be divided equally.

The first systematic study of expected value appears in Huygens' book. Like Pascal, Huygens find the value of a gamble by assuming that the answer is obvious for certain symmetric situations and uses this to deduce the expected for the general situation. He does this in steps. His first proposition is

> Prop. I. If I expect a or b, either of which, with equal probability, may fall to me, then my Expectation is worth $(a+b)/2$, that is, the half Sum of a and b.[3]

Huygens proved this as follows: Assume that two player A and B play a game in which each player puts up a stake of $(a+b)/2$ with an equal chance of winning the total stake. Then the value of the game to each player is $(a+b)/2$. For example, if the game had to be called off clearly each player should just get back his original stake. Now, by symmetry, this value is not changed if we add the condition that the winner of the game has to pay the loser an amount b as a consolation prize. Then for player A the value is still $(a+b)/2$. But what are his possible outcomes for the modified game? If he wins he gets the total stake $a+b$ and must pay B an

---

[2]Quoted in F. N. David, *Games, Gods and Gambling* (London: Griffin, 1962), p. 231.

[3]C. Huygens, *Calculating in Games of Chance*, translation attributed to John Arbuthnot (London, 1692), p. 34.

amount b so ends up with a. If he loses he gets an amount b from player B. Thus player A wins a or b with equal chances and the value to him is $(a+b)/2$.

Huygens illustrated this proof in terms of an example. If you are offered a game in which you have an equal chance of winning 2 or 8, the expected value is 5, since this game is equivalent to the game in which each player stakes 5 and agrees to pay the loser 3 — a game in which the value is obviously 5.

Huygens' second proposition is

> Prop. II. If I expect a, b, or c, either of which, with equal facility, may happen, then the Value of my Expectation is $(a+b+c)/3$, or the third of the Sum of a, b, and c.[4]

His argument here is similar. Three players, A, B, and C, each stake

$$(a+b+c)/3$$

in a game they have an equal chance of winning. The value of this game to player A is clearly the amount he has staked. Further, this value is not changed if A enters into an agreement with B that if one of them wins he pays the other a consolation prize of b and with C that if one of them wins he pays the other a consolation prize of c. By symmetry these agreements do not change the value of the game. In this modified game, if A wins he wins the total stake $a+b+c$ minus the consolation prizes $b+c$ giving him a final winning of a. If B wins, A wins b and if C wins, A wins c. Thus A finds himself in a game with value $(a+b+c)/3$ and with outcomes a, b, and c occurring with equal chance. This proves Proposition II.

More generally, this reasoning shows that if there are n outcomes

$$a_1,\ a_2,\ \ldots,\ a_n\ ,$$

all occurring with the same probability, the expected value is

$$\frac{a_1 + a_2 + \cdots + a_n}{n}\ .$$

In his third proposition Huygens considered the case where you win a or b but with unequal probabilities. He assumed there are p chances of winning a, and q chances of winning b, all having the same probability. He then showed that the expected value is

$$E = \frac{p}{p+q} \cdot a + \frac{q}{p+q} \cdot b\ .$$

This follows by considering an equivalent gamble with $p+q$ outcomes all occurring with the same probability and with a payoff of a in p of the outcomes and b in q of the outcomes. This allowed Huygens to compute the expected value for experiments with unequal probabilities, at least when these probablities are rational numbers.

Thus, instead of defining the expected value as a weighted average, Huygens assumed that the expected value of certain symmetric gambles are known and deduced the other values from these. Although this requires a good deal of clever

---
[4]ibid., p. 35.

6.1. EXPECTED VALUE

manipulation, Huygens ended up with values that agree with those given by our modern definition of expected value. One advantage of this method is that it gives a justification for the expected value in cases where it is not reasonable to assume that you can repeat the experiment a large number of times, as for example, in betting that at least two presidents died on the same day of the year. (In fact, three did; all were signers of the Declaration of Independence, and all three died on July 4.)

In his book, Huygens calculated the expected value of games using techniques similar to those which we used in computing the expected value for roulette at Monte Carlo. For example, his proposition XIV is:

> Prop. XIV. If I were playing with another by turns, with two Dice, on this Condition, that if I throw 7 I gain, and if he throws 6 he gains allowing him the first Throw: To find the proportion of my Hazard to his.[5]

A modern description of this game is as follows. Huygens and his opponent take turns rolling a die. The game is over if Huygens rolls a 7 or his opponent rolls a 6. His opponent rolls first. What is the probability that Huygens wins the game?

To solve this problem Huygens let x be his chance of winning when his opponent threw first and y his chance of winning when he threw first. Then on the first roll his opponent wins on 5 out of the 36 possibilities. Thus,

$$x = \frac{31}{36} \cdot y \ .$$

But when Huygens rolls he wins on 6 out of the 36 possible outcomes, and in the other 30, he is led back to where his chances are x. Thus

$$y = \frac{6}{36} + \frac{30}{36} \cdot x \ .$$

From these two equations Huygens found that $x = 31/61$.

Another early use of expected value appeared in Pascal's argument to show that a rational person should believe in the existence of God.[6] Pascal said that we have to make a wager whether to believe or not to believe. Let p denote the probability that God does not exist. His discussion suggests that we are playing a game with two strategies, believe and not believe, with payoffs as shown in Table 6.4.

Here $-u$ represents the cost to you of passing up some worldly pleasures as a consequence of believing that God exists. If you do not believe, and God is a vengeful God, you will lose x. If God exists and you do believe you will gain v. Now to determine which strategy is best you should compare the two expected values

$$p(-u) + (1-p)v \quad \text{and} \quad p0 + (1-p)(-x),$$

---

[5]ibid., p. 47.
[6]Quoted in I. Hacking, *The Emergence of Probability* (Cambridge: Cambridge Univ. Press, 1975).

	God does not exist	God exists
	p	$1-p$
believe	$-u$	v
not believe	0	$-x$

Table 6.4: Payoffs.

Age	Survivors
0	100
6	64
16	40
26	25
36	16
46	10
56	6
66	3
76	1

Table 6.5: Graunt's mortality data.

and choose the larger of the two. In general, the choice will depend upon the value of p. But Pascal assumed that the value of v is infinite and so the strategy of believing is best no matter what probability you assign for the existence of God. This example is considered by some to be the beginning of decision theory. Decision analyses of this kind appear today in many fields, and, in particular, are an important part of medical diagnostics and corporate business decisions.

Another early use of expected value was to decide the price of annuities. The study of statistics has its origins in the use of the bills of mortality kept in the parishes in London from 1603. These records kept a weekly tally of christenings and burials. From these John Graunt made estimates for the population of London and also provided the first mortality data,[7] shown in Table 6.5.

As Hacking observes, Graunt apparently constructed this table by assuming that after the age of 6 there is a constant probability of about 5/8 of surviving for another decade.[8] For example, of the 64 people who survive to age 6, 5/8 of 64 or 40 survive to 16, 5/8 of these 40 or 25 survive to 26, and so forth. Of course, he rounded off his figures to the nearest whole person.

Clearly, a constant mortality rate cannot be correct throughout the whole range, and later tables provided by Halley were more realistic in this respect.[9]

---

[7] ibid., p. 108.
[8] ibid., p. 109.
[9] E. Halley, "An Estimate of The Degrees of Mortality of Mankind," *Phil. Trans. Royal. Soc.*,

6.1. EXPECTED VALUE

A *terminal annuity* provides a fixed amount of money during a period of n years. To determine the price of a terminal annuity one needs only to know the appropriate interest rate. A *life annuity* provides a fixed amount during each year of the buyer's life. The appropriate price for a life annuity is the expected value of the terminal annuity evaluated for the random lifetime of the buyer. Thus, the work of Huygens in introducing expected value and the work of Graunt and Halley in determining mortality tables led to a more rational method for pricing annuities. This was one of the first serious uses of probability theory outside the gambling houses.

Although expected value plays a role now in every branch of science, it retains its importance in the casino. In 1962, Edward Thorp's book *Beat the Dealer*[10] provided the reader with a strategy for playing the popular casino game of blackjack that would assure the player a positive expected winning. This book forevermore changed the belief of the casinos that they could not be beat.

Exercises

1 A card is drawn at random from a deck consisting of cards numbered 2 through 10. A player wins 1 dollar if the number on the card is odd and loses 1 dollar if the number if even. What is the expected value of his winnings?

2 A card is drawn at random from a deck of playing cards. If it is red, the player wins 1 dollar; if it is black, the player loses 2 dollars. Find the expected value of the game.

3 In a class there are 20 students: 3 are 5' 6", 5 are 5'8", 4 are 5'10", 4 are 6', and 4 are 6' 2". A student is chosen at random. What is the student's expected height?

4 In Las Vegas the roulette wheel has a 0 and a 00 and then the numbers 1 to 36 marked on equal slots; the wheel is spun and a ball stops randomly in one slot. When a player bets 1 dollar on a number, he receives 36 dollars if the ball stops on this number, for a net gain of 35 dollars; otherwise, he loses his dollar bet. Find the expected value for his winnings.

5 In a second version of roulette in Las Vegas, a player bets on red or black. Half of the numbers from 1 to 36 are red, and half are black. If a player bets a dollar on black, and if the ball stops on a black number, he gets his dollar back and another dollar. If the ball stops on a red number or on 0 or 00 he loses his dollar. Find the expected winnings for this bet.

6 A die is rolled twice. Let X denote the sum of the two numbers that turn up, and Y the difference of the numbers (specifically, the number on the first roll minus the number on the second). Show that $E(XY) = E(X)E(Y)$. Are X and Y independent?

---

vol. 17 (1693), pp. 596–610; 654–656.
[10]E. Thorp, *Beat the Dealer* (New York: Random House, 1962).

*7 Show that, if X and Y are random variables taking on only two values each, and if $E(XY) = E(X)E(Y)$, then X and Y are independent.

8 A royal family has children until it has a boy or until it has three children, whichever comes first. Assume that each child is a boy with probability 1/2. Find the expected number of boys in this royal family and the expected number of girls.

9 If the first roll in a game of craps is neither a natural nor craps, the player can make an additional bet, equal to his original one, that he will make his point before a seven turns up. If his point is four or ten he is paid off at 2 : 1 odds; if it is a five or nine he is paid off at odds 3 : 2; and if it is a six or eight he is paid off at odds 6 : 5. Find the player's expected winnings if he makes this additional bet when he has the opportunity.

10 In Example 6.16 assume that Mr. Ace decides to buy the stock and hold it until it goes up 1 dollar and then sell and not buy again. Modify the program **StockSystem** to find the distribution of his profit under this system after a twenty-day period. Find the expected profit and the probability that he comes out ahead.

11 On September 26, 1980, the *New York Times* reported that a mysterious stranger strode into a Las Vegas casino, placed a single bet of 777,000 dollars on the "don't pass" line at the crap table, and walked away with more than 1.5 million dollars. In the "don't pass" bet, the bettor is essentially betting with the house. An exception occurs if the roller rolls a 12 on the first roll. In this case, the roller loses and the "don't pass" better just gets back the money bet instead of winning. Show that the "don't pass" bettor has a more favorable bet than the roller.

12 Recall that in the *martingale doubling system* (see Exercise 1.1.10), the player doubles his bet each time he loses and quits the first time he is ahead. Suppose that you are playing roulette in a *fair casino* where there are no 0's, and you bet on red each time. You then win with probability 1/2 each time. Assume that you start with a 1-dollar bet and employ the martingale system. Since you entered the casino with 100 dollars, you also quit in the unlikely event that black turns up six times in a row so that you are down 63 dollars and cannot make the required 64-dollar bet. Find your expected winnings under this system of play.

13 You have 80 dollars and play the following game. An urn contains two white balls and two black balls. You draw the balls out one at a time without replacement until all the balls are gone. On each draw, you bet half of your present fortune that you will draw a white ball. What is your final fortune?

14 In the hat check problem (see Example 3.12), it was assumed that N people check their hats and the hats are handed back at random. Let $X_j = 1$ if the

6.1. EXPECTED VALUE

jth person gets his or her hat and 0 otherwise. Find $E(X_j)$ and $E(X_j \cdot X_k)$ for j not equal to k. Are X_j and X_k independent?

15 A box contains two gold balls and three silver balls. You are allowed to choose successively balls from the box at random. You win 1 dollar each time you draw a gold ball and lose 1 dollar each time you draw a silver ball. After a draw, the ball is not replaced. Show that, if you draw until you are ahead by 1 dollar or until there are no more gold balls, this is a favorable game.

16 Gerolamo Cardano in his book, *The Gambling Scholar*, written in the early 1500s, considers the following carnival game. There are six dice. Each of the dice has five blank sides. The sixth side has a number between 1 and 6—a different number on each die. The six dice are rolled and the player wins a prize depending on the total of the numbers which turn up.

  (a) Find, as Cardano did, the expected total without finding its distribution.

  (b) Large prizes were given for large totals with a modest fee to play the game. Explain why this could be done.

17 Let X be the first time that a *failure* occurs in an infinite sequence of Bernoulli trials with probability p for success. Let $p_k = P(X = k)$ for $k = 1, 2, \ldots$. Show that $p_k = p^{k-1}q$ where $q = 1 - p$. Show that $\sum_k p_k = 1$. Show that $E(X) = 1/q$. What is the expected number of tosses of a coin required to obtain the first tail?

18 Exactly one of six similar keys opens a certain door. If you try the keys, one after another, what is the expected number of keys that you will have to try before success?

19 A multiple choice exam is given. A problem has four possible answers, and exactly one answer is correct. The student is allowed to choose a subset of the four possible answers as his answer. If his chosen subset contains the correct answer, the student receives three points, but he loses one point for each wrong answer in his chosen subset. Show that if he just guesses a subset uniformly and randomly his expected score is zero.

20 You are offered the following game to play: a fair coin is tossed until heads turns up for the first time (see Example 6.3). If this occurs on the first toss you receive 2 dollars, if it occurs on the second toss you receive $2^2 = 4$ dollars and, in general, if heads turns up for the first time on the nth toss you receive 2^n dollars.

  (a) Show that the expected value of your winnings does not exist (i.e., is given by a divergent sum) for this game. Does this mean that this game is favorable no matter how much you pay to play it?

  (b) Assume that you only receive 2^{10} dollars if any number greater than or equal to ten tosses are required to obtain the first head. Show that your expected value for this modified game is finite and find its value.

(c) Assume that you pay 10 dollars for each play of the original game. Write a program to simulate 100 plays of the game and see how you do.

(d) Now assume that the utility of n dollars is $\sqrt{n}$. Write an expression for the expected utility of the payment, and show that this expression has a finite value. Estimate this value. Repeat this exercise for the case that the utility function is $\log(n)$.

21 Let X be a random variable which is Poisson distributed with parameter λ. Show that $E(X) = \lambda$. *Hint*: Recall that

$$e^x = 1 + x + \frac{x^2}{2!} + \frac{x^3}{3!} + \cdots.$$

22 Recall that in Exercise 1.1.14, we considered a town with two hospitals. In the large hospital about 45 babies are born each day, and in the smaller hospital about 15 babies are born each day. We were interested in guessing which hospital would have on the average the largest number of days with the property that more than 60 percent of the children born on that day are boys. For each hospital find the expected number of days in a year that have the property that more than 60 percent of the children born on that day were boys.

23 An insurance company has 1,000 policies on men of age 50. The company estimates that the probability that a man of age 50 dies within a year is .01. Estimate the number of claims that the company can expect from beneficiaries of these men within a year.

24 Using the life table for 1981 in Appendix C, write a program to compute the expected lifetime for males and females of each possible age from 1 to 85. Compare the results for males and females. Comment on whether life insurance should be priced differently for males and females.

*25 A deck of ESP cards consists of 20 cards each of two types: say ten stars, ten circles (normally there are five types). The deck is shuffled and the cards turned up one at a time. You, the alleged percipient, are to name the symbol on each card *before* it is turned up.

Suppose that you are really just guessing at the cards. If you do not get to see each card after you have made your guess, then it is easy to calculate the expected number of correct guesses, namely ten.

If, on the other hand, you are guessing with information, that is, if you see each card after your guess, then, of course, you might expect to get a higher score. This is indeed the case, but calculating the correct expectation is no longer easy.

But it is easy to do a computer simulation of this guessing with information, so we can get a good idea of the expectation by simulation. (This is similar to the way that skilled blackjack players make blackjack into a favorable game by observing the cards that have already been played. See Exercise 29.)

(a) First, do a simulation of guessing without information, repeating the experiment at least 1000 times. Estimate the expected number of correct answers and compare your result with the theoretical expectation.

(b) What is the best strategy for guessing with information?

(c) Do a simulation of guessing with information, using the strategy in (b). Repeat the experiment at least 1000 times, and estimate the expectation in this case.

(d) Let S be the number of stars and C the number of circles in the deck. Let $h(S, C)$ be the expected winnings using the optimal guessing strategy in (b). Show that $h(S, C)$ satisfies the recursion relation

$$h(S,C) = \frac{S}{S+C} h(S-1, C) + \frac{C}{S+C} h(S, C-1) + \frac{\max(S,C)}{S+C} ,$$

and $h(0, 0) = h(-1, 0) = h(0, -1) = 0$. Using this relation, write a program to compute $h(S, C)$ and find $h(10, 10)$. Compare the computed value of $h(10, 10)$ with the result of your simulation in (c). For more about this exercise and Exercise 26 see Diaconis and Graham.[11]

*26 Consider the ESP problem as described in Exercise 25. You are again guessing with information, and you are using the optimal guessing strategy of guessing *star* if the remaining deck has more stars, *circle* if more circles, and tossing a coin if the number of stars and circles are equal. Assume that $S \geq C$, where S is the number of stars and C the number of circles.

We can plot the results of a typical game on a graph, where the horizontal axis represents the number of steps and the vertical axis represents the *difference* between the number of stars and the number of circles that have been turned up. A typical game is shown in Figure 6.6. In this particular game, the order in which the cards were turned up is $(C, S, S, S, S, C, C, S, S, C)$. Thus, in this particular game, there were six stars and four circles in the deck. This means, in particular, that every game played with this deck would have a graph which ends at the point $(10, 2)$. We define the line L to be the horizontal line which goes through the ending point on the graph (so its vertical coordinate is just the difference between the number of stars and circles in the deck).

(a) Show that, when the random walk is below the line L, the player guesses right when the graph goes up (star is turned up) and, when the walk is above the line, the player guesses right when the walk goes down (circle turned up). Show from this property that the subject is sure to have at least S correct guesses.

(b) When the walk is at a point (x, x) on the line L the number of stars and circles remaining is the same, and so the subject tosses a coin. Show that

---

[11] P. Diaconis and R. Graham, "The Analysis of Sequential Experiments with Feedback to Subjects," *Annals of Statistics*, vol. 9 (1981), pp. 3–23.

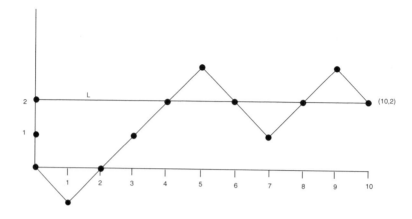

Figure 6.6: Random walk for ESP.

the probability that the walk reaches (x, x) is

$$\frac{\binom{S}{x}\binom{C}{x}}{\binom{S+C}{2x}}.$$

Hint: The outcomes of $2x$ cards is a hypergeometric distribution (see Section 5.1).

(c) Using the results of (a) and (b) show that the expected number of correct guesses under intelligent guessing is

$$S + \sum_{x=1}^{C} \frac{1}{2} \frac{\binom{S}{x}\binom{C}{x}}{\binom{S+C}{2x}}.$$

27 It has been said[12] that a Dr. B. Muriel Bristol declined a cup of tea stating that she preferred a cup into which milk had been poured first. The famous statistician R. A. Fisher carried out a test to see if she could tell whether milk was put in before or after the tea. Assume that for the test Dr. Bristol was given eight cups of tea—four in which the milk was put in before the tea and four in which the milk was put in after the tea.

(a) What is the expected number of correct guesses the lady would make if she had no information after each test and was just guessing?

(b) Using the result of Exercise 26 find the expected number of correct guesses if she was told the result of each guess and used an optimal guessing strategy.

28 In a popular computer game the computer picks an integer from 1 to n at random. The player is given k chances to guess the number. After each guess the computer responds "correct," "too small," or "too big."

---
[12] J. F. Box, *R. A. Fisher, The Life of a Scientist* (New York: John Wiley and Sons, 1978).

6.1. EXPECTED VALUE

(a) Show that if $n \leq 2^k - 1$, then there is a strategy that guarantees you will correctly guess the number in k tries.

(b) Show that if $n \geq 2^k - 1$, there is a strategy that assures you of identifying one of $2^k - 1$ numbers and hence gives a probability of $(2^k - 1)/n$ of winning. Why is this an optimal strategy? Illustrate your result in terms of the case $n = 9$ and $k = 3$.

29 In the casino game of blackjack the dealer is dealt two cards, one face up and one face down, and each player is dealt two cards, both face down. If the dealer is showing an ace the player can look at his down cards and then make a bet called an *insurance* bet. (Expert players will recognize why it is called insurance.) If you make this bet you will win the bet if the dealer's second card is a *ten card*: namely, a ten, jack, queen, or king. If you win, you are paid twice your insurance bet; otherwise you lose this bet. Show that, if the only cards you can see are the dealer's ace and your two cards and if your cards are not ten cards, then the insurance bet is an unfavorable bet. Show, however, that if you are playing two hands simultaneously, and you have no ten cards, then it is a favorable bet. (Thorp[13] has shown that the game of blackjack is favorable to the player if he or she can keep good enough track of the cards that have been played.)

30 Assume that, every time you buy a box of Wheaties, you receive a picture of one of the n players for the New York Yankees (see Exercise 3.2.34). Let X_k be the number of additional boxes you have to buy, after you have obtained $k-1$ different pictures, in order to obtain the next new picture. Thus $X_1 = 1$, X_2 is the number of boxes bought after this to obtain a picture different from the first pictured obtained, and so forth.

(a) Show that X_k has a geometric distribution with $p = (n - k + 1)/n$.

(b) Simulate the experiment for a team with 26 players (25 would be more accurate but we want an even number). Carry out a number of simulations and estimate the expected time required to get the first 13 players and the expected time to get the second 13. How do these expectations compare?

(c) Show that, if there are $2n$ players, the expected time to get the first half of the players is

$$2n\left(\frac{1}{2n} + \frac{1}{2n-1} + \cdots + \frac{1}{n+1}\right),$$

and the expected time to get the second half is

$$2n\left(\frac{1}{n} + \frac{1}{n-1} + \cdots + 1\right).$$

---
[13]E. Thorp, *Beat the Dealer* (New York: Random House, 1962).

(d) In Section 3.1 we showed that

$$1 + \frac{1}{2} + \frac{1}{3} + \cdots + \frac{1}{n} \sim \log n + .5772 + \frac{1}{2n}.$$

Use this to estimate the expression in (c). Compare these estimates with the exact values and also with your estimates obtained by simulation for the case $n = 26$.

*31 (Feller[14]) A large number, N, of people are subjected to a blood test. This can be administered in two ways: (1) Each person can be tested separately, in this case N test are required, (2) the blood samples of k persons can be pooled and analyzed together. If this test is *negative*, this one test suffices for the k people. If the test is *positive*, each of the k persons must be tested separately, and in all, $k + 1$ tests are required for the k people. Assume that the probability p that a test is positive is the same for all people and that these events are independent.

(a) Find the probability that the test for a pooled sample of k people will be positive.

(b) What is the expected value of the number X of tests necessary under plan (2)? (Assume that N is divisible by k.)

(c) For small p, show that the value of k which will minimize the expected number of tests under the second plan is approximately $1/\sqrt{p}$.

32 Write a program to add random numbers chosen from $[0, 1]$ until the first time the sum is greater than one. Have your program repeat this experiment a number of times to estimate the expected number of selections necessary in order that the sum of the chosen numbers first exceeds 1. On the basis of your experiments, what is your estimate for this number?

*33 The following related discrete problem also gives a good clue for the answer to Exercise 32. Randomly select with replacement $t_1, t_2, \ldots, t_r$ from the set $(1/n, 2/n, \ldots, n/n)$. Let X be the smallest value of r satisfying

$$t_1 + t_2 + \cdots + t_r > 1.$$

Then $E(X) = (1 + 1/n)^n$. To prove this, we can just as well choose $t_1, t_2, \ldots, t_r$ randomly with replacement from the set $(1, 2, \ldots, n)$ and let X be the smallest value of r for which

$$t_1 + t_2 + \cdots + t_r > n.$$

(a) Use Exercise 3.2.36 to show that

$$P(X \geq j+1) = \binom{n}{j}\left(\frac{1}{n}\right)^j.$$

---

[14] W. Feller, *Introduction to Probability Theory and Its Applications*, 3rd ed., vol. 1 (New York: John Wiley and Sons, 1968), p. 240.

(b) Show that

$$E(X) = \sum_{j=0}^{n} P(X \geq j+1) \ .$$

(c) From these two facts, find an expression for $E(X)$. This proof is due to Harris Schultz.[15]

*34 (Banach's Matchbox[16]) A man carries in each of his two front pockets a box of matches originally containing N matches. Whenever he needs a match, he chooses a pocket at random and removes one from that box. One day he reaches into a pocket and finds the box empty.

(a) Let p_r denote the probability that the other pocket contains r matches. Define a sequence of *counter* random variables as follows: Let $X_i = 1$ if the ith draw is from the left pocket, and 0 if it is from the right pocket. Interpret p_r in terms of $S_n = X_1 + X_2 + \cdots + X_n$. Find a binomial expression for p_r.

(b) Write a computer program to compute the p_r, as well as the probability that the other pocket contains at least r matches, for $N = 100$ and r from 0 to 50.

(c) Show that $(N-r)p_r = (1/2)(2N+1)p_{r+1} - (1/2)(r+1)p_{r+1}$.

(d) Evaluate $\sum_r p_r$.

(e) Use (c) and (d) to determine the expectation E of the distribution $\{p_r\}$.

(f) Use Stirling's formula to obtain an approximation for E. How many matches must each box contain to ensure a value of about 13 for the expectation E? (Take $\pi = 22/7$.)

35 A coin is tossed until the first time a head turns up. If this occurs on the nth toss and n is odd you win $2^n/n$, but if n is even then you lose $2^n/n$. Then if your expected winnings exist they are given by the convergent series

$$1 - \frac{1}{2} + \frac{1}{3} - \frac{1}{4} + \cdots$$

called the alternating *harmonic series*. It is tempting to say that this should be the expected value of the experiment. Show that if we were to do this, the expected value of an experiment would depend upon the order in which the outcomes are listed.

36 Suppose we have an urn containing c yellow balls and d green balls. We draw k balls, without replacement, from the urn. Find the expected number of yellow balls drawn. *Hint*: Write the number of yellow balls drawn as the sum of c random variables.

---

[15]H. Schultz, "An Expected Value Problem," *Two-Year Mathematics Journal*, vol. 10, no. 4 (1979), pp. 277–78.

[16]W. Feller, *Introduction to Probability Theory*, vol. 1, p. 166.

37 The reader is referred to Example 6.13 for an explanation of the various options available in Monte Carlo roulette.

(a) Compute the expected winnings of a 1 franc bet on red under option (a).

(b) Repeat part (a) for option (b).

(c) Compare the expected winnings for all three options.

*38 (from Pittel[17]) Telephone books, n in number, are kept in a stack. The probability that the book numbered i (where $1 \leq i \leq n$) is consulted for a given phone call is $p_i > 0$, where the p_i's sum to 1. After a book is used, it is placed at the top of the stack. Assume that the calls are independent and evenly spaced, and that the system has been employed indefinitely far into the past. Let d_i be the average depth of book i in the stack. Show that $d_i \leq d_j$ whenever $p_i \geq p_j$. Thus, on the average, the more popular books have a tendency to be closer to the top of the stack. *Hint*: Let p_{ij} denote the probability that book i is above book j. Show that $p_{ij} = p_{ij}(1 - p_j) + p_{ji}p_i$.

*39 (from Propp[18]) In the previous problem, let P be the probability that at the present time, each book is in its proper place, i.e., book i is ith from the top. Find a formula for P in terms of the p_i's. In addition, find the least upper bound on P, if the p_i's are allowed to vary. *Hint*: First find the probability that book 1 is in the right place. Then find the probability that book 2 is in the right place, given that book 1 is in the right place. Continue.

*40 (from H. Shultz and B. Leonard[19]) A sequence of random numbers in $[0, 1)$ is generated until the sequence is no longer monotone increasing. The numbers are chosen according to the uniform distribution. What is the expected length of the sequence? (In calculating the length, the term that destroys monotonicity is included.) *Hint*: Let a_1, a_2, ... be the sequence and let X denote the length of the sequence. Then

$$P(X > k) = P(a_1 < a_2 < \cdots < a_k) \ ,$$

and the probability on the right-hand side is easy to calculate. Furthermore, one can show that

$$E(X) = 1 + P(X > 1) + P(X > 2) + \cdots \ .$$

41 Let T be the random variable that counts the number of 2-unshuffles performed on an n-card deck until all of the labels on the cards are distinct. This random variable was discussed in Section 3.3. Using Equation 3.4 in that section, together with the formula

$$E(T) = \sum_{s=0}^{\infty} P(T > s)$$

---

[17]B. Pittel, Problem #1195, *Mathematics Magazine*, vol. 58, no. 3 (May 1985), pg. 183.

[18]J. Propp, Problem #1159, *Mathematics Magazine* vol. 57, no. 1 (Feb. 1984), pg. 50.

[19]H. Shultz and B. Leonard, "Unexpected Occurrences of the Number e," *Mathematics Magazine* vol. 62, no. 4 (October, 1989), pp. 269-271.

6.2. VARIANCE OF DISCRETE RANDOM VARIABLES

that was proved in Exercise 33, show that

$$E(T) = \sum_{s=0}^{\infty} \left(1 - \binom{2^s}{n} \frac{n!}{2^{sn}}\right) .$$

Show that for $n = 52$, this expression is approximately equal to 11.7. (As was stated in Chapter 3, this means that on the average, almost 12 riffle shuffles of a 52-card deck are required in order for the process to be considered random.)

6.2 Variance of Discrete Random Variables

The usefulness of the expected value as a prediction for the outcome of an experiment is increased when the outcome is not likely to deviate too much from the expected value. In this section we shall introduce a measure of this deviation, called the variance.

Variance

Definition 6.3 Let X be a numerically valued random variable with expected value $\mu = E(X)$. Then the *variance* of X, denoted by $V(X)$, is

$$V(X) = E((X - \mu)^2) .$$

$\square$

Note that, by Theorem 6.1, $V(X)$ is given by

$$V(X) = \sum_x (x - \mu)^2 m(x) , \qquad (6.1)$$

where m is the distribution function of X.

Standard Deviation

The *standard deviation* of X, denoted by $D(X)$, is $D(X) = \sqrt{V(X)}$. We often write σ for $D(X)$ and σ^2 for $V(X)$.

Example 6.17 Consider one roll of a die. Let X be the number that turns up. To find $V(X)$, we must first find the expected value of X. This is

$$\begin{aligned} \mu &= E(X) = 1\left(\frac{1}{6}\right) + 2\left(\frac{1}{6}\right) + 3\left(\frac{1}{6}\right) + 4\left(\frac{1}{6}\right) + 5\left(\frac{1}{6}\right) + 6\left(\frac{1}{6}\right) \\ &= \frac{7}{2} . \end{aligned}$$

To find the variance of X, we form the new random variable $(X - \mu)^2$ and compute its expectation. We can easily do this using the following table.

x	$m(x)$	$(x-7/2)^2$
1	1/6	25/4
2	1/6	9/4
3	1/6	1/4
4	1/6	1/4
5	1/6	9/4
6	1/6	25/4

Table 6.6: Variance calculation.

From this table we find $E((X-\mu)^2)$ is

$$V(X) = \frac{1}{6}\left(\frac{25}{4} + \frac{9}{4} + \frac{1}{4} + \frac{1}{4} + \frac{9}{4} + \frac{25}{4}\right)$$
$$= \frac{35}{12},$$

and the standard deviation $D(X) = \sqrt{35/12} \approx 1.707$. □

Calculation of Variance

We next prove a theorem that gives us a useful alternative form for computing the variance.

Theorem 6.6 If X is any random variable with $E(X) = \mu$, then

$$V(X) = E(X^2) - \mu^2 .$$

Proof. We have

$$V(X) = E((X-\mu)^2) = E(X^2 - 2\mu X + \mu^2)$$
$$= E(X^2) - 2\mu E(X) + \mu^2 = E(X^2) - \mu^2 .$$

□

Using Theorem 6.6, we can compute the variance of the outcome of a roll of a die by first computing

$$E(X^2) = 1\left(\frac{1}{6}\right) + 4\left(\frac{1}{6}\right) + 9\left(\frac{1}{6}\right) + 16\left(\frac{1}{6}\right) + 25\left(\frac{1}{6}\right) + 36\left(\frac{1}{6}\right)$$
$$= \frac{91}{6},$$

and,

$$V(X) = E(X^2) - \mu^2 = \frac{91}{6} - \left(\frac{7}{2}\right)^2 = \frac{35}{12},$$

in agreement with the value obtained directly from the definition of $V(X)$.

6.2. VARIANCE OF DISCRETE RANDOM VARIABLES

Properties of Variance

The variance has properties very different from those of the expectation. If c is any constant, $E(cX) = cE(X)$ and $E(X+c) = E(X)+c$. These two statements imply that the expectation is a linear function. However, the variance is not linear, as seen in the next theorem.

Theorem 6.7 If X is any random variable and c is any constant, then
$$V(cX) = c^2 V(X)$$
and
$$V(X+c) = V(X)\ .$$

Proof. Let $\mu = E(X)$. Then $E(cX) = c\mu$, and
$$\begin{aligned} V(cX) &= E((cX - c\mu)^2) = E(c^2(X-\mu)^2) \\ &= c^2 E((X-\mu)^2) = c^2 V(X)\ . \end{aligned}$$

To prove the second assertion, we note that, to compute $V(X+c)$, we would replace ω by $\omega + c$ and μ by $\mu + c$ in Equation 6.1. But the c's would cancel, leaving $V(X)$. □

We turn now to some general properties of the variance. Recall that if X and Y are any two random variables, $E(X+Y) = E(X)+E(Y)$. This is not always true for the case of the variance. For example, let X be a random variable with $V(X) \neq 0$, and define $Y = -X$. Then $V(X) = V(Y)$, so that $V(X)+V(Y) = 2V(X)$. But $X+Y$ is always 0 and hence has variance 0. Thus $V(X+Y) \neq V(X)+V(Y)$.

In the important case of mutually independent random variables, however, *the variance of the sum is the sum of the variances.*

Theorem 6.8 Let X and Y be two *independent* random variables. Then
$$V(X+Y) = V(X)+V(Y)\ .$$

Proof. Let $E(X) = a$ and $E(Y) = b$. Then
$$\begin{aligned} V(X+Y) &= E((X+Y)^2) - (a+b)^2 \\ &= E(X^2) + 2E(XY) + E(Y^2) - a^2 - 2ab - b^2\ . \end{aligned}$$

Since X and Y are independent, $E(XY) = E(X)E(Y) = ab$. Thus,
$$V(X+Y) = E(X^2) - a^2 + E(Y^2) - b^2 = V(X) + V(Y)\ .$$

□

It is easy to extend this proof, by mathematical induction, to show that *the variance of the sum of any number of mutually independent random variables is the sum of the individual variances.* Thus we have the following theorem.

Theorem 6.9 Let X_1, X_2, ..., X_n be an independent trials process with $E(X_j) = \mu$ and $V(X_j) = \sigma^2$. Let
$$S_n = X_1 + X_2 + \cdots + X_n$$
be the sum, and
$$A_n = \frac{S_n}{n}$$
be the average. Then
$$\begin{aligned} E(S_n) &= n\mu \, , \\ V(S_n) &= n\sigma^2 \, , \\ E(A_n) &= \mu \, , \\ V(A_n) &= \frac{\sigma^2}{n} \, . \end{aligned}$$

Proof. Since all the random variables X_j have the same expected value, we have
$$E(S_n) = E(X_1) + \cdots + E(X_n) = n\mu \, ,$$
and
$$V(S_n) = V(X_1) + \cdots + V(X_n) = n\sigma^2 \, .$$

We have seen that, if we multiply a random variable X with mean μ and variance σ^2 by a constant c, the new random variable has expected value $c\mu$ and variance $c^2\sigma^2$. Thus,
$$E(A_n) = E\left(\frac{S_n}{n}\right) = \frac{n\mu}{n} = \mu \, ,$$
and
$$V(A_n) = V\left(\frac{S_n}{n}\right) = \frac{V(S_n)}{n^2} = \frac{n\sigma^2}{n^2} = \frac{\sigma^2}{n} \, .$$

Finally, the standard deviation of A_n is given by
$$\sigma(A_n) = \frac{\sigma}{\sqrt{n}} \, .$$
$\square$

The last statement in the above proof implies that in an independent trials process, if the individual summands have finite variance, then the standard deviation of the average goes to 0 as $n \to \infty$. Since the standard deviation tells us something about the spread of the distribution around the mean, we see that for large values of n, the value of A_n is usually very close to the mean of A_n, which equals μ, as shown above. This statement is made precise in Chapter 8, where it is called the Law of Large Numbers. For example, let X represent the roll of a fair die. In Figure 6.7, we show the distribution of a random variable A_n corresponding to X, for $n = 10$ and $n = 100$.

6.2. VARIANCE OF DISCRETE RANDOM VARIABLES

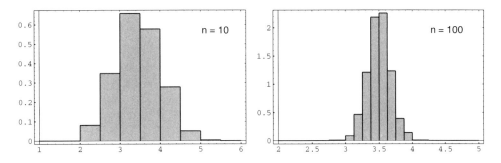

Figure 6.7: Empirical distribution of A_n.

Example 6.18 Consider n rolls of a die. We have seen that, if X_j is the outcome if the jth roll, then $E(X_j) = 7/2$ and $V(X_j) = 35/12$. Thus, if S_n is the sum of the outcomes, and $A_n = S_n/n$ is the average of the outcomes, we have $E(A_n) = 7/2$ and $V(A_n) = (35/12)/n$. Therefore, as n increases, the expected value of the average remains constant, but the variance tends to 0. If the variance is a measure of the expected deviation from the mean this would indicate that, for large n, we can expect the average to be very near the expected value. This is in fact the case, and we shall justify it in Chapter 8. $\square$

Bernoulli Trials

Consider next the general Bernoulli trials process. As usual, we let $X_j = 1$ if the jth outcome is a success and 0 if it is a failure. If p is the probability of a success, and $q = 1 - p$, then

$$
\begin{aligned}
E(X_j) &= 0q + 1p = p, \\
E(X_j^2) &= 0^2 q + 1^2 p = p,
\end{aligned}
$$

and

$$V(X_j) = E(X_j^2) - (E(X_j))^2 = p - p^2 = pq.$$

Thus, for Bernoulli trials, if $S_n = X_1 + X_2 + \cdots + X_n$ is the number of successes, then $E(S_n) = np$, $V(S_n) = npq$, and $D(S_n) = \sqrt{npq}$. If $A_n = S_n/n$ is the average number of successes, then $E(A_n) = p$, $V(A_n) = pq/n$, and $D(A_n) = \sqrt{pq/n}$. We see that the expected proportion of successes remains p and the variance tends to 0. This suggests that the frequency interpretation of probability is a correct one. We shall make this more precise in Chapter 8.

Example 6.19 Let T denote the number of trials until the first success in a Bernoulli trials process. Then T is geometrically distributed. What is the variance of T? In Example 4.15, we saw that

$$m_T = \begin{pmatrix} 1 & 2 & 3 & \cdots \\ p & qp & q^2 p & \cdots \end{pmatrix}.$$

In Example 6.4, we showed that

$$E(T) = 1/p .$$

Thus,

$$V(T) = E(T^2) - 1/p^2 ,$$

so we need only find

$$\begin{aligned} E(T^2) &= 1p + 4qp + 9q^2 p + \cdots \\ &= p(1 + 4q + 9q^2 + \cdots) . \end{aligned}$$

To evaluate this sum, we start again with

$$1 + x + x^2 + \cdots = \frac{1}{1-x} .$$

Differentiating, we obtain

$$1 + 2x + 3x^2 + \cdots = \frac{1}{(1-x)^2} .$$

Multiplying by x,

$$x + 2x^2 + 3x^3 + \cdots = \frac{x}{(1-x)^2} .$$

Differentiating again gives

$$1 + 4x + 9x^2 + \cdots = \frac{1+x}{(1-x)^3} .$$

Thus,

$$E(T^2) = p\frac{1+q}{(1-q)^3} = \frac{1+q}{p^2}$$

and

$$\begin{aligned} V(T) &= E(T^2) - (E(T))^2 \\ &= \frac{1+q}{p^2} - \frac{1}{p^2} = \frac{q}{p^2} . \end{aligned}$$

For example, the variance for the number of tosses of a coin until the first head turns up is $(1/2)/(1/2)^2 = 2$. The variance for the number of rolls of a die until the first six turns up is $(5/6)/(1/6)^2 = 30$. Note that, as p decreases, the variance increases rapidly. This corresponds to the increased spread of the geometric distribution as p decreases (noted in Figure 5.1). □

Poisson Distribution

Just as in the case of expected values, it is easy to guess the variance of the Poisson distribution with parameter λ. We recall that the variance of a binomial distribution with parameters n and p equals npq. We also recall that the Poisson distribution could be obtained as a limit of binomial distributions, if n goes to ∞ and p goes to 0 in such a way that their product is kept fixed at the value λ. In this case, $npq = \lambda q$ approaches λ, since q goes to 1. So, given a Poisson distribution with parameter λ, we should guess that its variance is λ. The reader is asked to show this in Exercise 30.

6.2. VARIANCE OF DISCRETE RANDOM VARIABLES

Exercises

1 A number is chosen at random from the set $S = \{-1, 0, 1\}$. Let X be the number chosen. Find the expected value, variance, and standard deviation of X.

2 A random variable X has the distribution

$$p_X = \begin{pmatrix} 0 & 1 & 2 & 4 \\ 1/3 & 1/3 & 1/6 & 1/6 \end{pmatrix}.$$

Find the expected value, variance, and standard deviation of X.

3 You place a 1-dollar bet on the number 17 at Las Vegas, and your friend places a 1-dollar bet on black (see Exercises 1.1.6 and 1.1.7). Let X be your winnings and Y be her winnings. Compare $E(X)$, $E(Y)$, and $V(X)$, $V(Y)$. What do these computations tell you about the nature of your winnings if you and your friend make a sequence of bets, with you betting each time on a number and your friend betting on a color?

4 X is a random variable with $E(X) = 100$ and $V(X) = 15$. Find

(a) $E(X^2)$.
(b) $E(3X + 10)$.
(c) $E(-X)$.
(d) $V(-X)$.
(e) $D(-X)$.

5 In a certain manufacturing process, the (Fahrenheit) temperature never varies by more than 2° from 62°. The temperature is, in fact, a random variable F with distribution

$$P_F = \begin{pmatrix} 60 & 61 & 62 & 63 & 64 \\ 1/10 & 2/10 & 4/10 & 2/10 & 1/10 \end{pmatrix}.$$

(a) Find $E(F)$ and $V(F)$.
(b) Define $T = F - 62$. Find $E(T)$ and $V(T)$, and compare these answers with those in part (a).
(c) It is decided to report the temperature readings on a Celsius scale, that is, $C = (5/9)(F - 32)$. What is the expected value and variance for the readings now?

6 Write a computer program to calculate the mean and variance of a distribution which you specify as data. Use the program to compare the variances for the following densities, both having expected value 0:

$$p_X = \begin{pmatrix} -2 & -1 & 0 & 1 & 2 \\ 3/11 & 2/11 & 1/11 & 2/11 & 3/11 \end{pmatrix};$$

$$p_Y = \begin{pmatrix} -2 & -1 & 0 & 1 & 2 \\ 1/11 & 2/11 & 5/11 & 2/11 & 1/11 \end{pmatrix}.$$

7 A coin is tossed three times. Let X be the number of heads that turn up. Find $V(X)$ and $D(X)$.

8 A random sample of 2400 people are asked if they favor a government proposal to develop new nuclear power plants. If 40 percent of the people in the country are in favor of this proposal, find the expected value and the standard deviation for the number S_{2400} of people in the sample who favored the proposal.

9 A die is loaded so that the probability of a face coming up is proportional to the number on that face. The die is rolled with outcome X. Find $V(X)$ and $D(X)$.

10 Prove the following facts about the standard deviation.

(a) $D(X + c) = D(X)$.

(b) $D(cX) = |c|D(X)$.

11 A number is chosen at random from the integers $1, 2, 3, \ldots, n$. Let X be the number chosen. Show that $E(X) = (n+1)/2$ and $V(X) = (n-1)(n+1)/12$. *Hint*: The following identity may be useful:

$$1^2 + 2^2 + \cdots + n^2 = \frac{(n)(n+1)(2n+1)}{6} .$$

12 Let X be a random variable with $\mu = E(X)$ and $\sigma^2 = V(X)$. Define $X^* = (X-\mu)/\sigma$. The random variable X^* is called the *standardized random variable* associated with X. Show that this standardized random variable has expected value 0 and variance 1.

13 Peter and Paul play Heads or Tails (see Example 1.4). Let W_n be Peter's winnings after n matches. Show that $E(W_n) = 0$ and $V(W_n) = n$.

14 Find the expected value and the variance for the number of boys and the number of girls in a royal family that has children until there is a boy or until there are three children, whichever comes first.

15 Suppose that n people have their hats returned at random. Let $X_i = 1$ if the ith person gets his or her own hat back and 0 otherwise. Let $S_n = \sum_{i=1}^{n} X_i$. Then S_n is the total number of people who get their own hats back. Show that

(a) $E(X_i^2) = 1/n$.

(b) $E(X_i \cdot X_j) = 1/n(n-1)$ for $i \neq j$.

(c) $E(S_n^2) = 2$ (using (a) and (b)).

(d) $V(S_n) = 1$.

6.2. VARIANCE OF DISCRETE RANDOM VARIABLES

16 Let S_n be the number of successes in n independent trials. Use the program **BinomialProbabilities** (Section 3.2) to compute, for given n, p, and j, the probability
$$P(-j\sqrt{npq} < S_n - np < j\sqrt{npq}) \ .$$

(a) Let $p = .5$, and compute this probability for $j = 1, 2, 3$ and $n = 10, 30, 50$. Do the same for $p = .2$.

(b) Show that the *standardized random variable* $S_n^* = (S_n - np)/\sqrt{npq}$ has expected value 0 and variance 1. What do your results from (a) tell you about this standardized quantity S_n^*?

17 Let X be the outcome of a chance experiment with $E(X) = \mu$ and $V(X) = \sigma^2$. When μ and σ^2 are unknown, the statistician often estimates them by repeating the experiment n times with outcomes $x_1, x_2, \ldots, x_n$, estimating μ by the *sample mean*
$$\bar{x} = \frac{1}{n} \sum_{i=1}^{n} x_i \ ,$$
and σ^2 by the *sample variance*
$$s^2 = \frac{1}{n} \sum_{i=1}^{n} (x_i - \bar{x})^2 \ .$$

Then s is the *sample standard deviation*. These formulas should remind the reader of the definitions of the theoretical mean and variance. (Many statisticians define the sample variance with the coefficient $1/n$ replaced by $1/(n-1)$. If this alternative definition is used, the expected value of s^2 is equal to σ^2. See Exercise 18, part (d).)

Write a computer program that will roll a die n times and compute the sample mean and sample variance. Repeat this experiment several times for $n = 10$ and $n = 1000$. How well do the sample mean and sample variance estimate the true mean $7/2$ and variance $35/12$?

18 Show that, for the sample mean $\bar{x}$ and sample variance s^2 as defined in Exercise 17,

(a) $E(\bar{x}) = \mu$.

(b) $E\big((\bar{x} - \mu)^2\big) = \sigma^2/n$.

(c) $E(s^2) = \frac{n-1}{n}\sigma^2$. *Hint:* For (c) write
$$\sum_{i=1}^{n}(x_i - \bar{x})^2 = \sum_{i=1}^{n}\big((x_i - \mu) - (\bar{x} - \mu)\big)^2$$
$$= \sum_{i=1}^{n}(x_i - \mu)^2 - 2(\bar{x} - \mu)\sum_{i=1}^{n}(x_i - \mu) + n(\bar{x} - \mu)^2$$
$$- \sum_{i=1}^{n}(x_i - \mu)^2 \quad n(\bar{x} - \mu)^2,$$

and take expectations of both sides, using part (b) when necessary.

(d) Show that if, in the definition of s^2 in Exercise 17, we replace the coefficient $1/n$ by the coefficient $1/(n-1)$, then $E(s^2) = \sigma^2$. (This shows why many statisticians use the coefficient $1/(n-1)$. The number s^2 is used to estimate the unknown quantity σ^2. If an estimator has an average value which equals the quantity being estimated, then the estimator is said to be *unbiased*. Thus, the statement $E(s^2) = \sigma^2$ says that s^2 is an unbiased estimator of σ^2.)

19 Let X be a random variable taking on values $a_1, a_2, \ldots, a_r$ with probabilities $p_1, p_2, \ldots, p_r$ and with $E(X) = \mu$. Define the *spread* of X as follows:

$$\bar{\sigma} = \sum_{i=1}^{r} |a_i - \mu| p_i \ .$$

This, like the standard deviation, is a way to quantify the amount that a random variable is spread out around its mean. Recall that the variance of a sum of mutually independent random variables is the sum of the individual variances. The square of the spread corresponds to the variance in a manner similar to the correspondence between the spread and the standard deviation. Show by an example that it is not necessarily true that the square of the spread of the sum of two independent random variables is the sum of the squares of the individual spreads.

20 We have two instruments that measure the distance between two points. The measurements given by the two instruments are random variables X_1 and X_2 that are independent with $E(X_1) = E(X_2) = \mu$, where μ is the true distance. From experience with these instruments, we know the values of the variances σ_1^2 and σ_2^2. These variances are not necessarily the same. From two measurements, we estimate μ by the weighted average $\bar{\mu} = wX_1 + (1-w)X_2$. Here w is chosen in $[0,1]$ to minimize the variance of $\bar{\mu}$.

(a) What is $E(\bar{\mu})$?

(b) How should w be chosen in $[0,1]$ to minimize the variance of $\bar{\mu}$?

21 Let X be a random variable with $E(X) = \mu$ and $V(X) = \sigma^2$. Show that the function $f(x)$ defined by

$$f(x) = \sum_{\omega} (X(\omega) - x)^2 p(\omega)$$

has its minimum value when $x = \mu$.

22 Let X and Y be two random variables defined on the finite sample space Ω. Assume that X, Y, $X+Y$, and $X-Y$ all have the same distribution. Prove that $P(X = Y = 0) = 1$.

6.2. VARIANCE OF DISCRETE RANDOM VARIABLES

23 If X and Y are any two random variables, then the *covariance* of X and Y is defined by $\text{Cov}(X, Y) = E((X - E(X))(Y - E(Y)))$. Note that $\text{Cov}(X, X) = V(X)$. Show that, if X and Y are independent, then $\text{Cov}(X, Y) = 0$; and show, by an example, that we can have $\text{Cov}(X, Y) = 0$ and X and Y not independent.

*24 A professor wishes to make up a true-false exam with n questions. She assumes that she can design the problems in such a way that a student will answer the jth problem correctly with probability p_j, and that the answers to the various problems may be considered independent experiments. Let S_n be the number of problems that a student will get correct. The professor wishes to choose p_j so that $E(S_n) = .7n$ and so that the variance of S_n is as large as possible. Show that, to achieve this, she should choose $p_j = .7$ for all j; that is, she should make all the problems have the same difficulty.

25 (Lamperti[20]) An urn contains exactly 5000 balls, of which an unknown number X are white and the rest red, where X is a random variable with a probability distribution on the integers 0, 1, 2, ..., 5000.

(a) Suppose we know that $E(X) = \mu$. Show that this is enough to allow us to calculate the probability that a ball drawn at random from the urn will be white. What is this probability?

(b) Suppose the variance of X is σ^2. What is the probability of drawing two white balls in part (b)?

26 We draw a ball from the urn, examine its color, replace it, and then draw another. Under what conditions, if any, are the results of the two drawings independent; that is, does

$$P(\text{white, white}) = P(\text{white})^2 \ ?$$

27 For a sequence of Bernoulli trials, let X_1 be the number of trials until the first success. For $j \geq 2$, let X_j be the number of trials after the $(j-1)$st success until the jth success. It can be shown that $X_1, X_2, \ldots$ is an independent trials process.

(a) What is the common distribution, expected value, and variance for X_j?

(b) Let $T_n = X_1 + X_2 + \cdots + X_n$. Then T_n is the time until the nth success. Find $E(T_n)$ and $V(T_n)$.

(c) Use the results of (b) to find the expected value and variance for the number of tosses of a coin until the nth occurrence of a head.

28 Referring to Exercise 6.1.30, find the variance for the number of boxes of Wheaties bought before getting half of the players' pictures and the variance for the number of additional boxes needed to get the second half of the players' pictures.

---

[20] Private communication.

29 In Example 5.3, assume that the book in question has 1000 pages. Let X be the number of pages with no mistakes. Show that $E(X) = 905$ and $V(X) = 86$. Using these results, show that the probability is $\leq .05$ that there will be more than 924 pages without errors or fewer than 866 pages without errors.

30 Let X be Poisson distributed with parameter λ. Show that $V(X) = \lambda$.

6.3 Continuous Random Variables

In this section we consider the properties of the expected value and the variance of a continuous random variable. These quantities are defined just as for discrete random variables and share the same properties.

Expected Value

Definition 6.4 Let X be a real-valued random variable with density function $f(x)$. The *expected value* $\mu = E(X)$ is defined by

$$\mu = E(X) = \int_{-\infty}^{+\infty} x f(x)\, dx\ ,$$

provided the integral

$$\int_{-\infty}^{+\infty} |x| f(x)\, dx$$

is finite. □

The reader should compare this definition with the corresponding one for discrete random variables in Section 6.1. Intuitively, we can interpret $E(X)$, as we did in the previous sections, as the value that we should expect to obtain if we perform a large number of independent experiments and average the resulting values of X.

We can summarize the properties of $E(X)$ as follows (cf. Theorem 6.2).

Theorem 6.10 If X and Y are real-valued random variables and c is any constant, then

$$\begin{aligned} E(X + Y) &= E(X) + E(Y)\ , \\ E(cX) &= cE(X)\ . \end{aligned}$$

The proof is very similar to the proof of Theorem 6.2, and we omit it. □

More generally, if $X_1, X_2, \ldots, X_n$ are n real-valued random variables, and $c_1, c_2, \ldots, c_n$ are n constants, then

$$E(c_1 X_1 + c_2 X_2 + \cdots + c_n X_n) = c_1 E(X_1) + c_2 E(X_2) + \cdots + c_n E(X_n)\ .$$

6.3. CONTINUOUS RANDOM VARIABLES

Example 6.20 Let X be uniformly distributed on the interval $[0, 1]$. Then

$$E(X) = \int_0^1 x \, dx = 1/2 \ .$$

It follows that if we choose a large number N of random numbers from $[0, 1]$ and take the average, then we can expect that this average should be close to the expected value of $1/2$. □

Example 6.21 Let $Z = (x, y)$ denote a point chosen uniformly and randomly from the unit disk, as in the dart game in Example 2.8 and let $X = (x^2 + y^2)^{1/2}$ be the distance from Z to the center of the disk. The density function of X can easily be shown to equal $f(x) = 2x$, so by the definition of expected value,

$$\begin{aligned} E(X) &= \int_0^1 x f(x) \, dx \\ &= \int_0^1 x(2x) \, dx \\ &= \frac{2}{3} \ . \end{aligned}$$

□

Example 6.22 In the example of the couple meeting at the Inn (Example 2.16), each person arrives at a time which is uniformly distributed between 5:00 and 6:00 PM. The random variable Z under consideration is the length of time the first person has to wait until the second one arrives. It was shown that

$$f_Z(z) = 2(1-z) \ ,$$

for $0 \leq z \leq 1$. Hence,

$$\begin{aligned} E(Z) &= \int_0^1 z f_Z(z) \, dz \\ &= \int_0^1 2z(1-z) \, dz \\ &= \left[z^2 - \frac{2}{3} z^3 \right]_0^1 \\ &= \frac{1}{3} \ . \end{aligned}$$

□

Expectation of a Function of a Random Variable

Suppose that X is a real-valued random variable and $\phi(x)$ is a continuous function from **R** to **R**. The following theorem is the continuous analogue of Theorem 6.1.

Theorem 6.11 If X is a real-valued random variable and if $\phi : \mathbf{R} \to \mathbf{R}$ is a continuous real-valued function with domain $[a, b]$, then

$$E(\phi(X)) = \int_{-\infty}^{+\infty} \phi(x) f_X(x)\, dx \; ,$$

provided the integral exists. □

For a proof of this theorem, see Ross.[21]

Expectation of the Product of Two Random Variables

In general, it is not true that $E(XY) = E(X)E(Y)$, since the integral of a product is not the product of integrals. But if X and Y are independent, then the expectations multiply.

Theorem 6.12 Let X and Y be independent real-valued continuous random variables with finite expected values. Then we have

$$E(XY) = E(X)E(Y) \; .$$

Proof. We will prove this only in the case that the ranges of X and Y are contained in the intervals $[a, b]$ and $[c, d]$, respectively. Let the density functions of X and Y be denoted by $f_X(x)$ and $f_Y(y)$, respectively. Since X and Y are independent, the joint density function of X and Y is the product of the individual density functions. Hence

$$\begin{aligned} E(XY) &= \int_a^b \int_c^d xy f_X(x) f_Y(y)\, dy\, dx \\ &= \int_a^b x f_X(x)\, dx \int_c^d y f_Y(y)\, dy \\ &= E(X)E(Y) \; . \end{aligned}$$

The proof in the general case involves using sequences of bounded random variables that approach X and Y, and is somewhat technical, so we will omit it. □

In the same way, one can show that if $X_1, X_2, \ldots, X_n$ are n mutually independent real-valued random variables, then

$$E(X_1 X_2 \cdots X_n) = E(X_1)\, E(X_2) \cdots E(X_n) \; .$$

Example 6.23 Let $Z = (X, Y)$ be a point chosen at random in the unit square. Let $A = X^2$ and $B = Y^2$. Then Theorem 4.3 implies that A and B are independent. Using Theorem 6.11, the expectations of A and B are easy to calculate:

$$\begin{aligned} E(A) = E(B) &= \int_0^1 x^2\, dx \\ &= \frac{1}{3} \; . \end{aligned}$$

---

[21]S. Ross, *A First Course in Probability*, (New York: Macmillan, 1984), pgs. 241-245.

6.3. CONTINUOUS RANDOM VARIABLES

Using Theorem 6.12, the expectation of AB is just the product of $E(A)$ and $E(B)$, or 1/9. The usefulness of this theorem is demonstrated by noting that it is quite a bit more difficult to calculate $E(AB)$ from the definition of expectation. One finds that the density function of AB is

$$f_{AB}(t) = \frac{-\log(t)}{4\sqrt{t}},$$

so

$$\begin{aligned} E(AB) &= \int_0^1 t f_{AB}(t)\, dt \\ &= \frac{1}{9}. \end{aligned}$$

$\square$

Example 6.24 Again let $Z = (X, Y)$ be a point chosen at random in the unit square, and let $W = X + Y$. Then Y and W are not independent, and we have

$$\begin{aligned} E(Y) &= \frac{1}{2}, \\ E(W) &= 1, \\ E(YW) &= E(XY + Y^2) = E(X)E(Y) + \frac{1}{3} = \frac{7}{12} \neq E(Y)E(W). \end{aligned}$$

$\square$

We turn now to the variance.

Variance

Definition 6.5 Let X be a real-valued random variable with density function $f(x)$. The *variance* $\sigma^2 = V(X)$ is defined by

$$\sigma^2 = V(X) = E((X - \mu)^2).$$

$\square$

The next result follows easily from Theorem 6.1. There is another way to calculate the variance of a continuous random variable, which is usually slightly easier. It is given in Theorem 6.15.

Theorem 6.13 If X is a real-valued random variable with $E(X) = \mu$, then

$$\sigma^2 = \int_{-\infty}^{\infty} (x - \mu)^2 f(x)\, dx.$$

$\square$

The properties listed in the next three theorems are all proved in exactly the same way that the corresponding theorems for discrete random variables were proved in Section 6.2.

Theorem 6.14 If X is a real-valued random variable defined on Ω and c is any constant, then (cf. Theorem 6.7)
$$\begin{aligned} V(cX) &= c^2 V(X) , \\ V(X+c) &= V(X) . \end{aligned}$$
$\square$

Theorem 6.15 If X is a real-valued random variable with $E(X) = \mu$, then (cf. Theorem 6.6)
$$V(X) = E(X^2) - \mu^2 .$$
$\square$

Theorem 6.16 If X and Y are independent real-valued random variables on Ω, then (cf. Theorem 6.8)
$$V(X+Y) = V(X) + V(Y) .$$
$\square$

Example 6.25 (continuation of Example 6.20) If X is uniformly distributed on $[0,1]$, then, using Theorem 6.15, we have
$$V(X) = \int_0^1 \left(x - \frac{1}{2}\right)^2 dx = \frac{1}{12} .$$
$\square$

Example 6.26 Let X be an exponentially distributed random variable with parameter λ. Then the density function of X is
$$f_X(x) = \lambda e^{-\lambda x} .$$

From the definition of expectation and integration by parts, we have
$$\begin{aligned} E(X) &= \int_0^\infty x f_X(x)\, dx \\ &= \lambda \int_0^\infty x e^{-\lambda x}\, dx \\ &= -x e^{-\lambda x} \Big|_0^\infty + \int_0^\infty e^{-\lambda x}\, dx \\ &= 0 + \frac{e^{-\lambda x}}{-\lambda} \Big|_0^\infty = \frac{1}{\lambda} . \end{aligned}$$

6.3. CONTINUOUS RANDOM VARIABLES

Similarly, using Theorems 6.11 and 6.15, we have

$$
\begin{aligned}
V(X) &= \int_0^\infty x^2 f_X(x)\,dx - \frac{1}{\lambda^2} \\
&= \lambda \int_0^\infty x^2 e^{-\lambda x}\,dx - \frac{1}{\lambda^2} \\
&= \left. -x^2 e^{-\lambda x} \right|_0^\infty + 2\int_0^\infty x e^{-\lambda x}\,dx - \frac{1}{\lambda^2} \\
&= \left. -x^2 e^{-\lambda x} \right|_0^\infty - \left.\frac{2x e^{-\lambda x}}{\lambda}\right|_0^\infty - \left.\frac{2}{\lambda^2} e^{-\lambda x}\right|_0^\infty - \frac{1}{\lambda^2} = \frac{2}{\lambda^2} - \frac{1}{\lambda^2} = \frac{1}{\lambda^2}\ .
\end{aligned}
$$

In this case, both $E(X)$ and $V(X)$ are finite if $\lambda > 0$. $\square$

Example 6.27 Let Z be a standard normal random variable with density function

$$f_Z(x) = \frac{1}{\sqrt{2\pi}} e^{-x^2/2}\ .$$

Since this density function is symmetric with respect to the y-axis, then it is easy to show that

$$\int_{-\infty}^\infty x f_Z(x)\,dx$$

has value 0. The reader should recall however, that the expectation is defined to be the above integral only if the integral

$$\int_{-\infty}^\infty |x| f_Z(x)\,dx$$

is finite. This integral equals

$$2\int_0^\infty x f_Z(x)\,dx\ ,$$

which one can easily show is finite. Thus, the expected value of Z is 0.

To calculate the variance of Z, we begin by applying Theorem 6.15:

$$V(Z) = \int_{-\infty}^{+\infty} x^2 f_Z(x)\,dx - \mu^2\ .$$

If we write x^2 as $x \cdot x$, and integrate by parts, we obtain

$$\frac{1}{\sqrt{2\pi}}\left(-x e^{-x^2/2}\right)\bigg|_{-\infty}^{+\infty} + \frac{1}{\sqrt{2\pi}}\int_{-\infty}^{+\infty} e^{-x^2/2}\,dx\ .$$

The first summand above can be shown to equal 0, since as $x \to \pm\infty$, $e^{-x^2/2}$ gets small more quickly than x gets large. The second summand is just the standard normal density integrated over its domain, so the value of this summand is 1. Therefore, the variance of the standard normal density equals 1.

Now let X be a (not necessarily standard) normal random variable with parameters μ and σ. Then the density function of X is

$$f_X(x) = \frac{1}{\sqrt{2\pi}\sigma} e^{-(x-\mu)^2/2\sigma^2} .$$

We can write $X = \sigma Z + \mu$, where Z is a standard normal random variable. Since $E(Z) = 0$ and $V(Z) = 1$ by the calculation above, Theorems 6.10 and 6.14 imply that

$$\begin{aligned} E(X) &= E(\sigma Z + \mu) = \mu , \\ V(X) &= V(\sigma Z + \mu) = \sigma^2 . \end{aligned}$$

$\square$

Example 6.28 Let X be a continuous random variable with the Cauchy density function

$$f_X(x) = \frac{a}{\pi} \frac{1}{a^2 + x^2} .$$

Then the expectation of X does not exist, because the integral

$$\frac{a}{\pi} \int_{-\infty}^{+\infty} \frac{|x|\, dx}{a^2 + x^2}$$

diverges. Thus the variance of X also fails to exist. Densities whose variance is not defined, like the Cauchy density, behave quite differently in a number of important respects from those whose variance is finite. We shall see one instance of this difference in Section 8.2. $\square$

Independent Trials

Corollary 6.1 If $X_1, X_2, \ldots, X_n$ is an independent trials process of real-valued random variables, with $E(X_i) = \mu$ and $V(X_i) = \sigma^2$, and if

$$\begin{aligned} S_n &= X_1 + X_2 + \cdots + X_n , \\ A_n &= \frac{S_n}{n} , \end{aligned}$$

then

$$\begin{aligned} E(S_n) &= n\mu , \\ E(A_n) &= \mu , \\ V(S_n) &= n\sigma^2 , \\ V(A_n) &= \frac{\sigma^2}{n} . \end{aligned}$$

It follows that if we set

$$S_n^* = \frac{S_n - n\mu}{\sqrt{n\sigma^2}} ,$$

6.3. CONTINUOUS RANDOM VARIABLES

then

$$E(S_n^*) = 0,$$
$$V(S_n^*) = 1.$$

We say that S_n^* is a *standardized version of* S_n (see Exercise 12 in Section 6.2). □

Queues

Example 6.29 Let us consider again the queueing problem, that is, the problem of the customers waiting in a queue for service (see Example 5.7). We suppose again that customers join the queue in such a way that the time between arrivals is an exponentially distributed random variable X with density function

$$f_X(t) = \lambda e^{-\lambda t}.$$

Then the expected value of the time between arrivals is simply $1/\lambda$ (see Example 6.26), as was stated in Example 5.7. The reciprocal λ of this expected value is often referred to as the *arrival rate*. The *service time* of an individual who is first in line is defined to be the amount of time that the person stays at the head of the line before leaving. We suppose that the customers are served in such a way that the service time is another exponentially distributed random variable Y with density function

$$f_X(t) = \mu e^{-\mu t}.$$

Then the expected value of the service time is

$$E(X) = \int_0^\infty t f_X(t)\, dt = \frac{1}{\mu}.$$

The reciprocal μ if this expected value is often referred to as the *service rate*.

We expect on grounds of our everyday experience with queues that if the service rate is greater than the arrival rate, then the average queue size will tend to stabilize, but if the service rate is less than the arrival rate, then the queue will tend to increase in length without limit (see Figure 5.7). The simulations in Example 5.7 tend to bear out our everyday experience. We can make this conclusion more precise if we introduce the *traffic intensity* as the product

$$\rho = \text{(arrival rate)(average service time)} = \frac{\lambda}{\mu} = \frac{1/\mu}{1/\lambda}.$$

The traffic intensity is also the ratio of the average service time to the average time between arrivals. If the traffic intensity is less than 1 the queue will perform reasonably, but if it is greater than 1 the queue will grow indefinitely large. In the critical case of $\rho = 1$, it can be shown that the queue will become large but there will always be times at which the queue is empty.[22]

---

[22]L. Kleinrock, *Queueing Systems*, vol. 2 (New York: John Wiley and Sons, 1975).

In the case that the traffic intensity is less than 1 we can consider the length of the queue as a random variable Z whose expected value is finite,

$$E(Z) = N \ .$$

The time spent in the queue by a single customer can be considered as a random variable W whose expected value is finite,

$$E(W) = T \ .$$

Then we can argue that, when a customer joins the queue, he expects to find N people ahead of him, and when he leaves the queue, he expects to find λT people behind him. Since, in equilibrium, these should be the same, we would expect to find that

$$N = \lambda T \ .$$

This last relationship is called *Little's law for queues*.[23] We will not prove it here. A proof may be found in Ross.[24] Note that in this case we are counting the waiting time of all customers, even those that do not have to wait at all. In our simulation in Section 4.2, we did not consider these customers.

If we knew the expected queue length then we could use Little's law to obtain the expected waiting time, since

$$T = \frac{N}{\lambda} \ .$$

The queue length is a random variable with a discrete distribution. We can estimate this distribution by simulation, keeping track of the queue lengths at the times at which a customer arrives. We show the result of this simulation (using the program **Queue**) in Figure 6.8.

We note that the distribution appears to be a geometric distribution. In the study of queueing theory it is shown that the distribution for the queue length in equilibrium is indeed a geometric distribution with

$$s_j = (1-\rho)\rho^j \qquad \text{for } j = 0, 1, 2, \ldots \ ,$$

if $\rho < 1$. The expected value of a random variable with this distribution is

$$N = \frac{\rho}{(1-\rho)}$$

(see Example 6.4). Thus by Little's result the expected waiting time is

$$T = \frac{\rho}{\lambda(1-\rho)} = \frac{1}{\mu - \lambda} \ ,$$

where μ is the service rate, λ the arrival rate, and ρ the traffic intensity.

---

[23]ibid., p. 17.
[24]S. M. Ross, *Applied Probability Models with Optimization Applications*, (San Francisco: Holden-Day, 1970)

6.3. CONTINUOUS RANDOM VARIABLES

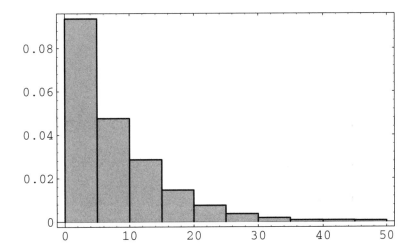

Figure 6.8: Distribution of queue lengths.

In our simulation, the arrival rate is 1 and the service rate is 1.1. Thus, the traffic intensity is $1/1.1 = 10/11$, the expected queue size is

$$\frac{10/11}{(1 - 10/11)} = 10 ,$$

and the expected waiting time is

$$\frac{1}{1.1 - 1} = 10 .$$

In our simulation the average queue size was 8.19 and the average waiting time was 7.37. In Figure 6.9, we show the histogram for the waiting times. This histogram suggests that the density for the waiting times is exponential with parameter $\mu - \lambda$, and this is the case. □

Exercises

1. Let X be a random variable with range $[-1, 1]$ and let $f_X(x)$ be the density function of X. Find $\mu(X)$ and $\sigma^2(X)$ if, for $|x| < 1$,

   (a) $f_X(x) = 1/2$.
   (b) $f_X(x) = |x|$.
   (c) $f_X(x) = 1 - |x|$.
   (d) $f_X(x) = (3/2)x^2$.

2. Let X be a random variable with range $[-1, 1]$ and f_X its density function. Find $\mu(X)$ and $\sigma^2(X)$ if, for $|x| > 1$, $f_X(x) = 0$, and for $|x| < 1$,

   (a) $f_X(x) = (3/4)(1 - x^2)$.

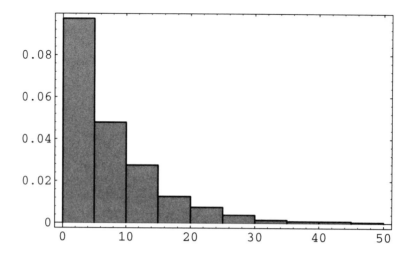

Figure 6.9: Distribution of queue waiting times.

(b) $f_X(x) = (\pi/4)\cos(\pi x/2)$.

(c) $f_X(x) = (x+1)/2$.

(d) $f_X(x) = (3/8)(x+1)^2$.

3 The lifetime, measure in hours, of the ACME super light bulb is a random variable T with density function $f_T(t) = \lambda^2 t e^{-\lambda t}$, where $\lambda = .05$. What is the expected lifetime of this light bulb? What is its variance?

4 Let X be a random variable with range $[-1,1]$ and density function $f_X(x) = ax + b$ if $|x| < 1$.

   (a) Show that if $\int_{-1}^{+1} f_X(x)\,dx = 1$, then $b = 1/2$.
   (b) Show that if $f_X(x) \geq 0$, then $-1/2 \leq a \leq 1/2$.
   (c) Show that $\mu = (2/3)a$, and hence that $-1/3 \leq \mu \leq 1/3$.
   (d) Show that $\sigma^2(X) = (2/3)b - (4/9)a^2 = 1/3 - (4/9)a^2$.

5 Let X be a random variable with range $[-1,1]$ and density function $f_X(x) = ax^2 + bx + c$ if $|x| < 1$ and 0 otherwise.

   (a) Show that $2a/3 + 2c = 1$ (see Exercise 4).
   (b) Show that $2b/3 = \mu(X)$.
   (c) Show that $2a/5 + 2c/3 = \sigma^2(X)$.
   (d) Find a, b, and c if $\mu(X) = 0$, $\sigma^2(X) = 1/15$, and sketch the graph of f_X.
   (e) Find a, b, and c if $\mu(X) = 0$, $\sigma^2(X) = 1/2$, and sketch the graph of f_X.

6 Let T be a random variable with range $[0, \infty]$ and f_T its density function. Find $\mu(T)$ and $\sigma^2(T)$ if, for $t < 0$, $f_T(t) = 0$, and for $t > 0$,

6.3. CONTINUOUS RANDOM VARIABLES

(a) $f_T(t) = 3e^{-3t}$.

(b) $f_T(t) = 9te^{-3t}/2$.

(c) $f_T(t) = 3/(1+t)^4$.

7 Let X be a random variable with density function f_X. Show, using elementary calculus, that the function

$$\phi(a) = E((X-a)^2)$$

takes its minimum value when $a = \mu(X)$, and in that case $\phi(a) = \sigma^2(X)$.

8 Let X be a random variable with mean μ and variance σ^2. Let $Y = aX^2 + bX + c$. Find the expected value of Y.

9 Let X, Y, and Z be independent random variables, each with mean μ and variance σ^2.

(a) Find the expected value and variance of $S = X + Y + Z$.

(b) Find the expected value and variance of $A = (1/3)(X + Y + Z)$.

(c) Find the expected value of S^2 and A^2.

10 Let X and Y be independent random variables with uniform density functions on $[0, 1]$. Find

(a) $E(|X - Y|)$.

(b) $E(\max(X, Y))$.

(c) $E(\min(X, Y))$.

(d) $E(X^2 + Y^2)$.

(e) $E((X + Y)^2)$.

11 The Pilsdorff Beer Company runs a fleet of trucks along the 100 mile road from Hangtown to Dry Gulch. The trucks are old, and are apt to break down at any point along the road with equal probability. Where should the company locate a garage so as to minimize the expected distance from a typical breakdown to the garage? In other words, if X is a random variable giving the location of the breakdown, measured, say, from Hangtown, and b gives the location of the garage, what choice of b minimizes $E(|X - b|)$? Now suppose X is not distributed uniformly over $[0, 100]$, but instead has density function $f_X(x) = 2x/10{,}000$. Then what choice of b minimizes $E(|X - b|)$?

12 Find $E(X^Y)$, where X and Y are independent random variables which are uniform on $[0, 1]$. Then verify your answer by simulation.

13 Let X be a random variable that takes on nonnegative values and has distribution function $F(x)$. Show that

$$E(X) = \int_0^\infty (1 - F(x))\, dx \ .$$

Hint: Integrate by parts.

Illustrate this result by calculating $E(X)$ by this method if X has an exponential distribution $F(x) = 1 - e^{-\lambda x}$ for $x \geq 0$, and $F(x) = 0$ otherwise.

14 Let X be a continuous random variable with density function $f_X(x)$. Show that if
$$\int_{-\infty}^{+\infty} x^2 f_X(x)\, dx < \infty\ ,$$
then
$$\int_{-\infty}^{+\infty} |x| f_X(x)\, dx < \infty\ .$$

Hint: Except on the interval $[-1, 1]$, the first integrand is greater than the second integrand.

15 Let X be a random variable distributed uniformly over $[0, 20]$. Define a new random variable Y by $Y = \lfloor X \rfloor$ (the greatest integer in X). Find the expected value of Y. Do the same for $Z = \lfloor X + .5 \rfloor$. Compute $E(|X - Y|)$ and $E(|X - Z|)$. (Note that Y is the value of X rounded off to the nearest smallest integer, while Z is the value of X rounded off to the nearest integer. Which method of rounding off is better? Why?)

16 Assume that the lifetime of a diesel engine part is a random variable X with density f_X. When the part wears out, it is replaced by another with the same density. Let $N(t)$ be the number of parts that are used in time t. We want to study the random variable $N(t)/t$. Since parts are replaced on the average every $E(X)$ time units, we expect about $t/E(X)$ parts to be used in time t. That is, we expect that
$$\lim_{t \to \infty} E\left(\frac{N(t)}{t}\right) = \frac{1}{E(X)}\ .$$

This result is correct but quite difficult to prove. Write a program that will allow you to specify the density f_X, and the time t, and simulate this experiment to find $N(t)/t$. Have your program repeat the experiment 500 times and plot a bar graph for the random outcomes of $N(t)/t$. From this data, estimate $E(N(t)/t)$ and compare this with $1/E(X)$. In particular, do this for $t = 100$ with the following two densities:

(a) $f_X = e^{-t}$.

(b) $f_X = te^{-t}$.

17 Let X and Y be random variables. The *covariance* Cov(X, Y) is defined by (see Exercise 6.2.23)
$$\operatorname{cov}(X, Y) = E((X - \mu(X))(Y - \mu(Y)))\ .$$

(a) Show that $\operatorname{cov}(X, Y) = E(XY) - E(X)E(Y)$.

(b) Using (a), show that $\text{cov}(X,Y) = 0$, if X and Y are independent. (Caution: the converse is *not* always true.)

(c) Show that $V(X+Y) = V(X) + V(Y) + 2\text{cov}(X,Y)$.

18 Let X and Y be random variables with positive variance. The *correlation* of X and Y is defined as

$$\rho(X,Y) = \frac{\text{cov}(X,Y)}{\sqrt{V(X)V(Y)}} \ .$$

(a) Using Exercise 17(c), show that

$$0 \le V\left(\frac{X}{\sigma(X)} + \frac{Y}{\sigma(Y)}\right) = 2(1 + \rho(X,Y)) \ .$$

(b) Now show that

$$0 \le V\left(\frac{X}{\sigma(X)} - \frac{Y}{\sigma(Y)}\right) = 2(1 - \rho(X,Y)) \ .$$

(c) Using (a) and (b), show that

$$-1 \le \rho(X,Y) \le 1 \ .$$

19 Let X and Y be independent random variables with uniform densities in $[0,1]$. Let $Z = X + Y$ and $W = X - Y$. Find

(a) $\rho(X,Y)$ (see Exercise 18).

(b) $\rho(X,Z)$.

(c) $\rho(Y,W)$.

(d) $\rho(Z,W)$.

*20 When studying certain physiological data, such as heights of fathers and sons, it is often natural to assume that these data (e.g., the heights of the fathers and the heights of the sons) are described by random variables with normal densities. These random variables, however, are not independent but rather are correlated. For example, a two-dimensional standard normal density for correlated random variables has the form

$$f_{X,Y}(x,y) = \frac{1}{2\pi\sqrt{1-\rho^2}} \cdot e^{-(x^2 - 2\rho xy + y^2)/2(1-\rho^2)} \ .$$

(a) Show that X and Y each have standard normal densities.

(b) Show that the correlation of X and Y (see Exercise 18) is ρ.

*21 For correlated random variables X and Y it is natural to ask for the expected value for X given Y. For example, Galton calculated the expected value of the height of a son given the height of the father. He used this to show that tall men can be expected to have sons who are less tall on the average. Similarly, students who do very well on one exam can be expected to do less well on the next exam, and so forth. This is called *regression on the mean*. To define this conditional expected value, we first define a conditional density of X given $Y = y$ by

$$f_{X|Y}(x|y) = \frac{f_{X,Y}(x,y)}{f_Y(y)},$$

where $f_{X,Y}(x,y)$ is the joint density of X and Y, and f_Y is the density for Y. Then the conditional expected value of X given Y is

$$E(X|Y = y) = \int_a^b x f_{X|Y}(x|y)\, dx.$$

For the normal density in Exercise 20, show that the conditional density of $f_{X|Y}(x|y)$ is normal with mean ρy and variance $1 - \rho^2$. From this we see that if X and Y are positively correlated ($0 < \rho < 1$), and if $y > E(Y)$, then the expected value for X given $Y = y$ will be less than y (i.e., we have regression on the mean).

22 A point Y is chosen at random from $[0, 1]$. A second point X is then chosen from the interval $[0, Y]$. Find the density for X. *Hint*: Calculate $f_{X|Y}$ as in Exercise 21 and then use

$$f_X(x) = \int_x^1 f_{X|Y}(x|y) f_Y(y)\, dy.$$

Can you also derive your result geometrically?

*23 Let X and V be two standard normal random variables. Let ρ be a real number between -1 and 1.

(a) Let $Y = \rho X + \sqrt{1 - \rho^2} V$. Show that $E(Y) = 0$ and $Var(Y) = 1$. We shall see later (see Example 7.5 and Example 10.17), that the sum of two independent normal random variables is again normal. Thus, assuming this fact, we have shown that Y is standard normal.

(b) Using Exercises 17 and 18, show that the correlation of X and Y is ρ.

(c) In Exercise 20, the joint density function $f_{X,Y}(x,y)$ for the random variable (X, Y) is given. Now suppose that we want to know the set of points (x, y) in the xy-plane such that $f_{X,Y}(x, y) = C$ for some constant C. This set of points is called a set of constant density. Roughly speaking, a set of constant density is a set of points where the outcomes (X, Y) are equally likely to fall. Show that for a given C, the set of points of constant density is a curve whose equation is

$$x^2 - 2\rho xy + y^2 = D,$$

where D is a constant which depends upon C. (This curve is an ellipse.)

6.3. CONTINUOUS RANDOM VARIABLES

(d) One can plot the ellipse in part (c) by using the parametric equations

$$x = \frac{r\cos\theta}{\sqrt{2(1-\rho)}} + \frac{r\sin\theta}{\sqrt{2(1+\rho)}},$$

$$y = \frac{r\cos\theta}{\sqrt{2(1-\rho)}} - \frac{r\sin\theta}{\sqrt{2(1+\rho)}}.$$

Write a program to plot 1000 pairs (X, Y) for $\rho = -1/2, 0, 1/2$. For each plot, have your program plot the above parametric curves for $r = 1, 2, 3$.

*24 Following Galton, let us assume that the fathers and sons have heights that are dependent normal random variables. Assume that the average height is 68 inches, standard deviation is 2.7 inches, and the correlation coefficient is .5 (see Exercises 20 and 21). That is, assume that the heights of the fathers and sons have the form $2.7X + 68$ and $2.7Y + 68$, respectively, where X and Y are correlated standardized normal random variables, with correlation coefficient .5.

(a) What is the expected height for the son of a father whose height is 72 inches?

(b) Plot a scatter diagram of the heights of 1000 father and son pairs. *Hint*: You can choose standardized pairs as in Exercise 23 and then plot $(2.7X + 68, 2.7Y + 68)$.

*25 When we have pairs of data (x_i, y_i) that are outcomes of the pairs of dependent random variables X, Y we can estimate the coorelation coefficient ρ by

$$\bar{r} = \frac{\sum_i (x_i - \bar{x})(y_i - \bar{y})}{(n-1)s_X s_Y},$$

where $\bar{x}$ and $\bar{y}$ are the sample means for X and Y, respectively, and s_X and s_Y are the sample standard deviations for X and Y (see Exercise 6.2.17). Write a program to compute the sample means, variances, and correlation for such dependent data. Use your program to compute these quantities for Galton's data on heights of parents and children given in Appendix B.

Plot the equal density ellipses as defined in Exercise 23 for $r = 4, 6$, and 8, and on the same graph print the values that appear in the table at the appropriate points. For example, print 12 at the point $(70.5, 68.2)$, indicating that there were 12 cases where the parent's height was 70.5 and the child's was 68.12. See if Galton's data is consistent with the equal density ellipses.

26 (from Hamming[25]) Suppose you are standing on the bank of a straight river.

(a) Choose, at random, a direction which will keep you on dry land, and walk 1 km in that direction. Let P denote your position. What is the expected distance from P to the river?

---

[25] R. W. Hamming, *The Art of Probability for Scientists and Engineers* (Redwood City: Addison-Wesley, 1991), p. 192.

(b) Now suppose you proceed as in part (a), but when you get to P, you pick a random direction (from among *all* directions) and walk 1 km. What is the probability that you will reach the river before the second walk is completed?

27 (from Hamming[26]) A game is played as follows: A random number X is chosen uniformly from $[0,1]$. Then a sequence $Y_1, Y_2, \ldots$ of random numbers is chosen independently and uniformly from $[0,1]$. The game ends the first time that $Y_i > X$. You are then paid $(i-1)$ dollars. What is a fair entrance fee for this game?

28 A long needle of length L much bigger than 1 is dropped on a grid with horizontal and vertical lines one unit apart. Show that the average number a of lines crossed is approximately

$$a = \frac{4L}{\pi}.$$

---

[26]ibid., pg. 205.

Chapter 7

Sums of Independent Random Variables

7.1 Sums of Discrete Random Variables

In this chapter we turn to the important question of determining the distribution of a sum of independent random variables in terms of the distributions of the individual constituents. In this section we consider only sums of discrete random variables, reserving the case of continuous random variables for the next section.

We consider here only random variables whose values are integers. Their distribution functions are then defined on these integers. We shall find it convenient to assume here that these distribution functions are defined for *all* integers, by defining them to be 0 where they are not otherwise defined.

Convolutions

Suppose X and Y are two independent discrete random variables with distribution functions $m_1(x)$ and $m_2(x)$. Let $Z = X + Y$. We would like to determine the distribution function $m_3(x)$ of Z. To do this, it is enough to determine the probability that Z takes on the value z, where z is an arbitrary integer. Suppose that $X = k$, where k is some integer. Then $Z = z$ if and only if $Y = z - k$. So the event $Z = z$ is the union of the pairwise disjoint events

$$(X = k) \text{ and } (Y = z - k),$$

where k runs over the integers. Since these events are pairwise disjoint, we have

$$P(Z = z) = \sum_{k=-\infty}^{\infty} P(X = k) \cdot P(Y = z - k).$$

Thus, we have found the distribution function of the random variable Z. This leads to the following definition.

Definition 7.1 Let X and Y be two independent integer-valued random variables, with distribution functions $m_1(x)$ and $m_2(x)$ respectively. Then the *convolution* of $m_1(x)$ and $m_2(x)$ is the distribution function $m_3 = m_1 * m_2$ given by

$$m_3(j) = \sum_k m_1(k) \cdot m_2(j-k) \;,$$

for $j = \ldots, -2, -1, 0, 1, 2, \ldots$. The function $m_3(x)$ is the distribution function of the random variable $Z = X + Y$. □

It is easy to see that the convolution operation is commutative, and it is straightforward to show that it is also associative.

Now let $S_n = X_1 + X_2 + \cdots + X_n$ be the sum of n independent random variables of an independent trials process with common distribution function m defined on the integers. Then the distribution function of S_1 is m. We can write

$$S_n = S_{n-1} + X_n \;.$$

Thus, since we know the distribution function of X_n is m, we can find the distribution function of S_n by induction.

Example 7.1 A die is rolled twice. Let X_1 and X_2 be the outcomes, and let $S_2 = X_1 + X_2$ be the sum of these outcomes. Then X_1 and X_2 have the common distribution function:

$$m = \begin{pmatrix} 1 & 2 & 3 & 4 & 5 & 6 \\ 1/6 & 1/6 & 1/6 & 1/6 & 1/6 & 1/6 \end{pmatrix}.$$

The distribution function of S_2 is then the convolution of this distribution with itself. Thus,

$$\begin{aligned}
P(S_2 = 2) &= m(1)m(1) \\
&= \frac{1}{6} \cdot \frac{1}{6} = \frac{1}{36}, \\
P(S_2 = 3) &= m(1)m(2) + m(2)m(1) \\
&= \frac{1}{6} \cdot \frac{1}{6} + \frac{1}{6} \cdot \frac{1}{6} = \frac{2}{36}, \\
P(S_2 = 4) &= m(1)m(3) + m(2)m(2) + m(3)m(1) \\
&= \frac{1}{6} \cdot \frac{1}{6} + \frac{1}{6} \cdot \frac{1}{6} + \frac{1}{6} \cdot \frac{1}{6} = \frac{3}{36}.
\end{aligned}$$

Continuing in this way we would find $P(S_2 = 5) = 4/36$, $P(S_2 = 6) = 5/36$, $P(S_2 = 7) = 6/36$, $P(S_2 = 8) = 5/36$, $P(S_2 = 9) = 4/36$, $P(S_2 = 10) = 3/36$, $P(S_2 = 11) = 2/36$, and $P(S_2 = 12) = 1/36$.

The distribution for S_3 would then be the convolution of the distribution for S_2 with the distribution for X_3. Thus

$$P(S_3 = 3) = P(S_2 = 2)P(X_3 = 1)$$

7.1. SUMS OF DISCRETE RANDOM VARIABLES

$$= \frac{1}{36} \cdot \frac{1}{6} = \frac{1}{216} ,$$
$$P(S_3 = 4) = P(S_2 = 3)P(X_3 = 1) + P(S_2 = 2)P(X_3 = 2)$$
$$= \frac{2}{36} \cdot \frac{1}{6} + \frac{1}{36} \cdot \frac{1}{6} = \frac{3}{216} ,$$

and so forth.

This is clearly a tedious job, and a program should be written to carry out this calculation. To do this we first write a program to form the convolution of two densities p and q and return the density r. We can then write a program to find the density for the sum S_n of n independent random variables with a common density p, at least in the case that the random variables have a finite number of possible values.

Running this program for the example of rolling a die n times for $n = 10$, 20, 30 results in the distributions shown in Figure 7.1. We see that, as in the case of Bernoulli trials, the distributions become bell-shaped. We shall discuss in Chapter 9 a very general theorem called the *Central Limit Theorem* that will explain this phenomenon. □

Example 7.2 A well-known method for evaluating a bridge hand is: an ace is assigned a value of 4, a king 3, a queen 2, and a jack 1. All other cards are assigned a value of 0. The *point count* of the hand is then the sum of the values of the cards in the hand. (It is actually more complicated than this, taking into account voids in suits, and so forth, but we consider here this simplified form of the point count.) If a card is dealt at random to a player, then the point count for this card has distribution

$$p_X = \begin{pmatrix} 0 & 1 & 2 & 3 & 4 \\ 36/52 & 4/52 & 4/52 & 4/52 & 4/52 \end{pmatrix} .$$

Let us regard the total hand of 13 cards as 13 independent trials with this common distribution. (Again this is not quite correct because we assume here that we are always choosing a card from a full deck.) Then the distribution for the point count C for the hand can be found from the program **NFoldConvolution** by using the distribution for a single card and choosing $n = 13$. A player with a point count of 13 or more is said to have an *opening bid*. The probability of having an opening bid is then

$$P(C \geq 13) .$$

Since we have the distribution of C, it is easy to compute this probability. Doing this we find that

$$P(C \geq 13) = .2845 ,$$

so that about one in four hands should be an opening bid according to this simplified model. A more realistic discussion of this problem can be found in Epstein, *The Theory of Gambling and Statistical Logic*.[1] □

---
[1] R. A. Epstein, *The Theory of Gambling and Statistical Logic*, rev. ed. (New York: Academic Press, 1977).

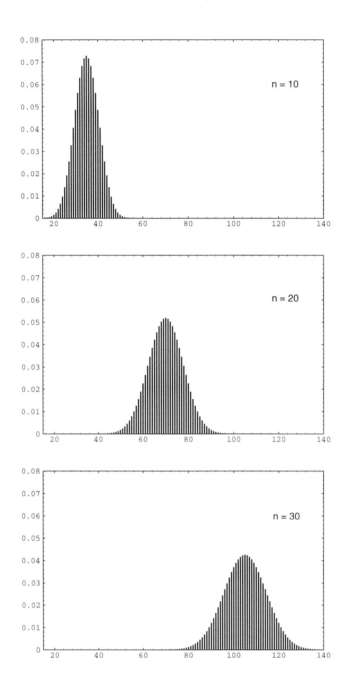

Figure 7.1: Density of S_n for rolling a die n times.

7.1. SUMS OF DISCRETE RANDOM VARIABLES

For certain special distributions it is possible to find an expression for the distribution that results from convoluting the distribution with itself n times.

The convolution of two binomial distributions, one with parameters m and p and the other with parameters n and p, is a binomial distribution with parameters $(m+n)$ and p. This fact follows easily from a consideration of the experiment which consists of first tossing a coin m times, and then tossing it n more times.

The convolution of k geometric distributions with common parameter p is a negative binomial distribution with parameters p and k. This can be seen by considering the experiment which consists of tossing a coin until the kth head appears.

Exercises

1 A die is rolled three times. Find the probability that the sum of the outcomes is

(a) greater than 9.

(b) an odd number.

2 The price of a stock on a given trading day changes according to the distribution
$$p_X = \begin{pmatrix} -1 & 0 & 1 & 2 \\ 1/4 & 1/2 & 1/8 & 1/8 \end{pmatrix}.$$

Find the distribution for the change in stock price after two (independent) trading days.

3 Let X_1 and X_2 be independent random variables with common distribution
$$p_X = \begin{pmatrix} 0 & 1 & 2 \\ 1/8 & 3/8 & 1/2 \end{pmatrix}.$$

Find the distribution of the sum $X_1 + X_2$.

4 In one play of a certain game you win an amount X with distribution
$$p_X = \begin{pmatrix} 1 & 2 & 3 \\ 1/4 & 1/4 & 1/2 \end{pmatrix}.$$

Using the program **NFoldConvolution** find the distribution for your total winnings after ten (independent) plays. Plot this distribution.

5 Consider the following two experiments: the first has outcome X taking on the values 0, 1, and 2 with equal probabilities; the second results in an (independent) outcome Y taking on the value 3 with probability 1/4 and 4 with probability 3/4. Find the distribution of

(a) $Y + X$.

(b) $Y - X$.

6 People arrive at a queue according to the following scheme: During each minute of time either 0 or 1 person arrives. The probability that 1 person arrives is p and that no person arrives is $q = 1 - p$. Let C_r be the number of customers arriving in the first r minutes. Consider a Bernoulli trials process with a success if a person arrives in a unit time and failure if no person arrives in a unit time. Let T_r be the number of failures before the rth success.

(a) What is the distribution for T_r?

(b) What is the distribution for C_r?

(c) Find the mean and variance for the number of customers arriving in the first r minutes.

7 (a) A die is rolled three times with outcomes X_1, X_2, and X_3. Let Y_3 be the maximum of the values obtained. Show that
$$P(Y_3 \le j) = P(X_1 \le j)^3 \ .$$
Use this to find the distribution of Y_3. Does Y_3 have a bell-shaped distribution?

(b) Now let Y_n be the maximum value when n dice are rolled. Find the distribution of Y_n. Is this distribution bell-shaped for large values of n?

8 A baseball player is to play in the World Series. Based upon his season play, you estimate that if he comes to bat four times in a game the number of hits he will get has a distribution
$$p_X = \begin{pmatrix} 0 & 1 & 2 & 3 & 4 \\ .4 & .2 & .2 & .1 & .1 \end{pmatrix} .$$
Assume that the player comes to bat four times in each game of the series.

(a) Let X denote the number of hits that he gets in a series. Using the program **NFoldConvolution**, find the distribution of X for each of the possible series lengths: four-game, five-game, six-game, seven-game.

(b) Using one of the distribution found in part (a), find the probability that his batting average exceeds .400 in a four-game series. (The batting average is the number of hits divided by the number of times at bat.)

(c) Given the distribution p_X, what is his long-term batting average?

9 Prove that you cannot load two dice in such a way that the probabilities for any sum from 2 to 12 are the same. (Be sure to consider the case where one or more sides turn up with probability zero.)

10 (Lévy[2]) Assume that n is an integer, not prime. Show that you can find two distributions a and b on the nonnegative integers such that the convolution of

---

[2]See M. Krasner and B. Ranulae, "Sur une Proprieté des Polynomes de la Division du Circle"; and the following note by J. Hadamard, in *C. R. Acad. Sci.*, vol. 204 (1937), pp. 397–399.

7.2. SUMS OF CONTINUOUS RANDOM VARIABLES

a and b is the equiprobable distribution on the set 0, 1, 2, ..., $n-1$. If n is prime this is not possible, but the proof is not so easy. (Assume that neither a nor b is concentrated at 0.)

11 Assume that you are playing craps with dice that are loaded in the following way: faces two, three, four, and five all come up with the same probability $(1/6) + r$. Faces one and six come up with probability $(1/6) - 2r$, with $0 < r < .02$. Write a computer program to find the probability of winning at craps with these dice, and using your program find which values of r make craps a favorable game for the player with these dice.

7.2 Sums of Continuous Random Variables

In this section we consider the continuous version of the problem posed in the previous section: How are sums of independent random variables distributed?

Convolutions

Definition 7.2 Let X and Y be two continuous random variables with density functions $f(x)$ and $g(y)$, respectively. Assume that both $f(x)$ and $g(y)$ are defined for all real numbers. Then the *convolution* $f * g$ of f and g is the function given by

$$(f * g)(z) = \int_{-\infty}^{+\infty} f(z-y)g(y)\, dy$$
$$= \int_{-\infty}^{+\infty} g(z-x)f(x)\, dx \ .$$

$\square$

This definition is analogous to the definition, given in Section 7.1, of the convolution of two distribution functions. Thus it should not be surprising that if X and Y are independent, then the density of their sum is the convolution of their densities. This fact is stated as a theorem below, and its proof is left as an exercise (see Exercise 1).

Theorem 7.1 Let X and Y be two independent random variables with density functions $f_X(x)$ and $f_Y(y)$ defined for all x. Then the sum $Z = X + Y$ is a random variable with density function $f_Z(z)$, where f_Z is the convolution of f_X and f_Y. $\square$

To get a better understanding of this important result, we will look at some examples.

Sum of Two Independent Uniform Random Variables

Example 7.3 Suppose we choose independently two numbers at random from the interval $[0, 1]$ with uniform probability density. What is the density of their sum?

Let X and Y be random variables describing our choices and $Z = X + Y$ their sum. Then we have

$$f_X(x) = f_Y(x) = \begin{cases} 1 & \text{if } 0 \le x \le 1, \\ 0 & \text{otherwise;} \end{cases}$$

and the density function for the sum is given by

$$f_Z(z) = \int_{-\infty}^{+\infty} f_X(z-y) f_Y(y)\, dy\ .$$

Since $f_Y(y) = 1$ if $0 \le y \le 1$ and 0 otherwise, this becomes

$$f_Z(z) = \int_0^1 f_X(z-y)\, dy\ .$$

Now the integrand is 0 unless $0 \le z - y \le 1$ (i.e., unless $z - 1 \le y \le z$) and then it is 1. So if $0 \le z \le 1$, we have

$$f_Z(z) = \int_0^z dy = z\ ,$$

while if $1 < z \le 2$, we have

$$f_Z(z) = \int_{z-1}^1 dy = 2 - z\ ,$$

and if $z < 0$ or $z > 2$ we have $f_Z(z) = 0$ (see Figure 7.2). Hence,

$$f_Z(z) = \begin{cases} z, & \text{if } 0 \le z \le 1, \\ 2 - z, & \text{if } 1 < z \le 2, \\ 0, & \text{otherwise.} \end{cases}$$

Note that this result agrees with that of Example 2.4. $\square$

Sum of Two Independent Exponential Random Variables

Example 7.4 Suppose we choose two numbers at random from the interval $[0, \infty)$ with an *exponential* density with parameter λ. What is the density of their sum?

Let X, Y, and $Z = X + Y$ denote the relevant random variables, and f_X, f_Y, and f_Z their densities. Then

$$f_X(x) = f_Y(x) = \begin{cases} \lambda e^{-\lambda x}, & \text{if } x \ge 0, \\ 0, & \text{otherwise;} \end{cases}$$

7.2. SUMS OF CONTINUOUS RANDOM VARIABLES

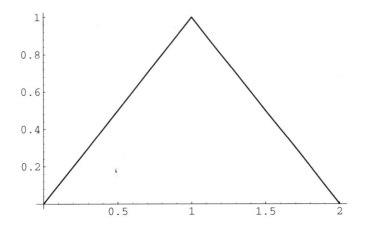

Figure 7.2: Convolution of two uniform densities.

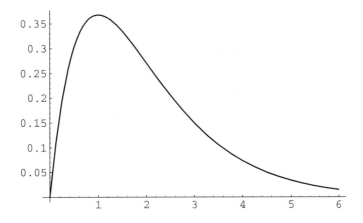

Figure 7.3: Convolution of two exponential densities with $\lambda = 1$.

and so, if $z > 0$,

$$\begin{aligned}
f_Z(z) &= \int_{-\infty}^{+\infty} f_X(z-y) f_Y(y)\, dy \\
&= \int_0^z \lambda e^{-\lambda(z-y)} \lambda e^{-\lambda y}\, dy \\
&= \int_0^z \lambda^2 e^{-\lambda z}\, dy \\
&= \lambda^2 z e^{-\lambda z},
\end{aligned}$$

while if $z < 0$, $f_Z(z) = 0$ (see Figure 7.3). Hence,

$$f_Z(z) = \begin{cases} \lambda^2 z e^{-\lambda z}, & \text{if } z \geq 0, \\ 0, & \text{otherwise.} \end{cases}$$

□

Sum of Two Independent Normal Random Variables

Example 7.5 It is an interesting and important fact that the convolution of two normal densities with means μ_1 and μ_2 and variances σ_1 and σ_2 is again a normal density, with mean $\mu_1 + \mu_2$ and variance $\sigma_1^2 + \sigma_2^2$. We will show this in the special case that both random variables are standard normal. The general case can be done in the same way, but the calculation is messier. Another way to show the general result is given in Example 10.17.

Suppose X and Y are two independent random variables, each with the standard *normal* density (see Example 5.8). We have

$$f_X(x) = f_Y(y) = \frac{1}{\sqrt{2\pi}} e^{-x^2/2},$$

and so

$$\begin{aligned} f_Z(z) &= f_X * f_Y(z) \\ &= \frac{1}{2\pi} \int_{-\infty}^{+\infty} e^{-(z-y)^2/2} e^{-y^2/2} \, dy \\ &= \frac{1}{2\pi} e^{-z^2/4} \int_{-\infty}^{+\infty} e^{-(y-z/2)^2} \, dy \\ &= \frac{1}{2\pi} e^{-z^2/4} \sqrt{\pi} \left[\frac{1}{\sqrt{\pi}} \int_{-\infty}^{\infty} e^{-(y-z/2)^2} \, dy \right] . \end{aligned}$$

The expression in the brackets equals 1, since it is the integral of the normal density function with $\mu = 0$ and $\sigma = \sqrt{2}$. So, we have

$$f_Z(z) = \frac{1}{\sqrt{4\pi}} e^{-z^2/4} .$$

$\square$

Sum of Two Independent Cauchy Random Variables

Example 7.6 Choose two numbers at random from the interval $(-\infty, +\infty)$ with the Cauchy density with parameter $a = 1$ (see Example 5.10). Then

$$f_X(x) = f_Y(x) = \frac{1}{\pi(1 + x^2)},$$

and $Z = X + Y$ has density

$$f_Z(z) = \frac{1}{\pi^2} \int_{-\infty}^{+\infty} \frac{1}{1 + (z-y)^2} \frac{1}{1 + y^2} \, dy .$$

7.2. SUMS OF CONTINUOUS RANDOM VARIABLES

This integral requires some effort, and we give here only the result (see Section 10.3, or Dwass[3]):

$$f_Z(z) = \frac{2}{\pi(4+z^2)}.$$

Now, suppose that we ask for the density function of the *average*

$$A = (1/2)(X+Y)$$

of X and Y. Then $A = (1/2)Z$. Exercise 5.2.19 shows that if U and V are two continuous random variables with density functions $f_U(x)$ and $f_V(x)$, respectively, and if $V = aU$, then

$$f_V(x) = \left(\frac{1}{a}\right) f_U\left(\frac{x}{a}\right).$$

Thus, we have

$$f_A(z) = 2f_Z(2z) = \frac{1}{\pi(1+z^2)}.$$

Hence, the density function for the average of two random variables, each having a Cauchy density, is again a random variable with a Cauchy density; this remarkable property is a peculiarity of the Cauchy density. One consequence of this is if the error in a certain measurement process had a Cauchy density and you averaged a number of measurements, the average could not be expected to be any more accurate than any one of your individual measurements! □

Rayleigh Density

Example 7.7 Suppose X and Y are two independent standard normal random variables. Now suppose we locate a point P in the xy-plane with coordinates (X, Y) and ask: What is the density of the square of the distance of P from the origin? (We have already simulated this problem in Example 5.9.) Here, with the preceding notation, we have

$$f_X(x) = f_Y(x) = \frac{1}{\sqrt{2\pi}} e^{-x^2/2}.$$

Moreover, if X^2 denotes the square of X, then (see Theorem 5.1 and the discussion following)

$$f_{X^2}(r) = \begin{cases} \frac{1}{2\sqrt{r}}(f_X(\sqrt{r}) + f_X(-\sqrt{r})) & \text{if } r > 0, \\ 0 & \text{otherwise.} \end{cases}$$

$$= \begin{cases} \frac{1}{\sqrt{2\pi r}}(e^{-r/2}) & \text{if } r > 0, \\ 0 & \text{otherwise.} \end{cases}$$

---

[3]M. Dwass, "On the Convolution of Cauchy Distributions," *American Mathematical Monthly*, vol. 92, no. 1, (1985), pp. 55–57; see also R. Nelson, letters to the Editor, ibid., p. 679.

This is a gamma density with $\lambda = 1/2$, $\beta = 1/2$ (see Example 7.4). Now let $R^2 = X^2 + Y^2$. Then

$$\begin{aligned}
f_{R^2}(r) &= \int_{-\infty}^{+\infty} f_{X^2}(r-s) f_{Y^2}(s)\, ds \\
&= \frac{1}{4\pi} \int_{-\infty}^{+\infty} e^{-(r-s)/2} \frac{r-s}{2}^{-1/2} e^{-s} \frac{s}{2}^{-1/2} ds , \\
&= \begin{cases} \frac{1}{2} e^{-r^2/2}, & \text{if } r \geq 0, \\ 0, & \text{otherwise.} \end{cases}
\end{aligned}$$

Hence, R^2 has a gamma density with $\lambda = 1/2$, $\beta = 1$. We can interpret this result as giving the density for the square of the distance of P from the center of a target if its coordinates are normally distributed.

The density of the random variable R is obtained from that of R^2 in the usual way (see Theorem 5.1), and we find

$$f_R(r) = \begin{cases} \frac{1}{2} e^{-r^2/2} \cdot 2r = r e^{-r^2/2}, & \text{if } r \geq 0, \\ 0, & \text{otherwise.} \end{cases}$$

Physicists will recognize this as a Rayleigh density. Our result here agrees with our simulation in Example 5.9. □

Chi-Squared Density

More generally, the same method shows that the sum of the squares of n independent normally distributed random variables with mean 0 and standard deviation 1 has a gamma density with $\lambda = 1/2$ and $\beta = n/2$. Such a density is called a *chi-squared density* with n degrees of freedom. This density was introduced in Chapter 4.3. In Example 5.10, we used this density to test the hypothesis that two traits were independent.

Another important use of the chi-squared density is in comparing experimental data with a theoretical discrete distribution, to see whether the data supports the theoretical model. More specifically, suppose that we have an experiment with a finite set of outcomes. If the set of outcomes is countable, we group them into finitely many sets of outcomes. We propose a theoretical distribution which we think will model the experiment well. We obtain some data by repeating the experiment a number of times. Now we wish to check how well the theoretical distribution fits the data.

Let X be the random variable which represents a theoretical outcome in the model of the experiment, and let $m(x)$ be the distribution function of X. In a manner similar to what was done in Example 5.10, we calculate the value of the expression

$$V = \sum_x \frac{(o_x - n \cdot m(x))^2}{n \cdot m(x)} ,$$

where the sum runs over all possible outcomes x, n is the number of data points, and o_x denotes the number of outcomes of type x observed in the data. Then

7.2. SUMS OF CONTINUOUS RANDOM VARIABLES

Outcome	Observed Frequency
1	15
2	8
3	7
4	5
5	7
6	18

Table 7.1: Observed data.

for moderate or large values of n, the quantity V is approximately chi-squared distributed, with $\nu-1$ degrees of freedom, where ν represents the number of possible outcomes. The proof of this is beyond the scope of this book, but we will illustrate the reasonableness of this statement in the next example. If the value of V is very large, when compared with the appropriate chi-squared density function, then we would tend to reject the hypothesis that the model is an appropriate one for the experiment at hand. We now give an example of this procedure.

Example 7.8 Suppose we are given a single die. We wish to test the hypothesis that the die is fair. Thus, our theoretical distribution is the uniform distribution on the integers between 1 and 6. So, if we roll the die n times, the expected number of data points of each type is $n/6$. Thus, if o_i denotes the actual number of data points of type i, for $1 \leq i \leq 6$, then the expression

$$V = \sum_{i=1}^{6} \frac{(o_i - n/6)^2}{n/6}$$

is approximately chi-squared distributed with 5 degrees of freedom.

Now suppose that we actually roll the die 60 times and obtain the data in Table 7.1. If we calculate V for this data, we obtain the value 13.6. The graph of the chi-squared density with 5 degrees of freedom is shown in Figure 7.4. One sees that values as large as 13.6 are rarely taken on by V if the die is fair, so we would reject the hypothesis that the die is fair. (When using this test, a statistician will reject the hypothesis if the data gives a value of V which is larger than 95% of the values one would expect to obtain if the hypothesis is true.)

In Figure 7.5, we show the results of rolling a die 60 times, then calculating V, and then repeating this experiment 1000 times. The program that performs these calculations is called **DieTest**. We have superimposed the chi-squared density with 5 degrees of freedom; one can see that the data values fit the curve fairly well, which supports the statement that the chi-squared density is the correct one to use. □

So far we have looked at several important special cases for which the convolution integral can be evaluated explicitly. In general, the convolution of two continuous densities cannot be evaluated explicitly, and we must resort to numerical methods. Fortunately, these prove to be remarkably effective, at least for bounded densities.

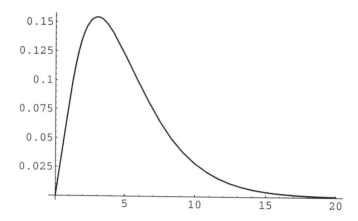

Figure 7.4: Chi-squared density with 5 degrees of freedom.

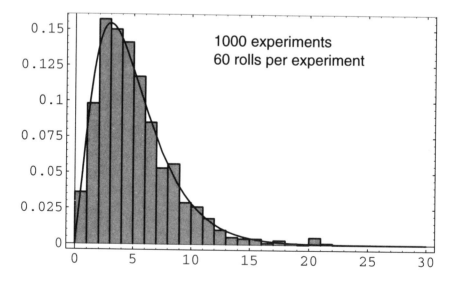

Figure 7.5: Rolling a fair die.

7.2. SUMS OF CONTINUOUS RANDOM VARIABLES

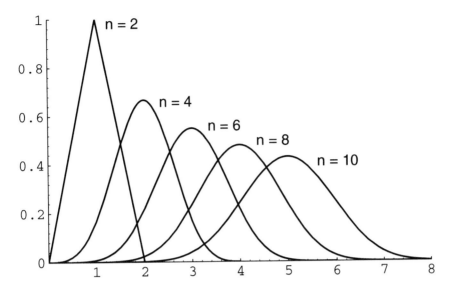

Figure 7.6: Convolution of n uniform densities.

Independent Trials

We now consider briefly the distribution of the sum of n independent random variables, all having the same density function. If X_1, X_2, ..., X_n are these random variables and $S_n = X_1 + X_2 + \cdots + X_n$ is their sum, then we will have

$$f_{S_n}(x) = (f_{X_1} * f_{X_2} * \cdots * f_{X_n})(x) ,$$

where the right-hand side is an n-fold convolution. It is possible to calculate this density for general values of n in certain simple cases.

Example 7.9 Suppose the X_i are uniformly distributed on the interval $[0, 1]$. Then

$$f_{X_i}(x) = \begin{cases} 1, & \text{if } 0 \leq x \leq 1, \\ 0, & \text{otherwise,} \end{cases}$$

and $f_{S_n}(x)$ is given by the formula[4]

$$f_{S_n}(x) = \begin{cases} \frac{1}{(n-1)!} \sum_{0 \leq j \leq x} (-1)^j \binom{n}{j}(x-j)^{n-1}, & \text{if } 0 < x < n, \\ 0, & \text{otherwise.} \end{cases}$$

The density $f_{S_n}(x)$ for $n = 2, 4, 6, 8, 10$ is shown in Figure 7.6.

If the X_i are distributed normally, with mean 0 and variance 1, then (cf. Example 7.5)

$$f_{X_i}(x) = \frac{1}{\sqrt{2\pi}} e^{-x^2/2} ,$$

---
[4]J. B. Uspensky, *Introduction to Mathematical Probability* (New York: McGraw-Hill, 1937), p. 277.

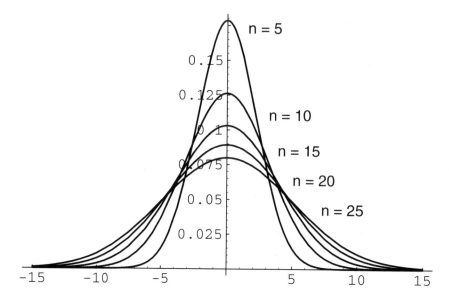

Figure 7.7: Convolution of n standard normal densities.

and
$$f_{S_n}(x) = \frac{1}{\sqrt{2\pi n}} e^{-x^2/2n} .$$

Here the density f_{S_n} for $n = 5, 10, 15, 20, 25$ is shown in Figure 7.7.

If the X_i are all exponentially distributed, with mean $1/\lambda$, then
$$f_{X_i}(x) = \lambda e^{-\lambda x} ,$$
and
$$f_{S_n}(x) = \frac{\lambda e^{-\lambda x}(\lambda x)^{n-1}}{(n-1)!} .$$

In this case the density f_{S_n} for $n = 2, 4, 6, 8, 10$ is shown in Figure 7.8. □

Exercises

1. Let X and Y be independent real-valued random variables with density functions $f_X(x)$ and $f_Y(y)$, respectively. Show that the density function of the sum $X + Y$ is the convolution of the functions $f_X(x)$ and $f_Y(y)$. *Hint*: Let $\bar{X}$ be the joint random variable (X, Y). Then the joint density function of $\bar{X}$ is $f_X(x)f_Y(y)$, since X and Y are independent. Now compute the probability that $X + Y \leq z$, by integrating the joint density function over the appropriate region in the plane. This gives the cumulative distribution function of Z. Now differentiate this function with respect to z to obtain the density function of z.

2. Let X and Y be independent random variables defined on the space Ω, with density functions f_X and f_Y, respectively. Suppose that $Z = X + Y$. Find the density f_Z of Z if

7.2. SUMS OF CONTINUOUS RANDOM VARIABLES

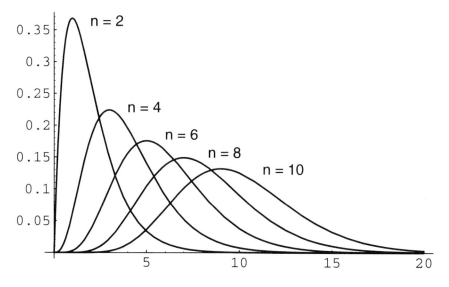

Figure 7.8: Convolution of n exponential densities with $\lambda = 1$.

(a)
$$f_X(x) = f_Y(x) = \begin{cases} 1/2, & \text{if } -1 \leq x \leq +1, \\ 0, & \text{otherwise.} \end{cases}$$

(b)
$$f_X(x) = f_Y(x) = \begin{cases} 1/2, & \text{if } 3 \leq x \leq 5, \\ 0, & \text{otherwise.} \end{cases}$$

(c)
$$f_X(x) = \begin{cases} 1/2, & \text{if } -1 \leq x \leq 1, \\ 0, & \text{otherwise.} \end{cases}$$

$$f_Y(x) = \begin{cases} 1/2, & \text{if } 3 \leq x \leq 5, \\ 0, & \text{otherwise.} \end{cases}$$

(d) What can you say about the set $E = \{\, z : f_Z(z) > 0 \,\}$ in each case?

3 Suppose again that $Z = X + Y$. Find f_Z if

(a)
$$f_X(x) = f_Y(x) = \begin{cases} x/2, & \text{if } 0 < x < 2, \\ 0, & \text{otherwise.} \end{cases}$$

(b)
$$f_X(x) = f_Y(x) = \begin{cases} (1/2)(x-3), & \text{if } 3 < x < 5, \\ 0, & \text{otherwise.} \end{cases}$$

(c)
$$f_X(x) = \begin{cases} 1/2, & \text{if } 0 < x < 2, \\ 0, & \text{otherwise.} \end{cases}$$

$$f_Y(x) = \begin{cases} x/2, & \text{if } 0 < x < 2, \\ 0, & \text{otherwise.} \end{cases}$$

(d) What can you say about the set $E = \{\, z : f_Z(z) > 0 \,\}$ in each case?

4 Let X, Y, and Z be independent random variables with

$$f_X(x) = f_Y(x) = f_Z(x) = \begin{cases} 1, & \text{if } 0 < x < 1, \\ 0, & \text{otherwise.} \end{cases}$$

Suppose that $W = X + Y + Z$. Find f_W directly, and compare your answer with that given by the formula in Example 7.9. *Hint*: See Example 7.3.

5 Suppose that X and Y are independent and $Z = X + Y$. Find f_Z if

(a)
$$f_X(x) = \begin{cases} \lambda e^{-\lambda x}, & \text{if } x > 0, \\ 0, & \text{otherwise.} \end{cases}$$

$$f_Y(x) = \begin{cases} \mu e^{-\mu x}, & \text{if } x > 0, \\ 0, & \text{otherwise.} \end{cases}$$

(b)
$$f_X(x) = \begin{cases} \lambda e^{-\lambda x}, & \text{if } x > 0, \\ 0, & \text{otherwise.} \end{cases}$$

$$f_Y(x) = \begin{cases} 1, & \text{if } 0 < x < 1, \\ 0, & \text{otherwise.} \end{cases}$$

6 Suppose again that $Z = X + Y$. Find f_Z if

$$f_X(x) = \frac{1}{\sqrt{2\pi}\sigma_1} e^{-(x-\mu_1)^2/2\sigma_1^2}$$

$$f_Y(x) = \frac{1}{\sqrt{2\pi}\sigma_2} e^{-(x-\mu_2)^2/2\sigma_2^2}.$$

*7 Suppose that $R^2 = X^2 + Y^2$. Find f_{R^2} and f_R if

$$f_X(x) = \frac{1}{\sqrt{2\pi}\sigma_1} e^{-(x-\mu_1)^2/2\sigma_1^2}$$

$$f_Y(x) = \frac{1}{\sqrt{2\pi}\sigma_2} e^{-(x-\mu_2)^2/2\sigma_2^2}.$$

8 Suppose that $R^2 = X^2 + Y^2$. Find f_{R^2} and f_R if

$$f_X(x) = f_Y(x) = \begin{cases} 1/2, & \text{if } -1 \leq x \leq 1, \\ 0, & \text{otherwise.} \end{cases}$$

9 Assume that the service time for a customer at a bank is exponentially distributed with mean service time 2 minutes. Let X be the total service time for 10 customers. Estimate the probability that $X > 22$ minutes.

7.2. SUMS OF CONTINUOUS RANDOM VARIABLES

10 Let $X_1, X_2, \ldots, X_n$ be n independent random variables each of which has an exponential density with mean μ. Let M be the *minimum* value of the X_j. Show that the density for M is exponential with mean μ/n. *Hint*: Use cumulative distribution functions.

11 A company buys 100 lightbulbs, each of which has an exponential lifetime of 1000 hours. What is the expected time for the first of these bulbs to burn out? (See Exercise 10.)

12 An insurance company assumes that the time between claims from each of its homeowners' policies is exponentially distributed with mean μ. It would like to estimate μ by averaging the times for a number of policies, but this is not very practical since the time between claims is about 30 years. At Galambos'[5] suggestion the company puts its customers in groups of 50 and observes the time of the first claim within each group. Show that this provides a practical way to estimate the value of μ.

13 Particles are subject to collisions that cause them to split into two parts with each part a fraction of the parent. Suppose that this fraction is uniformly distributed between 0 and 1. Following a single particle through several splittings we obtain a fraction of the original particle $Z_n = X_1 \cdot X_2 \cdot \ldots \cdot X_n$ where each X_j is uniformly distributed between 0 and 1. Show that the density for the random variable Z_n is

$$f_n(z) = \frac{1}{(n-1)!}(-\log z)^{n-1}.$$

Hint: Show that $Y_k = -\log X_k$ is exponentially distributed. Use this to find the density function for $S_n = Y_1 + Y_2 + \cdots + Y_n$, and from this the cumulative distribution and density of $Z_n = e^{-S_n}$.

14 Assume that X_1 and X_2 are independent random variables, each having an exponential density with parameter λ. Show that $Z = X_1 - X_2$ has density

$$f_Z(z) = (1/2)\lambda e^{-\lambda|z|}.$$

15 Suppose we want to test a coin for fairness. We flip the coin n times and record the number of times X_0 that the coin turns up tails and the number of times $X_1 = n - X_0$ that the coin turns up heads. Now we set

$$Z = \sum_{i=0}^{1} \frac{(X_i - n/2)^2}{n/2}.$$

Then for a fair coin Z has approximately a chi-squared distribution with $2 - 1 = 1$ degree of freedom. Verify this by computer simulation first for a fair coin ($p = 1/2$) and then for a biased coin ($p = 1/3$).

---

[5] J. Galambos, *Introductory Probability Theory* (New York: Marcel Dekker, 1984), p. 159.

16 Verify your answers in Exercise 2(a) by computer simulation: Choose X and Y from $[-1, 1]$ with uniform density and calculate $Z = X + Y$. Repeat this experiment 500 times, recording the outcomes in a bar graph on $[-2, 2]$ with 40 bars. Does the density f_Z calculated in Exercise 2(a) describe the shape of your bar graph? Try this for Exercises 2(b) and Exercise 2(c), too.

17 Verify your answers to Exercise 3 by computer simulation.

18 Verify your answer to Exercise 4 by computer simulation.

19 The *support* of a function $f(x)$ is defined to be the set

$$\{x \;:\; f(x) > 0\}\;.$$

Suppose that X and Y are two continuous random variables with density functions $f_X(x)$ and $f_Y(y)$, respectively, and suppose that the supports of these density functions are the intervals $[a, b]$ and $[c, d]$, respectively. Find the support of the density function of the random variable $X + Y$.

20 Let $X_1, X_2, \ldots, X_n$ be a sequence of independent random variables, all having a common density function f_X with support $[a, b]$ (see Exercise 19). Let $S_n = X_1 + X_2 + \cdots + X_n$, with density function f_{S_n}. Show that the support of f_{S_n} is the interval $[na, nb]$. *Hint*: Write $f_{S_n} = f_{S_{n-1}} * f_X$. Now use Exercise 19 to establish the desired result by induction.

21 Let $X_1, X_2, \ldots, X_n$ be a sequence of independent random variables, all having a common density function f_X. Let $A = S_n/n$ be their average. Find f_A if

(a) $f_X(x) = (1/\sqrt{2\pi})e^{-x^2/2}$ (normal density).

(b) $f_X(x) = e^{-x}$ (exponential density).

Hint: Write $f_A(x)$ in terms of $f_{S_n}(x)$.

Chapter 8

Law of Large Numbers

8.1 Law of Large Numbers for Discrete Random Variables

We are now in a position to prove our first fundamental theorem of probability. We have seen that an intuitive way to view the probability of a certain outcome is as the frequency with which that outcome occurs in the long run, when the experiment is repeated a large number of times. We have also defined probability mathematically as a value of a distribution function for the random variable representing the experiment. The Law of Large Numbers, which is a theorem proved about the mathematical model of probability, shows that this model is consistent with the frequency interpretation of probability. This theorem is sometimes called the *law of averages*. To find out what would happen if this law were not true, see the article by Robert M. Coates.[1]

Chebyshev Inequality

To discuss the Law of Large Numbers, we first need an important inequality called the *Chebyshev Inequality*.

Theorem 8.1 (Chebyshev Inequality) Let X be a discrete random variable with expected value $\mu = E(X)$, and let $\epsilon > 0$ be any positive real number. Then

$$P(|X - \mu| \geq \epsilon) \leq \frac{V(X)}{\epsilon^2} .$$

Proof. Let $m(x)$ denote the distribution function of X. Then the probability that X differs from μ by at least ϵ is given by

$$P(|X - \mu| \geq \epsilon) = \sum_{|x - \mu| \geq \epsilon} m(x) .$$

---

[1] R. M. Coates, "The Law," *The World of Mathematics*, ed. James R. Newman (New York: Simon and Schuster, 1956).

We know that
$$V(X) = \sum_x (x-\mu)^2 m(x) \ ,$$
and this is clearly at least as large as
$$\sum_{|x-\mu|\geq\epsilon} (x-\mu)^2 m(x) \ ,$$
since all the summands are positive and we have restricted the range of summation in the second sum. But this last sum is at least
$$\sum_{|x-\mu|\geq\epsilon} \epsilon^2 m(x) = \epsilon^2 \sum_{|x-\mu|\geq\epsilon} m(x)$$
$$= \epsilon^2 P(|X-\mu|\geq\epsilon) \ .$$

So,
$$P(|X-\mu|\geq\epsilon) \leq \frac{V(X)}{\epsilon^2} \ .$$
$\square$

Note that X in the above theorem can be any discrete random variable, and ϵ any positive number.

Example 8.1 Let X by any random variable with $E(X) = \mu$ and $V(X) = \sigma^2$. Then, if $\epsilon = k\sigma$, Chebyshev's Inequality states that
$$P(|X-\mu|\geq k\sigma) \leq \frac{\sigma^2}{k^2\sigma^2} = \frac{1}{k^2} \ .$$

Thus, for any random variable, the probability of a deviation from the mean of more than k standard deviations is $\leq 1/k^2$. If, for example, $k=5$, $1/k^2 = .04$. $\square$

Chebyshev's Inequality is the best possible inequality in the sense that, for any $\epsilon > 0$, it is possible to give an example of a random variable for which Chebyshev's Inequality is in fact an equality. To see this, given $\epsilon > 0$, choose X with distribution
$$p_X = \begin{pmatrix} -\epsilon & +\epsilon \\ 1/2 & 1/2 \end{pmatrix} \ .$$
Then $E(X) = 0$, $V(X) = \epsilon^2$, and
$$P(|X-\mu|\geq\epsilon) = \frac{V(X)}{\epsilon^2} = 1 \ .$$

We are now prepared to state and prove the Law of Large Numbers.

8.1. DISCRETE RANDOM VARIABLES

Law of Large Numbers

Theorem 8.2 (Law of Large Numbers) Let $X_1, X_2, \ldots, X_n$ be an independent trials process, with finite expected value $\mu = E(X_j)$ and finite variance $\sigma^2 = V(X_j)$. Let $S_n = X_1 + X_2 + \cdots + X_n$. Then for any $\epsilon > 0$,

$$P\left(\left|\frac{S_n}{n} - \mu\right| \geq \epsilon\right) \to 0$$

as $n \to \infty$. Equivalently,

$$P\left(\left|\frac{S_n}{n} - \mu\right| < \epsilon\right) \to 1$$

as $n \to \infty$.

Proof. Since $X_1, X_2, \ldots, X_n$ are independent and have the same distributions, we can apply Theorem 6.9. We obtain

$$V(S_n) = n\sigma^2 \ ,$$

and

$$V(\frac{S_n}{n}) = \frac{\sigma^2}{n} \ .$$

Also we know that

$$E(\frac{S_n}{n}) = \mu \ .$$

By Chebyshev's Inequality, for any $\epsilon > 0$,

$$P\left(\left|\frac{S_n}{n} - \mu\right| \geq \epsilon\right) \leq \frac{\sigma^2}{n\epsilon^2} \ .$$

Thus, for fixed ϵ,

$$P\left(\left|\frac{S_n}{n} - \mu\right| \geq \epsilon\right) \to 0$$

as $n \to \infty$, or equivalently,

$$P\left(\left|\frac{S_n}{n} - \mu\right| < \epsilon\right) \to 1$$

as $n \to \infty$. $\square$

Law of Averages

Note that S_n/n is an average of the individual outcomes, and one often calls the Law of Large Numbers the "law of averages." It is a striking fact that we can start with a random experiment about which little can be predicted and, by taking averages, obtain an experiment in which the outcome can be predicted with a high degree of certainty. The Law of Large Numbers, as we have stated it, is often called the "Weak Law of Large Numbers" to distinguish it from the "Strong Law of Large Numbers" described in Exercise 15.

Consider the important special case of Bernoulli trials with probability p for success. Let $X_j = 1$ if the jth outcome is a success and 0 if it is a failure. Then $S_n = X_1 + X_2 + \cdots + X_n$ is the number of successes in n trials and $\mu = E(X_1) = p$. The Law of Large Numbers states that for any $\epsilon > 0$

$$P\left(\left|\frac{S_n}{n} - p\right| < \epsilon\right) \to 1$$

as $n \to \infty$. The above statement says that, in a large number of repetitions of a Bernoulli experiment, we can expect the proportion of times the event will occur to be near p. This shows that our mathematical model of probability agrees with our frequency interpretation of probability.

Coin Tossing

Let us consider the special case of tossing a coin n times with S_n the number of heads that turn up. Then the random variable S_n/n represents the fraction of times heads turns up and will have values between 0 and 1. The Law of Large Numbers predicts that the outcomes for this random variable will, for large n, be near $1/2$.

In Figure 8.1, we have plotted the distribution for this example for increasing values of n. We have marked the outcomes between .45 and .55 by dots at the top of the spikes. We see that as n increases the distribution gets more and more concentrated around .5 and a larger and larger percentage of the total area is contained within the interval $(.45, .55)$, as predicted by the Law of Large Numbers.

Die Rolling

Example 8.2 Consider n rolls of a die. Let X_j be the outcome of the jth roll. Then $S_n = X_1 + X_2 + \cdots + X_n$ is the sum of the first n rolls. This is an independent trials process with $E(X_j) = 7/2$. Thus, by the Law of Large Numbers, for any $\epsilon > 0$

$$P\left(\left|\frac{S_n}{n} - \frac{7}{2}\right| \geq \epsilon\right) \to 0$$

as $n \to \infty$. An equivalent way to state this is that, for any $\epsilon > 0$,

$$P\left(\left|\frac{S_n}{n} - \frac{7}{2}\right| < \epsilon\right) \to 1$$

as $n \to \infty$. $\square$

Numerical Comparisons

It should be emphasized that, although Chebyshev's Inequality proves the Law of Large Numbers, it is actually a very crude inequality for the probabilities involved. However, its strength lies in the fact that it is true for any random variable at all, and it allows us to prove a very powerful theorem.

In the following example, we compare the estimates given by Chebyshev's Inequality with the actual values.

8.1. DISCRETE RANDOM VARIABLES

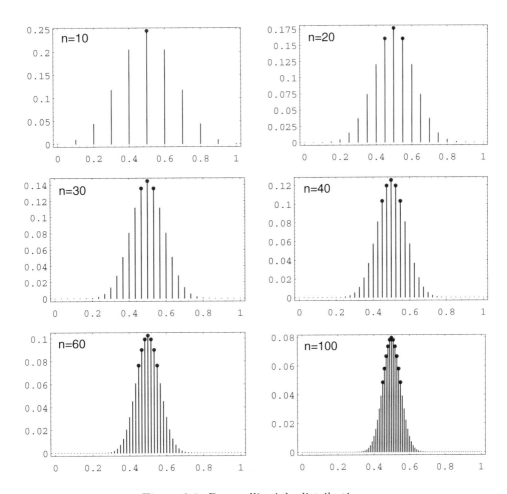

Figure 8.1: Bernoulli trials distributions.

Example 8.3 Let $X_1, X_2, \ldots, X_n$ be a Bernoulli trials process with probability .3 for success and .7 for failure. Let $X_j = 1$ if the jth outcome is a success and 0 otherwise. Then, $E(X_j) = .3$ and $V(X_j) = (.3)(.7) = .21$. If

$$A_n = \frac{S_n}{n} = \frac{X_1 + X_2 + \cdots + X_n}{n}$$

is the *average* of the X_i, then $E(A_n) = .3$ and $V(A_n) = V(S_n)/n^2 = .21/n$. Chebyshev's Inequality states that if, for example, $\epsilon = .1$,

$$P(|A_n - .3| \geq .1) \leq \frac{.21}{n(.1)^2} = \frac{21}{n} .$$

Thus, if $n = 100$,

$$P(|A_{100} - .3| \geq .1) \leq .21 ,$$

or if $n = 1000$,

$$P(|A_{1000} - .3| \geq .1) \leq .021 .$$

These can be rewritten as

$$P(.2 < A_{100} < .4) \geq .79 ,$$
$$P(.2 < A_{1000} < .4) \geq .979 .$$

These values should be compared with the actual values, which are (to six decimal places)

$$P(.2 < A_{100} < .4) \approx .962549$$
$$P(.2 < A_{1000} < .4) \approx 1 .$$

The program **Law** can be used to carry out the above calculations in a systematic way. □

Historical Remarks

The Law of Large Numbers was first proved by the Swiss mathematician James Bernoulli in the fourth part of his work *Ars Conjectandi* published posthumously in 1713.[2] As often happens with a first proof, Bernoulli's proof was much more difficult than the proof we have presented using Chebyshev's inequality. Chebyshev developed his inequality to prove a general form of the Law of Large Numbers (see Exercise 12). The inequality itself appeared much earlier in a work by Bienaymé, and in discussing its history Maistrov remarks that it was referred to as the Bienaymé-Chebyshev Inequality for a long time.[3]

In *Ars Conjectandi* Bernoulli provides his reader with a long discussion of the meaning of his theorem with lots of examples. In modern notation he has an event

---

[2] J. Bernoulli, *The Art of Conjecturing IV*, trans. Bing Sung, Technical Report No. 2, Dept. of Statistics, Harvard Univ., 1966

[3] L. E. Maistrov, *Probability Theory: A Historical Approach*, trans. and ed. Samual Kotz, (New York: Academic Press, 1974), p. 202

8.1. DISCRETE RANDOM VARIABLES

that occurs with probability p but he does not know p. He wants to estimate p by the fraction $\bar{p}$ of the times the event occurs when the experiment is repeated a number of times. He discusses in detail the problem of estimating, by this method, the proportion of white balls in an urn that contains an unknown number of white and black balls. He would do this by drawing a sequence of balls from the urn, replacing the ball drawn after each draw, and estimating the unknown proportion of white balls in the urn by the proportion of the balls drawn that are white. He shows that, by choosing n large enough he can obtain any desired accuracy and reliability for the estimate. He also provides a lively discussion of the applicability of his theorem to estimating the probability of dying of a particular disease, of different kinds of weather occurring, and so forth.

In speaking of the number of trials necessary for making a judgement, Bernoulli observes that the "man on the street" believes the "law of averages."

> Further, it cannot escape anyone that for judging in this way about any event at all, it is not enough to use one or two trials, but rather a great number of trials is required. And sometimes the stupidest man—by some instinct of nature *per se* and by no previous instruction (this is truly amazing)— knows for sure that the more observations of this sort that are taken, the less the danger will be of straying from the mark.[4]

But he goes on to say that he must contemplate another possibility.

> Something futher must be contemplated here which perhaps no one has thought about till now. It certainly remains to be inquired whether after the number of observations has been increased, the probability is increased of attaining the true ratio between the number of cases in which some event can happen and in which it cannot happen, so that this probability finally exceeds any given degree of certainty; or whether the problem has, so to speak, its own asymptote—that is, whether some degree of certainty is given which one can never exceed.[5]

Bernoulli recognized the importance of this theorem, writing:

> Therefore, this is the problem which I now set forth and make known after I have already pondered over it for twenty years. Both its novelty and its very great usefullness, coupled with its just as great difficulty, can exceed in weight and value all the remaining chapters of this thesis.[6]

Bernoulli concludes his long proof with the remark:

> Whence, finally, this one thing seems to follow: that if observations of all events were to be continued throughout all eternity, (and hence the ultimate probability would tend toward perfect certainty), everything in

---
[4] Bernoulli, op. cit., p. 38.
[5] ibid., p. 39.
[6] ibid., p. 42.

the world would be perceived to happen in fixed ratios and according to a constant law of alternation, so that even in the most accidental and fortuitous occurrences we would be bound to recognize, as it were, a certain necessity and, so to speak, a certain fate.

I do now know whether Plato wished to aim at this in his doctrine of the universal return of things, according to which he predicted that all things will return to their original state after countless ages have past.[7]

Exercises

1 A fair coin is tossed 100 times. The expected number of heads is 50, and the standard deviation for the number of heads is $(100 \cdot 1/2 \cdot 1/2)^{1/2} = 5$. What does Chebyshev's Inequality tell you about the probability that the number of heads that turn up deviates from the expected number 50 by three or more standard deviations (i.e., by at least 15)?

2 Write a program that uses the function binomial(n, p, x) to compute the exact probability that you estimated in Exercise 1. Compare the two results.

3 Write a program to toss a coin 10,000 times. Let S_n be the number of heads in the first n tosses. Have your program print out, after every 1000 tosses, $S_n - n/2$. On the basis of this simulation, is it correct to say that you can expect heads about half of the time when you toss a coin a large number of times?

4 A 1-dollar bet on craps has an expected winning of $-.0141$. What does the Law of Large Numbers say about your winnings if you make a large number of 1-dollar bets at the craps table? Does it assure you that your losses will be small? Does it assure you that if n is very large you will lose?

5 Let X be a random variable with $E(X) = 0$ and $V(X) = 1$. What integer value k will assure us that $P(|X| \geq k) \leq .01$?

6 Let S_n be the number of successes in n Bernoulli trials with probability p for success on each trial. Show, using Chebyshev's Inequality, that for any $\epsilon > 0$

$$P\left(\left|\frac{S_n}{n} - p\right| \geq \epsilon\right) \leq \frac{p(1-p)}{n\epsilon^2} \ .$$

7 Find the maximum possible value for $p(1-p)$ if $0 < p < 1$. Using this result and Exercise 6, show that the estimate

$$P\left(\left|\frac{S_n}{n} - p\right| \geq \epsilon\right) \leq \frac{1}{4n\epsilon^2}$$

is valid for any p.

---

[7]ibid., pp. 65–66.

8.1. DISCRETE RANDOM VARIABLES

8 A fair coin is tossed a large number of times. Does the Law of Large Numbers assure us that, if n is large enough, with probability $> .99$ the number of heads that turn up will not deviate from $n/2$ by more than 100?

9 In Exercise 6.2.15, you showed that, for the hat check problem, the number S_n of people who get their own hats back has $E(S_n) = V(S_n) = 1$. Using Chebyshev's Inequality, show that $P(S_n \geq 11) \leq .01$ for any $n \geq 11$.

10 Let X by any random variable which takes on values 0, 1, 2, ..., n and has $E(X) = V(X) = 1$. Show that, for any integer k,

$$P(X \geq k+1) \leq \frac{1}{k^2} .$$

11 We have two coins: one is a fair coin and the other is a coin that produces heads with probability 3/4. One of the two coins is picked at random, and this coin is tossed n times. Let S_n be the number of heads that turns up in these n tosses. Does the Law of Large Numbers allow us to predict the proportion of heads that will turn up in the long run? After we have observed a large number of tosses, can we tell which coin was chosen? How many tosses suffice to make us 95 percent sure?

12 (Chebyshev[8]) Assume that $X_1, X_2, \ldots, X_n$ are independent random variables with possibly different distributions and let S_n be their sum. Let $m_k = E(X_k)$, $\sigma_k^2 = V(X_k)$, and $M_n = m_1 + m_2 + \cdots + m_n$. Assume that $\sigma_k^2 < R$ for all k. Prove that, for any $\epsilon > 0$,

$$P\left(\left|\frac{S_n}{n} - \frac{M_n}{n}\right| < \epsilon\right) \to 1$$

as $n \to \infty$.

13 A fair coin is tossed repeatedly. Before each toss, you are allowed to decide whether to bet on the outcome. Can you describe a betting system with infinitely many bets which will enable you, in the long run, to win more than half of your bets? (Note that we are disallowing a betting system that says to bet until you are ahead, then quit.) Write a computer program that implements this betting system. As stated above, your program must decide whether to bet on a particular outcome before that outcome is determined. For example, you might select only outcomes that come after there have been three tails in a row. See if you can get more than 50% heads by your "system."

*14 Prove the following analogue of Chebyshev's Inequality:

$$P(|X - E(X)| \geq \epsilon) \leq \frac{1}{\epsilon} E(|X - E(X)|) .$$

---
[8] P. L. Chebyshev, "On Mean Values," *J. Math. Pure. Appl.*, vol. 12 (1867), pp. 177–184.

*15 We have proved a theorem often called the "Weak Law of Large Numbers." Most people's intuition and our computer simulations suggest that, if we toss a coin a sequence of times, the proportion of heads will really approach 1/2; that is, if S_n is the number of heads in n times, then we will have

$$A_n = \frac{S_n}{n} \to \frac{1}{2}$$

as $n \to \infty$. Of course, we cannot be sure of this since we are not able to toss the coin an infinite number of times, and, if we could, the coin could come up heads every time. However, the "Strong Law of Large Numbers," proved in more advanced courses, states that

$$P\left(\frac{S_n}{n} \to \frac{1}{2}\right) = 1 \ .$$

Describe a sample space Ω that would make it possible for us to talk about the event

$$E = \left\{\omega : \frac{S_n}{n} \to \frac{1}{2}\right\} \ .$$

Could we assign the equiprobable measure to this space? (See Example 2.18.)

*16 In this problem, you will construct a sequence of random variables which satisfies the Weak Law of Large Numbers, but not the Strong Law of Large Numbers (see Exercise 15). For each positive integer n, let the random variable X_n be defined by

$$P(X_n = \pm n2^n) = f(n) \ ,$$
$$P(X_n = 0) = 1 - 2f(n) \ ,$$

where $f(n)$ is a function that will be chosen later (and which satisfies $0 \leq f(n) \leq 1/2$ for all positive integers n). Let $S_n = X_1 + X_2 + \cdots + X_n$.

(a) Show that $\mu(S_n) = 0$ for all n.

(b) Show that if $X_n > 0$, then $S_n \geq 2^n$.

(c) Use part (b) to show that $S_n/n \to 0$ as $n \to \infty$ if and only if there exists an n_0 such that $X_k = 0$ for all $k \geq n_0$. Show that this happens with probability 0 if we require that $f(n) < 1/2$ for all n. This shows that the sequence $\{X_n\}$ does not satisfy the Strong Law of Large Numbers.

(d) We now turn our attention to the Weak Law of Large Numbers. Given a positive ϵ, we wish to estimate

$$P\left(\left|\frac{S_n}{n}\right| \geq \epsilon\right) \ .$$

Suppose that $X_k = 0$ for $m < k \leq n$. Show that

$$|S_n| \leq 2^{2m} \ .$$

(e) Show that if we define $g(n) = (1/2)\log_2(\epsilon n)$, then
$$2^{2m} < \epsilon n \ .$$

This shows that if $X_k = 0$ for $g(n) < k \le n$, then
$$|S_n| < \epsilon n \ ,$$

or
$$\left|\frac{S_n}{n}\right| < \epsilon \ .$$

We wish to show that the probability of this event tends to 1 as $n \to \infty$, or equivalently, that the probability of the complementary event tends to 0 as $n \to \infty$. The complementary event is the event that $X_k \ne 0$ for some k with $g(n) < k \le n$. Show that the probability of this event equals
$$1 - \prod_{k=\lceil g(n) \rceil}^{n} \bigl(1 - 2f(n)\bigr) \ ,$$

and show that this expression is less than
$$1 - \prod_{k=\lceil g(n) \rceil}^{\infty} \bigl(1 - 2f(n)\bigr) \ .$$

(f) Show that by making $f(n) \to 0$ rapidly enough, the expression in part (e) can be made to approach 1 as $n \to \infty$. This shows that the sequence $\{X_n\}$ satisfies the Weak Law of Large Numbers.

*17 Let us toss a biased coin that comes up heads with probability p and assume the validity of the Strong Law of Large Numbers as described in Exercise 15. Then, with probability 1,
$$\frac{S_n}{n} \to p$$

as $n \to \infty$. If $f(x)$ is a continuous function on the unit interval, then we also have
$$f\left(\frac{S_n}{n}\right) \to f(p) \ .$$

Finally, we could hope that
$$E\left(f\left(\frac{S_n}{n}\right)\right) \to E(f(p)) = f(p) \ .$$

Show that, if all this is correct, as in fact it is, we would have proven that any continuous function on the unit interval is a limit of polynomial functions. This is a sketch of a probabilistic proof of an important theorem in mathematics called the *Weierstrass approximation theorem*.

8.2 Law of Large Numbers for Continuous Random Variables

In the previous section we discussed in some detail the Law of Large Numbers for discrete probability distributions. This law has a natural analogue for continuous probability distributions, which we consider somewhat more briefly here.

Chebyshev Inequality

Just as in the discrete case, we begin our discussion with the Chebyshev Inequality.

Theorem 8.3 (Chebyshev Inequality) Let X be a continuous random variable with density function $f(x)$. Suppose X has a finite expected value $\mu = E(X)$ and finite variance $\sigma^2 = V(X)$. Then for any positive number $\epsilon > 0$ we have

$$P(|X - \mu| \geq \epsilon) \leq \frac{\sigma^2}{\epsilon^2} \ .$$

$\square$

The proof is completely analogous to the proof in the discrete case, and we omit it.

Note that this theorem says nothing if $\sigma^2 = V(X)$ is infinite.

Example 8.4 Let X be any continuous random variable with $E(X) = \mu$ and $V(X) = \sigma^2$. Then, if $\epsilon = k\sigma = k$ standard deviations for some integer k, then

$$P(|X - \mu| \geq k\sigma) \leq \frac{\sigma^2}{k^2\sigma^2} = \frac{1}{k^2} \ ,$$

just as in the discrete case. $\square$

Law of Large Numbers

With the Chebyshev Inequality we can now state and prove the Law of Large Numbers for the continuous case.

Theorem 8.4 (Law of Large Numbers) Let $X_1, X_2, \ldots, X_n$ be an independent trials process with a continuous density function f, finite expected value μ, and finite variance σ^2. Let $S_n = X_1 + X_2 + \cdots + X_n$ be the sum of the X_i. Then for any real number $\epsilon > 0$ we have

$$\lim_{n \to \infty} P\left(\left|\frac{S_n}{n} - \mu\right| \geq \epsilon\right) = 0 \ ,$$

or equivalently,

$$\lim_{n \to \infty} P\left(\left|\frac{S_n}{n} - \mu\right| < \epsilon\right) = 1 \ .$$

$\square$

8.2. CONTINUOUS RANDOM VARIABLES

Note that this theorem is not necessarily true if σ^2 is infinite (see Example 8.8).

As in the discrete case, the Law of Large Numbers says that the average value of n independent trials tends to the expected value as $n \to \infty$, in the precise sense that, given $\epsilon > 0$, the probability that the average value and the expected value differ by more than ϵ tends to 0 as $n \to \infty$.

Once again, we suppress the proof, as it is identical to the proof in the discrete case.

Uniform Case

Example 8.5 Suppose we choose at random n numbers from the interval $[0, 1]$ with uniform distribution. Then if X_i describes the ith choice, we have

$$\mu = E(X_i) = \int_0^1 x\,dx = \frac{1}{2}\ ,$$

$$\sigma^2 = V(X_i) = \int_0^1 x^2\,dx - \mu^2$$

$$= \frac{1}{3} - \frac{1}{4} = \frac{1}{12}\ .$$

Hence,

$$E\left(\frac{S_n}{n}\right) = \frac{1}{2}\ ,$$

$$V\left(\frac{S_n}{n}\right) = \frac{1}{12n}\ ,$$

and for any $\epsilon > 0$,

$$P\left(\left|\frac{S_n}{n} - \frac{1}{2}\right| \geq \epsilon\right) \leq \frac{1}{12n\epsilon^2}\ .$$

This says that if we choose n numbers at random from $[0, 1]$, then the chances are better than $1 - 1/(12n\epsilon^2)$ that the difference $|S_n/n - 1/2|$ is less than ϵ. Note that ϵ plays the role of the amount of error we are willing to tolerate: If we choose $\epsilon = 0.1$, say, then the chances that $|S_n/n - 1/2|$ is less than 0.1 are better than $1 - 100/(12n)$. For $n = 100$, this is about .92, but if $n = 1000$, this is better than .99 and if $n = 10{,}000$, this is better than .999.

We can illustrate what the Law of Large Numbers says for this example graphically. The density for $A_n = S_n/n$ is determined by

$$f_{A_n}(x) = nf_{S_n}(nx)\ .$$

We have seen in Section 7.2, that we can compute the density $f_{S_n}(x)$ for the sum of n uniform random variables. In Figure 8.2 we have used this to plot the density for A_n for various values of n. We have shaded in the area for which A_n would lie between .45 and .55. We see that as we increase n, we obtain more and more of the total area inside the shaded region. The Law of Large Numbers tells us that we can obtain as much of the total area as we please inside the shaded region by choosing n large enough (see also Figure 8.1). □

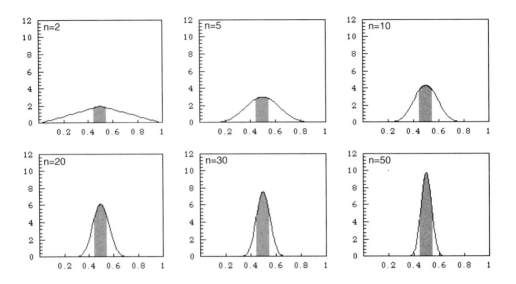

Figure 8.2: Illustration of Law of Large Numbers — uniform case.

Normal Case

Example 8.6 Suppose we choose n real numbers at random, using a normal distribution with mean 0 and variance 1. Then

$$\mu = E(X_i) = 0 ,$$
$$\sigma^2 = V(X_i) = 1 .$$

Hence,

$$E\left(\frac{S_n}{n}\right) = 0 ,$$
$$V\left(\frac{S_n}{n}\right) = \frac{1}{n} ,$$

and, for any $\epsilon > 0$,

$$P\left(\left|\frac{S_n}{n} - 0\right| \geq \epsilon\right) \leq \frac{1}{n\epsilon^2} .$$

In this case it is possible to compare the Chebyshev estimate for $P(|S_n/n - \mu| \geq \epsilon)$ in the Law of Large Numbers with exact values, since we know the density function for S_n/n exactly (see Example 7.9). The comparison is shown in Table 8.1, for $\epsilon = .1$. The data in this table was produced by the program **LawContinuous**. We see here that the Chebyshev estimates are in general *not* very accurate. □

8.2. CONTINUOUS RANDOM VARIABLES

n	$P(|S_n/n| \geq .1)$	Chebyshev
100	.31731	1.00000
200	.15730	.50000
300	.08326	.33333
400	.04550	.25000
500	.02535	.20000
600	.01431	.16667
700	.00815	.14286
800	.00468	.12500
900	.00270	.11111
1000	.00157	.10000

Table 8.1: Chebyshev estimates.

Monte Carlo Method

Here is a somewhat more interesting example.

Example 8.7 Let $g(x)$ be a continuous function defined for $x \in [0,1]$ with values in $[0,1]$. In Section 2.1, we showed how to estimate the area of the region under the graph of $g(x)$ by the Monte Carlo method, that is, by choosing a large number of random values for x and y with uniform distribution and seeing what fraction of the points $P(x, y)$ fell inside the region under the graph (see Example 2.2).

Here is a better way to estimate the same area (see Figure 8.3). Let us choose a large number of independent values X_n at random from $[0, 1]$ with uniform density, set $Y_n = g(X_n)$, and find the average value of the Y_n. Then this average is our estimate for the area. To see this, note that if the density function for X_n is uniform,

$$\begin{aligned} \mu &= E(Y_n) = \int_0^1 g(x) f(x)\, dx \\ &= \int_0^1 g(x)\, dx \\ &= \text{average value of } g(x)\ , \end{aligned}$$

while the variance is

$$\sigma^2 = E((Y_n - \mu)^2) = \int_0^1 (g(x) - \mu)^2\, dx < 1\ ,$$

since for all x in $[0, 1]$, $g(x)$ is in $[0, 1]$, hence μ is in $[0, 1]$, and so $|g(x) - \mu| \leq 1$. Now let $A_n = (1/n)(Y_1 + Y_2 + \cdots + Y_n)$. Then by Chebyshev's Inequality, we have

$$P(|A_n - \mu| \geq \epsilon) \leq \frac{\sigma^2}{n\epsilon^2} < \frac{1}{n\epsilon^2}\ .$$

This says that to get within ϵ of the true value for $\mu = \int_0^1 g(x)\, dx$ with probability at least p, we should choose n so that $1/n\epsilon^2 \leq 1 - p$ (i.e., so that $n \geq 1/\epsilon^2(1-p)$). Note that this method tells us how large to take n to get a desired accuracy. □

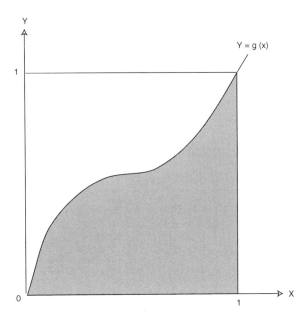

Figure 8.3: Area problem.

The Law of Large Numbers requires that the variance σ^2 of the original underlying density be finite: $\sigma^2 < \infty$. In cases where this fails to hold, the Law of Large Numbers may fail, too. An example follows.

Cauchy Case

Example 8.8 Suppose we choose n numbers from $(-\infty, +\infty)$ with a Cauchy density with parameter $a = 1$. We know that for the Cauchy density the expected value and variance are undefined (see Example 6.28). In this case, the density function for
$$A_n = \frac{S_n}{n}$$
is given by (see Example 7.6)
$$f_{A_n}(x) = \frac{1}{\pi(1+x^2)} \ ,$$
that is, *the density function for A_n is the same for all n.* In this case, as n increases, the density function does not change at all, and the Law of Large Numbers does not hold. □

Exercises

1 Let X be a continuous random variable with mean $\mu = 10$ and variance $\sigma^2 = 100/3$. Using Chebyshev's Inequality, find an upper bound for the following probabilities.

8.2. CONTINUOUS RANDOM VARIABLES

(a) $P(|X - 10| \geq 2)$.

(b) $P(|X - 10| \geq 5)$.

(c) $P(|X - 10| \geq 9)$.

(d) $P(|X - 10| \geq 20)$.

2 Let X be a continuous random variable with values uniformly distributed over the interval $[0, 20]$.

(a) Find the mean and variance of X.

(b) Calculate $P(|X - 10| \geq 2)$, $P(|X - 10| \geq 5)$, $P(|X - 10| \geq 9)$, and $P(|X - 10| \geq 20)$ exactly. How do your answers compare with those of Exercise 1? How good is Chebyshev's Inequality in this case?

3 Let X be the random variable of Exercise 2.

(a) Calculate the function $f(x) = P(|X - 10| \geq x)$.

(b) Now graph the function $f(x)$, and on the same axes, graph the Chebyshev function $g(x) = 100/(3x^2)$. Show that $f(x) \leq g(x)$ for all $x > 0$, but that $g(x)$ is not a very good approximation for $f(x)$.

4 Let X be a continuous random variable with values exponentially distributed over $[0, \infty)$ with parameter $\lambda = 0.1$.

(a) Find the mean and variance of X.

(b) Using Chebyshev's Inequality, find an upper bound for the following probabilities: $P(|X - 10| \geq 2)$, $P(|X - 10| \geq 5)$, $P(|X - 10| \geq 9)$, and $P(|X - 10| \geq 20)$.

(c) Calculate these probabilities exactly, and compare with the bounds in (b).

5 Let X be a continuous random variable with values normally distributed over $(-\infty, +\infty)$ with mean $\mu = 0$ and variance $\sigma^2 = 1$.

(a) Using Chebyshev's Inequality, find upper bounds for the following probabilities: $P(|X| \geq 1)$, $P(|X| \geq 2)$, and $P(|X| \geq 3)$.

(b) The area under the normal curve between -1 and 1 is .6827, between -2 and 2 is .9545, and between -3 and 3 it is .9973 (see the table in Appendix A). Compare your bounds in (a) with these exact values. How good is Chebyshev's Inequality in this case?

6 If X is normally distributed, with mean μ and variance σ^2, find an upper bound for the following probabilities, using Chebyshev's Inequality.

(a) $P(|X - \mu| \geq \sigma)$.

(b) $P(|X - \mu| \geq 2\sigma)$.

(c) $P(|X - \mu| \geq 3\sigma)$.

(d) $P(|X - \mu| \geq 4\sigma)$.

Now find the exact value using the program **NormalArea** or the normal table in Appendix A, and compare.

7 If X is a random variable with mean $\mu \neq 0$ and variance σ^2, define the *relative deviation* D of X from its mean by

$$D = \left|\frac{X - \mu}{\mu}\right|.$$

(a) Show that $P(D \geq a) \leq \sigma^2/(\mu^2 a^2)$.

(b) If X is the random variable of Exercise 1, find an upper bound for $P(D \geq .2)$, $P(D \geq .5)$, $P(D \geq .9)$, and $P(D \geq 2)$.

8 Let X be a continuous random variable and define the *standardized version* X^* of X by:

$$X^* = \frac{X - \mu}{\sigma}.$$

(a) Show that $P(|X^*| \geq a) \leq 1/a^2$.

(b) If X is the random variable of Exercise 1, find bounds for $P(|X^*| \geq 2)$, $P(|X^*| \geq 5)$, and $P(|X^*| \geq 9)$.

9 (a) Suppose a number X is chosen at random from $[0, 20]$ with uniform probability. Find a lower bound for the probability that X lies between 8 and 12, using Chebyshev's Inequality.

(b) Now suppose 20 real numbers are chosen independently from $[0, 20]$ with uniform probability. Find a lower bound for the probability that their average lies between 8 and 12.

(c) Now suppose 100 real numbers are chosen independently from $[0, 20]$. Find a lower bound for the probability that their average lies between 8 and 12.

10 A student's score on a particular calculus final is a random variable with values of $[0, 100]$, mean 70, and variance 25.

(a) Find a lower bound for the probability that the student's score will fall between 65 and 75.

(b) If 100 students take the final, find a lower bound for the probability that the class average will fall between 65 and 75.

11 The Pilsdorff beer company runs a fleet of trucks along the 100 mile road from Hangtown to Dry Gulch, and maintains a garage halfway in between. Each of the trucks is apt to break down at a point X miles from Hangtown, where X is a random variable uniformly distributed over $[0, 100]$.

(a) Find a lower bound for the probability $P(|X - 50| \leq 10)$.

8.2. CONTINUOUS RANDOM VARIABLES

(b) Suppose that in one bad week, 20 trucks break down. Find a lower bound for the probability $P(|A_{20} - 50| \leq 10)$, where A_{20} is the average of the distances from Hangtown at the time of breakdown.

12 A share of common stock in the Pilsdorff beer company has a price Y_n on the nth business day of the year. Finn observes that the price change $X_n = Y_{n+1} - Y_n$ appears to be a random variable with mean $\mu = 0$ and variance $\sigma^2 = 1/4$. If $Y_1 = 30$, find a lower bound for the following probabilities, under the assumption that the X_n's are mutually independent.

(a) $P(25 \leq Y_2 \leq 35)$.

(b) $P(25 \leq Y_{11} \leq 35)$.

(c) $P(25 \leq Y_{101} \leq 35)$.

13 Suppose one hundred numbers $X_1, X_2, \ldots, X_{100}$ are chosen independently at random from $[0, 20]$. Let $S = X_1 + X_2 + \cdots + X_{100}$ be the sum, $A = S/100$ the average, and $S^* = (S - 1000)/(10/\sqrt{3})$ the standardized sum. Find lower bounds for the probabilities

(a) $P(|S - 1000| \leq 100)$.

(b) $P(|A - 10| \leq 1)$.

(c) $P(|S^*| \leq \sqrt{3})$.

14 Let X be a continuous random variable normally distributed on $(-\infty, +\infty)$ with mean 0 and variance 1. Using the normal table provided in Appendix A, or the program **NormalArea**, find values for the function $f(x) = P(|X| \geq x)$ as x increases from 0 to 4.0 in steps of .25. Note that for $x \geq 0$ the table gives $NA(0, x) = P(0 \leq X \leq x)$ and thus $P(|X| \geq x) = 2(.5 - NA(0, x))$. Plot by hand the graph of $f(x)$ using these values, and the graph of the Chebyshev function $g(x) = 1/x^2$, and compare (see Exercise 3).

15 Repeat Exercise 14, but this time with mean 10 and variance 3. Note that the table in Appendix A presents values for a standard normal variable. Find the standardized version X^* for X, find values for $f^*(x) = P(|X^*| \geq x)$ as in Exercise 14, and then rescale these values for $f(x) = P(|X - 10| \geq x)$. Graph and compare this function with the Chebyshev function $g(x) = 3/x^2$.

16 Let $Z = X/Y$ where X and Y have normal densities with mean 0 and standard deviation 1. Then it can be shown that Z has a Cauchy density.

(a) Write a program to illustrate this result by plotting a bar graph of 1000 samples obtained by forming the ratio of two standard normal outcomes. Compare your bar graph with the graph of the Cauchy density. Depending upon which computer language you use, you may or may not need to tell the computer how to simulate a normal random variable. A method for doing this was described in Section 5.2.

(b) We have seen that the Law of Large Numbers does not apply to the Cauchy density (see Example 8.8). Simulate a large number of experiments with Cauchy density and compute the average of your results. Do these averages seem to be approaching a limit? If so can you explain why this might be?

17 Show that, if $X \geq 0$, then $P(X \geq a) \leq E(X)/a$.

18 (Lamperti[9]) Let X be a non-negative random variable. What is the best upper bound you can give for $P(X \geq a)$ if you know

(a) $E(X) = 20$.

(b) $E(X) = 20$ and $V(X) = 25$.

(c) $E(X) = 20$, $V(X) = 25$, and X is symmetric about its mean.

---

[9] Private communication.

Chapter 9

Central Limit Theorem

9.1 Central Limit Theorem for Bernoulli Trials

The second fundamental theorem of probability is the *Central Limit Theorem*. This theorem says that if S_n is the sum of n mutually independent random variables, then the distribution function of S_n is well-approximated by a certain type of continuous function known as a normal density function, which is given by the formula

$$f_{\mu,\sigma}(x) = \frac{1}{\sqrt{2\pi}\sigma} e^{-(x-\mu)^2/(2\sigma^2)} \ ,$$

as we have seen in Chapter 4.3. In this section, we will deal only with the case that $\mu = 0$ and $\sigma = 1$. We will call this particular normal density function the *standard normal density*, and we will denote it by $\phi(x)$:

$$\phi(x) = \frac{1}{\sqrt{2\pi}} e^{-x^2/2} \ .$$

A graph of this function is given in Figure 9.1. It can be shown that the area under any normal density equals 1.

The Central Limit Theorem tells us, quite generally, what happens when we have the sum of a large number of independent random variables each of which contributes a small amount to the total. In this section we shall discuss this theorem as it applies to the Bernoulli trials and in Section 9.2 we shall consider more general processes. We will discuss the theorem in the case that the individual random variables are identically distributed, but the theorem is true, under certain conditions, even if the individual random variables have different distributions.

Bernoulli Trials

Consider a Bernoulli trials process with probability p for success on each trial. Let $X_i = 1$ or 0 according as the ith outcome is a success or failure, and let $S_n = X_1 + X_2 + \cdots + X_n$. Then S_n is the number of successes in n trials. We know that S_n has as its distribution the binomial probabilities $b(n, p, j)$. In Section 3.2,

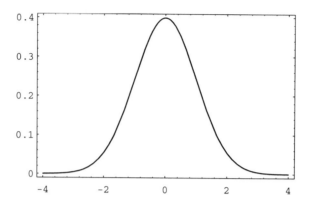

Figure 9.1: Standard normal density.

we plotted these distributions for $p = .3$ and $p = .5$ for various values of n (see Figure 3.5).

We note that the maximum values of the distributions appeared near the expected value np, which causes their spike graphs to drift off to the right as n increased. Moreover, these maximum values approach 0 as n increased, which causes the spike graphs to flatten out.

Standardized Sums

We can prevent the drifting of these spike graphs by subtracting the expected number of successes np from S_n, obtaining the new random variable $S_n - np$. Now the maximum values of the distributions will always be near 0.

To prevent the spreading of these spike graphs, we can normalize $S_n - np$ to have variance 1 by dividing by its standard deviation $\sqrt{npq}$ (see Exercise 6.2.12 and Exercise 6.2.16).

Definition 9.1 The *standardized sum* of S_n is given by

$$S_n^* = \frac{S_n - np}{\sqrt{npq}} \ .$$

S_n^* always has expected value 0 and variance 1. □

Suppose we plot a spike graph with the spikes placed at the possible values of S_n^*: $x_0, x_1, \ldots, x_n$, where

$$x_j = \frac{j - np}{\sqrt{npq}} \ . \tag{9.1}$$

We make the height of the spike at x_j equal to the distribution value $b(n, p, j)$. An example of this standardized spike graph, with $n = 270$ and $p = .3$, is shown in Figure 9.2. This graph is beautifully bell-shaped. We would like to fit a normal density to this spike graph. The obvious choice to try is the standard normal density, since it is centered at 0, just as the standardized spike graph is. In this figure, we

9.1. BERNOULLI TRIALS

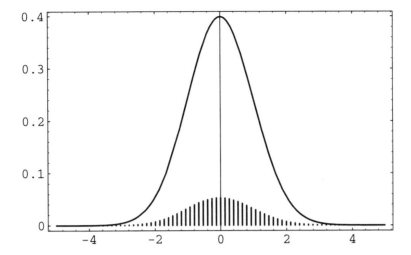

Figure 9.2: Normalized binomial distribution and standard normal density.

have drawn this standard normal density. The reader will note that a horrible thing has occurred: Even though the shapes of the two graphs are the same, the heights are quite different.

If we want the two graphs to fit each other, we must modify one of them; we choose to modify the spike graph. Since the shapes of the two graphs look fairly close, we will attempt to modify the spike graph without changing its shape. The reason for the differing heights is that the sum of the heights of the spikes equals 1, while the area under the standard normal density equals 1. If we were to draw a continuous curve through the top of the spikes, and find the area under this curve, we see that we would obtain, approximately, the sum of the heights of the spikes multiplied by the distance between consecutive spikes, which we will call ϵ. Since the sum of the heights of the spikes equals one, the area under this curve would be approximately ϵ. Thus, to change the spike graph so that the area under this curve has value 1, we need only multiply the heights of the spikes by $1/\epsilon$. It is easy to see from Equation 9.1 that

$$\epsilon = \frac{1}{\sqrt{npq}} \ .$$

In Figure 9.3 we show the standardized sum S_n^* for $n = 270$ and $p = .3$, after correcting the heights, together with the standard normal density. (This figure was produced with the program **CLTBernoulliPlot**.) The reader will note that the standard normal fits the height-corrected spike graph extremely well. In fact, one version of the Central Limit Theorem (see Theorem 9.1) says that as n increases, the standard normal density will do an increasingly better job of approximating the height-corrected spike graphs corresponding to a Bernoulli trials process with n summands.

Let us fix a value x on the x-axis and let n be a fixed positive integer. Then, using Equation 9.1, the point x_j that is closest to x has a subscript j given by the

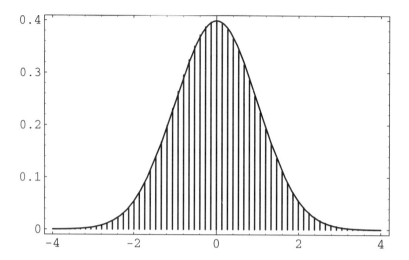

Figure 9.3: Corrected spike graph with standard normal density.

formula
$$j = \langle np + x\sqrt{npq}\rangle \ ,$$
where $\langle a \rangle$ means the integer nearest to a. Thus the height of the spike above x_j will be
$$\sqrt{npq}\, b(n,p,j) = \sqrt{npq}\, b(n,p,\langle np + x_j\sqrt{npq}\rangle) \ .$$

For large n, we have seen that the height of the spike is very close to the height of the normal density at x. This suggests the following theorem.

Theorem 9.1 (Central Limit Theorem for Binomial Distributions) For the binomial distribution $b(n,p,j)$ we have
$$\lim_{n\to\infty} \sqrt{npq}\, b(n,p,\langle np + x\sqrt{npq}\rangle) = \phi(x) \ ,$$
where $\phi(x)$ is the standard normal density.

The proof of this theorem can be carried out using Stirling's approximation from Section 3.1. We indicate this method of proof by considering the case $x = 0$. In this case, the theorem states that
$$\lim_{n\to\infty} \sqrt{npq}\, b(n,p,\langle np\rangle) = \frac{1}{\sqrt{2\pi}} = .3989\ldots \ .$$

In order to simplify the calculation, we assume that np is an integer, so that $\langle np\rangle = np$. Then
$$\sqrt{npq}\, b(n,p,np) = \sqrt{npq}\, p^{np} q^{nq} \frac{n!}{(np)!\,(nq)!} \ .$$

Recall that Stirling's formula (see Theorem 3.3) states that
$$n! \sim \sqrt{2\pi n}\, n^n e^{-n} \qquad \text{as} \quad n \to \infty \ .$$

9.1. BERNOULLI TRIALS

Using this, we have

$$\sqrt{npq}\, b(n,p,np) \sim \frac{\sqrt{npq}\, p^{np} q^{nq} \sqrt{2\pi n}\, n^n e^{-n}}{\sqrt{2\pi np}\sqrt{2\pi nq}\, (np)^{np}(nq)^{nq} e^{-np} e^{-nq}},$$

which simplifies to $1/\sqrt{2\pi}$. □

Approximating Binomial Distributions

We can use Theorem 9.1 to find approximations for the values of binomial distribution functions. If we wish to find an approximation for $b(n,p,j)$, we set

$$j = np + x\sqrt{npq}$$

and solve for x, obtaining

$$x = \frac{j - np}{\sqrt{npq}}.$$

Theorem 9.1 then says that

$$\sqrt{npq}\, b(n,p,j)$$

is approximately equal to $\phi(x)$, so

$$b(n,p,j) \approx \frac{\phi(x)}{\sqrt{npq}}$$
$$= \frac{1}{\sqrt{npq}} \phi\left(\frac{j-np}{\sqrt{npq}}\right).$$

Example 9.1 Let us estimate the probability of exactly 55 heads in 100 tosses of a coin. For this case $np = 100 \cdot 1/2 = 50$ and $\sqrt{npq} = \sqrt{100 \cdot 1/2 \cdot 1/2} = 5$. Thus $x_{55} = (55 - 50)/5 = 1$ and

$$P(S_{100} = 55) \sim \frac{\phi(1)}{5} = \frac{1}{5}\left(\frac{1}{\sqrt{2\pi}} e^{-1/2}\right)$$
$$= .0484\,.$$

To four decimal places, the actual value is .0485, and so the approximation is very good. □

The program **CLTBernoulliLocal** illustrates this approximation for any choice of n, p, and j. We have run this program for two examples. The first is the probability of exactly 50 heads in 100 tosses of a coin; the estimate is .0798, while the actual value, to four decimal places, is .0796. The second example is the probability of exactly eight sixes in 36 rolls of a die; here the estimate is .1093, while the actual value, to four decimal places, is .1196.

The individual binomial probabilities tend to 0 as n tends to infinity. In most applications we are not interested in the probability that a specific outcome occurs, but rather in the probability that the outcome lies in a given interval, say the interval $[a, b]$. In order to find this probability, we add the heights of the spike graphs for values of j between a and b. This is the same as asking for the probability that the standardized sum S_n^* lies between a^* and b^*, where a^* and b^* are the standardized values of a and b. But as n tends to infinity the sum of these areas could be expected to approach the area under the standard normal density between a^* and b^*. The *Central Limit Theorem* states that this does indeed happen.

Theorem 9.2 (Central Limit Theorem for Bernoulli Trials) Let S_n be the number of successes in n Bernoulli trials with probability p for success, and let a and b be two fixed real numbers. Define

$$a^* = \frac{a - np}{\sqrt{npq}}$$

and

$$b^* = \frac{b - np}{\sqrt{npq}} .$$

Then

$$\lim_{n \to \infty} P(a \leq S_n \leq b) = \int_{a^*}^{b^*} \phi(x)\, dx .$$

$\square$

This theorem can be proved by adding together the approximations to $b(n, p, k)$ given in Theorem 9.1. It is also a special case of the more general Central Limit Theorem (see Section 10.3).

We know from calculus that the integral on the right side of this equation is equal to the area under the graph of the standard normal density $\phi(x)$ between a and b. We denote this area by $\mathrm{NA}(a^*, b^*)$. Unfortunately, there is no simple way to integrate the function $e^{-x^2/2}$, and so we must either use a table of values or else a numerical integration program. (See Figure 9.4 for values of $\mathrm{NA}(0, z)$. A more extensive table is given in Appendix A.)

It is clear from the symmetry of the standard normal density that areas such as that between -2 and 3 can be found from this table by adding the area from 0 to 2 (same as that from -2 to 0) to the area from 0 to 3.

Approximation of Binomial Probabilities

Suppose that S_n is binomially distributed with parameters n and p. We have seen that the above theorem shows how to estimate a probability of the form

$$P(i \leq S_n \leq j) , \tag{9.2}$$

where i and j are integers between 0 and n. As we have seen, the binomial distribution can be represented as a spike graph, with spikes at the integers between 0 and n, and with the height of the kth spike given by $b(n, p, k)$. For moderate-sized

9.1. BERNOULLI TRIALS

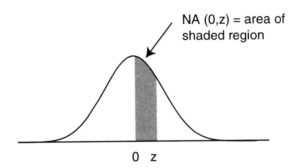

z	NA(z)	z	NA(z)	z	NA(z)	z	NA(z)
.0	.0000	1.0	.3413	2.0	.4772	3.0	.4987
.1	.0398	1.1	.3643	2.1	.4821	3.1	.4990
.2	.0793	1.2	.3849	2.2	.4861	3.2	.4993
.3	.1179	1.3	.4032	2.3	.4893	3.3	.4995
.4	.1554	1.4	.4192	2.4	.4918	3.4	.4997
.5	.1915	1.5	.4332	2.5	.4938	3.5	.4998
.6	.2257	1.6	.4452	2.6	.4953	3.6	.4998
.7	.2580	1.7	.4554	2.7	.4965	3.7	.4999
.8	.2881	1.8	.4641	2.8	.4974	3.8	.4999
.9	.3159	1.9	.4713	2.9	.4981	3.9	.5000

Figure 9.4: Table of values of $NA(0, z)$, the normal area from 0 to z.

values of n, if we standardize this spike graph, and change the heights of its spikes, in the manner described above, the sum of the heights of the spikes is approximated by the area under the standard normal density between i^* and j^*. It turns out that a slightly more accurate approximation is afforded by the area under the standard normal density between the standardized values corresponding to $(i - 1/2)$ and $(j + 1/2)$; these values are

$$i^* = \frac{i - 1/2 - np}{\sqrt{npq}}$$

and

$$j^* = \frac{j + 1/2 - np}{\sqrt{npq}} \ .$$

Thus,

$$P(i \leq S_n \leq j) \approx \mathrm{NA}\left(\frac{i - \frac{1}{2} - np}{\sqrt{npq}}, \frac{j + \frac{1}{2} - np}{\sqrt{npq}}\right) \ .$$

We now illustrate this idea with some examples.

Example 9.2 A coin is tossed 100 times. Estimate the probability that the number of heads lies between 40 and 60 (the word "between" in mathematics means inclusive of the endpoints). The expected number of heads is $100 \cdot 1/2 = 50$, and the standard deviation for the number of heads is $\sqrt{100 \cdot 1/2 \cdot 1/2} = 5$. Thus, since $n = 100$ is reasonably large, we have

$$\begin{aligned}
P(40 \leq S_n \leq 60) &\approx P\left(\frac{39.5 - 50}{5} \leq S_n^* \leq \frac{60.5 - 50}{5}\right) \\
&= P(-2.1 \leq S_n^* \leq 2.1) \\
&\approx \mathrm{NA}(-2.1, 2.1) \\
&= 2\mathrm{NA}(0, 2.1) \\
&\approx .9642 \ .
\end{aligned}$$

The actual value is .96480, to five decimal places.

Note that in this case we are asking for the probability that the outcome will not deviate by more than two standard deviations from the expected value. Had we asked for the probability that the number of successes is between 35 and 65, this would have represented three standard deviations from the mean, and, using our 1/2 correction, our estimate would be the area under the standard normal curve between -3.1 and 3.1, or $2\mathrm{NA}(0, 3.1) = .9980$. The actual answer in this case, to five places, is .99821. □

It is important to work a few problems by hand to understand the conversion from a given inequality to an inequality relating to the standardized variable. After this, one can then use a computer program that carries out this conversion, including the 1/2 correction. The program **CLTBernoulliGlobal** is such a program for estimating probabilities of the form $P(a \leq S_n \leq b)$.

9.1. BERNOULLI TRIALS

Example 9.3 Dartmouth College would like to have 1050 freshmen. This college cannot accommodate more than 1060. Assume that each applicant accepts with probability .6 and that the acceptances can be modeled by Bernoulli trials. If the college accepts 1700, what is the probability that it will have too many acceptances?

If it accepts 1700 students, the expected number of students who matriculate is $.6 \cdot 1700 = 1020$. The standard deviation for the number that accept is $\sqrt{1700 \cdot .6 \cdot .4} \approx 20$. Thus we want to estimate the probability

$$\begin{aligned} P(S_{1700} > 1060) &= P(S_{1700} \geq 1061) \\ &= P\left(S^*_{1700} \geq \frac{1060.5 - 1020}{20}\right) \\ &= P(S^*_{1700} \geq 2.025) \: . \end{aligned}$$

From Table 9.4, if we interpolate, we would estimate this probability to be $.5 - .4784 = .0216$. Thus, the college is fairly safe using this admission policy. □

Applications to Statistics

There are many important questions in the field of statistics that can be answered using the Central Limit Theorem for independent trials processes. The following example is one that is encountered quite frequently in the news. Another example of an application of the Central Limit Theorem to statistics is given in Section 9.2.

Example 9.4 One frequently reads that a poll has been taken to estimate the proportion of people in a certain population who favor one candidate over another in a race with two candidates. (This model also applies to races with more than two candidates A and B, and to ballot propositions.) Clearly, it is not possible for pollsters to ask everyone for their preference. What is done instead is to pick a subset of the population, called a sample, and ask everyone in the sample for their preference. Let p be the actual proportion of people in the population who are in favor of candidate A and let $q = 1 - p$. If we choose a sample of size n from the population, the preferences of the people in the sample can be represented by random variables $X_1, X_2, \ldots, X_n$, where $X_i = 1$ if person i is in favor of candidate A, and $X_i = 0$ if person i is in favor of candidate B. Let $S_n = X_1 + X_2 + \cdots + X_n$. If each subset of size n is chosen with the same probability, then S_n is hypergeometrically distributed. If n is small relative to the size of the population (which is typically true in practice), then S_n is approximately binomially distributed, with parameters n and p.

The pollster wants to estimate the value p. An estimate for p is provided by the value $\bar{p} = S_n/n$, which is the proportion of people in the sample who favor candidate B. The Central Limit Theorem says that the random variable $\bar{p}$ is approximately normally distributed. (In fact, our version of the Central Limit Theorem says that the distribution function of the random variable

$$S^*_n = \frac{S_n - np}{\sqrt{npq}}$$

is approximated by the standard normal density.) But we have

$$\bar{p} = \frac{S_n - np}{\sqrt{npq}}\sqrt{\frac{pq}{n}} + p \ ,$$

i.e., $\bar{p}$ is just a linear function of S_n^*. Since the distribution of S_n^* is approximated by the standard normal density, the distribution of the random variable $\bar{p}$ must also be bell-shaped. We also know how to write the mean and standard deviation of $\bar{p}$ in terms of p and n. The mean of $\bar{p}$ is just p, and the standard deviation is

$$\sqrt{\frac{pq}{n}} \ .$$

Thus, it is easy to write down the standardized version of $\bar{p}$; it is

$$\bar{p}^* = \frac{\bar{p} - p}{\sqrt{pq/n}} \ .$$

Since the distribution of the standardized version of $\bar{p}$ is approximated by the standard normal density, we know, for example, that 95% of its values will lie within two standard deviations of its mean, and the same is true of $\bar{p}$. So we have

$$P\left(p - 2\sqrt{\frac{pq}{n}} < \bar{p} < p + 2\sqrt{\frac{pq}{n}}\right) \approx .954 \ .$$

Now the pollster does not know p or q, but he can use $\bar{p}$ and $\bar{q} = 1 - \bar{p}$ in their place without too much danger. With this idea in mind, the above statement is equivalent to the statement

$$P\left(\bar{p} - 2\sqrt{\frac{\bar{p}\bar{q}}{n}} < p < \bar{p} + 2\sqrt{\frac{\bar{p}\bar{q}}{n}}\right) \approx .954 \ .$$

The resulting interval

$$\left(\bar{p} - \frac{2\sqrt{\bar{p}\bar{q}}}{\sqrt{n}}, \ \bar{p} + \frac{2\sqrt{\bar{p}\bar{q}}}{\sqrt{n}}\right)$$

is called the *95 percent confidence interval* for the unknown value of p. The name is suggested by the fact that if we use this method to estimate p in a large number of samples we should expect that in about 95 percent of the samples the true value of p is contained in the confidence interval obtained from the sample. In Exercise 11 you are asked to write a program to illustrate that this does indeed happen.

The pollster has control over the value of n. Thus, if he wants to create a 95% confidence interval with length 6%, then he should choose a value of n so that

$$\frac{2\sqrt{\bar{p}\bar{q}}}{\sqrt{n}} \leq .03 \ .$$

Using the fact that $\bar{p}\bar{q} \leq 1/4$, no matter what the value of $\bar{p}$ is, it is easy to show that if he chooses a value of n so that

$$\frac{1}{\sqrt{n}} \leq .03 \ ,$$

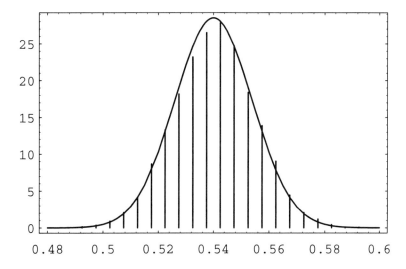

Figure 9.5: Polling simulation.

he will be safe. This is equivalent to choosing

$$n \geq 1111 \ .$$

So if the pollster chooses n to be 1200, say, and calculates $\bar{p}$ using his sample of size 1200, then 19 times out of 20 (i.e., 95% of the time), his confidence interval, which is of length 6%, will contain the true value of p. This type of confidence interval is typically reported in the news as follows: this survey has a 3% margin of error. In fact, most of the surveys that one sees reported in the paper will have sample sizes around 1000. A somewhat surprising fact is that the size of the population has apparently no effect on the sample size needed to obtain a 95% confidence interval for p with a given margin of error. To see this, note that the value of n that was needed depended only on the number .03, which is the margin of error. In other words, whether the population is of size 100,000 or 100,000,000, the pollster needs only to choose a sample of size 1200 or so to get the same accuracy of estimate of p. (We did use the fact that the sample size was small relative to the population size in the statement that S_n is approximately binomially distributed.)

In Figure 9.5, we show the results of simulating the polling process. The population is of size 100,000, and for the population, $p = .54$. The sample size was chosen to be 1200. The spike graph shows the distribution of $\bar{p}$ for 10,000 randomly chosen samples. For this simulation, the program kept track of the number of samples for which $\bar{p}$ was within 3% of .54. This number was 9648, which is close to 95% of the number of samples used.

Another way to see what the idea of confidence intervals means is shown in Figure 9.6. In this figure, we show 100 confidence intervals, obtained by computing $\bar{p}$ for 100 different samples of size 1200 from the same population as before. The reader can see that most of these confidence intervals (96, to be exact) contain the true value of p.

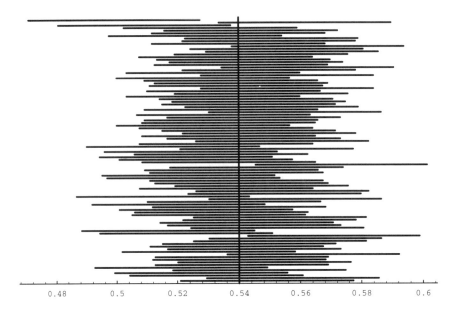

Figure 9.6: Confidence interval simulation.

The Gallup Poll has used these polling techniques in every Presidential election since 1936 (and in innumerable other elections as well). Table 9.1[1] shows the results of their efforts. The reader will note that most of the approximations to p are within 3% of the actual value of p. The sample sizes for these polls were typically around 1500. (In the table, both the predicted and actual percentages for the winning candidate refer to the percentage of the vote among the "major" political parties. In most elections, there were two major parties, but in several elections, there were three.)

This technique also plays an important role in the evaluation of the effectiveness of drugs in the medical profession. For example, it is sometimes desired to know what proportion of patients will be helped by a new drug. This proportion can be estimated by giving the drug to a subset of the patients, and determining the proportion of this sample who are helped by the drug. □

Historical Remarks

The Central Limit Theorem for Bernoulli trials was first proved by Abraham de Moivre and appeared in his book, *The Doctrine of Chances*, first published in 1718.[2]

De Moivre spent his years from age 18 to 21 in prison in France because of his Protestant background. When he was released he left France for England, where he worked as a tutor to the sons of noblemen. Newton had presented a copy of his *Principia Mathematica* to the Earl of Devonshire. The story goes that, while

---

[1] The Gallup Poll Monthly, November 1992, No. 326, p. 33. Supplemented with the help of Lydia K. Saab, The Gallup Organization.

[2] A. de Moivre, *The Doctrine of Chances*, 3d ed. (London: Millar, 1756).

9.1. BERNOULLI TRIALS

Year	Winning Candidate	Gallup Final Survey	Election Result	Deviation
1936	Roosevelt	55.7%	62.5%	6.8%
1940	Roosevelt	52.0%	55.0%	3.0%
1944	Roosevelt	51.5%	53.3%	1.8%
1948	Truman	44.5%	49.9%	5.4%
1952	Eisenhower	51.0%	55.4%	4.4%
1956	Eisenhower	59.5%	57.8%	1.7%
1960	Kennedy	51.0%	50.1%	0.9%
1964	Johnson	64.0%	61.3%	2.7%
1968	Nixon	43.0%	43.5%	0.5%
1972	Nixon	62.0%	61.8%	0.2%
1976	Carter	48.0%	50.0%	2.0%
1980	Reagan	47.0%	50.8%	3.8%
1984	Reagan	59.0%	59.1%	0.1%
1988	Bush	56.0%	53.9%	2.1%
1992	Clinton	49.0%	43.2%	5.8%
1996	Clinton	52.0%	50.1%	1.9%

Table 9.1: Gallup Poll accuracy record.

de Moivre was tutoring at the Earl's house, he came upon Newton's work and found that it was beyond him. It is said that he then bought a copy of his own and tore it into separate pages, learning it page by page as he walked around London to his tutoring jobs. De Moivre frequented the coffeehouses in London, where he started his probability work by calculating odds for gamblers. He also met Newton at such a coffeehouse and they became fast friends. De Moivre dedicated his book to Newton.

The Doctrine of Chances provides the techniques for solving a wide variety of gambling problems. In the midst of these gambling problems de Moivre rather modestly introduces his proof of the Central Limit Theorem, writing

> A Method of approximating the Sum of the Terms of the Binomial $(a + b)^n$ expanded into a Series, from whence are deduced some practical Rules to estimate the Degree of Assent which is to be given to Experiments.[3]

De Moivre's proof used the approximation to factorials that we now call Stirling's formula. De Moivre states that he had obtained this formula before Stirling but without determining the exact value of the constant $\sqrt{2\pi}$. While he says it is not really necessary to know this exact value, he concedes that knowing it "has spread a singular Elegancy on the Solution."

The complete proof and an interesting discussion of the life of de Moivre can be found in the book *Games, Gods and Gambling* by F. N. David.[4]

---

[3] ibid., p. 243.
[4] F. N. David, *Games, Gods and Gambling* (London: Griffin, 1962).

Exercises

1 Let S_{100} be the number of heads that turn up in 100 tosses of a fair coin. Use the Central Limit Theorem to estimate

(a) $P(S_{100} \leq 45)$.

(b) $P(45 < S_{100} < 55)$.

(c) $P(S_{100} > 63)$.

(d) $P(S_{100} < 57)$.

2 Let S_{200} be the number of heads that turn up in 200 tosses of a fair coin. Estimate

(a) $P(S_{200} = 100)$.

(b) $P(S_{200} = 90)$.

(c) $P(S_{200} = 80)$.

3 A true-false examination has 48 questions. June has probability 3/4 of answering a question correctly. April just guesses on each question. A passing score is 30 or more correct answers. Compare the probability that June passes the exam with the probability that April passes it.

4 Let S be the number of heads in 1,000,000 tosses of a fair coin. Use (a) Chebyshev's inequality, and (b) the Central Limit Theorem, to estimate the probability that S lies between 499,500 and 500,500. Use the same two methods to estimate the probability that S lies between 499,000 and 501,000, and the probability that S lies between 498,500 and 501,500.

5 A rookie is brought to a baseball club on the assumption that he will have a .300 batting average. (Batting average is the ratio of the number of hits to the number of times at bat.) In the first year, he comes to bat 300 times and his batting average is .267. Assume that his at bats can be considered Bernoulli trials with probability .3 for success. Could such a low average be considered just bad luck or should he be sent back to the minor leagues? Comment on the assumption of Bernoulli trials in this situation.

6 Once upon a time, there were two railway trains competing for the passenger traffic of 1000 people leaving from Chicago at the same hour and going to Los Angeles. Assume that passengers are equally likely to choose each train. How many seats must a train have to assure a probability of .99 or better of having a seat for each passenger?

7 Assume that, as in Example 9.3, Dartmouth admits 1750 students. What is the probability of too many acceptances?

8 A club serves dinner to members only. They are seated at 12-seat tables. The manager observes over a long period of time that 95 percent of the time there are between six and nine full tables of members, and the remainder of the

9.1. BERNOULLI TRIALS

time the numbers are equally likely to fall above or below this range. Assume that each member decides to come with a given probability p, and that the decisions are independent. How many members are there? What is p?

9 Let S_n be the number of successes in n Bernoulli trials with probability .8 for success on each trial. Let $A_n = S_n/n$ be the average number of successes. In each case give the value for the limit, and give a reason for your answer.

(a) $\lim_{n\to\infty} P(A_n = .8)$.

(b) $\lim_{n\to\infty} P(.7n < S_n < .9n)$.

(c) $\lim_{n\to\infty} P(S_n < .8n + .8\sqrt{n})$.

(d) $\lim_{n\to\infty} P(.79 < A_n < .81)$.

10 Find the probability that among 10,000 random digits the digit 3 appears not more than 931 times.

11 Write a computer program to simulate 10,000 Bernoulli trials with probability .3 for success on each trial. Have the program compute the 95 percent confidence interval for the probability of success based on the proportion of successes. Repeat the experiment 100 times and see how many times the true value of .3 is included within the confidence limits.

12 A balanced coin is flipped 400 times. Determine the number x such that the probability that the number of heads is between $200 - x$ and $200 + x$ is approximately .80.

13 A noodle machine in Spumoni's spaghetti factory makes about 5 percent defective noodles even when properly adjusted. The noodles are then packed in crates containing 1900 noodles each. A crate is examined and found to contain 115 defective noodles. What is the approximate probability of finding at least this many defective noodles if the machine is properly adjusted?

14 A restaurant feeds 400 customers per day. On the average 20 percent of the customers order apple pie.

(a) Give a range (called a 95 percent confidence interval) for the number of pieces of apple pie ordered on a given day such that you can be 95 percent sure that the actual number will fall in this range.

(b) How many customers must the restaurant have, on the average, to be at least 95 percent sure that the number of customers ordering pie on that day falls in the 19 to 21 percent range?

15 Recall that if X is a random variable, the *cumulative distribution function* of X is the function $F(x)$ defined by

$$F(x) = P(X \leq x) .$$

(a) Let S_n be the number of successes in n Bernoulli trials with probability p for success. Write a program to plot the cumulative distribution for S_n.

(b) Modify your program in (a) to plot the cumulative distribution $F_n^*(x)$ of the standardized random variable

$$S_n^* = \frac{S_n - np}{\sqrt{npq}} \ .$$

(c) Define the *normal distribution* $N(x)$ to be the area under the normal curve up to the value x. Modify your program in (b) to plot the normal distribution as well, and compare it with the cumulative distribution of S_n^*. Do this for $n = 10, 50$, and 100.

16 In Example 3.11, we were interested in testing the hypothesis that a new form of aspirin is effective 80 percent of the time rather than the 60 percent of the time as reported for standard aspirin. The new aspirin is given to n people. If it is effective in m or more cases, we accept the claim that the new drug is effective 80 percent of the time and if not we reject the claim. Using the Central Limit Theorem, show that you can choose the number of trials n and the critical value m so that the probability that we reject the hypothesis when it is true is less than .01 and the probability that we accept it when it is false is also less than .01. Find the smallest value of n that will suffice for this.

17 In an opinion poll it is assumed that an unknown proportion p of the people are in favor of a proposed new law and a proportion $1 - p$ are against it. A sample of n people is taken to obtain their opinion. The proportion $\bar{p}$ in favor in the sample is taken as an estimate of p. Using the Central Limit Theorem, determine how large a sample will ensure that the estimate will, with probability .95, be correct to within .01.

18 A description of a poll in a certain newspaper says that one can be 95% confident that error due to sampling will be no more than plus or minus 3 percentage points. A poll in the New York Times taken in Iowa says that "according to statistical theory, in 19 out of 20 cases the results based on such samples will differ by no more than 3 percentage points in either direction from what would have been obtained by interviewing all adult Iowans." These are both attempts to explain the concept of confidence intervals. Do both statements say the same thing? If not, which do you think is the more accurate description?

9.2 Central Limit Theorem for Discrete Independent Trials

We have illustrated the Central Limit Theorem in the case of Bernoulli trials, but this theorem applies to a much more general class of chance processes. In particular, it applies to any independent trials process such that the individual trials have finite variance. For such a process, both the normal approximation for individual terms and the Central Limit Theorem are valid.

9.2. DISCRETE INDEPENDENT TRIALS

Let $S_n = X_1 + X_2 + \cdots + X_n$ be the sum of n independent discrete random variables of an independent trials process with common distribution function $m(x)$ defined on the integers, with mean μ and variance σ^2. We have seen in Section 7.2 that the distributions for such independent sums have shapes resembling the normal curve, but the largest values drift to the right and the curves flatten out (see Figure 7.6). We can prevent this just as we did for Bernoulli trials.

Standardized Sums

Consider the standardized random variable

$$S_n^* = \frac{S_n - n\mu}{\sqrt{n\sigma^2}} \ .$$

This standardizes S_n to have expected value 0 and variance 1. If $S_n = j$, then S_n^* has the value x_j with

$$x_j = \frac{j - n\mu}{\sqrt{n\sigma^2}} \ .$$

We can construct a spike graph just as we did for Bernoulli trials. Each spike is centered at some x_j. The distance between successive spikes is

$$b = \frac{1}{\sqrt{n\sigma^2}} \ ,$$

and the height of the spike is

$$h = \sqrt{n\sigma^2} P(S_n = j) \ .$$

The case of Bernoulli trials is the special case for which $X_j = 1$ if the jth outcome is a success and 0 otherwise; then $\mu = p$ and $\sigma^2 = \sqrt{pq}$.

We now illustrate this process for two different discrete distributions. The first is the distribution m, given by

$$m = \begin{pmatrix} 1 & 2 & 3 & 4 & 5 \\ .2 & .2 & .2 & .2 & .2 \end{pmatrix} \ .$$

In Figure 9.7 we show the standardized sums for this distribution for the cases $n = 2$ and $n = 10$. Even for $n = 2$ the approximation is surprisingly good.

For our second discrete distribution, we choose

$$m = \begin{pmatrix} 1 & 2 & 3 & 4 & 5 \\ .4 & .3 & .1 & .1 & .1 \end{pmatrix} \ .$$

This distribution is quite asymmetric and the approximation is not very good for $n - 3$, but by $n = 10$ we again have an excellent approximation (see Figure 9.8). Figures 9.7 and 9.8 were produced by the program **CLTIndTrialsPlot**.

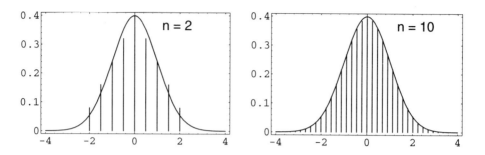

Figure 9.7: Distribution of standardized sums.

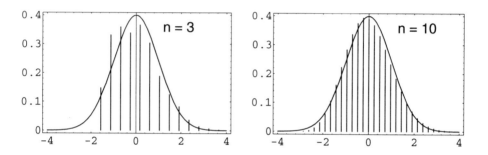

Figure 9.8: Distribution of standardized sums.

Approximation Theorem

As in the case of Bernoulli trials, these graphs suggest the following approximation theorem for the individual probabilities.

Theorem 9.3 Let $X_1, X_2, \ldots, X_n$ be an independent trials process and let $S_n = X_1 + X_2 + \cdots + X_n$. Assume that the greatest common divisor of the differences of all the values that the X_j can take on is 1. Let $E(X_j) = \mu$ and $V(X_j) = \sigma^2$. Then for n large,

$$P(S_n = j) \sim \frac{\phi(x_j)}{\sqrt{n\sigma^2}} ,$$

where $x_j = (j - n\mu)/\sqrt{n\sigma^2}$, and $\phi(x)$ is the standard normal density. □

The program **CLTIndTrialsLocal** implements this approximation. When we run this program for 6 rolls of a die, and ask for the probability that the sum of the rolls equals 21, we obtain an actual value of .09285, and a normal approximation value of .09537. If we run this program for 24 rolls of a die, and ask for the probability that the sum of the rolls is 72, we obtain an actual value of .01724 and a normal approximation value of .01705. These results show that the normal approximations are quite good.

Central Limit Theorem for a Discrete Independent Trials Process

The Central Limit Theorem for a discrete independent trials process is as follows.

Theorem 9.4 (Central Limit Theorem) Let $S_n = X_1 + X_2 + \cdots + X_n$ be the sum of n discrete independent random variables with common distribution having expected value μ and variance σ^2. Then, for $a < b$,

$$\lim_{n \to \infty} P\left(a < \frac{S_n - n\mu}{\sqrt{n\sigma^2}} < b\right) = \frac{1}{\sqrt{2\pi}} \int_a^b e^{-x^2/2} \, dx \ .$$

$\Box$

We will give the proofs of Theorems 9.3 and Theorem 9.4 in Section 10.3. Here we consider several examples.

Examples

Example 9.5 A die is rolled 420 times. What is the probability that the sum of the rolls lies between 1400 and 1550?

The sum is a random variable

$$S_{420} = X_1 + X_2 + \cdots + X_{420} \ ,$$

where each X_j has distribution

$$m_X = \begin{pmatrix} 1 & 2 & 3 & 4 & 5 & 6 \\ 1/6 & 1/6 & 1/6 & 1/6 & 1/6 & 1/6 \end{pmatrix}$$

We have seen that $\mu = E(X) = 7/2$ and $\sigma^2 = V(X) = 35/12$. Thus, $E(S_{420}) = 420 \cdot 7/2 = 1470$, $\sigma^2(S_{420}) = 420 \cdot 35/12 = 1225$, and $\sigma(S_{420}) = 35$. Therefore,

$$\begin{aligned} P(1400 \leq S_{420} \leq 1550) &\approx P\left(\frac{1399.5 - 1470}{35} \leq S_{240}^* \leq \frac{1550.5 - 1470}{35}\right) \\ &= P(-2.01 \leq S_{420}^* \leq 2.30) \\ &\approx \text{NA}(-2.01, 2.30) = .9670 \ . \end{aligned}$$

We note that the program **CLTIndTrialsGlobal** could be used to calculate these probabilities. $\Box$

Example 9.6 A student's grade point average is the average of his grades in 30 courses. The grades are based on 100 possible points and are recorded as integers. Assume that, in each course, the instructor makes an error in grading of k with probability $|p/k|$, where $k = \pm 1, \pm 2, \pm 3, \pm 4, \pm 5$. The probability of no error is then $1 - (137/30)p$. (The parameter p represents the inaccuracy of the instructor's grading.) Thus, in each course, there are two grades for the student, namely the "correct" grade and the recorded grade. So there are two average grades for the student, namely the average of the correct grades and the average of the recorded grades.

We wish to estimate the probability that these two average grades differ by less than .05 for a given student. We now assume that $p = 1/20$. We also assume that the total error is the sum S_{30} of 30 independent random variables each with distribution

$$m_X : \begin{Bmatrix} -5 & -4 & -3 & -2 & -1 & 0 & 1 & 2 & 3 & 4 & 5 \\ \frac{1}{100} & \frac{1}{80} & \frac{1}{60} & \frac{1}{40} & \frac{1}{20} & \frac{463}{600} & \frac{1}{20} & \frac{1}{40} & \frac{1}{60} & \frac{1}{80} & \frac{1}{100} \end{Bmatrix}.$$

One can easily calculate that $E(X) = 0$ and $\sigma^2(X) = 1.5$. Then we have

$$P\left(-.05 \leq \tfrac{S_{30}}{30} \leq .05\right) = P(-1.5 \leq S_{30} \leq 1.5)$$

$$= P\left(\tfrac{-1.5}{\sqrt{30 \cdot 1.5}} \leq S_{30}^* \leq \tfrac{1.5}{\sqrt{30 \cdot 1.5}}\right)$$

$$= P(-.224 \leq S_{30}^* \leq .224)$$

$$\approx \mathrm{NA}(-.224, .224) = .1772 .$$

This means that there is only a 17.7% chance that a given student's grade point average is accurate to within .05. (Thus, for example, if two candidates for valedictorian have recorded averages of 97.1 and 97.2, there is an appreciable probability that their correct averages are in the reverse order.) For a further discussion of this example, see the article by R. M. Kozelka.[5] □

A More General Central Limit Theorem

In Theorem 9.4, the discrete random variables that were being summed were assumed to be independent and identically distributed. It turns out that the assumption of identical distributions can be substantially weakened. Much work has been done in this area, with an important contribution being made by J. W. Lindeberg. Lindeberg found a condition on the sequence $\{X_n\}$ which guarantees that the distribution of the sum S_n is asymptotically normally distributed. Feller showed that Lindeberg's condition is necessary as well, in the sense that if the condition does not hold, then the sum S_n is not asymptotically normally distributed. For a precise statement of Lindeberg's Theorem, we refer the reader to Feller.[6] A sufficient condition that is stronger (but easier to state) than Lindeberg's condition, and is weaker than the condition in Theorem 9.4, is given in the following theorem.

---

[5]R. M. Kozelka, "Grade-Point Averages and the Central Limit Theorem," *American Math. Monthly*, vol. 86 (Nov 1979), pp. 773-777.

[6]W. Feller, *Introduction to Probability Theory and its Applications*, vol. 1, 3rd ed. (New York: John Wiley & Sons, 1968), p. 254.

9.2. DISCRETE INDEPENDENT TRIALS

Theorem 9.5 (Central Limit Theorem) Let X_1, X_2, ..., X_n, ... be a sequence of independent discrete random variables, and let $S_n = X_1 + X_2 + \cdots + X_n$. For each n, denote the mean and variance of X_n by μ_n and σ_n^2, respectively. Define the mean and variance of S_n to be m_n and s_n^2, respectively, and assume that $s_n \to \infty$. If there exists a constant A, such that $|X_n| \leq A$ for all n, then for $a < b$,

$$\lim_{n \to \infty} P\left(a < \frac{S_n - m_n}{s_n} < b\right) = \frac{1}{\sqrt{2\pi}} \int_a^b e^{-x^2/2}\, dx \ .$$

$\square$

The condition that $|X_n| \leq A$ for all n is sometimes described by saying that the sequence $\{X_n\}$ is uniformly bounded. The condition that $s_n \to \infty$ is necessary (see Exercise 15).

We illustrate this theorem by generating a sequence of n random distributions on the interval $[a, b]$. We then convolute these distributions to find the distribution of the sum of n experiments governed by these distributions. Finally, we standardized the distribution for the sum to have mean 0 and standard deviation 1 and compare it with the normal density. The program **CLTGeneral** carries out this procedure.

In Figure 9.9 we show the result of running this program for $[a, b] = [-2, 4]$, and $n = 1, 4$, and 10. We see that our first random distribution is quite asymmetric. By the time we choose the sum of ten such experiments we have a very good fit to the normal curve.

The above theorem essentially says that anything that can be thought of as being made up as the sum of many small independent pieces is approximately normally distributed. This brings us to one of the most important questions that was asked about genetics in the 1800's.

The Normal Distribution and Genetics

When one looks at the distribution of heights of adults of one sex in a given population, one cannot help but notice that this distribution looks like the normal distribution. An example of this is shown in Figure 9.10. This figure shows the distribution of heights of 9593 women between the ages of 21 and 74. These data come from the Health and Nutrition Examination Survey I (HANES I). For this survey, a sample of the U.S. civilian population was chosen. The survey was carried out between 1971 and 1974.

A natural question to ask is "How does this come about?". Francis Galton, an English scientist in the 19th century, studied this question, and other related questions, and constructed probability models that were of great importance in explaining the genetic effects on such attributes as height. In fact, one of the most important ideas in statistics, the idea of regression to the mean, was invented by Galton in his attempts to understand these genetic effects.

Galton was faced with an apparent contradiction. On the one hand, he knew that the normal distribution arises in situations in which many small independent effects are being summed. On the other hand, he also knew that many quantitative attributes, such as height, are strongly influenced by genetic factors: tall parents

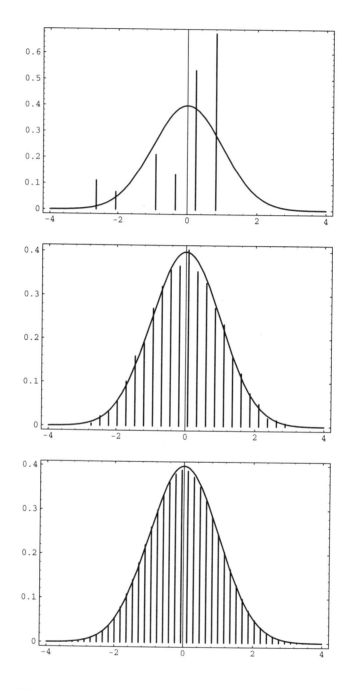

Figure 9.9: Sums of randomly chosen random variables.

9.2. DISCRETE INDEPENDENT TRIALS

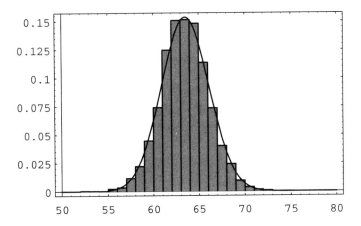

Figure 9.10: Distribution of heights of adult women.

tend to have tall offspring. Thus in this case, there seem to be two large effects, namely the parents. Galton was certainly aware of the fact that non-genetic factors played a role in determining the height of an individual. Nevertheless, unless these non-genetic factors overwhelm the genetic ones, thereby refuting the hypothesis that heredity is important in determining height, it did not seem possible for sets of parents of given heights to have offspring whose heights were normally distributed.

One can express the above problem symbolically as follows. Suppose that we choose two specific positive real numbers x and y, and then find all pairs of parents one of whom is x units tall and the other of whom is y units tall. We then look at all of the offspring of these pairs of parents. One can postulate the existence of a function $f(x, y)$ which denotes the genetic effect of the parents' heights on the heights of the offspring. One can then let W denote the effects of the non-genetic factors on the heights of the offspring. Then, for a given set of heights $\{x, y\}$, the random variable which represents the heights of the offspring is given by

$$H = f(x, y) + W ,$$

where f is a deterministic function, i.e., it gives one output for a pair of inputs $\{x, y\}$. If we assume that the effect of f is large in comparison with the effect of W, then the variance of W is small. But since f is deterministic, the variance of H equals the variance of W, so the variance of H is small. However, Galton observed from his data that the variance of the heights of the offspring of a given pair of parent heights is not small. This would seem to imply that inheritance plays a small role in the determination of the height of an individual. Later in this section, we will describe the way in which Galton got around this problem.

We will now consider the modern explanation of why certain traits, such as heights, are approximately normally distributed. In order to do so, we need to introduce some terminology from the field of genetics. The cells in a living organism that are not directly involved in the transmission of genetic material to offspring are called somatic cells, and the remaining cells are called germ cells. Organisms of a given species have their genetic information encoded in sets of physical entities, called chromosomes. The chromosomes are paired in each somatic cell. For example, human beings have 23 pairs of chromosomes in each somatic cell. The sex cells contain one chromosome from each pair. In sexual reproduction, two sex cells, one from each parent, contribute their chromosomes to create the set of chromosomes for the offspring.

Chromosomes contain many subunits, called genes. Genes consist of molecules of DNA, and one gene has, encoded in its DNA, information that leads to the regulation of proteins. In the present context, we will consider those genes containing information that has an effect on some physical trait, such as height, of the organism. The pairing of the chromosomes gives rise to a pairing of the genes on the chromosomes.

In a given species, each gene can be any one of several forms. These various forms are called alleles. One should think of the different alleles as potentially producing different effects on the physical trait in question. Of the two alleles that are found in a given gene pair in an organism, one of the alleles came from one parent and the other allele came from the other parent. The possible types of pairs of alleles (without regard to order) are called genotypes.

If we assume that the height of a human being is largely controlled by a specific gene, then we are faced with the same difficulty that Galton was. We are assuming that each parent has a pair of alleles which largely controls their heights. Since each parent contributes one allele of this gene pair to each of its offspring, there are four possible allele pairs for the offspring at this gene location. The assumption is that these pairs of alleles largely control the height of the offspring, and we are also assuming that genetic factors outweigh non-genetic factors. It follows that among the offspring we should see several modes in the height distribution of the offspring, one mode corresponding to each possible pair of alleles. This distribution does not correspond to the observed distribution of heights.

An alternative hypothesis, which does explain the observation of normally distributed heights in offspring of a given sex, is the multiple-gene hypothesis. Under this hypothesis, we assume that there are many genes that affect the height of an individual. These genes may differ in the amount of their effects. Thus, we can represent each gene pair by a random variable X_i, where the value of the random variable is the allele pair's effect on the height of the individual. Thus, for example, if each parent has two different alleles in the gene pair under consideration, then the offspring has one of four possible pairs of alleles at this gene location. Now the height of the offspring is a random variable, which can be expressed as

$$H = X_1 + X_2 + \cdots + X_n + W ,$$

if there are n genes that affect height. (Here, as before, the random variable W denotes non-genetic effects.) Although n is fixed, if it is fairly large, then Theorem 9.5 implies that the sum $X_1 + X_2 + \cdots + X_n$ is approximately normally distributed. Now, if we assume that the X_i's have a significantly larger cumulative effect than W does, then H is approximately normally distributed.

Another observed feature of the distribution of heights of adults of one sex in a population is that the variance does not seem to increase or decrease from one

9.2. DISCRETE INDEPENDENT TRIALS

generation to the next. This was known at the time of Galton, and his attempts to explain this led him to the idea of regression to the mean. This idea will be discussed further in the historical remarks at the end of the section. (The reason that we only consider one sex is that human heights are clearly sex-linked, and in general, if we have two populations that are each normally distributed, then their union need not be normally distributed.)

Using the multiple-gene hypothesis, it is easy to explain why the variance should be constant from generation to generation. We begin by assuming that for a specific gene location, there are k alleles, which we will denote by A_1, A_2, ..., A_k. We assume that the offspring are produced by random mating. By this we mean that given any offspring, it is equally likely that it came from any pair of parents in the preceding generation. There is another way to look at random mating that makes the calculations easier. We consider the set S of all of the alleles (at the given gene location) in all of the germ cells of all of the individuals in the parent generation. In terms of the set S, by random mating we mean that each pair of alleles in S is equally likely to reside in any particular offspring. (The reader might object to this way of thinking about random mating, as it allows two alleles from the same parent to end up in an offspring; but if the number of individuals in the parent population is large, then whether or not we allow this event does not affect the probabilities very much.)

For $1 \leq i \leq k$, we let p_i denote the proportion of alleles in the parent population that are of type A_i. It is clear that this is the same as the proportion of alleles in the germ cells of the parent population, assuming that each parent produces roughly the same number of germs cells. Consider the distribution of alleles in the offspring. Since each germ cell is equally likely to be chosen for any particular offspring, the distribution of alleles in the offspring is the same as in the parents.

We next consider the distribution of genotypes in the two generations. We will prove the following fact: the distribution of genotypes in the offspring generation depends only upon the distribution of alleles in the parent generation (in particular, it does not depend upon the distribution of genotypes in the parent generation). Consider the possible genotypes; there are $k(k+1)/2$ of them. Under our assumptions, the genotype A_iA_i will occur with frequency p_i^2, and the genotype A_iA_j, with $i \neq j$, will occur with frequency $2p_ip_j$. Thus, the frequencies of the genotypes depend only upon the allele frequencies in the parent generation, as claimed.

This means that if we start with a certain generation, and a certain distribution of alleles, then in all generations after the one we started with, both the allele distribution and the genotype distribution will be fixed. This last statement is known as the Hardy-Weinberg Law.

We can describe the consequences of this law for the distribution of heights among adults of one sex in a population. We recall that the height of an offspring was given by a random variable H, where

$$H = X_1 + X_2 + \cdots + X_n + W,$$

with the X_i's corresponding to the genes that affect height, and the random variable W denoting non-genetic effects. The Hardy-Weinberg Law states that for each X_i, the distribution in the offspring generation is the same as the distribution in the parent generation. Thus, if we assume that the distribution of W is roughly the same from generation to generation (or if we assume that its effects are small), then the distribution of H is the same from generation to generation. (In fact, dietary effects are part of W, and it is clear that in many human populations, diets have changed quite a bit from one generation to the next in recent times. This change is thought to be one of the reasons that humans, on the average, are getting taller. It is also the case that the effects of W are thought to be small relative to the genetic effects of the parents.)

Discussion

Generally speaking, the Central Limit Theorem contains more information than the Law of Large Numbers, because it gives us detailed information about the *shape* of the distribution of S_n^*; for large n the shape is approximately the same as the shape of the standard normal density. More specifically, the Central Limit Theorem says that if we standardize and height-correct the distribution of S_n, then the normal density function is a very good approximation to this distribution when n is large. Thus, we have a computable approximation for the distribution for S_n, which provides us with a powerful technique for generating answers for all sorts of questions about sums of independent random variables, even if the individual random variables have different distributions.

Historical Remarks

In the mid-1800's, the Belgian mathematician Quetelet[7] had shown empirically that the normal distribution occurred in real data, and had also given a method for fitting the normal curve to a given data set. Laplace[8] had shown much earlier that the sum of many independent identically distributed random variables is approximately normal. Galton knew that certain physical traits in a population appeared to be approximately normally distributed, but he did not consider Laplace's result to be a good explanation of how this distribution comes about. We give a quote from Galton that appears in the fascinating book by S. Stigler[9] on the history of statistics:

> First, let me point out a fact which Quetelet and all writers who have followed in his paths have unaccountably overlooked, and which has an intimate bearing on our work to-night. It is that, although characteristics of plants and animals conform to the law, the reason of their doing so is as yet totally unexplained. The essence of the law is that differences should be wholly due to the collective actions of a host of independent *petty* influences in various combinations...Now the processes of heredity...are not petty influences, but very important ones...The conclusion is...that the processes of heredity must work harmoniously with the law of deviation, and be themselves in some sense conformable to it.

---

[7]S. Stigler, *The History of Statistics,* (Cambridge: Harvard University Press, 1986), p. 203.
[8]ibid., p. 136
[9]ibid., p. 281.

9.2. DISCRETE INDEPENDENT TRIALS

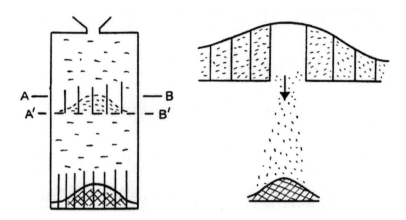

Figure 9.11: Two-stage version of the quincunx.

Galton invented a device known as a quincunx (now commonly called a Galton board), which we used in Example 3.10 to show how to physically obtain a binomial distribution. Of course, the Central Limit Theorem says that for large values of the parameter n, the binomial distribution is approximately normal. Galton used the quincunx to explain how inheritance affects the distribution of a trait among offspring.

We consider, as Galton did, what happens if we interrupt, at some intermediate height, the progress of the shot that is falling in the quincunx. The reader is referred to Figure 9.11. This figure is a drawing of Karl Pearson,[10] based upon Galton's notes. In this figure, the shot is being temporarily segregated into compartments at the line AB. (The line A'B' forms a platform on which the shot can rest.) If the line AB is not too close to the top of the quincunx, then the shot will be approximately normally distributed at this line. Now suppose that one compartment is opened, as shown in the figure. The shot from that compartment will fall, forming a normal distribution at the bottom of the quincunx. If now all of the compartments are opened, all of the shot will fall, producing the same distribution as would occur if the shot were not temporarily stopped at the line AB. But the action of stopping the shot at the line AB, and then releasing the compartments one at a time, is just the same as convoluting two normal distributions. The normal distributions at the bottom, corresponding to each compartment at the line AB, are being mixed, with their weights being the number of shot in each compartment. On the other hand, it is already known that if the shot are unimpeded, the final distribution is approximately normal. Thus, this device shows that the convolution of two normal distributions is again normal.

Galton also considered the quincunx from another perspective. He segregated into seven groups, by weight, a set of 490 sweet pea seeds. He gave 10 seeds from each of the seven group to each of seven friends, who grew the plants from the seeds. Galton found that each group produced seeds whose weights were normally

---

[10]Karl Pearson, *The Life, Letters and Labours of Francis Galton*, vol. IIIB, (Cambridge at the University Press 1930.) p. 466. Reprinted with permission.

distributed. (The sweet pea reproduces by self-pollination, so he did not need to consider the possibility of interaction between different groups.) In addition, he found that the variances of the weights of the offspring were the same for each group. This segregation into groups corresponds to the compartments at the line AB in the quincunx. Thus, the sweet peas were acting as though they were being governed by a convolution of normal distributions.

He now was faced with a problem. We have shown in Chapter 7, and Galton knew, that the convolution of two normal distributions produces a normal distribution with a larger variance than either of the original distributions. But his data on the sweet pea seeds showed that the variance of the offspring population was the same as the variance of the parent population. His answer to this problem was to postulate a mechanism that he called *reversion*, and is now called *regression to the mean*. As Stigler puts it:[11]

> The seven groups of progeny were normally distributed, but not about their parents' weight. Rather they were in every case distributed about a value that was closer to the average population weight than was that of the parent. Furthermore, this reversion followed "the simplest possible law," that is, it was linear. The average deviation of the progeny from the population average was in the same direction as that of the parent, but only a third as great. The mean progeny reverted to type, and the increased variation was just sufficient to maintain the population variability.

Galton illustrated reversion with the illustration shown in Figure 9.12.[12] The parent population is shown at the top of the figure, and the slanted lines are meant to correspond to the reversion effect. The offspring population is shown at the bottom of the figure.

Exercises

1 A die is rolled 24 times. Use the Central Limit Theorem to estimate the probability that

(a) the sum is greater than 84.

(b) the sum is equal to 84.

2 A random walker starts at 0 on the x-axis and at each time unit moves 1 step to the right or 1 step to the left with probability $1/2$. Estimate the probability that, after 100 steps, the walker is more than 10 steps from the starting position.

3 A piece of rope is made up of 100 strands. Assume that the breaking strength of the rope is the sum of the breaking strengths of the individual strands.

---

[11] ibid., p. 282.
[12] Karl Pearson, *The Life, Letters and Labours of Francis Galton*, vol. IIIA, (Cambridge at the University Press 1930.) p. 9. Reprinted with permission.

9.2. DISCRETE INDEPENDENT TRIALS

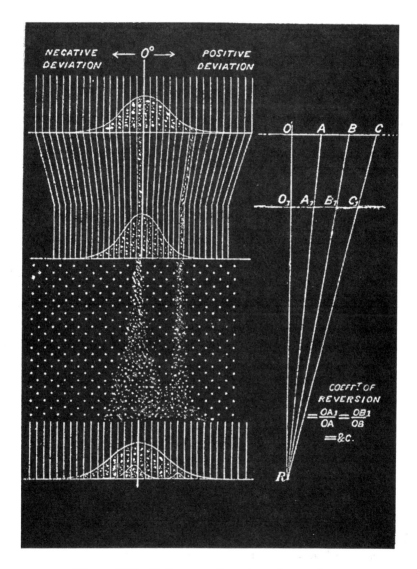

Figure 9.12: Galton's explanation of reversion.

Assume further that this sum may be considered to be the sum of an independent trials process with 100 experiments each having expected value of 10 pounds and standard deviation 1. Find the approximate probability that the rope will support a weight

(a) of 1000 pounds.

(b) of 970 pounds.

4 Write a program to find the average of 1000 random digits 0, 1, 2, 3, 4, 5, 6, 7, 8, or 9. Have the program test to see if the average lies within three standard deviations of the expected value of 4.5. Modify the program so that it repeats this simulation 1000 times and keeps track of the number of times the test is passed. Does your outcome agree with the Central Limit Theorem?

5 A die is thrown until the first time the total sum of the face values of the die is 700 or greater. Estimate the probability that, for this to happen,

(a) more than 210 tosses are required.

(b) less than 190 tosses are required.

(c) between 180 and 210 tosses, inclusive, are required.

6 A bank accepts rolls of pennies and gives 50 cents credit to a customer without counting the contents. Assume that a roll contains 49 pennies 30 percent of the time, 50 pennies 60 percent of the time, and 51 pennies 10 percent of the time.

(a) Find the expected value and the variance for the amount that the bank loses on a typical roll.

(b) Estimate the probability that the bank will lose more than 25 cents in 100 rolls.

(c) Estimate the probability that the bank will lose exactly 25 cents in 100 rolls.

(d) Estimate the probability that the bank will lose any money in 100 rolls.

(e) How many rolls does the bank need to collect to have a 99 percent chance of a net loss?

7 A surveying instrument makes an error of -2, -1, 0, 1, or 2 feet with equal probabilities when measuring the height of a 200-foot tower.

(a) Find the expected value and the variance for the height obtained using this instrument once.

(b) Estimate the probability that in 18 independent measurements of this tower, the average of the measurements is between 199 and 201, inclusive.

8 For Example 9.6 estimate $P(S_{30} = 0)$. That is, estimate the probability that the errors cancel out and the student's grade point average is correct.

9.3. CONTINUOUS INDEPENDENT TRIALS

9 Prove the Law of Large Numbers using the Central Limit Theorem.

10 Peter and Paul match pennies 10,000 times. Describe briefly what each of the following theorems tells you about Peter's fortune.

  (a) The Law of Large Numbers.

  (b) The Central Limit Theorem.

11 A tourist in Las Vegas was attracted by a certain gambling game in which the customer stakes 1 dollar on each play; a win then pays the customer 2 dollars plus the return of her stake, although a loss costs her only her stake. Las Vegas insiders, and alert students of probability theory, know that the probability of winning at this game is 1/4. When driven from the tables by hunger, the tourist had played this game 240 times. Assuming that no near miracles happened, about how much poorer was the tourist upon leaving the casino? What is the probability that she lost no money?

12 We have seen that, in playing roulette at Monte Carlo (Example 6.13), betting 1 dollar on red or 1 dollar on 17 amounts to choosing between the distributions

$$m_X = \begin{pmatrix} -1 & -1/2 & 1 \\ 18/37 & 1/37 & 18/37 \end{pmatrix}$$

or

$$m_X = \begin{pmatrix} -1 & 35 \\ 36/37 & 1/37 \end{pmatrix}$$

You plan to choose one of these methods and use it to make 100 1-dollar bets using the method chosen. Which gives you the higher probability of winning at least 20 dollars? Which gives you the higher probability of winning any money?

13 In Example 9.6 find the largest value of p that gives probability .954 that the first decimal place is correct.

14 It has been suggested that Example 9.6 is unrealistic, in the sense that the probabilities of errors are too low. Make up your own (reasonable) estimate for the distribution $m(x)$, and determine the probability that a student's grade point average is accurate to within .05. Also determine the probability that it is accurate to within .5.

15 Find a sequence of uniformly bounded discrete independent random variables $\{X_n\}$ such that the variance of their sum does not tend to ∞ as $n \to \infty$, and such that their sum is not asymptotically normally distributed.

9.3 Central Limit Theorem for Continuous Independent Trials

We have seen in Section 9.2 that the distribution function for the sum of a large number n of independent discrete random variables with mean μ and variance σ^2

tends to look like a normal density with mean $n\mu$ and variance $n\sigma^2$. What is remarkable about this result is that it holds for *any* distribution with finite mean and variance. We shall see in this section that the same result also holds true for continuous random variables having a common density function.

Let us begin by looking at some examples to see whether such a result is even plausible.

Standardized Sums

Example 9.7 Suppose we choose n random numbers from the interval $[0, 1]$ with uniform density. Let X_1, X_2, ..., X_n denote these choices, and $S_n = X_1 + X_2 + \cdots + X_n$ their sum.

We saw in Example 7.9 that the density function for S_n tends to have a normal shape, but is centered at $n/2$ and is flattened out. In order to compare the shapes of these density functions for different values of n, we proceed as in the previous section: we *standardize* S_n by defining

$$S_n^* = \frac{S_n - n\mu}{\sqrt{n}\sigma} \ .$$

Then we see that for all n we have

$$\begin{aligned} E(S_n^*) &= 0 \ , \\ V(S_n^*) &= 1 \ . \end{aligned}$$

The density function for S_n^* is just a standardized version of the density function for S_n (see Figure 9.13). □

Example 9.8 Let us do the same thing, but now choose numbers from the interval $[0, +\infty)$ with an exponential density with parameter λ. Then (see Example 6.26)

$$\begin{aligned} \mu &= E(X_i) = \frac{1}{\lambda} \ , \\ \sigma^2 &= V(X_j) = \frac{1}{\lambda^2} \ . \end{aligned}$$

Here we know the density function for S_n explicitly (see Section 7.2). We can use Corollary 5.1 to calculate the density function for S_n^*. We obtain

$$\begin{aligned} f_{S_n}(x) &= \frac{\lambda e^{-\lambda x}(\lambda x)^{n-1}}{(n-1)!} \ , \\ f_{S_n^*}(x) &= \frac{\sqrt{n}}{\lambda} f_{S_n}\left(\frac{\sqrt{n}x + n}{\lambda}\right) \ . \end{aligned}$$

The graph of the density function for S_n^* is shown in Figure 9.14. □

9.3. CONTINUOUS INDEPENDENT TRIALS

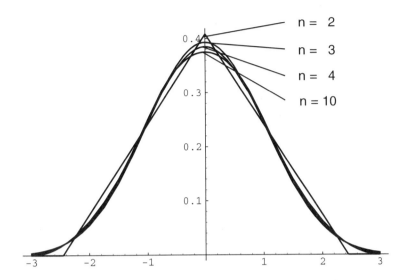

Figure 9.13: Density function for S_n^* (uniform case, $n = 2, 3, 4, 10$).

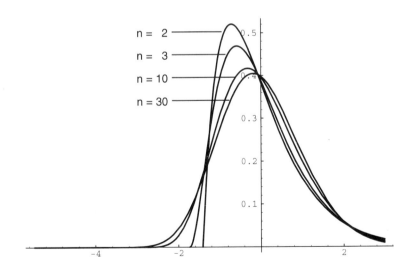

Figure 9.14: Density function for S_n^* (exponential case, $n = 2, 3, 10, 30$, $\lambda = 1$).

These examples make it seem plausible that the density function for the normalized random variable S_n^* for large n will look very much like the normal density with mean 0 and variance 1 in the continuous case as well as in the discrete case. The Central Limit Theorem makes this statement precise.

Central Limit Theorem

Theorem 9.6 (Central Limit Theorem) Let $S_n = X_1 + X_2 + \cdots + X_n$ be the sum of n independent continuous random variables with common density function p having expected value μ and variance σ^2. Let $S_n^* = (S_n - n\mu)/\sqrt{n}\sigma$. Then we have, for all $a < b$,

$$\lim_{n \to \infty} P(a < S_n^* < b) = \frac{1}{\sqrt{2\pi}} \int_a^b e^{-x^2/2}\, dx \ .$$

□

We shall give a proof of this theorem in Section 10.3. We will now look at some examples.

Example 9.9 Suppose a surveyor wants to measure a known distance, say of 1 mile, using a transit and some method of triangulation. He knows that because of possible motion of the transit, atmospheric distortions, and human error, any one measurement is apt to be slightly in error. He plans to make several measurements and take an average. He assumes that his measurements are independent random variables with a common distribution of mean $\mu = 1$ and standard deviation $\sigma = .0002$ (so, if the errors are approximately normally distributed, then his measurements are within 1 foot of the correct distance about 65% of the time). What can he say about the average?

He can say that if n is large, the average S_n/n has a density function that is approximately normal, with mean $\mu = 1$ mile, and standard deviation $\sigma = .0002/\sqrt{n}$ miles.

How many measurements should he make to be reasonably sure that his average lies within .0001 of the true value? The Chebyshev inequality says

$$P\left(\left|\frac{S_n}{n} - \mu\right| \geq .0001\right) \leq \frac{(.0002)^2}{n(10^{-8})} = \frac{4}{n} \ ,$$

so that we must have $n \geq 80$ before the probability that his error is less than .0001 exceeds .95.

We have already noticed that the estimate in the Chebyshev inequality is not always a good one, and here is a case in point. If we assume that n is large enough so that the density for S_n is approximately normal, then we have

$$P\left(\left|\frac{S_n}{n} - \mu\right| < .0001\right) = P(-.5\sqrt{n} < S_n^* < +.5\sqrt{n})$$

$$\approx \frac{1}{\sqrt{2\pi}} \int_{-.5\sqrt{n}}^{+.5\sqrt{n}} e^{-x^2/2}\, dx \ ,$$

9.3. CONTINUOUS INDEPENDENT TRIALS

and this last expression is greater than .95 if $.5\sqrt{n} \geq 2$. This says that it suffices to take $n = 16$ measurements for the same results. This second calculation is stronger, but depends on the assumption that $n = 16$ is large enough to establish the normal density as a good approximation to S_n^*, and hence to S_n. The Central Limit Theorem here says nothing about how large n has to be. In most cases involving sums of independent random variables, a good rule of thumb is that for $n \geq 30$, the approximation is a good one. In the present case, if we assume that the errors are approximately normally distributed, then the approximation is probably fairly good even for $n = 16$. $\square$

Estimating the Mean

Example 9.10 (Continuation of Example 9.9) Now suppose our surveyor is measuring an unknown distance with the same instruments under the same conditions. He takes 36 measurements and averages them. How sure can he be that his measurement lies within .0002 of the true value?

Again using the normal approximation, we get

$$
\begin{aligned}
P\left(\left|\frac{S_n}{n} - \mu\right| < .0002\right) &= P(|S_n^*| < .5\sqrt{n}) \\
&\approx \frac{2}{\sqrt{2\pi}} \int_{-3}^{3} e^{-x^2/2}\, dx \\
&\approx .997\ .
\end{aligned}
$$

This means that the surveyor can be 99.7 percent sure that his average is within .0002 of the true value. To improve his confidence, he can take more measurements, or require less accuracy, or improve the quality of his measurements (i.e., reduce the variance σ^2). In each case, the Central Limit Theorem gives quantitative information about the confidence of a measurement process, assuming always that the normal approximation is valid.

Now suppose the surveyor does not know the mean or standard deviation of his measurements, but assumes that they are independent. How should he proceed?

Again, he makes several measurements of a known distance and averages them. As before, the average error is approximately normally distributed, but now with unknown mean and variance. $\square$

Sample Mean

If he knows the variance σ^2 of the error distribution is .0002, then he can estimate the mean μ by taking the *average*, or *sample mean* of, say, 36 measurements:

$$\bar{\mu} = \frac{x_1 + x_2 + \cdots + x_n}{n}\ ,$$

where $n = 36$. Then, as before, $E(\bar{\mu}) = \mu$. Moreover, the preceding argument shows that

$$P(|\bar{\mu} - \mu| < .0002) \approx .997\ .$$

The interval $(\bar{\mu} - .0002, \bar{\mu} + .0002)$ is called *the 99.7% confidence interval* for μ (see Example 9.4).

Sample Variance

If he does not know the variance σ^2 of the error distribution, then he can estimate σ^2 by the *sample variance*:

$$\bar{\sigma}^2 = \frac{(x_1 - \bar{\mu})^2 + (x_2 - \bar{\mu})^2 + \cdots + (x_n - \bar{\mu})^2}{n},$$

where $n = 36$. The Law of Large Numbers, applied to the random variables $(X_i - \bar{\mu})^2$, says that for large n, the sample variance $\bar{\sigma}^2$ lies close to the variance σ^2, so that the surveyor can use $\bar{\sigma}^2$ in place of σ^2 in the argument above.

Experience has shown that, in most practical problems of this type, the sample variance is a good estimate for the variance, and can be used in place of the variance to determine confidence levels for the sample mean. This means that we can rely on the Law of Large Numbers for estimating the variance, and the Central Limit Theorem for estimating the mean.

We can check this in some special cases. Suppose we know that the error distribution is *normal*, with unknown mean and variance. Then we can take a sample of n measurements, find the sample mean $\bar{\mu}$ and sample variance $\bar{\sigma}^2$, and form

$$T_n^* = \frac{S_n - n\bar{\mu}}{\sqrt{n}\bar{\sigma}},$$

where $n = 36$. We expect T_n^* to be a good approximation for S_n^* for large n.

t-Density

The statistician W. S. Gosset[13] has shown that in this case T_n^* has a density function that is not normal but rather a *t-density* with n degrees of freedom. (The number n of degrees of freedom is simply a parameter which tells us which t-density to use.) In this case we can use the t-density in place of the normal density to determine confidence levels for μ. As n increases, the t-density approaches the normal density. Indeed, even for $n = 8$ the t-density and normal density are practically the same (see Figure 9.15).

Exercises

Notes on computer problems:

(a) Simulation: Recall (see Corollary 5.2) that

$$X = F^{-1}(rnd)$$

---

[13] W. S. Gosset discovered the distribution we now call the t-distribution while working for the Guinness Brewery in Dublin. He wrote under the pseudonym "Student." The results discussed here first appeared in Student, "The Probable Error of a Mean," *Biometrika*, vol. 6 (1908), pp. 1-24.

9.3. CONTINUOUS INDEPENDENT TRIALS

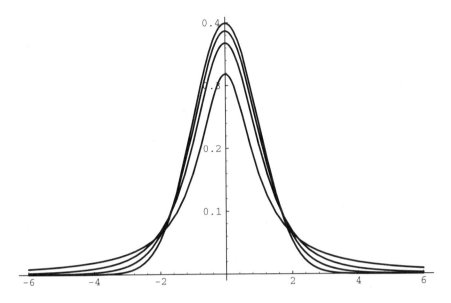

Figure 9.15: Graph of t–density for $n = 1, 3, 8$ and the normal density with $\mu = 0, \sigma = 1$.

will simulate a random variable with density $f(x)$ and distribution

$$F(X) = \int_{-\infty}^{x} f(t)\,dt \ .$$

In the case that $f(x)$ is a normal density function with mean μ and standard deviation σ, where neither F nor F^{-1} can be expressed in closed form, use instead

$$X = \sigma\sqrt{-2\log(rnd)}\cos 2\pi(rnd) + \mu \ .$$

(b) Bar graphs: you should aim for about 20 to 30 bars (of equal width) in your graph. You can achieve this by a good choice of the range $[x\text{min}, x\text{min}]$ and the number of bars (for instance, $[\mu - 3\sigma, \mu + 3\sigma]$ with 30 bars will work in many cases). Experiment!

1. Let X be a continuous random variable with mean $\mu(X)$ and variance $\sigma^2(X)$, and let $X^* = (X - \mu)/\sigma$ be its standardized version. Verify directly that $\mu(X^*) = 0$ and $\sigma^2(X^*) = 1$.

2. Let $\{X_k\}$, $1 \leq k \leq n$, be a sequence of independent random variables, all with mean 0 and variance 1, and let S_n, S_n^*, and A_n be their sum, standardized sum, and average, respectively. Verify directly that $S_n^* = S_n/\sqrt{n} = \sqrt{n}A_n$.

3. Let $\{X_k\}$, $1 \leq k \leq n$, be a sequence of random variables, all with mean μ and variance σ^2, and $Y_k = X_k^*$ be their standardized versions. Let S_n and T_n be the sum of the X_k and Y_k, and S_n^* and T_n^* their standardized version. Show that $S_n^* = T_n^* = T_n/\sqrt{n}$.

4 Suppose we choose independently 25 numbers at random (uniform density) from the interval $[0, 20]$. Write the normal densities that approximate the densities of their sum S_{25}, their standardized sum S_{25}^*, and their average A_{25}.

5 Write a program to choose independently 25 numbers at random from $[0, 20]$, compute their sum S_{25}, and repeat this experiment 1000 times. Make a bar graph for the density of S_{25} and compare it with the normal approximation of Exercise 4. How good is the fit? Now do the same for the standardized sum S_{25}^* and the average A_{25}.

6 In general, the Central Limit Theorem gives a better estimate than Chebyshev's inequality for the average of a sum. To see this, let A_{25} be the average calculated in Exercise 5, and let N be the normal approximation for A_{25}. Modify your program in Exercise 5 to provide a table of the function $F(x) = P(|A_{25} - 10| \geq x) =$ fraction of the total of 1000 trials for which $|A_{25} - 10| \geq x$. Do the same for the function $f(x) = P(|N - 10| \geq x)$. (You can use the normal table, Table 9.4, or the procedure **NormalArea** for this.) Now plot on the same axes the graphs of $F(x)$, $f(x)$, and the Chebyshev function $g(x) = 4/(3x^2)$. How do $f(x)$ and $g(x)$ compare as estimates for $F(x)$?

7 The Central Limit Theorem says the sums of independent random variables tend to look normal, no matter what crazy distribution the individual variables have. Let us test this by a computer simulation. Choose independently 25 numbers from the interval $[0, 1]$ with the probability density $f(x)$ given below, and compute their sum S_{25}. Repeat this experiment 1000 times, and make up a bar graph of the results. Now plot on the same graph the density $\phi(x) =$ normal $(x, \mu(S_{25}), \sigma(S_{25}))$. How well does the normal density fit your bar graph in each case?

(a) $f(x) = 1$.

(b) $f(x) = 2x$.

(c) $f(x) = 3x^2$.

(d) $f(x) = 4|x - 1/2|$.

(e) $f(x) = 2 - 4|x - 1/2|$.

8 Repeat the experiment described in Exercise 7 but now choose the 25 numbers from $[0, \infty)$, using $f(x) = e^{-x}$.

9 How large must n be before $S_n = X_1 + X_2 + \cdots + X_n$ is approximately normal? This number is often surprisingly small. Let us explore this question with a computer simulation. Choose n numbers from $[0, 1]$ with probability density $f(x)$, where $n = 3, 6, 12, 20$, and $f(x)$ is each of the densities in Exercise 7. Compute their sum S_n, repeat this experiment 1000 times, and make up a bar graph of 20 bars of the results. How large must n be before you get a good fit?

9.3. CONTINUOUS INDEPENDENT TRIALS

10 A surveyor is measuring the height of a cliff known to be about 1000 feet. He assumes his instrument is properly calibrated and that his measurement errors are independent, with mean $\mu = 0$ and variance $\sigma^2 = 10$. He plans to take n measurements and form the average. Estimate, using (a) Chebyshev's inequality and (b) the normal approximation, how large n should be if he wants to be 95 percent sure that his average falls within 1 foot of the true value. Now estimate, using (a) and (b), what value should σ^2 have if he wants to make only 10 measurements with the same confidence?

11 The price of one share of stock in the Pilsdorff Beer Company (see Exercise 8.2.12) is given by Y_n on the nth day of the year. Finn observes that the differences $X_n = Y_{n+1} - Y_n$ appear to be independent random variables with a common distribution having mean $\mu = 0$ and variance $\sigma^2 = 1/4$. If $Y_1 = 100$, estimate the probability that Y_{365} is

(a) ≥ 100.

(b) ≥ 110.

(c) ≥ 120.

12 Test your conclusions in Exercise 11 by computer simulation. First choose 364 numbers X_i with density $f(x) = \text{normal}(x, 0, 1/4)$. Now form the sum $Y_{365} = 100 + X_1 + X_2 + \cdots + X_{364}$, and repeat this experiment 200 times. Make up a bar graph on $[50, 150]$ of the results, superimposing the graph of the approximating normal density. What does this graph say about your answers in Exercise 11?

13 Physicists say that particles in a long tube are constantly moving back and forth along the tube, each with a velocity V_k (in cm/sec) at any given moment that is normally distributed, with mean $\mu = 0$ and variance $\sigma^2 = 1$. Suppose there are 10^{20} particles in the tube.

(a) Find the mean and variance of the average velocity of the particles.

(b) What is the probability that the average velocity is $\geq 10^{-9}$ cm/sec?

14 An astronomer makes n measurements of the distance between Jupiter and a particular one of its moons. Experience with the instruments used leads her to believe that for the proper units the measurements will be normally distributed with mean d, the true distance, and variance 16. She performs a series of n measurements. Let

$$A_n = \frac{X_1 + X_2 + \cdots + X_n}{n}$$

be the average of these measurements.

(a) Show that

$$P\left(A_n - \frac{8}{\sqrt{n}} \leq d \leq A_n + \frac{8}{\sqrt{n}}\right) \approx .95.$$

(b) When nine measurements were taken, the average of the distances turned out to be 23.2 units. Putting the observed values in (a) gives the *95 percent confidence interval* for the unknown distance d. Compute this interval.

(c) Why not say in (b) more simply that the probability is .95 that the value of d lies in the computed confidence interval?

(d) What changes would you make in the above procedure if you wanted to compute a 99 percent confidence interval?

15 Plot a bar graph similar to that in Figure 9.10 for the heights of the midparents in Galton's data as given in Appendix B and compare this bar graph to the appropriate normal curve.

Chapter 10

Generating Functions

10.1 Generating Functions for Discrete Distributions

So far we have considered in detail only the two most important attributes of a random variable, namely, the mean and the variance. We have seen how these attributes enter into the fundamental limit theorems of probability, as well as into all sorts of practical calculations. We have seen that the mean and variance of a random variable contain important information about the random variable, or, more precisely, about the distribution function of that variable. Now we shall see that the mean and variance do *not* contain *all* the available information about the density function of a random variable. To begin with, it is easy to give examples of different distribution functions which have the same mean and the same variance. For instance, suppose X and Y are random variables, with distributions

$$p_X = \begin{pmatrix} 1 & 2 & 3 & 4 & 5 & 6 \\ 0 & 1/4 & 1/2 & 0 & 0 & 1/4 \end{pmatrix},$$

$$p_Y = \begin{pmatrix} 1 & 2 & 3 & 4 & 5 & 6 \\ 1/4 & 0 & 0 & 1/2 & 1/4 & 0 \end{pmatrix}.$$

Then with these choices, we have $E(X) = E(Y) = 7/2$ and $V(X) = V(Y) = 9/4$, and yet certainly p_X and p_Y are quite different density functions.

This raises a question: If X is a random variable with range $\{x_1, x_2, \ldots\}$ of at most countable size, and distribution function $p = p_X$, and if we know its mean $\mu = E(X)$ and its variance $\sigma^2 = V(X)$, then what else do we need to know to determine p completely?

Moments

A nice answer to this question, at least in the case that X has finite range, can be given in terms of the *moments* of X, which are numbers defined as follows:

$$\begin{aligned}\mu_k &= k\text{th moment of } X \\ &= E(X^k) \\ &= \sum_{j=1}^{\infty}(x_j)^k p(x_j) ,\end{aligned}$$

provided the sum converges. Here $p(x_j) = P(X = x_j)$.

In terms of these moments, the mean μ and variance σ^2 of X are given simply by

$$\begin{aligned}\mu &= \mu_1, \\ \sigma^2 &= \mu_2 - \mu_1^2 ,\end{aligned}$$

so that a knowledge of the first two moments of X gives us its mean and variance. But a knowledge of *all* the moments of X determines its distribution function p completely.

Moment Generating Functions

To see how this comes about, we introduce a new variable t, and define a function $g(t)$ as follows:

$$\begin{aligned}g(t) &= E(e^{tX}) \\ &= \sum_{k=0}^{\infty}\frac{\mu_k t^k}{k!} \\ &= E\left(\sum_{k=0}^{\infty}\frac{X^k t^k}{k!}\right) \\ &= \sum_{j=1}^{\infty} e^{tx_j} p(x_j) .\end{aligned}$$

We call $g(t)$ the *moment generating function* for X, and think of it as a convenient bookkeeping device for describing the moments of X. Indeed, if we differentiate $g(t)$ n times and then set $t = 0$, we get μ_n:

$$\begin{aligned}\left.\frac{d^n}{dt^n}g(t)\right|_{t=0} &= g^{(n)}(0) \\ &= \left.\sum_{k=n}^{\infty}\frac{k!\,\mu_k t^{k-n}}{(k-n)!\,k!}\right|_{t=0} \\ &= \mu_n .\end{aligned}$$

It is easy to calculate the moment generating function for simple examples.

10.1. DISCRETE DISTRIBUTIONS

Examples

Example 10.1 Suppose X has range $\{1, 2, 3, \ldots, n\}$ and $p_X(j) = 1/n$ for $1 \leq j \leq n$ (uniform distribution). Then

$$\begin{aligned} g(t) &= \sum_{j=1}^{n} \frac{1}{n} e^{tj} \\ &= \frac{1}{n}(e^t + e^{2t} + \cdots + e^{nt}) \\ &= \frac{e^t(e^{nt} - 1)}{n(e^t - 1)} \ . \end{aligned}$$

If we use the expression on the right-hand side of the second line above, then it is easy to see that

$$\begin{aligned} \mu_1 &= g'(0) = \frac{1}{n}(1 + 2 + 3 + \cdots + n) = \frac{n+1}{2} , \\ \mu_2 &= g''(0) = \frac{1}{n}(1 + 4 + 9 + \cdots + n^2) = \frac{(n+1)(2n+1)}{6} , \end{aligned}$$

and that $\mu = \mu_1 = (n+1)/2$ and $\sigma^2 = \mu_2 - \mu_1^2 = (n^2 - 1)/12$. □

Example 10.2 Suppose now that X has range $\{0, 1, 2, 3, \ldots, n\}$ and $p_X(j) = \binom{n}{j} p^j q^{n-j}$ for $0 \leq j \leq n$ (binomial distribution). Then

$$\begin{aligned} g(t) &= \sum_{j=0}^{n} e^{tj} \binom{n}{j} p^j q^{n-j} \\ &= \sum_{j=0}^{n} \binom{n}{j} (pe^t)^j q^{n-j} \\ &= (pe^t + q)^n \ . \end{aligned}$$

Note that

$$\begin{aligned} \mu_1 = g'(0) &= n(pe^t + q)^{n-1} pe^t \big|_{t=0} = np , \\ \mu_2 = g''(0) &= n(n-1)p^2 + np , \end{aligned}$$

so that $\mu = \mu_1 = np$, and $\sigma^2 = \mu_2 - \mu_1^2 = np(1-p)$, as expected. □

Example 10.3 Suppose X has range $\{1, 2, 3, \ldots\}$ and $p_X(j) = q^{j-1}p$ for all j (geometric distribution). Then

$$\begin{aligned} g(t) &= \sum_{j=1}^{\infty} e^{tj} q^{j-1} p \\ &= \frac{pe^t}{1 - qe^t} \ . \end{aligned}$$

Here

$$\mu_1 = g'(0) = \left.\frac{pe^t}{(1-qe^t)^2}\right|_{t=0} = \frac{1}{p},$$

$$\mu_2 = g''(0) = \left.\frac{pe^t + pqe^{2t}}{(1-qe^t)^3}\right|_{t=0} = \frac{1+q}{p^2},$$

$\mu = \mu_1 = 1/p$, and $\sigma^2 = \mu_2 - \mu_1^2 = q/p^2$, as computed in Example 6.26. □

Example 10.4 Let X have range $\{0, 1, 2, 3, \ldots\}$ and let $p_X(j) = e^{-\lambda}\lambda^j/j!$ for all j (Poisson distribution with mean λ). Then

$$g(t) = \sum_{j=0}^{\infty} e^{tj}\frac{e^{-\lambda}\lambda^j}{j!}$$

$$= e^{-\lambda}\sum_{j=0}^{\infty} \frac{(\lambda e^t)^j}{j!}$$

$$= e^{-\lambda}e^{\lambda e^t} = e^{\lambda(e^t-1)}.$$

Then

$$\mu_1 = g'(0) = \left.e^{\lambda(e^t-1)}\lambda e^t\right|_{t=0} = \lambda,$$

$$\mu_2 = g''(0) = \left.e^{\lambda(e^t-1)}(\lambda^2 e^{2t} + \lambda e^t)\right|_{t=0} = \lambda^2 + \lambda,$$

$\mu = \mu_1 = \lambda$, and $\sigma^2 = \mu_2 - \mu_1^2 = \lambda$.

The variance of the Poisson distribution is easier to obtain in this way than directly from the definition (as was done in Exercise 6.2.30). □

Moment Problem

Using the moment generating function, we can now show, at least in the case of a discrete random variable with finite range, that its distribution function is completely determined by its moments.

Theorem 10.1 Let X be a discrete random variable with finite range

$$\{x_1, x_2, \ldots, x_n\},$$

and moments $\mu_k = E(X^k)$. Then the moment series

$$g(t) = \sum_{k=0}^{\infty} \frac{\mu_k t^k}{k!}$$

converges for all t to an infinitely differentiable function $g(t)$.

Proof. We know that

$$\mu_k = \sum_{j=1}^{n} (x_j)^k p(x_j).$$

10.1. DISCRETE DISTRIBUTIONS

If we set $M = \max |x_j|$, then we have

$$\begin{aligned} |\mu_k| &\leq \sum_{j=1}^{n} |x_j|^k p(x_j) \\ &\leq M^k \cdot \sum_{j=1}^{n} p(x_j) = M^k \ . \end{aligned}$$

Hence, for all N we have

$$\sum_{k=0}^{N} \left|\frac{\mu_k t^k}{k!}\right| \leq \sum_{k=0}^{N} \frac{(M|t|)^k}{k!} \leq e^{M|t|} \ ,$$

which shows that the moment series converges for all t. Since it is a power series, we know that its sum is infinitely differentiable.

This shows that the μ_k determine $g(t)$. Conversely, since $\mu_k = g^{(k)}(0)$, we see that $g(t)$ determines the μ_k. □

Theorem 10.2 Let X be a discrete random variable with finite range $\{x_1, x_2, \ldots, x_n\}$, distribution function p, and moment generating function g. Then g is uniquely determined by p, and conversely.

Proof. We know that p determines g, since

$$g(t) = \sum_{j=1}^{n} e^{tx_j} p(x_j) \ .$$

In this formula, we set $a_j = p(x_j)$ and, after choosing n convenient distinct values t_i of t, we set $b_i = g(t_i)$. Then we have

$$b_i = \sum_{j=1}^{n} e^{t_i x_j} a_j \ ,$$

or, in matrix notation

$$\mathbf{B} = \mathbf{MA} \ .$$

Here $\mathbf{B} = (b_i)$ and $\mathbf{A} = (a_j)$ are column n-vectors, and $\mathbf{M} = (e^{t_i x_j})$ is an $n \times n$ matrix.

We can solve this matrix equation for $\mathbf{A}$:

$$\mathbf{A} = \mathbf{M}^{-1}\mathbf{B} \ ,$$

provided only that the matrix $\mathbf{M}$ is *invertible* (i.e., provided that the determinant of $\mathbf{M}$ is different from 0). We can always arrange for this by choosing the values $t_i = i - 1$, since then the determinant of $\mathbf{M}$ is the *Vandermonde determinant*

$$\det \begin{pmatrix} 1 & 1 & 1 & \cdots & 1 \\ e^{tx_1} & e^{tx_2} & e^{tx_3} & \cdots & e^{tx_n} \\ e^{2tx_1} & e^{2tx_2} & e^{2tx_3} & \cdots & e^{2tx_n} \\ \cdots & & & & \\ e^{(n-1)tx_1} & e^{(n-1)tx_2} & e^{(n-1)tx_3} & \cdots & e^{(n-1)tx^n} \end{pmatrix}$$

of the e^{x_i}, with value $\prod_{i<j}(e^{x_i} - e^{x_j})$. This determinant is always different from 0 if the x_j are distinct. $\square$

If we delete the hypothesis that X have finite range in the above theorem, then the conclusion is no longer necessarily true.

Ordinary Generating Functions

In the special but important case where the x_j are all nonnegative integers, $x_j = j$, we can prove this theorem in a simpler way.

In this case, we have

$$g(t) = \sum_{j=0}^{n} e^{tj} p(j) ,$$

and we see that $g(t)$ is a *polynomial* in e^t. If we write $z = e^t$, and define the function h by

$$h(z) = \sum_{j=0}^{n} z^j p(j) ,$$

then $h(z)$ is a polynomial in z containing the same information as $g(t)$, and in fact

$$\begin{aligned} h(z) &= g(\log z) , \\ g(t) &= h(e^t) . \end{aligned}$$

The function $h(z)$ is often called the *ordinary generating function* for X. Note that $h(1) = g(0) = 1$, $h'(1) = g'(0) = \mu_1$, and $h''(1) = g''(0) - g'(0) = \mu_2 - \mu_1$. It follows from all this that if we know $g(t)$, then we know $h(z)$, and if we know $h(z)$, then we can find the $p(j)$ by Taylor's formula:

$$\begin{aligned} p(j) &= \text{coefficient of } z^j \text{ in } h(z) \\ &= \frac{h^{(j)}(0)}{j!} . \end{aligned}$$

For example, suppose we know that the moments of a certain discrete random variable X are given by

$$\begin{aligned} \mu_0 &= 1 , \\ \mu_k &= \frac{1}{2} + \frac{2^k}{4} , \quad \text{for } k \geq 1 . \end{aligned}$$

Then the moment generating function g of X is

$$\begin{aligned} g(t) &= \sum_{k=0}^{\infty} \frac{\mu_k t^k}{k!} \\ &= 1 + \frac{1}{2} \sum_{k=1}^{\infty} \frac{t^k}{k!} + \frac{1}{4} \sum_{k=1}^{\infty} \frac{(2t)^k}{k!} \\ &= \frac{1}{4} + \frac{1}{2} e^t + \frac{1}{4} e^{2t} . \end{aligned}$$

10.1. DISCRETE DISTRIBUTIONS

This is a polynomial in $z = e^t$, and

$$h(z) = \frac{1}{4} + \frac{1}{2}z + \frac{1}{4}z^2 \ .$$

Hence, X must have range $\{0, 1, 2\}$, and p must have values $\{1/4, 1/2, 1/4\}$.

Properties

Both the moment generating function g and the ordinary generating function h have many properties useful in the study of random variables, of which we can consider only a few here. In particular, if X is any discrete random variable and $Y = X + a$, then

$$\begin{aligned} g_Y(t) &= E(e^{tY}) \\ &= E(e^{t(X+a)}) \\ &= e^{ta} E(e^{tX}) \\ &= e^{ta} g_X(t) \ , \end{aligned}$$

while if $Y = bX$, then

$$\begin{aligned} g_Y(t) &= E(e^{tY}) \\ &= E(e^{tbX}) \\ &= g_X(bt) \ . \end{aligned}$$

In particular, if

$$X^* = \frac{X - \mu}{\sigma} \ ,$$

then (see Exercise 11)

$$g_{X^*}(t) = e^{-\mu t/\sigma} g_X\left(\frac{t}{\sigma}\right) \ .$$

If X and Y are *independent* random variables and $Z = X + Y$ is their sum, with p_X, p_Y, and p_Z the associated distribution functions, then we have seen in Chapter 7 that p_Z is the *convolution* of p_X and p_Y, and we know that convolution involves a rather complicated calculation. But for the generating functions we have instead the simple relations

$$\begin{aligned} g_Z(t) &= g_X(t) g_Y(t) \ , \\ h_Z(z) &= h_X(z) h_Y(z) \ , \end{aligned}$$

that is, g_Z is simply the *product* of g_X and g_Y, and similarly for h_Z.

To see this, first note that if X and Y are independent, then e^{tX} and e^{tY} are independent (see Exercise 5.2.38), and hence

$$E(e^{tX} e^{tY}) = E(e^{tX}) E(e^{tY}) \ .$$

It follows that

$$\begin{aligned}g_Z(t) &= E(e^{tZ}) = E(e^{t(X+Y)}) \\ &= E(e^{tX})E(e^{tY}) \\ &= g_X(t)g_Y(t) \ ,\end{aligned}$$

and, replacing t by $\log z$, we also get

$$h_Z(z) = h_X(z)h_Y(z) \ .$$

Example 10.5 If X and Y are independent discrete random variables with range $\{0, 1, 2, \ldots, n\}$ and binomial distribution

$$p_X(j) = p_Y(j) = \binom{n}{j}p^j q^{n-j} \ ,$$

and if $Z = X + Y$, then we know (cf. Section 7.1) that the range of X is

$$\{0, 1, 2, \ldots, 2n\}$$

and X has binomial distribution

$$p_Z(j) = (p_X * p_Y)(j) = \binom{2n}{j}p^j q^{2n-j} \ .$$

Here we can easily verify this result by using generating functions. We know that

$$\begin{aligned}g_X(t) = g_Y(t) &= \sum_{j=0}^{n} e^{tj}\binom{n}{j}p^j q^{n-j} \\ &= (pe^t + q)^n \ ,\end{aligned}$$

and

$$h_X(z) = h_Y(z) = (pz + q)^n \ .$$

Hence, we have

$$g_Z(t) = g_X(t)g_Y(t) = (pe^t + q)^{2n} \ ,$$

or, what is the same,

$$\begin{aligned}h_Z(z) &= h_X(z)h_Y(z) = (pz + q)^{2n} \\ &= \sum_{j=0}^{2n}\binom{2n}{j}(pz)^j q^{2n-j} \ ,\end{aligned}$$

from which we can see that the coefficient of z^j is just $p_Z(j) = \binom{2n}{j}p^j q^{2n-j}$. □

10.1. DISCRETE DISTRIBUTIONS

Example 10.6 If X and Y are independent discrete random variables with the non-negative integers $\{0, 1, 2, 3, \ldots\}$ as range, and with geometric distribution function

$$p_X(j) = p_Y(j) = q^j p ,$$

then

$$g_X(t) = g_Y(t) = \frac{p}{1 - qe^t} ,$$

and if $Z = X + Y$, then

$$\begin{aligned} g_Z(t) &= g_X(t)g_Y(t) \\ &= \frac{p^2}{1 - 2qe^t + q^2 e^{2t}} . \end{aligned}$$

If we replace e^t by z, we get

$$\begin{aligned} h_Z(z) &= \frac{p^2}{(1 - qz)^2} \\ &= p^2 \sum_{k=0}^{\infty} (k+1) q^k z^k , \end{aligned}$$

and we can read off the values of $p_Z(j)$ as the coefficient of z^j in this expansion for $h(z)$, even though $h(z)$ is not a polynomial in this case. The distribution p_Z is a negative binomial distribution (see Section 5.1). □

Here is a more interesting example of the power and scope of the method of generating functions.

Heads or Tails

Example 10.7 In the coin-tossing game discussed in Example 1.4, we now consider the question "When is Peter first in the lead?"

Let X_k describe the outcome of the kth trial in the game

$$X_k = \begin{cases} +1, & \text{if } k\text{th toss is heads,} \\ -1, & \text{if } k\text{th toss is tails.} \end{cases}$$

Then the X_k are independent random variables describing a Bernoulli process. Let $S_0 = 0$, and, for $n \geq 1$, let

$$S_n = X_1 + X_2 + \cdots + X_n .$$

Then S_n describes Peter's fortune after n trials, and Peter is first in the lead after n trials if $S_k \leq 0$ for $1 \leq k < n$ and $S_n = 1$.

Now this can happen when $n = 1$, in which case $S_1 = X_1 = 1$, or when $n > 1$, in which case $S_1 = X_1 = -1$. In the latter case, $S_k = 0$ for $k = n - 1$, and perhaps for other k between 1 and n. Let m be the *least* such value of k; then $S_m = 0$ and

$S_k < 0$ for $1 \leq k < m$. In this case Peter loses on the first trial, regains his initial position in the next $m - 1$ trials, and gains the lead in the next $n - m$ trials.

Let p be the probability that the coin comes up heads, and let $q = 1 - p$. Let r_n be the probability that Peter is first in the lead after n trials. Then from the discussion above, we see that

$$r_n = 0, \quad \text{if } n \text{ even},$$
$$r_1 = p \quad (= \text{probability of heads in a single toss}),$$
$$r_n = q(r_1 r_{n-2} + r_3 r_{n-4} + \cdots + r_{n-2} r_1), \quad \text{if } n > 1, n \text{ odd}.$$

Now let T describe the time (that is, the number of trials) required for Peter to take the lead. Then T is a random variable, and since $P(T = n) = r_n$, r is the distribution function for T.

We introduce the generating function $h_T(z)$ for T:

$$h_T(z) = \sum_{n=0}^{\infty} r_n z^n.$$

Then, by using the relations above, we can verify the relation

$$h_T(z) = pz + qz(h_T(z))^2.$$

If we solve this quadratic equation for $h_T(z)$, we get

$$h_T(z) = \frac{1 \pm \sqrt{1 - 4pqz^2}}{2qz} = \frac{2pz}{1 \mp \sqrt{1 - 4pqz^2}}.$$

Of these two solutions, we want the one that has a convergent power series in z (i.e., that is finite for $z = 0$). Hence we choose

$$h_T(z) = \frac{1 - \sqrt{1 - 4pqz^2}}{2qz} = \frac{2pz}{1 + \sqrt{1 - 4pqz^2}}.$$

Now we can ask: What is the probability that Peter is *ever* in the lead? This probability is given by (see Exercise 10)

$$\sum_{n=0}^{\infty} r_n = h_T(1) = \frac{1 - \sqrt{1 - 4pq}}{2q}$$
$$= \frac{1 - |p - q|}{2q}$$
$$= \begin{cases} p/q, & \text{if } p < q, \\ 1, & \text{if } p \geq q, \end{cases}$$

so that Peter is sure to be in the lead eventually if $p \geq q$.

How long will it take? That is, what is the expected value of T? This value is given by

$$E(T) = h_T'(1) = \begin{cases} 1/(p - q), & \text{if } p > q, \\ \infty, & \text{if } p = q. \end{cases}$$

10.1. DISCRETE DISTRIBUTIONS

This says that if $p > q$, then Peter can expect to be in the lead by about $1/(p-q)$ trials, but if $p = q$, he can expect to wait a long time.

A related problem, known as the Gambler's Ruin problem, is studied in Exercise 23 and in Section 12.2. □

Exercises

1. Find the generating functions, both ordinary $h(z)$ and moment $g(t)$, for the following discrete probability distributions.

   (a) The distribution describing a fair coin.

   (b) The distribution describing a fair die.

   (c) The distribution describing a die that always comes up 3.

   (d) The uniform distribution on the set $\{n, n+1, n+2, \ldots, n+k\}$.

   (e) The binomial distribution on $\{n, n+1, n+2, \ldots, n+k\}$.

   (f) The geometric distribution on $\{0, 1, 2, \ldots,\}$ with $p(j) = 2/3^{j+1}$.

2. For each of the distributions (a) through (d) of Exercise 1 calculate the first and second moments, μ_1 and μ_2, directly from their definition, and verify that $h(1) = 1$, $h'(1) = \mu_1$, and $h''(1) = \mu_2 - \mu_1$.

3. Let p be a probability distribution on $\{0, 1, 2\}$ with moments $\mu_1 = 1$, $\mu_2 = 3/2$.

   (a) Find its ordinary generating function $h(z)$.

   (b) Using (a), find its moment generating function.

   (c) Using (b), find its first six moments.

   (d) Using (a), find p_0, p_1, and p_2.

4. In Exercise 3, the probability distribution is completely determined by its first two moments. Show that this is always true for any probability distribution on $\{0, 1, 2\}$. *Hint*: Given μ_1 and μ_2, find $h(z)$ as in Exercise 3 and use $h(z)$ to determine p.

5. Let p and p' be the two distributions

$$p = \begin{pmatrix} 1 & 2 & 3 & 4 & 5 \\ 1/3 & 0 & 0 & 2/3 & 0 \end{pmatrix},$$

$$p' = \begin{pmatrix} 1 & 2 & 3 & 4 & 5 \\ 0 & 2/3 & 0 & 0 & 1/3 \end{pmatrix}.$$

   (a) Show that p and p' have the same first and second moments, but not the same third and fourth moments.

   (b) Find the ordinary and moment generating functions for p and p'.

6 Let p be the probability distribution

$$p = \begin{pmatrix} 0 & 1 & 2 \\ 0 & 1/3 & 2/3 \end{pmatrix},$$

and let $p_n = p * p * \cdots * p$ be the n-fold convolution of p with itself.

(a) Find p_2 by direct calculation (see Definition 7.1).

(b) Find the ordinary generating functions $h(z)$ and $h_2(z)$ for p and p_2, and verify that $h_2(z) = (h(z))^2$.

(c) Find $h_n(z)$ from $h(z)$.

(d) Find the first two moments, and hence the mean and variance, of p_n from $h_n(z)$. Verify that the mean of p_n is n times the mean of p.

(e) Find those integers j for which $p_n(j) > 0$ from $h_n(z)$.

7 Let X be a discrete random variable with values in $\{0, 1, 2, \ldots, n\}$ and moment generating function $g(t)$. Find, in terms of $g(t)$, the generating functions for

(a) $-X$.

(b) $X + 1$.

(c) $3X$.

(d) $aX + b$.

8 Let $X_1, X_2, \ldots, X_n$ be an independent trials process, with values in $\{0, 1\}$ and mean $\mu = 1/3$. Find the ordinary and moment generating functions for the distribution of

(a) $S_1 = X_1$. *Hint*: First find X_1 explicitly.

(b) $S_2 = X_1 + X_2$.

(c) $S_n = X_1 + X_2 + \cdots + X_n$.

(d) $A_n = S_n/n$.

(e) $S_n^* = (S_n - n\mu)/\sqrt{n\sigma^2}$.

9 Let X and Y be random variables with values in $\{1, 2, 3, 4, 5, 6\}$ with distribution functions p_X and p_Y given by

$$p_X(j) = a_j,$$
$$p_Y(j) = b_j.$$

(a) Find the ordinary generating functions $h_X(z)$ and $h_Y(z)$ for these distributions.

(b) Find the ordinary generating function $h_Z(z)$ for the distribution $Z = X + Y$.

(c) Show that $h_Z(z)$ cannot ever have the form
$$h_Z(z) = \frac{z^2 + z^3 + \cdots + z^{12}}{11}.$$

Hint: h_X and h_Y must have at least one nonzero root, but $h_Z(z)$ in the form given has no nonzero real roots.

It follows from this observation that there is no way to load two dice so that the probability that a given sum will turn up when they are tossed is the same for all sums (i.e., that all outcomes are equally likely).

10 Show that if
$$h(z) = \frac{1 - \sqrt{1 - 4pqz^2}}{2qz},$$
then
$$h(1) = \begin{cases} p/q, & \text{if } p \leq q, \\ 1, & \text{if } p \geq q, \end{cases}$$
and
$$h'(1) = \begin{cases} 1/(p-q), & \text{if } p > q, \\ \infty, & \text{if } p = q. \end{cases}$$

11 Show that if X is a random variable with mean μ and variance σ^2, and if $X^* = (X - \mu)/\sigma$ is the standardized version of X, then
$$g_{X^*}(t) = e^{-\mu t/\sigma} g_X\left(\frac{t}{\sigma}\right).$$

10.2 Branching Processes

Historical Background

In this section we apply the theory of generating functions to the study of an important chance process called a *branching process*.

Until recently it was thought that the theory of branching processes originated with the following problem posed by Francis Galton in the *Educational Times* in 1873.[1]

> Problem 4001: A large nation, of whom we will only concern ourselves with the adult males, N in number, and who each bear separate surnames, colonise a district. Their law of population is such that, in each generation, a_0 per cent of the adult males have no male children who reach adult life; a_1 have one such male child; a_2 have two; and so on up to a_5 who have five.
>
> Find (1) what proportion of the surnames will have become extinct after r generations; and (2) how many instances there will be of the same surname being held by m persons.

---
[1] D. G. Kendall, "Branching Processes Since 1873," *Journal of London Mathematics Society*, vol. 41 (1966), p. 386.

The first attempt at a solution was given by Reverend H. W. Watson. Because of a mistake in algebra, he incorrectly concluded that a family name would always die out with probability 1. However, the methods that he employed to solve the problems were, and still are, the basis for obtaining the correct solution.

Heyde and Seneta discovered an earlier communication by Bienaymé (1845) that anticipated Galton and Watson by 28 years. Bienaymé showed, in fact, that he was aware of the correct solution to Galton's problem. Heyde and Seneta in their book *I. J. Bienaymé: Statistical Theory Anticipated*,[2] give the following translation from Bienaymé's paper:

> If . . . the mean of the number of male children who replace the number of males of the preceding generation were less than unity, it would be easily realized that families are dying out due to the disappearance of the members of which they are composed. However, the analysis shows further that when this mean is equal to unity families tend to disappear, although less rapidly . . . .
>
> The analysis also shows clearly that if the mean ratio is greater than unity, the probability of the extinction of families with the passing of time no longer reduces to certainty. It only approaches a finite limit, which is fairly simple to calculate and which has the singular characteristic of being given by one of the roots of the equation (in which the number of generations is made infinite) which is not relevant to the question when the mean ratio is less than unity.[3]

Although Bienaymé does not give his reasoning for these results, he did indicate that he intended to publish a special paper on the problem. The paper was never written, or at least has never been found. In his communication Bienaymé indicated that he was motivated by the same problem that occurred to Galton. The opening paragraph of his paper as translated by Heyde and Seneta says,

> A great deal of consideration has been given to the possible multiplication of the numbers of mankind; and recently various very curious observations have been published on the fate which allegedly hangs over the aristocrary and middle classes; the families of famous men, etc. This fate, it is alleged, will inevitably bring about the disappearance of the so-called *families fermées*.[4]

A much more extensive discussion of the history of branching processes may be found in two papers by David G. Kendall.[5]

---

[2]C. C. Heyde and E. Seneta, *I. J. Bienaymé: Statistical Theory Anticipated* (New York: Springer Verlag, 1977).

[3]ibid., pp. 117–118.

[4]ibid., p. 118.

[5]D. G. Kendall, "Branching Processes Since 1873," pp. 385–406; and "The Genealogy of Genealogy: Branching Processes Before (and After) 1873," *Bulletin London Mathematics Society*, vol. 7 (1975), pp. 225–253.

10.2. BRANCHING PROCESSES

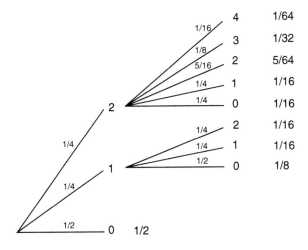

Figure 10.1: Tree diagram for Example 10.8.

Branching processes have served not only as crude models for population growth but also as models for certain physical processes such as chemical and nuclear chain reactions.

Problem of Extinction

We turn now to the first problem posed by Galton (i.e., the problem of finding the probability of extinction for a branching process). We start in the 0th generation with 1 male parent. In the first generation we shall have 0, 1, 2, 3, ... male offspring with probabilities $p_0, p_1, p_2, p_3, \ldots$. If in the first generation there are k offspring, then in the second generation there will be $X_1 + X_2 + \cdots + X_k$ offspring, where $X_1, X_2, \ldots, X_k$ are independent random variables, each with the common distribution $p_0, p_1, p_2, \ldots$. This description enables us to construct a tree, and a tree measure, for any number of generations.

Examples

Example 10.8 Assume that $p_0 = 1/2$, $p_1 = 1/4$, and $p_2 = 1/4$. Then the tree measure for the first two generations is shown in Figure 10.1.

Note that we use the theory of sums of independent random variables to assign branch probabilities. For example, if there are two offspring in the first generation, the probability that there will be two in the second generation is

$$\begin{aligned} P(X_1 + X_2 = 2) &= p_0 p_2 + p_1 p_1 + p_2 p_0 \\ &= \frac{1}{2} \cdot \frac{1}{4} + \frac{1}{4} \cdot \frac{1}{4} + \frac{1}{4} \cdot \frac{1}{2} = \frac{5}{16} . \end{aligned}$$

We now study the probability that our process dies out (i.e., that at some generation there are no offspring).

Let d_m be the probability that the process dies out by the mth generation. Of course, $d_0 = 0$. In our example, $d_1 = 1/2$ and $d_2 = 1/2 + 1/8 + 1/16 = 11/16$ (see Figure 10.1). Note that we must add the probabilities for all paths that lead to 0 by the mth generation. It is clear from the definition that

$$0 = d_0 \leq d_1 \leq d_2 \leq \cdots \leq 1 \ .$$

Hence, d_m converges to a limit d, $0 \leq d \leq 1$, and d is the probability that the process will ultimately die out. It is this value that we wish to determine. We begin by expressing the value d_m in terms of all possible outcomes on the first generation. If there are j offspring in the first generation, then to die out by the mth generation, each of these lines must die out in $m-1$ generations. Since they proceed independently, this probability is $(d_{m-1})^j$. Therefore

$$d_m = p_0 + p_1 d_{m-1} + p_2(d_{m-1})^2 + p_3(d_{m-1})^3 + \cdots \ . \tag{10.1}$$

Let $h(z)$ be the ordinary generating function for the p_i:

$$h(z) = p_0 + p_1 z + p_2 z^2 + \cdots \ .$$

Using this generating function, we can rewrite Equation 10.1 in the form

$$d_m = h(d_{m-1}) \ . \tag{10.2}$$

Since $d_m \to d$, by Equation 10.2 we see that the value d that we are looking for satisfies the equation

$$d = h(d) \ . \tag{10.3}$$

One solution of this equation is always $d = 1$, since

$$1 = p_0 + p_1 + p_2 + \cdots \ .$$

This is where Watson made his mistake. He assumed that 1 was the only solution to Equation 10.3. To examine this question more carefully, we first note that solutions to Equation 10.3 represent intersections of the graphs of

$$y = z$$

and

$$y = h(z) = p_0 + p_1 z + p_2 z^2 + \cdots \ .$$

Thus we need to study the graph of $y = h(z)$. We note that $h(0) = p_0$. Also,

$$h'(z) = p_1 + 2p_2 z + 3p_3 z^2 + \cdots \ , \tag{10.4}$$

and

$$h''(z) = 2p_2 + 3 \cdot 2p_3 z + 4 \cdot 3p_4 z^2 + \cdots \ .$$

From this we see that for $z \geq 0$, $h'(z) \geq 0$ and $h''(z) \geq 0$. Thus for nonnegative z, $h(z)$ is an increasing function and is concave upward. Therefore the graph of

10.2. BRANCHING PROCESSES

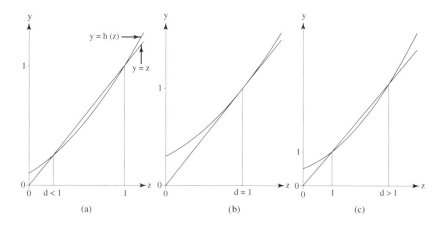

Figure 10.2: Graphs of $y = z$ and $y = h(z)$.

$y = h(z)$ can intersect the line $y = z$ in at most two points. Since we know it must intersect the line $y = z$ at $(1,1)$, we know that there are just three possibilities, as shown in Figure 10.2.

In case (a) the equation $d = h(d)$ has roots $\{d, 1\}$ with $0 \le d < 1$. In the second case (b) it has only the one root $d = 1$. In case (c) it has two roots $\{1, d\}$ where $1 < d$. Since we are looking for a solution $0 \le d \le 1$, we see in cases (b) and (c) that our only solution is 1. In these cases we can conclude that the process will die out with probability 1. However in case (a) we are in doubt. We must study this case more carefully.

From Equation 10.4 we see that

$$h'(1) = p_1 + 2p_2 + 3p_3 + \cdots = m \; ,$$

where m is the expected number of offspring produced by a single parent. In case (a) we have $h'(1) > 1$, in (b) $h'(1) = 1$, and in (c) $h'(1) < 1$. Thus our three cases correspond to $m > 1$, $m = 1$, and $m < 1$. We assume now that $m > 1$. Recall that $d_0 = 0$, $d_1 = h(d_0) = p_0$, $d_2 = h(d_1)$, ..., and $d_n = h(d_{n-1})$. We can construct these values geometrically, as shown in Figure 10.3.

We can see geometrically, as indicated for d_0, d_1, d_2, and d_3 in Figure 10.3, that the points $(d_i, h(d_i))$ will always lie above the line $y = z$. Hence, they must converge to the first intersection of the curves $y = z$ and $y = h(z)$ (i.e., to the root $d < 1$). This leads us to the following theorem. $\square$

Theorem 10.3 Consider a branching process with generating function $h(z)$ for the number of offspring of a given parent. Let d be the smallest root of the equation $z = h(z)$. If the mean number m of offspring produced by a single parent is ≤ 1, then $d = 1$ and the process dies out with probability 1. If $m > 1$ then $d < 1$ and the process dies out with probability d. $\square$

We shall often want to know the probability that a branching process dies out by a particular generation, as well as the limit of these probabilities. Let d_n be

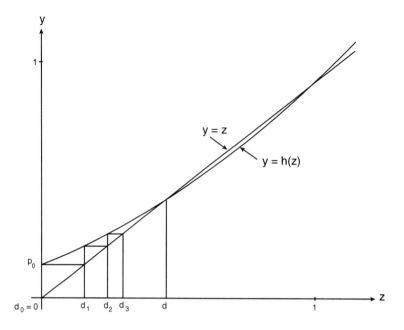

Figure 10.3: Geometric determination of d.

the probability of dying out by the nth generation. Then we know that $d_1 = p_0$. We know further that $d_n = h(d_{n-1})$ where $h(z)$ is the generating function for the number of offspring produced by a single parent. This makes it easy to compute these probabilities.

The program **Branch** calculates the values of d_n. We have run this program for 12 generations for the case that a parent can produce at most two offspring and the probabilities for the number produced are $p_0 = .2$, $p_1 = .5$, and $p_2 = .3$. The results are given in Table 10.1.

We see that the probability of dying out by 12 generations is about .6. We shall see in the next example that the probability of eventually dying out is 2/3, so that even 12 generations is not enough to give an accurate estimate for this probability.

We now assume that at most two offspring can be produced. Then

$$h(z) = p_0 + p_1 z + p_2 z^2 \ .$$

In this simple case the condition $z = h(z)$ yields the equation

$$d = p_0 + p_1 d + p_2 d^2 \ ,$$

which is satisfied by $d = 1$ and $d = p_0/p_2$. Thus, in addition to the root $d = 1$ we have the second root $d = p_0/p_2$. The mean number m of offspring produced by a single parent is

$$m = p_1 + 2p_2 = 1 - p_0 - p_2 + 2p_2 = 1 - p_0 + p_2 \ .$$

Thus, if $p_0 > p_2$, $m < 1$ and the second root is > 1. If $p_0 = p_2$, we have a double root $d = 1$. If $p_0 < p_2$, $m > 1$ and the second root d is less than 1 and represents the probability that the process will die out.

10.2. BRANCHING PROCESSES

Generation	Probability of dying out
1	.2
2	.312
3	.385203
4	.437116
5	.475879
6	.505878
7	.529713
8	.549035
9	.564949
10	.578225
11	.589416
12	.598931

Table 10.1: Probability of dying out.

$$p_0 = .2092$$
$$p_1 = .2584$$
$$p_2 = .2360$$
$$p_3 = .1593$$
$$p_4 = .0828$$
$$p_5 = .0357$$
$$p_6 = .0133$$
$$p_7 = .0042$$
$$p_8 = .0011$$
$$p_9 = .0002$$
$$p_{10} = .0000$$

Table 10.2: Distribution of number of female children.

Example 10.9 Keyfitz[6] compiled and analyzed data on the continuation of the female family line among Japanese women. His estimates at the basic probability distribution for the number of female children born to Japanese women of ages 45–49 in 1960 are given in Table 10.2.

The expected number of girls in a family is then 1.837 so the probability d of extinction is less than 1. If we run the program **Branch**, we can estimate that d is in fact only about .324. □

Distribution of Offspring

So far we have considered only the first of the two problems raised by Galton, namely the probability of extinction. We now consider the second problem, that is, the distribution of the number Z_n of offspring in the nth generation. The exact form of the distribution is not known except in very special cases. We shall see,

---

[6]N. Keyfitz, *Introduction to the Mathematics of Population*, rev. ed. (Reading, PA: Addison Wesley, 1977).

however, that we can describe the limiting behavior of Z_n as $n \to \infty$.

We first show that the generating function $h_n(z)$ of the distribution of Z_n can be obtained from $h(z)$ for any branching process.

We recall that the value of the generating function at the value z for any random variable X can be written as

$$h(z) = E(z^X) = p_0 + p_1 z + p_2 z^2 + \cdots .$$

That is, $h(z)$ is the expected value of an experiment which has outcome z^j with probability p_j.

Let $S_n = X_1 + X_2 + \cdots + X_n$ where each X_j has the same integer-valued distribution (p_j) with generating function $k(z) = p_0 + p_1 z + p_2 z^2 + \cdots$. Let $k_n(z)$ be the generating function of S_n. Then using one of the properties of ordinary generating functions discussed in Section 10.1, we have

$$k_n(z) = (k(z))^n ,$$

since the X_j's are independent and all have the same distribution.

Consider now the branching process Z_n. Let $h_n(z)$ be the generating function of Z_n. Then

$$\begin{aligned} h_{n+1}(z) &= E(z^{Z_{n+1}}) \\ &= \sum_k E(z^{Z_{n+1}} | Z_n = k) P(Z_n = k) . \end{aligned}$$

If $Z_n = k$, then $Z_{n+1} = X_1 + X_2 + \cdots + X_k$ where $X_1, X_2, \ldots, X_k$ are independent random variables with common generating function $h(z)$. Thus

$$E(z^{Z_{n+1}} | Z_n = k) = E(z^{X_1 + X_2 + \cdots + X_k}) = (h(z))^k ,$$

and

$$h_{n+1}(z) = \sum_k (h(z))^k P(Z_n = k) .$$

But

$$h_n(z) = \sum_k P(Z_n = k) z^k .$$

Thus,

$$h_{n+1}(z) = h_n(h(z)) . \tag{10.5}$$

Hence the generating function for Z_2 is $h_2(z) = h(h(z))$, for Z_3 is

$$h_3(z) = h(h(h(z))) ,$$

and so forth. From this we see also that

$$h_{n+1}(z) = h(h_n(z)) . \tag{10.6}$$

If we differentiate Equation 10.6 and use the chain rule we have

$$h'_{n+1}(z) = h'(h_n(z)) h'_n(z) .$$

10.2. BRANCHING PROCESSES

Putting $z = 1$ and using the fact that $h_n(1) = 1$ and $h'_n(1) = m_n =$ the mean number of offspring in the n'th generation, we have

$$m_{n+1} = m \cdot m_n \ .$$

Thus, $m_2 = m \cdot m = m^2$, $m_3 = m \cdot m^2 = m^3$, and in general

$$m_n = m^n \ .$$

Thus, for a branching process with $m > 1$, the mean number of offspring grows exponentially at a rate m.

Examples

Example 10.10 For the branching process of Example 10.8 we have

$$\begin{aligned} h(z) &= 1/2 + (1/4)z + (1/4)z^2 \ , \\ h_2(z) &= h(h(z)) = 1/2 + (1/4)[1/2 + (1/4)z + (1/4)z^2] \\ &\quad + (1/4)[1/2 + (1/4)z + (1/4)z^2]^2 \\ &= 11/16 + (1/8)z + (9/64)z^2 + (1/32)z^3 + (1/64)z^4 \ . \end{aligned}$$

The probabilities for the number of offspring in the second generation agree with those obtained directly from the tree measure (see Figure 1). $\square$

It is clear that even in the simple case of at most two offspring, we cannot easily carry out the calculation of $h_n(z)$ by this method. However, there is one special case in which this can be done.

Example 10.11 Assume that the probabilities $p_1, p_2, \ldots$ form a geometric series: $p_k = bc^{k-1}$, $k = 1, 2, \ldots$, with $0 < b \leq 1 - c$ and

$$\begin{aligned} p_0 &= 1 - p_1 - p_2 - \cdots \\ &= 1 - b - bc - bc^2 - \cdots \\ &= 1 - \frac{b}{1 - c} \ . \end{aligned}$$

Then the generating function $h(z)$ for this distribution is

$$\begin{aligned} h(z) &= p_0 + p_1 z + p_2 z^2 + \cdots \\ &= 1 - \frac{b}{1-c} + bz + bcz^2 + bc^2 z^3 + \cdots \\ &= 1 - \frac{b}{1-c} + \frac{bz}{1-cz} \ . \end{aligned}$$

From this we find

$$h'(z) = \frac{bcz}{(1-cz)^2} + \frac{b}{1-cz} = \frac{b}{(1-cz)^2}$$

and
$$m = h'(1) = \frac{b}{(1-c)^2}.$$

We know that if $m \leq 1$ the process will surely die out and $d = 1$. To find the probability d when $m > 1$ we must find a root $d < 1$ of the equation
$$z = h(z),$$
or
$$z = 1 - \frac{b}{1-c} + \frac{bz}{1-cz}.$$

This leads us to a quadratic equation. We know that $z = 1$ is one solution. The other is found to be
$$d = \frac{1-b-c}{c(1-c)}.$$

It is easy to verify that $d < 1$ just when $m > 1$.

It is possible in this case to find the distribution of Z_n. This is done by first finding the generating function $h_n(z)$.[7] The result for $m \neq 1$ is:

$$h_n(z) = 1 - m^n \left[\frac{1-d}{m^n - d}\right] + \frac{m^n \left[\frac{1-d}{m^n-d}\right]^2 z}{1 - \left[\frac{m^n-1}{m^n-d}\right] z}.$$

The coefficients of the powers of z give the distribution for Z_n:

$$P(Z_n = 0) = 1 - m^n \frac{1-d}{m^n - d} = \frac{d(m^n - 1)}{m^n - d}$$

and
$$P(Z_n = j) = m^n \left(\frac{1-d}{m^n - d}\right)^2 \cdot \left(\frac{m^n - 1}{m^n - d}\right)^{j-1},$$

for $j \geq 1$. $\square$

Example 10.12 Let us re-examine the Keyfitz data to see if a distribution of the type considered in Example 10.11 could reasonably be used as a model for this population. We would have to estimate from the data the parameters b and c for the formula $p_k = bc^{k-1}$. Recall that

$$m = \frac{b}{(1-c)^2} \tag{10.7}$$

and the probability d that the process dies out is

$$d = \frac{1-b-c}{c(1-c)}. \tag{10.8}$$

Solving Equation 10.7 and 10.8 for b and c gives

$$c = \frac{m-1}{m-d}$$

---
[7]T. E. Harris, *The Theory of Branching Processes* (Berlin: Springer, 1963), p. 9.

10.2. BRANCHING PROCESSES

		Geometric
p_j	Data	Model
0	.2092	.1816
1	.2584	.3666
2	.2360	.2028
3	.1593	.1122
4	.0828	.0621
5	.0357	.0344
6	.0133	.0190
7	.0042	.0105
8	.0011	.0058
9	.0002	.0032
10	.0000	.0018

Table 10.3: Comparison of observed and expected frequencies.

and
$$b = m\left(\frac{1-d}{m-d}\right)^2.$$

We shall use the value 1.837 for m and .324 for d that we found in the Keyfitz example. Using these values, we obtain $b = .3666$ and $c = .5533$. Note that $(1-c)^2 < b < 1-c$, as required. In Table 10.3 we give for comparison the probabilities p_0 through p_8 as calculated by the geometric distribution versus the empirical values.

The geometric model tends to favor the larger numbers of offspring but is similar enough to show that this modified geometric distribution might be appropriate to use for studies of this kind.

Recall that if $S_n = X_1 + X_2 + \cdots + X_n$ is the sum of independent random variables with the same distribution then the Law of Large Numbers states that S_n/n converges to a constant, namely $E(X_1)$. It is natural to ask if there is a similar limiting theorem for branching processes.

Consider a branching process with Z_n representing the number of offspring after n generations. Then we have seen that the expected value of Z_n is m^n. Thus we can scale the random variable Z_n to have expected value 1 by considering the random variable
$$W_n = \frac{Z_n}{m^n}.$$

In the theory of branching processes it is proved that this random variable W_n will tend to a limit as n tends to infinity. However, unlike the case of the Law of Large Numbers where this limit is a constant, for a branching process the limiting value of the random variables W_n is itself a random variable.

Although we cannot prove this theorem here we can illustrate it by simulation. This requires a little care. When a branching process survives, the number of offspring is apt to get very large. If in a given generation there are 1000 offspring, the offspring of the next generation are the result of 1000 chance events, and it will take a while to simulate these 1000 experiments. However, since the final result is

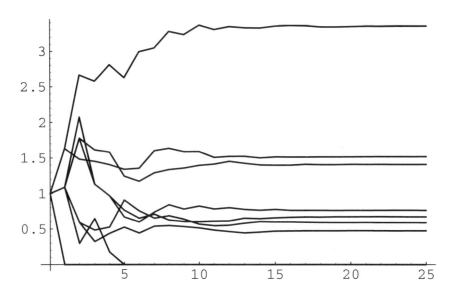

Figure 10.4: Simulation of Z_n/m^n for the Keyfitz example.

the sum of 1000 independent experiments we can use the Central Limit Theorem to replace these 1000 experiments by a single experiment with normal density having the appropriate mean and variance. The program **BranchingSimulation** carries out this process.

We have run this program for the Keyfitz example, carrying out 10 simulations and graphing the results in Figure 10.4.

The expected number of female offspring per female is 1.837, so that we are graphing the outcome for the random variables $W_n = Z_n/(1.837)^n$. For three of the simulations the process died out, which is consistent with the value $d = .3$ that we found for this example. For the other seven simulations the value of W_n tends to a limiting value which is different for each simulation. $\square$

Example 10.13 We now examine the random variable Z_n more closely for the case $m < 1$ (see Example 10.11). Fix a value $t > 0$; let $[tm^n]$ be the integer part of tm^n. Then

$$P(Z_n = [tm^n]) = m^n \left(\frac{1-d}{m^n - d}\right)^2 \left(\frac{m^n - 1}{m^n - d}\right)^{[tm^n]-1}$$

$$= \frac{1}{m^n}\left(\frac{1-d}{1 - d/m^n}\right)^2 \left(\frac{1 - 1/m^n}{1 - d/m^n}\right)^{tm^n + a},$$

where $|a| \leq 2$. Thus, as $n \to \infty$,

$$m^n P(Z_n = [tm^n]) \to (1-d)^2 \frac{e^{-t}}{e^{-td}} = (1-d)^2 e^{-t(1-d)} .$$

For $t = 0$,

$$P(Z_n = 0) \to d .$$

10.2. BRANCHING PROCESSES

We can compare this result with the Central Limit Theorem for sums S_n of integer-valued independent random variables (see Theorem 9.3), which states that if t is an integer and $u = (t - n\mu)/\sqrt{\sigma^2 n}$, then as $n \to \infty$,

$$\sqrt{\sigma^2 n}\, P(S_n = u\sqrt{\sigma^2 n} + \mu n) \to \frac{1}{\sqrt{2\pi}} e^{-u^2/2} \ .$$

We see that the form of these statements are quite similar. It is possible to prove a limit theorem for a general class of branching processes that states that under suitable hypotheses, as $n \to \infty$,

$$m^n P(Z_n = [tm^n]) \to k(t) \ ,$$

for $t > 0$, and

$$P(Z_n = 0) \to d \ .$$

However, unlike the Central Limit Theorem for sums of independent random variables, the function $k(t)$ will depend upon the basic distribution that determines the process. Its form is known for only a very few examples similar to the one we have considered here. □

Chain Letter Problem

Example 10.14 An interesting example of a branching process was suggested by Free Huizinga.[8] In 1978, a chain letter called the "Circle of Gold," believed to have started in California, found its way across the country to the theater district of New York. The chain required a participant to buy a letter containing a list of 12 names for 100 dollars. The buyer gives 50 dollars to the person from whom the letter was purchased and then sends 50 dollars to the person whose name is at the top of the list. The buyer then crosses off the name at the top of the list and adds her own name at the bottom in each letter before it is sold again.

Let us first assume that the buyer may sell the letter only to a single person. If you buy the letter you will want to compute your expected winnings. (We are ignoring here the fact that the passing on of chain letters through the mail is a federal offense with certain obvious resulting penalties.) Assume that each person involved has a probability p of selling the letter. Then you will receive 50 dollars with probability p and another 50 dollars if the letter is sold to 12 people, since then your name would have risen to the top of the list. This occurs with probability p^{12}, and so your expected winnings are $-100 + 50p + 50p^{12}$. Thus the chain in this situation is a highly unfavorable game.

It would be more reasonable to allow each person involved to make a copy of the list and try to sell the letter to at least 2 other people. Then you would have a chance of recovering your 100 dollars on these sales, and if any of the letters is sold 12 times you will receive a bonus of 50 dollars for each of these cases. We can consider this as a branching process with 12 generations. The members of the first

---
[8]Private communication.

generation are the letters you sell. The second generation consists of the letters sold by members of the first generation, and so forth.

Let us assume that the probabilities that each individual sells letters to 0, 1, or 2 others are p_0, p_1, and p_2, respectively. Let Z_1, Z_2, ..., Z_{12} be the number of letters in the first 12 generations of this branching process. Then your expected winnings are
$$50(E(Z_1) + E(Z_{12})) = 50m + 50m^{12},$$
where $m = p_1 + 2p_2$ is the expected number of letters you sold. Thus to be favorable we just have
$$50m + 50m^{12} > 100,$$
or
$$m + m^{12} > 2.$$
But this will be true if and only if $m > 1$. We have seen that this will occur in the quadratic case if and only if $p_2 > p_0$. Let us assume for example that $p_0 = .2$, $p_1 = .5$, and $p_2 = .3$. Then $m = 1.1$ and the chain would be a favorable game. Your expected profit would be
$$50(1.1 + 1.1^{12}) - 100 \approx 112.$$

The probability that you receive at least one payment from the 12th generation is $1 - d_{12}$. We find from our program **Branch** that $d_{12} = .599$. Thus, $1 - d_{12} = .401$ is the probability that you receive some bonus. The maximum that you could receive from the chain would be $50(2 + 2^{12}) = 204{,}900$ if everyone were to successfully sell two letters. Of course you can not always expect to be so lucky. (What is the probability of this happening?)

To simulate this game, we need only simulate a branching process for 12 generations. Using a slightly modified version of our program **BranchingSimulation** we carried out twenty such simulations, giving the results shown in Table 10.4.

Note that we were quite lucky on a few runs, but we came out ahead only a little less than half the time. The process died out by the twelfth generation in 12 out of the 20 experiments, in good agreement with the probability $d_{12} = .599$ that we calculated using the program **Branch**.

Let us modify the assumptions about our chain letter to let the buyer sell the letter to as many people as she can instead of to a maximum of two. We shall assume, in fact, that a person has a large number N of acquaintances and a small probability p of persuading any one of them to buy the letter. Then the distribution for the number of letters that she sells will be a binomial distribution with mean $m = Np$. Since N is large and p is small, we can assume that the probability p_j that an individual sells the letter to j people is given by the Poisson distribution
$$p_j = \frac{e^{-m} m^j}{j!}.$$

10.2. BRANCHING PROCESSES

Z_1	Z_2	Z_3	Z_4	Z_5	Z_6	Z_7	Z_8	Z_9	Z_{10}	Z_{11}	Z_{12}	Profit
1	0	0	0	0	0	0	0	0	0	0	0	-50
1	1	2	3	2	3	2	1	2	3	3	6	250
0	0	0	0	0	0	0	0	0	0	0	0	-100
2	4	4	2	3	4	4	3	2	2	1	1	50
1	2	3	5	4	3	3	3	5	8	6	6	250
0	0	0	0	0	0	0	0	0	0	0	0	-100
2	3	2	2	2	1	2	3	3	3	4	6	300
1	2	1	1	1	1	2	1	0	0	0	0	-50
0	0	0	0	0	0	0	0	0	0	0	0	-100
1	0	0	0	0	0	0	0	0	0	0	0	-50
2	3	2	3	3	3	5	9	12	12	13	15	750
1	1	1	0	0	0	0	0	0	0	0	0	-50
1	2	2	3	3	0	0	0	0	0	0	0	-50
1	1	1	1	2	2	3	4	4	6	4	5	200
1	1	0	0	0	0	0	0	0	0	0	0	-50
1	0	0	0	0	0	0	0	0	0	0	0	-50
1	0	0	0	0	0	0	0	0	0	0	0	-50
1	1	2	3	3	4	2	3	3	3	3	2	50
1	2	4	6	6	9	10	13	16	17	15	18	850
1	0	0	0	0	0	0	0	0	0	0	0	-50

Table 10.4: Simulation of chain letter (finite distribution case).

Z_1	Z_2	Z_3	Z_4	Z_5	Z_6	Z_7	Z_8	Z_9	Z_{10}	Z_{11}	Z_{12}	Profit
1	2	6	7	7	8	11	9	7	6	6	5	200
1	0	0	0	0	0	0	0	0	0	0	0	-50
1	0	0	0	0	0	0	0	0	0	0	0	-50
1	1	1	0	0	0	0	0	0	0	0	0	-50
0	0	0	0	0	0	0	0	0	0	0	0	-100
1	1	1	1	1	1	2	4	9	7	9	7	300
2	3	3	4	2	0	0	0	0	0	0	0	0
1	0	0	0	0	0	0	0	0	0	0	0	-50
2	1	0	0	0	0	0	0	0	0	0	0	0
3	3	4	7	11	17	14	11	11	10	16	25	1300
0	0	0	0	0	0	0	0	0	0	0	0	-100
1	2	2	1	1	3	1	0	0	0	0	0	-50
0	0	0	0	0	0	0	0	0	0	0	0	-100
2	3	1	0	0	0	0	0	0	0	0	0	0
3	1	0	0	0	0	0	0	0	0	0	0	50
1	0	0	0	0	0	0	0	0	0	0	0	-50
3	4	4	7	10	11	9	11	12	14	13	10	550
1	3	3	4	9	5	7	9	8	8	6	3	100
1	0	4	6	6	9	10	13	0	0	0	0	-50
1	0	0	0	0	0	0	0	0	0	0	0	-50

Table 10.5: Simulation of chain letter (Poisson case).

The generating function for the Poisson distribution is

$$\begin{aligned} h(z) &= \sum_{j=0}^{\infty} \frac{e^{-m} m^j z^j}{j!} \\ &= e^{-m} \sum_{j=0}^{\infty} \frac{m^j z^j}{j!} \\ &= e^{-m} e^{mz} = e^{m(z-1)} \ . \end{aligned}$$

The expected number of letters that an individual passes on is m, and again to be favorable we must have $m > 1$. Let us assume again that $m = 1.1$. Then we can find again the probability $1 - d_{12}$ of a bonus from **Branch**. The result is .232. Although the expected winnings are the same, the variance is larger in this case, and the buyer has a better chance for a reasonably large profit. We again carried out 20 simulations using the Poisson distribution with mean 1.1. The results are shown in Table 10.5.

We note that, as before, we came out ahead less than half the time, but we also had one large profit. In only 6 of the 20 cases did we receive any profit. This is again in reasonable agreement with our calculation of a probability .232 for this happening. □

Exercises

1 Let $Z_1, Z_2, \ldots, Z_N$ describe a branching process in which each parent has j offspring with probability p_j. Find the probability d that the process eventually dies out if

   (a) $p_0 = 1/2$, $p_1 = 1/4$, and $p_2 = 1/4$.
   (b) $p_0 = 1/3$, $p_1 = 1/3$, and $p_2 = 1/3$.
   (c) $p_0 = 1/3$, $p_1 = 0$, and $p_2 = 2/3$.
   (d) $p_j = 1/2^{j+1}$, for $j = 0, 1, 2, \ldots$.
   (e) $p_j = (1/3)(2/3)^j$, for $j = 0, 1, 2, \ldots$.
   (f) $p_j = e^{-2} 2^j / j!$, for $j = 0, 1, 2, \ldots$ (estimate d numerically).

2 Let $Z_1, Z_2, \ldots, Z_N$ describe a branching process in which each parent has j offspring with probability p_j. Find the probability d that the process dies out if

   (a) $p_0 = 1/2$, $p_1 = p_2 = 0$, and $p_3 = 1/2$.
   (b) $p_0 = p_1 = p_2 = p_3 = 1/4$.
   (c) $p_0 = t$, $p_1 = 1 - 2t$, $p_2 = 0$, and $p_3 = t$, where $t \leq 1/2$.

3 In the chain letter problem (see Example 10.14) find your expected profit if

   (a) $p_0 = 1/2$, $p_1 = 0$, and $p_2 = 1/2$.
   (b) $p_0 = 1/6$, $p_1 = 1/2$, and $p_2 = 1/3$.

   Show that if $p_0 > 1/2$, you cannot expect to make a profit.

4 Let $S_N = X_1 + X_2 + \cdots + X_N$, where the X_i's are independent random variables with common distribution having generating function $f(z)$. Assume that N is an integer valued random variable independent of all of the X_j and having generating function $g(z)$. Show that the generating function for S_N is $h(z) = g(f(z))$. *Hint*: Use the fact that

$$h(z) = E(z^{S_N}) = \sum_k E(z^{S_N} | N = k) P(N = k) \ .$$

5 We have seen that if the generating function for the offspring of a single parent is $f(z)$, then the generating function for the number of offspring after two generations is given by $h(z) = f(f(z))$. Explain how this follows from the result of Exercise 4.

6 Consider a queueing process (see Example 5.7) such that in each minute either 0 or 1 customers arrive with probabilities p or $q = 1 - p$, respectively. (The number p is called the *arrival rate*.) When a customer starts service she finishes in the next minute with probability r. The number r is called the *service rate*.) Thus when a customer begins being served she will finish being served in j minutes with probability $(1 - r)^{j-1} r$, for $j = 1, 2, 3, \ldots$.

(a) Find the generating function $f(z)$ for the number of customers who arrive in one minute and the generating function $g(z)$ for the length of time that a person spends in service once she begins service.

(b) Consider a *customer branching process* by considering the offspring of a customer to be the customers who arrive while she is being served. Using Exercise 4, show that the generating function for our customer branching process is $h(z) = g(f(z))$.

(c) If we start the branching process with the arrival of the first customer, then the length of time until the branching process dies out will be the *busy period* for the server. Find a condition in terms of the arrival rate and service rate that will assure that the server will ultimately have a time when he is not busy.

7 Let N be the expected total number of offspring in a branching process. Let m be the mean number of offspring of a single parent. Show that

$$N = 1 + \left(\sum p_k \cdot k\right) N = 1 + mN$$

and hence that N is finite if and only if $m < 1$ and in that case $N = 1/(1-m)$.

8 Consider a branching process such that the number of offspring of a parent is j with probability $1/2^{j+1}$ for $j = 0, 1, 2, \ldots$.

(a) Using the results of Example 10.11 show that the probability that there are j offspring in the nth generation is

$$p_j^{(n)} = \begin{cases} \frac{1}{n(n+1)}\left(\frac{n}{n+1}\right)^j, & \text{if } j \geq 1, \\ \frac{n}{n+1}, & \text{if } j = 0. \end{cases}$$

(b) Show that the probability that the process dies out exactly at the nth generation is $1/n(n+1)$.

(c) Show that the expected lifetime is infinite even though $d = 1$.

10.3 Generating Functions for Continuous Densities

In the previous section, we introduced the concepts of moments and moment generating functions for discrete random variables. These concepts have natural analogues for continuous random variables, provided some care is taken in arguments involving convergence.

Moments

If X is a continuous random variable defined on the probability space Ω, with density function f_X, then we define the nth moment of X by the formula

$$\mu_n = E(X^n) = \int_{-\infty}^{+\infty} x^n f_X(x)\, dx\ ,$$

10.3. CONTINUOUS DENSITIES

provided the integral

$$\mu_n = E(X^n) = \int_{-\infty}^{+\infty} |x|^n f_X(x)\,dx \ ,$$

is finite. Then, just as in the discrete case, we see that $\mu_0 = 1$, $\mu_1 = \mu$, and $\mu_2 - \mu_1^2 = \sigma^2$.

Moment Generating Functions

Now we define the *moment generating function* $g(t)$ for X by the formula

$$\begin{aligned} g(t) &= \sum_{k=0}^{\infty} \frac{\mu_k t^k}{k!} = \sum_{k=0}^{\infty} \frac{E(X^k) t^k}{k!} \\ &= E(e^{tX}) = \int_{-\infty}^{+\infty} e^{tx} f_X(x)\,dx \ , \end{aligned}$$

provided this series converges. Then, as before, we have

$$\mu_n = g^{(n)}(0) \ .$$

Examples

Example 10.15 Let X be a continuous random variable with range $[0,1]$ and density function $f_X(x) = 1$ for $0 \le x \le 1$ (uniform density). Then

$$\mu_n = \int_0^1 x^n\,dx = \frac{1}{n+1} \ ,$$

and

$$\begin{aligned} g(t) &= \sum_{k=0}^{\infty} \frac{t^k}{(k+1)!} \\ &= \frac{e^t - 1}{t} \ . \end{aligned}$$

Here the series converges for all t. Alternatively, we have

$$\begin{aligned} g(t) &= \int_{-\infty}^{+\infty} e^{tx} f_X(x)\,dx \\ &= \int_0^1 e^{tx}\,dx = \frac{e^t - 1}{t} \ . \end{aligned}$$

Then (by L'Hôpital's rule)

$$\begin{aligned} \mu_0 &= g(0) = \lim_{t \to 0} \frac{e^t - 1}{t} = 1 \ , \\ \mu_1 &= g'(0) = \lim_{t \to 0} \frac{te^t - e^t + 1}{t^2} = \frac{1}{2} \ , \\ \mu_2 &= g''(0) = \lim_{t \to 0} \frac{t^3 e^t - 2t^2 e^t + 2te^t - 2t}{t^4} = \frac{1}{3} \ . \end{aligned}$$

In particular, we verify that $\mu = g'(0) = 1/2$ and

$$\sigma^2 = g''(0) - (g'(0))^2 = \frac{1}{3} - \frac{1}{4} = \frac{1}{12}$$

as before (see Example 6.25). □

Example 10.16 Let X have range $[0, \infty)$ and density function $f_X(x) = \lambda e^{-\lambda x}$ (exponential density with parameter λ). In this case

$$\begin{aligned}\mu_n &= \int_0^\infty x^n \lambda e^{-\lambda x}\, dx = \lambda(-1)^n \frac{d^n}{d\lambda^n} \int_0^\infty e^{-\lambda x}\, dx \\ &= \lambda(-1)^n \frac{d^n}{d\lambda^n}[\frac{1}{\lambda}] = \frac{n!}{\lambda^n} ,\end{aligned}$$

and

$$\begin{aligned}g(t) &= \sum_{k=0}^\infty \frac{\mu_k t^k}{k!} \\ &= \sum_{k=0}^\infty [\frac{t}{\lambda}]^k = \frac{\lambda}{\lambda - t} .\end{aligned}$$

Here the series converges only for $|t| < \lambda$. Alternatively, we have

$$\begin{aligned}g(t) &= \int_0^\infty e^{tx} \lambda e^{-\lambda x}\, dx \\ &= \frac{\lambda e^{(t-\lambda)x}}{t - \lambda}\Big|_0^\infty = \frac{\lambda}{\lambda - t} .\end{aligned}$$

Now we can verify directly that

$$\mu_n = g^{(n)}(0) = \frac{\lambda n!}{(\lambda - t)^{n+1}}\Big|_{t=0} = \frac{n!}{\lambda^n} .$$

□

Example 10.17 Let X have range $(-\infty, +\infty)$ and density function

$$f_X(x) = \frac{1}{\sqrt{2\pi}} e^{-x^2/2}$$

(normal density). In this case we have

$$\begin{aligned}\mu_n &= \frac{1}{\sqrt{2\pi}} \int_{-\infty}^{+\infty} x^n e^{-x^2/2}\, dx \\ &= \begin{cases} \frac{(2m)!}{2^m m!}, & \text{if } n = 2m, \\ 0, & \text{if } n = 2m + 1. \end{cases}\end{aligned}$$

10.3. CONTINUOUS DENSITIES

(These moments are calculated by integrating once by parts to show that $\mu_n = (n-1)\mu_{n-2}$, and observing that $\mu_0 = 1$ and $\mu_1 = 0$.) Hence,

$$\begin{aligned} g(t) &= \sum_{n=0}^{\infty} \frac{\mu_n t^n}{n!} \\ &= \sum_{m=0}^{\infty} \frac{t^{2m}}{2^m m!} = e^{t^2/2} \ . \end{aligned}$$

This series converges for all values of t. Again we can verify that $g^{(n)}(0) = \mu_n$.

Let X be a normal random variable with parameters μ and σ. It is easy to show that the moment generating function of X is given by

$$e^{t\mu + (\sigma^2/2)t^2} \ .$$

Now suppose that X and Y are two independent normal random variables with parameters μ_1, σ_1, and μ_2, σ_2, respectively. Then, the product of the moment generating functions of X and Y is

$$e^{t(\mu_1+\mu_2)+((\sigma_1^2+\sigma_2^2)/2)t^2} \ .$$

This is the moment generating function for a normal random variable with mean $\mu_1 + \mu_2$ and variance $\sigma_1^2 + \sigma_2^2$. Thus, the sum of two independent normal random variables is again normal. (This was proved for the special case that both summands are standard normal in Example 7.5.) $\square$

In general, the series defining $g(t)$ will not converge for all t. But in the important special case where X is bounded (i.e., where the range of X is contained in a finite interval), we can show that the series does converge for all t.

Theorem 10.4 Suppose X is a continuous random variable with range contained in the interval $[-M, M]$. Then the series

$$g(t) = \sum_{k=0}^{\infty} \frac{\mu_k t^k}{k!}$$

converges for all t to an infinitely differentiable function $g(t)$, and $g^{(n)}(0) = \mu_n$.

Proof. We have

$$\mu_k = \int_{-M}^{+M} x^k f_X(x)\, dx \ ,$$

so

$$\begin{aligned} |\mu_k| &\leq \int_{-M}^{+M} |x|^k f_X(x)\, dx \\ &\leq M^k \int_{-M}^{+M} f_X(x)\, dx = M^k \ . \end{aligned}$$

Hence, for all N we have

$$\sum_{k=0}^{N} \left| \frac{\mu_k t^k}{k!} \right| \leq \sum_{k=0}^{N} \frac{(M|t|)^k}{k!} \leq e^{M|t|} ,$$

which shows that the power series converges for all t. We know that the sum of a convergent power series is always differentiable. □

Moment Problem

Theorem 10.5 If X is a bounded random variable, then the moment generating function $g_X(t)$ of x determines the density function $f_X(x)$ uniquely.

Sketch of the Proof. We know that

$$\begin{aligned} g_X(t) &= \sum_{k=0}^{\infty} \frac{\mu_k t^k}{k!} \\ &= \int_{-\infty}^{+\infty} e^{tx} f(x) \, dx . \end{aligned}$$

If we replace t by $i\tau$, where τ is real and $i = \sqrt{-1}$, then the series converges for all τ, and we can define the function

$$k_X(\tau) = g_X(i\tau) = \int_{-\infty}^{+\infty} e^{i\tau x} f_X(x) \, dx .$$

The function $k_X(\tau)$ is called the *characteristic function* of X, and is defined by the above equation even when the series for g_X does not converge. This equation says that k_X is the *Fourier transform* of f_X. It is known that the Fourier transform has an inverse, given by the formula

$$f_X(x) = \frac{1}{2\pi} \int_{-\infty}^{+\infty} e^{-i\tau x} k_X(\tau) \, d\tau ,$$

suitably interpreted.[9] Here we see that the characteristic function k_X, and hence the moment generating function g_X, determines the density function f_X uniquely under our hypotheses. □

Sketch of the Proof of the Central Limit Theorem

With the above result in mind, we can now sketch a proof of the Central Limit Theorem for bounded continuous random variables (see Theorem 9.6). To this end, let X be a continuous random variable with density function f_X, mean $\mu = 0$ and variance $\sigma^2 = 1$, and moment generating function $g(t)$ defined by its series for all t.

---

[9]H. Dym and H. P. McKean, *Fourier Series and Integrals* (New York: Academic Press, 1972).

10.3. CONTINUOUS DENSITIES

Let $X_1, X_2, \ldots, X_n$ be an independent trials process with each X_i having density f_X, and let $S_n = X_1 + X_2 + \cdots + X_n$, and $S_n^* = (S_n - n\mu)/\sqrt{n\sigma^2} = S_n/\sqrt{n}$. Then each X_i has moment generating function $g(t)$, and since the X_i are independent, the sum S_n, just as in the discrete case (see Section 10.1), has moment generating function

$$g_n(t) = (g(t))^n \;,$$

and the standardized sum S_n^* has moment generating function

$$g_n^*(t) = \left(g\left(\frac{t}{\sqrt{n}}\right)\right)^n \;.$$

We now show that, as $n \to \infty$, $g_n^*(t) \to e^{t^2/2}$, where $e^{t^2/2}$ is the moment generating function of the normal density $n(x) = (1/\sqrt{2\pi})e^{-x^2/2}$ (see Example 10.17).

To show this, we set $u(t) = \log g(t)$, and

$$\begin{aligned} u_n^*(t) &= \log g_n^*(t) \\ &= n \log g\left(\frac{t}{\sqrt{n}}\right) = nu\left(\frac{t}{\sqrt{n}}\right) \;, \end{aligned}$$

and show that $u_n^*(t) \to t^2/2$ as $n \to \infty$. First we note that

$$\begin{aligned} u(0) &= \log g_n(0) = 0 \;, \\ u'(0) &= \frac{g'(0)}{g(0)} = \frac{\mu_1}{1} = 0 \;, \\ u''(0) &= \frac{g''(0)g(0) - (g'(0))^2}{(g(0))^2} \\ &= \frac{\mu_2 - \mu_1^2}{1} = \sigma^2 = 1 \;. \end{aligned}$$

Now by using L'Hôpital's rule twice, we get

$$\begin{aligned} \lim_{n \to \infty} u_n^*(t) &= \lim_{s \to \infty} \frac{u(t/\sqrt{s})}{s^{-1}} \\ &= \lim_{s \to \infty} \frac{u'(t/\sqrt{s})t}{2s^{-1/2}} \\ &= \lim_{s \to \infty} u''\left(\frac{t}{\sqrt{s}}\right)\frac{t^2}{2} = \sigma^2 \frac{t^2}{2} = \frac{t^2}{2} \;. \end{aligned}$$

Hence, $g_n^*(t) \to e^{t^2/2}$ as $n \to \infty$. Now to complete the proof of the Central Limit Theorem, we must show that if $g_n^*(t) \to e^{t^2/2}$, then under our hypotheses the distribution functions $F_n^*(x)$ of the S_n^* must converge to the distribution function $F_N^*(x)$ of the normal variable N; that is, that

$$F_n^*(a) = P(S_n^* \le a) \to \frac{1}{\sqrt{2\pi}} \int_{-\infty}^a e^{-x^2/2} \, dx \;,$$

and furthermore, that the density functions $f_n^*(x)$ of the S_n^* must converge to the density function for N; that is, that

$$f_n^*(x) \to \frac{1}{\sqrt{2\pi}} e^{-x^2/2} \;,$$

as $n \to \infty$.

Since the densities, and hence the distributions, of the S_n^* are uniquely determined by their moment generating functions under our hypotheses, these conclusions are certainly plausible, but their proofs involve a detailed examination of characteristic functions and Fourier transforms, and we shall not attempt them here.

In the same way, we can prove the Central Limit Theorem for bounded discrete random variables with integer values (see Theorem 9.4). Let X be a discrete random variable with density function $p(j)$, mean $\mu = 0$, variance $\sigma^2 = 1$, and moment generating function $g(t)$, and let $X_1, X_2, \ldots, X_n$ form an independent trials process with common density p. Let $S_n = X_1 + X_2 + \cdots + X_n$ and $S_n^* = S_n/\sqrt{n}$, with densities p_n and p_n^*, and moment generating functions $g_n(t)$ and $g_n^*(t) = \left(g(\frac{t}{\sqrt{n}})\right)^n$. Then we have

$$g_n^*(t) \to e^{t^2/2} ,$$

just as in the continuous case, and this implies in the same way that the distribution functions $F_n^*(x)$ converge to the normal distribution; that is, that

$$F_n^*(a) = P(S_n^* \leq a) \to \frac{1}{\sqrt{2\pi}} \int_{-\infty}^{a} e^{-x^2/2}\, dx ,$$

as $n \to \infty$.

The corresponding statement about the distribution functions p_n^*, however, requires a little extra care (see Theorem 9.3). The trouble arises because the distribution $p(x)$ is not defined for all x, but only for integer x. It follows that the distribution $p_n^*(x)$ is defined only for x of the form $j/\sqrt{n}$, and these values change as n changes.

We can fix this, however, by introducing the function $\bar{p}(x)$, defined by the formula

$$\bar{p}(x) = \begin{cases} p(j), & \text{if } j - 1/2 \leq x < j + 1/2, \\ 0, & \text{otherwise.} \end{cases}$$

Then $\bar{p}(x)$ is defined for all x, $\bar{p}(j) = p(j)$, and the graph of $\bar{p}(x)$ is the step function for the distribution $p(j)$ (see Figure 3 of Section 9.1).

In the same way we introduce the step function $\bar{p}_n(x)$ and $\bar{p}_n^*(x)$ associated with the distributions p_n and p_n^*, and their moment generating functions $\bar{g}_n(t)$ and $\bar{g}_n^*(t)$. If we can show that $\bar{g}_n^*(t) \to e^{t^2/2}$, then we can conclude that

$$\bar{p}_n^*(x) \to \frac{1}{\sqrt{2\pi}} e^{t^2/2} ,$$

as $n \to \infty$, for all x, a conclusion strongly suggested by Figure 9.3.

Now $\bar{g}(t)$ is given by

$$\begin{aligned}
\bar{g}(t) &= \int_{-\infty}^{+\infty} e^{tx} \bar{p}(x)\, dx \\
&= \sum_{j=-N}^{+N} \int_{j-1/2}^{j+1/2} e^{tx} p(j)\, dx
\end{aligned}$$

10.3. CONTINUOUS DENSITIES

$$\begin{aligned}
&= \sum_{j=-N}^{+N} p(j)e^{tj}\frac{e^{t/2}-e^{-t/2}}{2t/2} \\
&= g(t)\frac{\sinh(t/2)}{t/2},
\end{aligned}$$

where we have put

$$\sinh(t/2) = \frac{e^{t/2}-e^{-t/2}}{2}.$$

In the same way, we find that

$$\begin{aligned}
\bar{g}_n(t) &= g_n(t)\frac{\sinh(t/2)}{t/2}, \\
\bar{g}_n^*(t) &= g_n^*(t)\frac{\sinh(t/2\sqrt{n})}{t/2\sqrt{n}}.
\end{aligned}$$

Now, as $n \to \infty$, we know that $g_n^*(t) \to e^{t^2/2}$, and, by L'Hôpital's rule,

$$\lim_{n\to\infty} \frac{\sinh(t/2\sqrt{n})}{t/2\sqrt{n}} = 1.$$

It follows that

$$\bar{g}_n^*(t) \to e^{t^2/2},$$

and hence that

$$\bar{p}_n^*(x) \to \frac{1}{\sqrt{2\pi}}e^{-x^2/2},$$

as $n \to \infty$. The astute reader will note that in this sketch of the proof of Theorem 9.3, we never made use of the hypothesis that the greatest common divisor of the differences of all the values that the X_i can take on is 1. This is a technical point that we choose to ignore. A complete proof may be found in Gnedenko and Kolmogorov.[10]

Cauchy Density

The characteristic function of a continuous density is a useful tool even in cases when the moment series does not converge, or even in cases when the moments themselves are not finite. As an example, consider the Cauchy density with parameter $a = 1$ (see Example 5.10)

$$f(x) = \frac{1}{\pi(1+x^2)}.$$

If X and Y are independent random variables with Cauchy density $f(x)$, then the average $Z = (X+Y)/2$ also has Cauchy density $f(x)$, that is,

$$f_Z(x) = f(x).$$

---

[10]B. V. Gnedenko and A. N. Kolmogorov, *Limit Distributions for Sums of Independent Random Variables* (Reading: Addison-Wesley, 1968), p. 233.

This is hard to check directly, but easy to check by using characteristic functions. Note first that

$$\mu_2 = E(X^2) = \int_{-\infty}^{+\infty} \frac{x^2}{\pi(1+x^2)} \, dx = \infty$$

so that μ_2 is infinite. Nevertheless, we can define the characteristic function $k_X(\tau)$ of x by the formula

$$k_X(\tau) = \int_{-\infty}^{+\infty} e^{i\tau x} \frac{1}{\pi(1+x^2)} \, dx \ .$$

This integral is easy to do by contour methods, and gives us

$$k_X(\tau) = k_Y(\tau) = e^{-|\tau|} \ .$$

Hence,

$$k_{X+Y}(\tau) = (e^{-|\tau|})^2 = e^{-2|\tau|} \ ,$$

and since

$$k_Z(\tau) = k_{X+Y}(\tau/2) \ ,$$

we have

$$k_Z(\tau) = e^{-2|\tau/2|} = e^{-|\tau|} \ .$$

This shows that $k_Z = k_X = k_Y$, and leads to the conclusions that $f_Z = f_X = f_Y$.

It follows from this that if $X_1, X_2, \ldots, X_n$ is an independent trials process with common Cauchy density, and if

$$A_n = \frac{X_1 + X_2 + \cdots + X_n}{n}$$

is the average of the X_i, then A_n has the same density as do the X_i. This means that the Law of Large Numbers fails for this process; the distribution of the average A_n is exactly the same as for the individual terms. Our proof of the Law of Large Numbers fails in this case because the variance of X_i is not finite.

Exercises

1 Let X be a continuous random variable with values in $[0,2]$ and density f_X. Find the moment generating function $g(t)$ for X if

(a) $f_X(x) = 1/2$.

(b) $f_X(x) = (1/2)x$.

(c) $f_X(x) = 1 - (1/2)x$.

(d) $f_X(x) = |1 - x|$.

(e) $f_X(x) = (3/8)x^2$.

Hint: Use the integral definition, as in Examples 10.15 and 10.16.

2 For each of the densities in Exercise 1 calculate the first and second moments, μ_1 and μ_2, directly from their definition and verify that $g(0) = 1$, $g'(0) = \mu_1$, and $g''(0) = \mu_2$.

10.3. CONTINUOUS DENSITIES

3 Let X be a continuous random variable with values in $[0, \infty)$ and density f_X. Find the moment generating functions for X if

(a) $f_X(x) = 2e^{-2x}$.

(b) $f_X(x) = e^{-2x} + (1/2)e^{-x}$.

(c) $f_X(x) = 4xe^{-2x}$.

(d) $f_X(x) = \lambda(\lambda x)^{n-1}e^{-\lambda x}/(n-1)!$.

4 For each of the densities in Exercise 3, calculate the first and second moments, μ_1 and μ_2, directly from their definition and verify that $g(0) = 1$, $g'(0) = \mu_1$, and $g''(0) = \mu_2$.

5 Find the characteristic function $k_X(\tau)$ for each of the random variables X of Exercise 1.

6 Let X be a continuous random variable whose characteristic function $k_X(\tau)$ is
$$k_X(\tau) = e^{-|\tau|}, \quad -\infty < \tau < +\infty\ .$$
Show directly that the density f_X of X is
$$f_X(x) = \frac{1}{\pi(1+x^2)}\ .$$

7 Let X be a continuous random variable with values in $[0,1]$, uniform density function $f_X(x) \equiv 1$ and moment generating function $g(t) = (e^t - 1)/t$. Find in terms of $g(t)$ the moment generating function for

(a) $-X$.

(b) $1 + X$.

(c) $3X$.

(d) $aX + b$.

8 Let $X_1, X_2, \ldots, X_n$ be an independent trials process with uniform density. Find the moment generating function for

(a) X_1.

(b) $S_2 = X_1 + X_2$.

(c) $S_n = X_1 + X_2 + \cdots + X_n$.

(d) $A_n = S_n/n$.

(e) $S_n^* = (S_n - n\mu)/\sqrt{n\sigma^2}$.

9 Let $X_1, X_2, \ldots, X_n$ be an independent trials process with normal density of mean 1 and variance 2. Find the moment generating function for

(a) X_1.

(b) $S_2 = X_1 + X_2$.

(c) $S_n = X_1 + X_2 + \cdots + X_n$.

(d) $A_n = S_n/n$.

(e) $S_n^* = (S_n - n\mu)/\sqrt{n\sigma^2}$.

10 Let $X_1, X_2, \ldots, X_n$ be an independent trials process with density

$$f(x) = \frac{1}{2} e^{-|x|}, \qquad -\infty < x < +\infty .$$

(a) Find the mean and variance of $f(x)$.

(b) Find the moment generating function for X_1, S_n, A_n, and S_n^*.

(c) What can you say about the moment generating function of S_n^* as $n \to \infty$?

(d) What can you say about the moment generating function of A_n as $n \to \infty$?

Chapter 11

Markov Chains

11.1 Introduction

Most of our study of probability has dealt with independent trials processes. These processes are the basis of classical probability theory and much of statistics. We have discussed two of the principal theorems for these processes: the Law of Large Numbers and the Central Limit Theorem.

We have seen that when a sequence of chance experiments forms an independent trials process, the possible outcomes for each experiment are the same and occur with the same probability. Further, knowledge of the outcomes of the previous experiments does not influence our predictions for the outcomes of the next experiment. The distribution for the outcomes of a single experiment is sufficient to construct a tree and a tree measure for a sequence of n experiments, and we can answer any probability question about these experiments by using this tree measure.

Modern probability theory studies chance processes for which the knowledge of previous outcomes influences predictions for future experiments. In principle, when we observe a sequence of chance experiments, all of the past outcomes could influence our predictions for the next experiment. For example, this should be the case in predicting a student's grades on a sequence of exams in a course. But to allow this much generality would make it very difficult to prove general results.

In 1907, A. A. Markov began the study of an important new type of chance process. In this process, the outcome of a given experiment can affect the outcome of the next experiment. This type of process is called a Markov chain.

Specifying a Markov Chain

We describe a Markov chain as follows: We have a set of *states*, $S = \{s_1, s_2, \ldots, s_r\}$. The process starts in one of these states and moves successively from one state to another. Each move is called a *step*. If the chain is currently in state s_i, then it moves to state s_j at the next step with a probability denoted by p_{ij}, and this probability does not depend upon which states the chain was in before the current state.

The probabilities p_{ij} are called *transition probabilities*. The process can remain in the state it is in, and this occurs with probability p_{ii}. An initial probability distribution, defined on S, specifies the starting state. Usually this is done by specifying a particular state as the starting state.

R. A. Howard[1] provides us with a picturesque description of a Markov chain as a frog jumping on a set of lily pads. The frog starts on one of the pads and then jumps from lily pad to lily pad with the appropriate transition probabilities.

Example 11.1 According to Kemeny, Snell, and Thompson,[2] the Land of Oz is blessed by many things, but not by good weather. They never have two nice days in a row. If they have a nice day, they are just as likely to have snow as rain the next day. If they have snow or rain, they have an even chance of having the same the next day. If there is change from snow or rain, only half of the time is this a change to a nice day. With this information we form a Markov chain as follows. We take as states the kinds of weather R, N, and S. From the above information we determine the transition probabilities. These are most conveniently represented in a square array as

$$\mathbf{P} = \begin{array}{c} \\ R \\ N \\ S \end{array} \begin{pmatrix} R & N & S \\ 1/2 & 1/4 & 1/4 \\ 1/2 & 0 & 1/2 \\ 1/4 & 1/4 & 1/2 \end{pmatrix} .$$

$\square$

Transition Matrix

The entries in the first row of the matrix $\mathbf{P}$ in Example 11.1 represent the probabilities for the various kinds of weather following a rainy day. Similarly, the entries in the second and third rows represent the probabilities for the various kinds of weather following nice and snowy days, respectively. Such a square array is called the *matrix of transition probabilities*, or the *transition matrix*.

We consider the question of determining the probability that, given the chain is in state i today, it will be in state j two days from now. We denote this probability by $p_{ij}^{(2)}$. In Example 11.1, we see that if it is rainy today then the event that it is snowy two days from now is the disjoint union of the following three events: 1) it is rainy tomorrow and snowy two days from now, 2) it is nice tomorrow and snowy two days from now, and 3) it is snowy tomorrow and snowy two days from now. The probability of the first of these events is the product of the conditional probability that it is rainy tomorrow, given that it is rainy today, and the conditional probability that it is snowy two days from now, given that it is rainy tomorrow. Using the transition matrix $\mathbf{P}$, we can write this product as $p_{11}p_{13}$. The other two

---

[1] R. A. Howard, *Dynamic Probabilistic Systems,* vol. 1 (New York: John Wiley and Sons, 1971).
[2] J. G. Kemeny, J. L. Snell, G. L. Thompson, *Introduction to Finite Mathematics,* 3rd ed. (Englewood Cliffs, NJ: Prentice-Hall, 1974).

11.1. INTRODUCTION

events also have probabilities that can be written as products of entries of **P**. Thus, we have

$$p_{13}^{(2)} = p_{11}p_{13} + p_{12}p_{23} + p_{13}p_{33} \ .$$

This equation should remind the reader of a dot product of two vectors; we are dotting the first row of **P** with the third column of **P**. This is just what is done in obtaining the 1, 3-entry of the product of **P** with itself. In general, if a Markov chain has r states, then

$$p_{ij}^{(2)} = \sum_{k=1}^{r} p_{ik}p_{kj} \ .$$

The following general theorem is easy to prove by using the above observation and induction.

Theorem 11.1 Let **P** be the transition matrix of a Markov chain. The ijth entry $p_{ij}^{(n)}$ of the matrix $\mathbf{P}^n$ gives the probability that the Markov chain, starting in state s_i, will be in state s_j after n steps.

Proof. The proof of this theorem is left as an exercise (Exercise 17). □

Example 11.2 (Example 11.1 continued) Consider again the weather in the Land of Oz. We know that the powers of the transition matrix give us interesting information about the process as it evolves. We shall be particularly interested in the state of the chain after a large number of steps. The program **MatrixPowers** computes the powers of **P**.

We have run the program **MatrixPowers** for the Land of Oz example to compute the successive powers of **P** from 1 to 6. The results are shown in Table 11.1. We note that after six days our weather predictions are, to three-decimal-place accuracy, independent of today's weather. The probabilities for the three types of weather, R, N, and S, are .4, .2, and .4 no matter where the chain started. This is an example of a type of Markov chain called a *regular* Markov chain. For this type of chain, it is true that long-range predictions are independent of the starting state. Not all chains are regular, but this is an important class of chains that we shall study in detail later. □

We now consider the long-term behavior of a Markov chain when it starts in a state chosen by a probability distribution on the set of states, which we will call a *probability vector*. A probability vector with r components is a row vector whose entries are non-negative and sum to 1. If **u** is a probability vector which represents the initial state of a Markov chain, then we think of the ith component of **u** as representing the probability that the chain starts in state s_i.

With this interpretation of random starting states, it is easy to prove the following theorem.

Theorem 11.2 Let **P** be the transition matrix of a Markov chain, and let **u** be the probability vector which represents the starting distribution. Then the probability

$$\mathbf{P}^1 = \begin{array}{c} \\ \text{Rain} \\ \text{Nice} \\ \text{Snow} \end{array} \begin{pmatrix} \text{Rain} & \text{Nice} & \text{Snow} \\ .500 & .250 & .250 \\ .500 & .000 & .500 \\ .250 & .250 & .500 \end{pmatrix}$$

$$\mathbf{P}^2 = \begin{array}{c} \\ \text{Rain} \\ \text{Nice} \\ \text{Snow} \end{array} \begin{pmatrix} \text{Rain} & \text{Nice} & \text{Snow} \\ .438 & .188 & .375 \\ .375 & .250 & .375 \\ .375 & .188 & .438 \end{pmatrix}$$

$$\mathbf{P}^3 = \begin{array}{c} \\ \text{Rain} \\ \text{Nice} \\ \text{Snow} \end{array} \begin{pmatrix} \text{Rain} & \text{Nice} & \text{Snow} \\ .406 & .203 & .391 \\ .406 & .188 & .406 \\ .391 & .203 & .406 \end{pmatrix}$$

$$\mathbf{P}^4 = \begin{array}{c} \\ \text{Rain} \\ \text{Nice} \\ \text{Snow} \end{array} \begin{pmatrix} \text{Rain} & \text{Nice} & \text{Snow} \\ .402 & .199 & .398 \\ .398 & .203 & .398 \\ .398 & .199 & .402 \end{pmatrix}$$

$$\mathbf{P}^5 = \begin{array}{c} \\ \text{Rain} \\ \text{Nice} \\ \text{Snow} \end{array} \begin{pmatrix} \text{Rain} & \text{Nice} & \text{Snow} \\ .400 & .200 & .399 \\ .400 & .199 & .400 \\ .399 & .200 & .400 \end{pmatrix}$$

$$\mathbf{P}^6 = \begin{array}{c} \\ \text{Rain} \\ \text{Nice} \\ \text{Snow} \end{array} \begin{pmatrix} \text{Rain} & \text{Nice} & \text{Snow} \\ .400 & .200 & .400 \\ .400 & .200 & .400 \\ .400 & .200 & .400 \end{pmatrix}$$

Table 11.1: Powers of the Land of Oz transition matrix.

11.1. INTRODUCTION

that the chain is in state s_i after n steps is the ith entry in the vector

$$\mathbf{u}^{(n)} = \mathbf{u}\mathbf{P}^n \ .$$

Proof. The proof of this theorem is left as an exercise (Exercise 18). □

We note that if we want to examine the behavior of the chain under the assumption that it starts in a certain state s_i, we simply choose $\mathbf{u}$ to be the probability vector with ith entry equal to 1 and all other entries equal to 0.

Example 11.3 In the Land of Oz example (Example 11.1) let the initial probability vector $\mathbf{u}$ equal $(1/3, 1/3, 1/3)$. Then we can calculate the distribution of the states after three days using Theorem 11.2 and our previous calculation of $\mathbf{P}^3$. We obtain

$$\mathbf{u}^{(3)} = \mathbf{u}\mathbf{P}^3 = (1/3, \ 1/3, \ 1/3) \begin{pmatrix} .406 & .203 & .391 \\ .406 & .188 & .406 \\ .391 & .203 & .406 \end{pmatrix}$$
$$= (.401, \ .188, \ .401) \ .$$

□

Examples

The following examples of Markov chains will be used throughout the chapter for exercises.

Example 11.4 The President of the United States tells person A his or her intention to run or not to run in the next election. Then A relays the news to B, who in turn relays the message to C, and so forth, always to some new person. We assume that there is a probability a that a person will change the answer from yes to no when transmitting it to the next person and a probability b that he or she will change it from no to yes. We choose as states the message, either yes or no. The transition matrix is then

$$\mathbf{P} = \begin{matrix} \\ \text{yes} \\ \text{no} \end{matrix} \begin{pmatrix} \text{yes} & \text{no} \\ 1-a & a \\ b & 1-b \end{pmatrix} .$$

The initial state represents the President's choice. □

Example 11.5 Each time a certain horse runs in a three-horse race, he has probability 1/2 of winning, 1/4 of coming in second, and 1/4 of coming in third, independent of the outcome of any previous race. We have an independent trials process, but it can also be considered from the point of view of Markov chain theory. The transition matrix is

$$\mathbf{P} = \begin{array}{c} \\ \text{W} \\ \text{P} \\ \text{S} \end{array} \begin{pmatrix} \text{W} & \text{P} & \text{S} \\ .5 & .25 & .25 \\ .5 & .25 & .25 \\ .5 & .25 & .25 \end{pmatrix}.$$

□

Example 11.6 In the Dark Ages, Harvard, Dartmouth, and Yale admitted only male students. Assume that, at that time, 80 percent of the sons of Harvard men went to Harvard and the rest went to Yale, 40 percent of the sons of Yale men went to Yale, and the rest split evenly between Harvard and Dartmouth; and of the sons of Dartmouth men, 70 percent went to Dartmouth, 20 percent to Harvard, and 10 percent to Yale. We form a Markov chain with transition matrix

$$\mathbf{P} = \begin{array}{c} \\ \text{H} \\ \text{Y} \\ \text{D} \end{array} \begin{pmatrix} \text{H} & \text{Y} & \text{D} \\ .8 & .2 & 0 \\ .3 & .4 & .3 \\ .2 & .1 & .7 \end{pmatrix}.$$

□

Example 11.7 Modify Example 11.6 by assuming that the son of a Harvard man always went to Harvard. The transition matrix is now

$$\mathbf{P} = \begin{array}{c} \\ \text{H} \\ \text{Y} \\ \text{D} \end{array} \begin{pmatrix} \text{H} & \text{Y} & \text{D} \\ 1 & 0 & 0 \\ .3 & .4 & .3 \\ .2 & .1 & .7 \end{pmatrix}.$$

□

Example 11.8 (Ehrenfest Model) The following is a special case of a model, called the Ehrenfest model,[3] that has been used to explain diffusion of gases. The general model will be discussed in detail in Section 11.5. We have two urns that, between them, contain four balls. At each step, one of the four balls is chosen at random and moved from the urn that it is in into the other urn. We choose, as states, the number of balls in the first urn. The transition matrix is then

$$\mathbf{P} = \begin{array}{c} \\ 0 \\ 1 \\ 2 \\ 3 \\ 4 \end{array} \begin{pmatrix} 0 & 1 & 2 & 3 & 4 \\ 0 & 1 & 0 & 0 & 0 \\ 1/4 & 0 & 3/4 & 0 & 0 \\ 0 & 1/2 & 0 & 1/2 & 0 \\ 0 & 0 & 3/4 & 0 & 1/4 \\ 0 & 0 & 0 & 1 & 0 \end{pmatrix}.$$

□

---

[3]P. and T. Ehrenfest, "Über zwei bekannte Einwände gegen das Boltzmannsche H-Theorem," *Physikalishce Zeitschrift*, vol. 8 (1907), pp. 311-314.

11.1. INTRODUCTION

Example 11.9 (Gene Model) The simplest type of inheritance of traits in animals occurs when a trait is governed by a pair of genes, each of which may be of two types, say G and g. An individual may have a GG combination or Gg (which is genetically the same as gG) or gg. Very often the GG and Gg types are indistinguishable in appearance, and then we say that the G gene dominates the g gene. An individual is called *dominant* if he or she has GG genes, *recessive* if he or she has gg, and *hybrid* with a Gg mixture.

In the mating of two animals, the offspring inherits one gene of the pair from each parent, and the basic assumption of genetics is that these genes are selected at random, independently of each other. This assumption determines the probability of occurrence of each type of offspring. The offspring of two purely dominant parents must be dominant, of two recessive parents must be recessive, and of one dominant and one recessive parent must be hybrid.

In the mating of a dominant and a hybrid animal, each offspring must get a G gene from the former and has an equal chance of getting G or g from the latter. Hence there is an equal probability for getting a dominant or a hybrid offspring. Again, in the mating of a recessive and a hybrid, there is an even chance for getting either a recessive or a hybrid. In the mating of two hybrids, the offspring has an equal chance of getting G or g from each parent. Hence the probabilities are 1/4 for GG, 1/2 for Gg, and 1/4 for gg.

Consider a process of continued matings. We start with an individual of known genetic character and mate it with a hybrid. We assume that there is at least one offspring. An offspring is chosen at random and is mated with a hybrid and this process repeated through a number of generations. The genetic type of the chosen offspring in successive generations can be represented by a Markov chain. The states are dominant, hybrid, and recessive, and indicated by GG, Gg, and gg respectively.

The transition probabilities are

$$\mathbf{P} = \begin{array}{c} \text{GG} \\ \text{Gg} \\ \text{gg} \end{array} \begin{pmatrix} \text{GG} & \text{Gg} & \text{gg} \\ .5 & .5 & 0 \\ .25 & .5 & .25 \\ 0 & .5 & .5 \end{pmatrix}.$$

□

Example 11.10 Modify Example 11.9 as follows: Instead of mating the oldest offspring with a hybrid, we mate it with a dominant individual. The transition matrix is

$$\mathbf{P} = \begin{array}{c} \text{GG} \\ \text{Gg} \\ \text{gg} \end{array} \begin{pmatrix} \text{GG} & \text{Gg} & \text{gg} \\ 1 & 0 & 0 \\ .5 & .5 & 0 \\ 0 & 1 & 0 \end{pmatrix}.$$

□

Example 11.11 We start with two animals of opposite sex, mate them, select two of their offspring of opposite sex, and mate those, and so forth. To simplify the example, we will assume that the trait under consideration is independent of sex.

Here a state is determined by a pair of animals. Hence, the states of our process will be: $s_1 = (\text{GG}, \text{GG})$, $s_2 = (\text{GG}, \text{Gg})$, $s_3 = (\text{GG}, \text{gg})$, $s_4 = (\text{Gg}, \text{Gg})$, $s_5 = (\text{Gg}, \text{gg})$, and $s_6 = (\text{gg}, \text{gg})$.

We illustrate the calculation of transition probabilities in terms of the state s_2. When the process is in this state, one parent has GG genes, the other Gg. Hence, the probability of a dominant offspring is $1/2$. Then the probability of transition to s_1 (selection of two dominants) is $1/4$, transition to s_2 is $1/2$, and to s_4 is $1/4$. The other states are treated the same way. The transition matrix of this chain is:

$$\mathbf{P}^1 = \begin{pmatrix} & \text{GG,GG} & \text{GG,Gg} & \text{GG,gg} & \text{Gg,Gg} & \text{Gg,gg} & \text{gg,gg} \\ \text{GG,GG} & 1.000 & .000 & .000 & .000 & .000 & .000 \\ \text{GG,Gg} & .250 & .500 & .000 & .250 & .000 & .000 \\ \text{GG,gg} & .000 & .000 & .000 & 1.000 & .000 & .000 \\ \text{Gg,Gg} & .062 & .250 & .125 & .250 & .250 & .062 \\ \text{Gg,gg} & .000 & .000 & .000 & .250 & .500 & .250 \\ \text{gg,gg} & .000 & .000 & .000 & .000 & .000 & 1.000 \end{pmatrix}.$$

□

Example 11.12 (Stepping Stone Model) Our final example is another example that has been used in the study of genetics. It is called the *stepping stone* model.[4] In this model we have an n-by-n array of squares, and each square is initially any one of k different colors. For each step, a square is chosen at random. This square then chooses one of its eight neighbors at random and assumes the color of that neighbor. To avoid boundary problems, we assume that if a square S is on the left-hand boundary, say, but not at a corner, it is adjacent to the square T on the right-hand boundary in the same row as S, and S is also adjacent to the squares just above and below T. A similar assumption is made about squares on the upper and lower boundaries. (These adjacencies are much easier to understand if one imagines making the array into a cylinder by gluing the top and bottom edge together, and then making the cylinder into a doughnut by gluing the two circular boundaries together.) With these adjacencies, each square in the array is adjacent to exactly eight other squares.

A state in this Markov chain is a description of the color of each square. For this Markov chain the number of states is k^{n^2}, which for even a small array of squares is enormous. This is an example of a Markov chain that is easy to simulate but difficult to analyze in terms of its transition matrix. The program **SteppingStone** simulates this chain. We have started with a random initial configuration of two colors with $n = 20$ and show the result after the process has run for some time in Figure 11.2.

---

[4]S. Sawyer, "Results for The Stepping Stone Model for Migration in Population Genetics," *Annals of Probability*, vol. 4 (1979), pp. 699–728.

11.1. INTRODUCTION

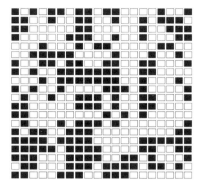

Figure 11.1: Initial state of the stepping stone model.

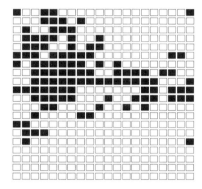

Figure 11.2: State of the stepping stone model after 10,000 steps.

This is an example of an *absorbing* Markov chain. This type of chain will be studied in Section 11.2. One of the theorems proved in that section, applied to the present example, implies that with probability 1, the stones will eventually all be the same color. By watching the program run, you can see that territories are established and a battle develops to see which color survives. At any time the probability that a particular color will win out is equal to the proportion of the array of this color. You are asked to prove this in Exercise 11.2.32. □

Exercises

1. It is raining in the Land of Oz. Determine a tree and a tree measure for the next three days' weather. Find $\mathbf{w}^{(1)}, \mathbf{w}^{(2)}$, and $\mathbf{w}^{(3)}$ and compare with the results obtained from $\mathbf{P}$, $\mathbf{P}^2$, and $\mathbf{P}^3$.

2. In Example 11.4, let $a = 0$ and $b = 1/2$. Find $\mathbf{P}$, $\mathbf{P}^2$, and $\mathbf{P}^3$. What would $\mathbf{P}^n$ be? What happens to $\mathbf{P}^n$ as n tends to infinity? Interpret this result.

3. In Example 11.5, find $\mathbf{P}$, $\mathbf{P}^2$, and $\mathbf{P}^3$. What is $\mathbf{P}^n$?

4 For Example 11.6, find the probability that the grandson of a man from Harvard went to Harvard.

5 In Example 11.7, find the probability that the grandson of a man from Harvard went to Harvard.

6 In Example 11.9, assume that we start with a hybrid bred to a hybrid. Find $\mathbf{w}^{(1)}$, $\mathbf{w}^{(2)}$, and $\mathbf{w}^{(3)}$. What would $\mathbf{w}^{(n)}$ be?

7 Find the matrices $\mathbf{P}^2$, $\mathbf{P}^3$, $\mathbf{P}^4$, and $\mathbf{P}^n$ for the Markov chain determined by the transition matrix $\mathbf{P} = \begin{pmatrix} 1 & 0 \\ 0 & 1 \end{pmatrix}$. Do the same for the transition matrix $\mathbf{P} = \begin{pmatrix} 0 & 1 \\ 1 & 0 \end{pmatrix}$. Interpret what happens in each of these processes.

8 A certain calculating machine uses only the digits 0 and 1. It is supposed to transmit one of these digits through several stages. However, at every stage, there is a probability p that the digit that enters this stage will be changed when it leaves and a probability $q = 1 - p$ that it won't. Form a Markov chain to represent the process of transmission by taking as states the digits 0 and 1. What is the matrix of transition probabilities?

9 For the Markov chain in Exercise 8, draw a tree and assign a tree measure assuming that the process begins in state 0 and moves through two stages of transmission. What is the probability that the machine, after two stages, produces the digit 0 (i.e., the correct digit)? What is the probability that the machine never changed the digit from 0? Now let $p = .1$. Using the program **MatrixPowers**, compute the 100th power of the transition matrix. Interpret the entries of this matrix. Repeat this with $p = .2$. Why do the 100th powers appear to be the same?

10 Modify the program **MatrixPowers** so that it prints out the average $\mathbf{A}_n$ of the powers $\mathbf{P}^n$, for $n = 1$ to N. Try your program on the Land of Oz example and compare $\mathbf{A}_n$ and $\mathbf{P}^n$.

11 Assume that a man's profession can be classified as professional, skilled laborer, or unskilled laborer. Assume that, of the sons of professional men, 80 percent are professional, 10 percent are skilled laborers, and 10 percent are unskilled laborers. In the case of sons of skilled laborers, 60 percent are skilled laborers, 20 percent are professional, and 20 percent are unskilled. Finally, in the case of unskilled laborers, 50 percent of the sons are unskilled laborers, and 25 percent each are in the other two categories. Assume that every man has at least one son, and form a Markov chain by following the profession of a randomly chosen son of a given family through several generations. Set up the matrix of transition probabilities. Find the probability that a randomly chosen grandson of an unskilled laborer is a professional man.

12 In Exercise 11, we assumed that every man has a son. Assume instead that the probability that a man has at least one son is .8. Form a Markov chain

11.2. ABSORBING MARKOV CHAINS

with four states. If a man has a son, the probability that this son is in a particular profession is the same as in Exercise 11. If there is no son, the process moves to state four which represents families whose male line has died out. Find the matrix of transition probabilities and find the probability that a randomly chosen grandson of an unskilled laborer is a professional man.

13 Write a program to compute $\mathbf{u}^{(n)}$ given $\mathbf{u}$ and $\mathbf{P}$. Use this program to compute $\mathbf{u}^{(10)}$ for the Land of Oz example, with $\mathbf{u} = (0, 1, 0)$, and with $\mathbf{u} = (1/3, 1/3, 1/3)$.

14 Using the program **MatrixPowers**, find $\mathbf{P}^1$ through $\mathbf{P}^6$ for Examples 11.9 and 11.10. See if you can predict the long-range probability of finding the process in each of the states for these examples.

15 Write a program to simulate the outcomes of a Markov chain after n steps, given the initial starting state and the transition matrix $\mathbf{P}$ as data (see Example 11.12). Keep this program for use in later problems.

16 Modify the program of Exercise 15 so that it keeps track of the proportion of times in each state in n steps. Run the modified program for different starting states for Example 11.1 and Example 11.8. Does the initial state affect the proportion of time spent in each of the states if n is large?

17 Prove Theorem 11.1.

18 Prove Theorem 11.2.

19 Consider the following process. We have two coins, one of which is fair, and the other of which has heads on both sides. We give these two coins to our friend, who chooses one of them at random (each with probability 1/2). During the rest of the process, she uses only the coin that she chose. She now proceeds to toss the coin many times, reporting the results. We consider this process to consist solely of what she reports to us.

(a) Given that she reports a head on the nth toss, what is the probability that a head is thrown on the $(n+1)$st toss?

(b) Consider this process as having two states, heads and tails. By computing the other three transition probabilities analogous to the one in part (a), write down a "transition matrix" for this process.

(c) Now assume that the process is in state "heads" on both the $(n-1)$st and the nth toss. Find the probability that a head comes up on the $(n+1)$st toss.

(d) Is this process a Markov chain?

11.2 Absorbing Markov Chains

The subject of Markov chains is best studied by considering special types of Markov chains. The first type that we shall study is called an *absorbing Markov chain*.

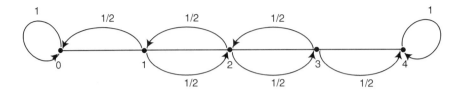

Figure 11.3: Drunkard's walk.

Definition 11.1 A state s_i of a Markov chain is called *absorbing* if it is impossible to leave it (i.e., $p_{ii} = 1$). A Markov chain is *absorbing* if it has at least one absorbing state, and if from every state it is possible to go to an absorbing state (not necessarily in one step). □

Definition 11.2 In an absorbing Markov chain, a state which is not absorbing is called *transient*. □

Drunkard's Walk

Example 11.13 A man walks along a four-block stretch of Park Avenue (see Figure 11.3). If he is at corner 1, 2, or 3, then he walks to the left or right with equal probability. He continues until he reaches corner 4, which is a bar, or corner 0, which is his home. If he reaches either home or the bar, he stays there.

We form a Markov chain with states 0, 1, 2, 3, and 4. States 0 and 4 are absorbing states. The transition matrix is then

$$\mathbf{P} = \begin{array}{c} 0 \\ 1 \\ 2 \\ 3 \\ 4 \end{array} \begin{pmatrix} 0 & 1 & 2 & 3 & 4 \\ 1 & 0 & 0 & 0 & 0 \\ 1/2 & 0 & 1/2 & 0 & 0 \\ 0 & 1/2 & 0 & 1/2 & 0 \\ 0 & 0 & 1/2 & 0 & 1/2 \\ 0 & 0 & 0 & 0 & 1 \end{pmatrix}.$$

The states 1, 2, and 3 are transient states, and from any of these it is possible to reach the absorbing states 0 and 4. Hence the chain is an absorbing chain. When a process reaches an absorbing state, we shall say that it is *absorbed*. □

The most obvious question that can be asked about such a chain is: What is the probability that the process will eventually reach an absorbing state? Other interesting questions include: (a) What is the probability that the process will end up in a given absorbing state? (b) On the average, how long will it take for the process to be absorbed? (c) On the average, how many times will the process be in each transient state? The answers to all these questions depend, in general, on the state from which the process starts as well as the transition probabilities.

Canonical Form

Consider an arbitrary absorbing Markov chain. Renumber the states so that the transient states come first. If there are r absorbing states and t transient states, the transition matrix will have the following *canonical form*

$$\mathbf{P} = \begin{array}{c} \\ \text{TR.} \\ \text{ABS.} \end{array} \begin{array}{c} \text{TR.} \quad \text{ABS.} \\ \left(\begin{array}{c|c} \mathbf{Q} & \mathbf{R} \\ \hline \mathbf{0} & \mathbf{I} \end{array} \right) \end{array}$$

Here $\mathbf{I}$ is an r-by-r indentity matrix, $\mathbf{0}$ is an r-by-t zero matrix, $\mathbf{R}$ is a nonzero t-by-r matrix, and $\mathbf{Q}$ is an t-by-t matrix. The first t states are transient and the last r states are absorbing.

In Section 11.1, we saw that the entry $p_{ij}^{(n)}$ of the matrix $\mathbf{P}^n$ is the probability of being in the state s_j after n steps, when the chain is started in state s_i. A standard matrix algebra argument shows that $\mathbf{P}^n$ is of the form

$$\mathbf{P}^n = \begin{array}{c} \\ \text{TR.} \\ \text{ABS.} \end{array} \begin{array}{c} \text{TR.} \quad \text{ABS.} \\ \left(\begin{array}{c|c} \mathbf{Q}^n & * \\ \hline \mathbf{0} & \mathbf{I} \end{array} \right) \end{array}$$

where the asterisk $*$ stands for the t-by-r matrix in the upper right-hand corner of $\mathbf{P}^n$. (This submatrix can be written in terms of $\mathbf{Q}$ and $\mathbf{R}$, but the expression is complicated and is not needed at this time.) The form of $\mathbf{P}^n$ shows that the entries of $\mathbf{Q}^n$ give the probabilities for being in each of the transient states after n steps for each possible transient starting state. For our first theorem we prove that the probability of being in the transient states after n steps approaches zero. Thus every entry of $\mathbf{Q}^n$ must approach zero as n approaches infinity (i.e, $\mathbf{Q}^n \to \mathbf{0}$).

In the following, if $\mathbf{u}$ and $\mathbf{v}$ are two vectors we say that $\mathbf{u} \leq \mathbf{v}$ if all components of $\mathbf{u}$ are less than or equal to the corresponding components of $\mathbf{v}$. Similarly, if $\mathbf{A}$ and $\mathbf{B}$ are matrices then $\mathbf{A} \leq \mathbf{B}$ if each entry of $\mathbf{A}$ is less than or equal to the corresponding entry of $\mathbf{B}$.

Probability of Absorption

Theorem 11.3 In an absorbing Markov chain, the probability that the process will be absorbed is 1 (i.e., $\mathbf{Q}^n \to \mathbf{0}$ as $n \to \infty$).

Proof. From each nonabsorbing state s_j it is possible to reach an absorbing state. Let m_j be the minimum number of steps required to reach an absorbing state, starting from s_j. Let p_j be the probability that, starting from s_j, the process will not reach an absorbing state in m_j steps. Then $p_j < 1$. Let m be the largest of the m_j and let p be the largest of p_j. The probability of not being absorbed in m steps is less than or equal to p, in $2n$ steps less than or equal to p^2, etc. Since $p < 1$ these probabilities tend to 0. Since the probability of not being absorbed in n steps is monotone decreasing, these probabilities also tend to 0, hence $\lim_{n \to \infty} \mathbf{Q}^n = 0$. $\square$

The Fundamental Matrix

Theorem 11.4 For an absorbing Markov chain the matrix $\mathbf{I} - \mathbf{Q}$ has an inverse $\mathbf{N}$ and $\mathbf{N} = \mathbf{I} + \mathbf{Q} + \mathbf{Q}^2 + \cdots$. The ij-entry n_{ij} of the matrix $\mathbf{N}$ is the expected number of times the chain is in state s_j, given that it starts in state s_i. The initial state is counted if $i = j$.

Proof. Let $(\mathbf{I} - \mathbf{Q})\mathbf{x} = 0$; that is $\mathbf{x} = \mathbf{Q}\mathbf{x}$. Then, iterating this we see that $\mathbf{x} = \mathbf{Q}^n \mathbf{x}$. Since $\mathbf{Q}^n \to 0$, we have $\mathbf{Q}^n \mathbf{x} \to 0$, so $\mathbf{x} = 0$. Thus $(\mathbf{I} - \mathbf{Q})^{-1} = \mathbf{N}$ exists. Note next that

$$(\mathbf{I} - \mathbf{Q})(\mathbf{I} + \mathbf{Q} + \mathbf{Q}^2 + \cdots + \mathbf{Q}^n) = \mathbf{I} - \mathbf{Q}^{n+1} .$$

Thus multiplying both sides by $\mathbf{N}$ gives

$$\mathbf{I} + \mathbf{Q} + \mathbf{Q}^2 + \cdots + \mathbf{Q}^n = \mathbf{N}(\mathbf{I} - \mathbf{Q}^{n+1}) .$$

Letting n tend to infinity we have

$$\mathbf{N} = \mathbf{I} + \mathbf{Q} + \mathbf{Q}^2 + \cdots .$$

Let s_i and s_j be two transient states, and assume throughout the remainder of the proof that i and j are fixed. Let $X^{(k)}$ be a random variable which equals 1 if the chain is in state s_j after k steps, and equals 0 otherwise. For each k, this random variable depends upon both i and j; we choose not to explicitly show this dependence in the interest of clarity. We have

$$P(X^{(k)} = 1) = q_{ij}^{(k)} ,$$

and

$$P(X^{(k)} = 0) = 1 - q_{ij}^{(k)} ,$$

where $q_{ij}^{(k)}$ is the ijth entry of $\mathbf{Q}^k$. These equations hold for $k = 0$ since $\mathbf{Q}^0 = \mathbf{I}$. Therefore, since $X^{(k)}$ is a 0-1 random variable, $E(X^{(k)}) = q_{ij}^{(k)}$.

The expected number of times the chain is in state s_j in the first n steps, given that it starts in state s_i, is clearly

$$E\left(X^{(0)} + X^{(1)} + \cdots + X^{(n)}\right) = q_{ij}^{(0)} + q_{ij}^{(1)} + \cdots + q_{ij}^{(n)} .$$

Letting n tend to infinity we have

$$E\left(X^{(0)} + X^{(1)} + \cdots\right) = q_{ij}^{(0)} + q_{ij}^{(1)} + \cdots = n_{ij} .$$

$\square$

11.2. ABSORBING MARKOV CHAINS

Definition 11.3 For an absorbing Markov chain $\mathbf{P}$, the matrix $\mathbf{N} = (\mathbf{I} - \mathbf{Q})^{-1}$ is called the *fundamental matrix* for $\mathbf{P}$. The entry n_{ij} of $\mathbf{N}$ gives the expected number of times that the process is in the transient state s_j if it is started in the transient state s_i. □

Example 11.14 (Example 11.13 continued) In the Drunkard's Walk example, the transition matrix in canonical form is

$$\mathbf{P} = \begin{array}{c} \\ 1 \\ 2 \\ 3 \\ 0 \\ 4 \end{array} \begin{pmatrix} \begin{array}{ccc|cc} 1 & 2 & 3 & 0 & 4 \\ 0 & 1/2 & 0 & 1/2 & 0 \\ 1/2 & 0 & 1/2 & 0 & 0 \\ 0 & 1/2 & 0 & 0 & 1/2 \\ \hline 0 & 0 & 0 & 1 & 0 \\ 0 & 0 & 0 & 0 & 1 \end{array} \end{pmatrix} .$$

From this we see that the matrix $\mathbf{Q}$ is

$$\mathbf{Q} = \begin{pmatrix} 0 & 1/2 & 0 \\ 1/2 & 0 & 1/2 \\ 0 & 1/2 & 0 \end{pmatrix} ,$$

and

$$\mathbf{I} - \mathbf{Q} = \begin{pmatrix} 1 & -1/2 & 0 \\ -1/2 & 1 & -1/2 \\ 0 & -1/2 & 1 \end{pmatrix} .$$

Computing $(\mathbf{I} - \mathbf{Q})^{-1}$, we find

$$\mathbf{N} = (\mathbf{I} - \mathbf{Q})^{-1} = \begin{array}{c} \\ 1 \\ 2 \\ 3 \end{array} \begin{pmatrix} \begin{array}{ccc} 1 & 2 & 3 \\ 3/2 & 1 & 1/2 \\ 1 & 2 & 1 \\ 1/2 & 1 & 3/2 \end{array} \end{pmatrix} .$$

From the middle row of $\mathbf{N}$, we see that if we start in state 2, then the expected number of times in states 1, 2, and 3 before being absorbed are 1, 2, and 1. □

Time to Absorption

We now consider the question: Given that the chain starts in state s_i, what is the expected number of steps before the chain is absorbed? The answer is given in the next theorem.

Theorem 11.5 Let t_i be the expected number of steps before the chain is absorbed, given that the chain starts in state s_i, and let $\mathbf{t}$ be the column vector whose ith entry is t_i. Then

$$\mathbf{t} = \mathbf{Nc} ,$$

where $\mathbf{c}$ is a column vector all of whose entries are 1.

Proof. If we add all the entries in the ith row of $\mathbf{N}$, we will have the expected number of times in any of the transient states for a given starting state s_i, that is, the expected time required before being absorbed. Thus, t_i is the sum of the entries in the ith row of $\mathbf{N}$. If we write this statement in matrix form, we obtain the theorem. $\square$

Absorption Probabilities

Theorem 11.6 Let b_{ij} be the probability that an absorbing chain will be absorbed in the absorbing state s_j if it starts in the transient state s_i. Let $\mathbf{B}$ be the matrix with entries b_{ij}. Then $\mathbf{B}$ is an t-by-r matrix, and

$$\mathbf{B} = \mathbf{NR} ,$$

where $\mathbf{N}$ is the fundamental matrix and $\mathbf{R}$ is as in the canonical form.

Proof. We have

$$\begin{aligned}
\mathbf{B}_{ij} &= \sum_n \sum_k q_{ik}^{(n)} r_{kj} \\
&= \sum_k \sum_n q_{ik}^{(n)} r_{kj} \\
&= \sum_k n_{ik} r_{kj} \\
&= (\mathbf{NR})_{ij} .
\end{aligned}$$

This completes the proof. $\square$

Another proof of this is given in Exercise 34.

Example 11.15 (Example 11.14 continued) In the Drunkard's Walk example, we found that

$$\mathbf{N} = \begin{array}{c} \\ 1 \\ 2 \\ 3 \end{array} \begin{array}{c} 1 \quad 2 \quad 3 \\ \begin{pmatrix} 3/2 & 1 & 1/2 \\ 1 & 2 & 1 \\ 1/2 & 1 & 3/2 \end{pmatrix} \end{array} .$$

Hence,

$$\begin{aligned}
\mathbf{t} = \mathbf{Nc} &= \begin{pmatrix} 3/2 & 1 & 1/2 \\ 1 & 2 & 1 \\ 1/2 & 1 & 3/2 \end{pmatrix} \begin{pmatrix} 1 \\ 1 \\ 1 \end{pmatrix} \\
&= \begin{pmatrix} 3 \\ 4 \\ 3 \end{pmatrix} .
\end{aligned}$$

11.2. ABSORBING MARKOV CHAINS

Thus, starting in states 1, 2, and 3, the expected times to absorption are 3, 4, and 3, respectively.

From the canonical form,

$$\mathbf{R} = \begin{matrix} 1 \\ 2 \\ 3 \end{matrix}\begin{pmatrix} 0 & 4 \\ 1/2 & 0 \\ 0 & 0 \\ 0 & 1/2 \end{pmatrix}.$$

Hence,

$$\mathbf{B} = \mathbf{NR} = \begin{pmatrix} 3/2 & 1 & 1/2 \\ 1 & 2 & 1 \\ 1/2 & 1 & 3/2 \end{pmatrix} \cdot \begin{pmatrix} 1/2 & 0 \\ 0 & 0 \\ 0 & 1/2 \end{pmatrix}$$

$$= \begin{matrix} 1 \\ 2 \\ 3 \end{matrix}\begin{pmatrix} 0 & 4 \\ 3/4 & 1/4 \\ 1/2 & 1/2 \\ 1/4 & 3/4 \end{pmatrix}.$$

Here the first row tells us that, starting from state 1, there is probability 3/4 of absorption in state 0 and 1/4 of absorption in state 4. □

Computation

The fact that we have been able to obtain these three descriptive quantities in matrix form makes it very easy to write a computer program that determines these quantities for a given absorbing chain matrix.

The program **AbsorbingChain** calculates the basic descriptive quantities of an absorbing Markov chain.

We have run the program **AbsorbingChain** for the example of the drunkard's walk (Example 11.13) with 5 blocks. The results are as follows:

$$\mathbf{Q} = \begin{matrix} 1 \\ 2 \\ 3 \\ 4 \end{matrix}\begin{pmatrix} 1 & 2 & 3 & 4 \\ .00 & .50 & .00 & .00 \\ .50 & .00 & .50 & .00 \\ .00 & .50 & .00 & .50 \\ .00 & .00 & .50 & .00 \end{pmatrix};$$

$$\mathbf{R} = \begin{matrix} 1 \\ 2 \\ 3 \\ 4 \end{matrix}\begin{pmatrix} 0 & 5 \\ .50 & .00 \\ .00 & .00 \\ .00 & .00 \\ .00 & .50 \end{pmatrix};$$

$$\mathbf{N} = \begin{array}{c} 1 \\ 2 \\ 3 \\ 4 \end{array} \begin{pmatrix} 1 & 2 & 3 & 4 \\ 1.60 & 1.20 & .80 & .40 \\ 1.20 & 2.40 & 1.60 & .80 \\ .80 & 1.60 & 2.40 & 1.20 \\ .40 & .80 & 1.20 & 1.60 \end{pmatrix} ;$$

$$\mathbf{t} = \begin{array}{c} 1 \\ 2 \\ 3 \\ 4 \end{array} \begin{pmatrix} 4.00 \\ 6.00 \\ 6.00 \\ 4.00 \end{pmatrix} ;$$

$$\mathbf{B} = \begin{array}{c} 1 \\ 2 \\ 3 \\ 4 \end{array} \begin{pmatrix} 0 & 5 \\ .80 & .20 \\ .60 & .40 \\ .40 & .60 \\ .20 & .80 \end{pmatrix} .$$

Note that the probability of reaching the bar before reaching home, starting at x, is $x/5$ (i.e., proportional to the distance of home from the starting point). (See Exercise 24.)

Exercises

1 In Example 11.4, for what values of a and b do we obtain an absorbing Markov chain?

2 Show that Example 11.7 is an absorbing Markov chain.

3 Which of the genetics examples (Examples 11.9, 11.10, and 11.11) are absorbing?

4 Find the fundamental matrix $\mathbf{N}$ for Example 11.10.

5 For Example 11.11, verify that the following matrix is the inverse of $\mathbf{I} - \mathbf{Q}$ and hence is the fundamental matrix $\mathbf{N}$.

$$\mathbf{N} = \begin{pmatrix} 8/3 & 1/6 & 4/3 & 2/3 \\ 4/3 & 4/3 & 8/3 & 4/3 \\ 4/3 & 1/3 & 8/3 & 4/3 \\ 2/3 & 1/6 & 4/3 & 8/3 \end{pmatrix} .$$

Find $\mathbf{Nc}$ and $\mathbf{NR}$. Interpret the results.

6 In the Land of Oz example (Example 11.1), change the transition matrix by making R an absorbing state. This gives

$$\mathbf{P} = \begin{array}{c} R \\ N \\ S \end{array} \begin{pmatrix} R & N & S \\ 1 & 0 & 0 \\ 1/2 & 0 & 1/2 \\ 1/4 & 1/4 & 1/2 \end{pmatrix} .$$

11.2. ABSORBING MARKOV CHAINS

Find the fundamental matrix **N**, and also **Nc** and **NR**. Interpret the results.

7 In Example 11.8, make states 0 and 4 into absorbing states. Find the fundamental matrix **N**, and also **Nc** and **NR**, for the resulting absorbing chain. Interpret the results.

8 In Example 11.13 (Drunkard's Walk) of this section, assume that the probability of a step to the right is 2/3, and a step to the left is 1/3. Find **N**, **Nc**, and **NR**. Compare these with the results of Example 11.15.

9 A process moves on the integers 1, 2, 3, 4, and 5. It starts at 1 and, on each successive step, moves to an integer greater than its present position, moving with equal probability to each of the remaining larger integers. State five is an absorbing state. Find the expected number of steps to reach state five.

10 Using the result of Exercise 9, make a conjecture for the form of the fundamental matrix if the process moves as in that exercise, except that it now moves on the integers from 1 to n. Test your conjecture for several different values of n. Can you conjecture an estimate for the expected number of steps to reach state n, for large n? (See Exercise 11 for a method of determining this expected number of steps.)

*11 Let b_k denote the expected number of steps to reach n from $n - k$, in the process described in Exercise 9.

   (a) Define $b_0 = 0$. Show that for $k > 0$, we have
   $$b_k = 1 + \frac{1}{k}\left(b_{k-1} + b_{k-2} + \cdots + b_0\right).$$

   (b) Let
   $$f(x) = b_0 + b_1 x + b_2 x^2 + \cdots .$$
   Using the recursion in part (a), show that $f(x)$ satisfies the differential equation
   $$(1 - x)^2 y' - (1 - x)y + 1 = 0 .$$

   (c) Show that the general solution of the differential equation in part (b) is
   $$y = \frac{-\log(1-x)}{1-x} + \frac{c}{1-x},$$
   where c is a constant.

   (d) Use part (c) to show that
   $$b_k = 1 + \frac{1}{2} + \frac{1}{3} + \cdots + \frac{1}{k} .$$

12 Three tanks fight a three-way duel. Tank A has probability 1/2 of destroying the tank at which it fires, tank B has probability 1/3 of destroying the tank at which it fires, and tank C has probability 1/6 of destroying the tank at which

it fires. The tanks fire together and each tank fires at the strongest opponent not yet destroyed. Form a Markov chain by taking as states the subsets of the set of tanks. Find **N**, **Nc**, and **NR**, and interpret your results. *Hint*: Take as states ABC, AC, BC, A, B, C, and none, indicating the tanks that could survive starting in state ABC. You can omit AB because this state cannot be reached from ABC.

13 Smith is in jail and has 3 dollars; he can get out on bail if he has 8 dollars. A guard agrees to make a series of bets with him. If Smith bets A dollars, he wins A dollars with probability .4 and loses A dollars with probability .6. Find the probability that he wins 8 dollars before losing all of his money if

(a) he bets 1 dollar each time (timid strategy).

(b) he bets, each time, as much as possible but not more than necessary to bring his fortune up to 8 dollars (bold strategy).

(c) Which strategy gives Smith the better chance of getting out of jail?

14 With the situation in Exercise 13, consider the strategy such that for $i < 4$, Smith bets $\min(i, 4-i)$, and for $i \geq 4$, he bets according to the bold strategy, where i is his current fortune. Find the probability that he gets out of jail using this strategy. How does this probability compare with that obtained for the bold strategy?

15 Consider the game of tennis when *deuce* is reached. If a player wins the next point, he has *advantage*. On the following point, he either wins the game or the game returns to *deuce*. Assume that for any point, player A has probability .6 of winning the point and player B has probability .4 of winning the point.

(a) Set this up as a Markov chain with state 1: A wins; 2: B wins; 3: advantage A; 4: deuce; 5: advantage B.

(b) Find the absorption probabilities.

(c) At deuce, find the expected duration of the game and the probability that B will win.

Exercises 16 and 17 concern the inheritance of color-blindness, which is a sex-linked characteristic. There is a pair of genes, g and G, of which the former tends to produce color-blindness, the latter normal vision. The G gene is dominant. But a man has only one gene, and if this is g, he is color-blind. A man inherits one of his mother's two genes, while a woman inherits one gene from each parent. Thus a man may be of type G or g, while a woman may be type GG or Gg or gg. We will study a process of inbreeding similar to that of Example 11.11 by constructing a Markov chain.

16 List the states of the chain. *Hint*: There are six. Compute the transition probabilities. Find the fundamental matrix **N**, **Nc**, and **NR**.

11.2. ABSORBING MARKOV CHAINS

17 Show that in both Example 11.11 and the example just given, the probability of absorption in a state having genes of a particular type is equal to the proportion of genes of that type in the starting state. Show that this can be explained by the fact that a game in which your fortune is the number of genes of a particular type in the state of the Markov chain is a fair game.[5]

18 Assume that a student going to a certain four-year medical school in northern New England has, each year, a probability q of flunking out, a probability r of having to repeat the year, and a probability p of moving on to the next year (in the fourth year, moving on means graduating).

  (a) Form a transition matrix for this process taking as states F, 1, 2, 3, 4, and G where F stands for flunking out and G for graduating, and the other states represent the year of study.

  (b) For the case $q = .1$, $r = .2$, and $p = .7$ find the time a beginning student can expect to be in the second year. How long should this student expect to be in medical school?

  (c) Find the probability that this beginning student will graduate.

19 (E. Brown[6]) Mary and John are playing the following game: They have a three-card deck marked with the numbers 1, 2, and 3 and a spinner with the numbers 1, 2, and 3 on it. The game begins by dealing the cards out so that the dealer gets one card and the other person gets two. A move in the game consists of a spin of the spinner. The person having the card with the number that comes up on the spinner hands that card to the other person. The game ends when someone has all the cards.

  (a) Set up the transition matrix for this absorbing Markov chain, where the states correspond to the number of cards that Mary has.

  (b) Find the fundamental matrix.

  (c) On the average, how many moves will the game last?

  (d) If Mary deals, what is the probability that John will win the game?

20 Assume that an experiment has m equally probable outcomes. Show that the expected number of independent trials before the first occurrence of k consecutive occurrences of one of these outcomes is $(m^k - 1)/(m - 1)$. *Hint*: Form an absorbing Markov chain with states 1, 2, ..., k with state i representing the length of the current run. The expected time until a run of k is 1 more than the expected time until absorption for the chain started in state 1. It has been found that, in the decimal expansion of pi, starting with the 24,658,601st digit, there is a run of nine 7's. What would your result say about the expected number of digits necessary to find such a run if the digits are produced randomly?

---

[5]H. Gonshor, "An Application of Random Walk to a Problem in Population Genetics," *American Math Monthly*, vol. 94 (1987), pp. 668–671
[6]Private communication.

21 (Roberts[7]) A city is divided into 3 areas 1, 2, and 3. It is estimated that amounts u_1, u_2, and u_3 of pollution are emitted each day from these three areas. A fraction q_{ij} of the pollution from region i ends up the next day at region j. A fraction $q_i = 1 - \sum_j q_{ij} > 0$ goes into the atmosphere and escapes. Let $w_i^{(n)}$ be the amount of pollution in area i after n days.

  (a) Show that $\mathbf{w}^{(n)} = \mathbf{u} + \mathbf{u}\mathbf{Q} + \cdots + \mathbf{u}\mathbf{Q}^{n-1}$.

  (b) Show that $\mathbf{w}^{(n)} \to \mathbf{w}$, and show how to compute $\mathbf{w}$ from $\mathbf{u}$.

  (c) The government wants to limit pollution levels to a prescribed level by prescribing $\mathbf{w}$. Show how to determine the levels of pollution $\mathbf{u}$ which would result in a prescribed limiting value $\mathbf{w}$.

22 In the Leontief economic model,[8] there are n industries 1, 2, ..., n. The ith industry requires an amount $0 \leq q_{ij} \leq 1$ of goods (in dollar value) from company j to produce 1 dollar's worth of goods. The outside demand on the industries, in dollar value, is given by the vector $\mathbf{d} = (d_1, d_2, \ldots, d_n)$. Let $\mathbf{Q}$ be the matrix with entries q_{ij}.

  (a) Show that if the industries produce total amounts given by the vector $\mathbf{x} = (x_1, x_2, \ldots, x_n)$ then the amounts of goods of each type that the industries will need just to meet their internal demands is given by the vector $\mathbf{xQ}$.

  (b) Show that in order to meet the outside demand $\mathbf{d}$ and the internal demands the industries must produce total amounts given by a vector $\mathbf{x} = (x_1, x_2, \ldots, x_n)$ which satisfies the equation $\mathbf{x} = \mathbf{xQ} + \mathbf{d}$.

  (c) Show that if $\mathbf{Q}$ is the $\mathbf{Q}$-matrix for an absorbing Markov chain, then it is possible to meet any outside demand $\mathbf{d}$.

  (d) Assume that the row sums of $\mathbf{Q}$ are less than or equal to 1. Give an economic interpretation of this condition. Form a Markov chain by taking the states to be the industries and the transition probabilites to be the q_{ij}. Add one absorbing state 0. Define
  $$q_{i0} = 1 - \sum_j q_{ij} \ .$$
  Show that this chain will be absorbing if every company is either making a profit or ultimately depends upon a profit-making company.

  (e) Define $\mathbf{xc}$ to be the gross national product. Find an expression for the gross national product in terms of the demand vector $\mathbf{d}$ and the vector $\mathbf{t}$ giving the expected time to absorption.

23 A gambler plays a game in which on each play he wins one dollar with probability p and loses one dollar with probability $q = 1 - p$. The *Gambler's Ruin*

---
[7]F. Roberts, *Discrete Mathematical Models* (Englewood Cliffs, NJ: Prentice Hall, 1976).
[8]W. W. Leontief, *Input-Output Economics* (Oxford: Oxford University Press, 1966).

11.2. ABSORBING MARKOV CHAINS

problem is the problem of finding the probability w_x of winning an amount T before losing everything, starting with state x. Show that this problem may be considered to be an absorbing Markov chain with states $0, 1, 2, \ldots, T$ with 0 and T absorbing states. Suppose that a gambler has probability $p = .48$ of winning on each play. Suppose, in addition, that the gambler starts with 50 dollars and that $T = 100$ dollars. Simulate this game 100 times and see how often the gambler is ruined. This estimates w_{50}.

24 Show that w_x of Exercise 23 satisfies the following conditions:

(a) $w_x = pw_{x+1} + qw_{x-1}$ for $x = 1, 2, \ldots, T-1$.

(b) $w_0 = 0$.

(c) $w_T = 1$.

Show that these conditions determine w_x. Show that, if $p = q = 1/2$, then

$$w_x = \frac{x}{T}$$

satisfies (a), (b), and (c) and hence is the solution. If $p \neq q$, show that

$$w_x = \frac{(q/p)^x - 1}{(q/p)^T - 1}$$

satisfies these conditions and hence gives the probability of the gambler winning.

25 Write a program to compute the probability w_x of Exercise 24 for given values of x, p, and T. Study the probability that the gambler will ruin the bank in a game that is only slightly unfavorable, say $p = .49$, if the bank has significantly more money than the gambler.

*26 We considered the two examples of the Drunkard's Walk corresponding to the cases $n = 4$ and $n = 5$ blocks (see Example 11.13). Verify that in these two examples the expected time to absorption, starting at x, is equal to $x(n-x)$. See if you can prove that this is true in general. *Hint*: Show that if $f(x)$ is the expected time to absorption then $f(0) = f(n) = 0$ and

$$f(x) = (1/2)f(x-1) + (1/2)f(x+1) + 1$$

for $0 < x < n$. Show that if $f_1(x)$ and $f_2(x)$ are two solutions, then their difference $g(x)$ is a solution of the equation

$$g(x) = (1/2)g(x-1) + (1/2)g(x+1) .$$

Also, $g(0) = g(n) = 0$. Show that it is not possible for $g(x)$ to have a strict maximum or a strict minimum at the point i, where $1 \leq i \leq n-1$. Use this to show that $g(i) = 0$ for all i. This shows that there is at most one solution. Then verify that the function $f(x) = x(n-x)$ is a solution.

27 Consider an absorbing Markov chain with state space S. Let f be a function defined on S with the property that

$$f(i) = \sum_{j \in S} p_{ij} f(j) \ ,$$

or in vector form

$$\mathbf{f} = \mathbf{P}\mathbf{f} \ .$$

Then f is called a *harmonic function* for $\mathbf{P}$. If you imagine a game in which your fortune is $f(i)$ when you are in state i, then the harmonic condition means that the game is *fair* in the sense that your expected fortune after one step is the same as it was before the step.

(a) Show that for f harmonic
$$\mathbf{f} = \mathbf{P}^n \mathbf{f}$$
for all n.

(b) Show, using (a), that for f harmonic
$$\mathbf{f} = \mathbf{P}^\infty \mathbf{f} \ ,$$
where
$$\mathbf{P}^\infty = \lim_{n \to \infty} \mathbf{P}^n = \left(\begin{array}{c|c} \mathbf{0} & \mathbf{B} \\ \hline \mathbf{0} & \mathbf{I} \end{array} \right) \ .$$

(c) Using (b), prove that when you start in a transient state i your expected final fortune
$$\sum_k b_{ik} f(k)$$
is equal to your starting fortune $f(i)$. In other words, a fair game on a finite state space remains fair to the end. (Fair games in general are called *martingales*. Fair games on infinite state spaces need not remain fair with an unlimited number of plays allowed. For example, consider the game of Heads or Tails (see Example 1.4). Let Peter start with 1 penny and play until he has 2. Then Peter will be sure to end up 1 penny ahead.)

28 A coin is tossed repeatedly. We are interested in finding the expected number of tosses until a particular pattern, say B = HTH, occurs for the first time. If, for example, the outcomes of the tosses are HHTTHTH we say that the pattern B has occurred for the first time after 7 tosses. Let T^B be the time to obtain pattern B for the first time. Li[9] gives the following method for determining $E(T^B)$.

We are in a casino and, before each toss of the coin, a gambler enters, pays 1 dollar to play, and bets that the pattern B = HTH will occur on the next

---

[9]S-Y. R. Li, "A Martingale Approach to the Study of Occurrence of Sequence Patterns in Repeated Experiments," *Annals of Probability*, vol. 8 (1980), pp. 1171–1176.

11.2. ABSORBING MARKOV CHAINS

three tosses. If H occurs, he wins 2 dollars and bets this amount that the next outcome will be T. If he wins, he wins 4 dollars and bets this amount that H will come up next time. If he wins, he wins 8 dollars and the pattern has occurred. If at any time he loses, he leaves with no winnings.

Let A and B be two patterns. Let AB be the amount the gamblers win who arrive while the pattern A occurs and bet that B will occur. For example, if A = HT and B = HTH then AB = 2 + 4 = 6 since the first gambler bet on H and won 2 dollars and then bet on T and won 4 dollars more. The second gambler bet on H and lost. If A = HH and B = HTH, then AB = 2 since the first gambler bet on H and won but then bet on T and lost and the second gambler bet on H and won. If A = B = HTH then AB = BB = 8 + 2 = 10.

Now for each gambler coming in, the casino takes in 1 dollar. Thus the casino takes in T^B dollars. How much does it pay out? The only gamblers who go off with any money are those who arrive during the time the pattern B occurs and they win the amount BB. But since all the bets made are perfectly fair bets, it seems quite intuitive that the expected amount the casino takes in should equal the expected amount that it pays out. That is, $E(T^B) = $ BB.

Since we have seen that for B = HTH, BB = 10, the expected time to reach the pattern HTH for the first time is 10. If we had been trying to get the pattern B = HHH, then BB = 8 + 4 + 2 = 14 since all the last three gamblers are paid off in this case. Thus the expected time to get the pattern HHH is 14. To justify this argument, Li used a theorem from the theory of martingales (fair games).

We can obtain these expectations by considering a Markov chain whose states are the possible initial segments of the sequence HTH; these states are HTH, HT, H, and ∅, where ∅ is the empty set. Then, for this example, the transition matrix is

$$\begin{array}{c c} & \begin{array}{cccc} \text{HTH} & \text{HT} & \text{H} & \emptyset \end{array} \\ \begin{array}{c} \text{HTH} \\ \text{HT} \\ \text{H} \\ \emptyset \end{array} & \left(\begin{array}{cccc} 1 & 0 & 0 & 0 \\ .5 & 0 & 0 & .5 \\ 0 & .5 & .5 & 0 \\ 0 & 0 & .5 & .5 \end{array} \right) \end{array},$$

and if B = HTH, $E(T^B)$ is the expected time to absorption for this chain started in state ∅.

Show, using the associated Markov chain, that the values $E(T^B) = 10$ and $E(T^B) = 14$ are correct for the expected time to reach the patterns HTH and HHH, respectively.

29 We can use the gambling interpretation given in Exercise 28 to find the expected number of tosses required to reach pattern B when we start with pattern A. To be a meaningful problem, we assume that pattern A does not have pattern B as a subpattern. Let $E_A(T^B)$ be the expected time to reach pattern B starting with pattern A. We use our gambling scheme and assume that the first k coin tosses produced the pattern A. During this time, the gamblers made an amount AB. The total amount the gamblers will have made when the pattern B occurs is BB. Thus, the amount that the gamblers made after the pattern A has occurred is BB - AB. Again by the fair game argument, $E_A(T^B) = $ BB-AB.

For example, suppose that we start with pattern A = HT and are trying to get the pattern B = HTH. Then we saw in Exercise 28 that AB = 4 and BB = 10 so $E_A(T^B) = $ BB-AB= 6.

Verify that this gambling interpretation leads to the correct answer for all starting states in the examples that you worked in Exercise 28.

30 Here is an elegant method due to Guibas and Odlyzko[10] to obtain the expected time to reach a pattern, say HTH, for the first time. Let $f(n)$ be the number of sequences of length n which do not have the pattern HTH. Let $f_p(n)$ be the number of sequences that have the pattern for the first time after n tosses. To each element of $f(n)$, add the pattern HTH. Then divide the resulting sequences into three subsets: the set where HTH occurs for the first time at time $n+1$ (for this, the original sequence must have ended with HT); the set where HTH occurs for the first time at time $n+2$ (cannot happen for this pattern); and the set where the sequence HTH occurs for the first time at time $n+3$ (the original sequence ended with anything except HT). Doing this, we have
$$f(n) = f_p(n+1) + f_p(n+3) \ .$$
Thus,
$$\frac{f(n)}{2^n} = \frac{2f_p(n+1)}{2^{n+1}} + \frac{2^3 f_p(n+3)}{2^{n+3}} \ .$$

If T is the time that the pattern occurs for the first time, this equality states that
$$P(T > n) = 2P(T = n+1) + 8P(T = n+3) \ .$$

Show that if you sum this equality over all n you obtain
$$\sum_{n=0}^{\infty} P(T > n) = 2 + 8 = 10 \ .$$

Show that for any integer-valued random variable
$$E(T) = \sum_{n=0}^{\infty} P(T > n) \ ,$$

and conclude that $E(T) = 10$. Note that this method of proof makes very clear that $E(T)$ is, in general, equal to the expected amount the casino pays out and avoids the martingale system theorem used by Li.

---

[10]L. J. Guibas and A. M. Odlyzko, "String Overlaps, Pattern Matching, and Non-transitive Games," *Journal of Combinatorial Theory*, Series A, vol. 30 (1981), pp. 183–208.

11.2. ABSORBING MARKOV CHAINS

31 In Example 11.11, define $f(i)$ to be the proportion of G genes in state i. Show that f is a harmonic function (see Exercise 27). Why does this show that the probability of being absorbed in state (GG, GG) is equal to the proportion of G genes in the starting state? (See Exercise 17.)

32 Show that the stepping stone model (Example 11.12) is an absorbing Markov chain. Assume that you are playing a game with red and green squares, in which your fortune at any time is equal to the proportion of red squares at that time. Give an argument to show that this is a fair game in the sense that your expected winning after each step is just what it was before this step. *Hint*: Show that for every possible outcome in which your fortune will decrease by one there is another outcome of exactly the same probability where it will increase by one.

Use this fact and the results of Exercise 27 to show that the probability that a particular color wins out is equal to the proportion of squares that are initially of this color.

33 Consider a random walker who moves on the integers $0, 1, \ldots, N$, moving one step to the right with probability p and one step to the left with probability $q = 1 - p$. If the walker ever reaches 0 or N he stays there. (This is the Gambler's Ruin problem of Exercise 23.) If $p = q$ show that the function

$$f(i) = i$$

is a harmonic function (see Exercise 27), and if $p \neq q$ then

$$f(i) = \left(\frac{q}{p}\right)^i$$

is a harmonic function. Use this and the result of Exercise 27 to show that the probability b_{iN} of being absorbed in state N starting in state i is

$$b_{iN} = \begin{cases} \frac{i}{N}, & \text{if } p = q, \\ \frac{(\frac{q}{p})^i - 1}{(\frac{q}{p})^N - 1}, & \text{if } p \neq q. \end{cases}$$

For an alternative derivation of these results see Exercise 24.

34 Complete the following alternate proof of Theorem 11.6. Let s_i be a transient state and s_j be an absorbing state. If we compute b_{ij} in terms of the possibilities on the outcome of the first step, then we have the equation

$$b_{ij} = p_{ij} + \sum_k p_{ik} b_{kj} ,$$

where the summation is carried out over all transient states s_k. Write this in matrix form, and derive from this equation the statement

$$\mathbf{B} = \mathbf{NR} .$$

35 In Monte Carlo roulette (see Example 6.6), under option (c), there are six states (S, W, L, E, P_1, and P_2). The reader is referred to Figure 6.2, which contains a tree for this option. Form a Markov chain for this option, and use the program **AbsorbingChain** to find the probabilities that you win, lose, or break even for a 1 franc bet on red. Using these probabilities, find the expected winnings for this bet. For a more general discussion of Markov chains applied to roulette, see the article of H. Sagan referred to in Example 6.13.

36 We consider next a game called *Penney-ante* by its inventor W. Penney.[11] There are two players; the first player picks a pattern A of H's and T's, and then the second player, knowing the choice of the first player, picks a different pattern B. We assume that neither pattern is a subpattern of the other pattern. A coin is tossed a sequence of times, and the player whose pattern comes up first is the winner. To analyze the game, we need to find the probability p_A that pattern A will occur before pattern B and the probability $p_B = 1 - p_A$ that pattern B occurs before pattern A. To determine these probabilities we use the results of Exercises 28 and 29. Here you were asked to show that, the expected time to reach a pattern B for the first time is,

$$E(T^B) = BB ,$$

and, starting with pattern A, the expected time to reach pattern B is

$$E_A(T^B) = BB - AB .$$

(a) Show that the odds that the first player will win are given by John Conway's formula[12]:

$$\frac{p_A}{1 - p_A} = \frac{p_A}{p_B} = \frac{BB - BA}{AA - AB} .$$

Hint: Explain why

$$E(T^B) = E(T^{A \text{ or } B}) + p_A E_A(T^B)$$

and thus

$$BB = E(T^{A \text{ or } B}) + p_A(BB - AB) .$$

Interchange A and B to find a similar equation involving the p_B. Finally, note that

$$p_A + p_B = 1 .$$

Use these equations to solve for p_A and p_B.

(b) Assume that both players choose a pattern of the same length k. Show that, if $k = 2$, this is a fair game, but, if $k = 3$, the second player has an advantage no matter what choice the first player makes. (It has been shown that, for $k \geq 3$, if the first player chooses $a_1, a_2, \ldots, a_k$, then the optimal strategy for the second player is of the form $b, a_1, \ldots, a_{k-1}$ where b is the better of the two choices H or T.[13])

---
[11] W. Penney, "Problem: Penney-Ante," *Journal of Recreational Math*, vol. 2 (1969), p. 241.
[12] M. Gardner, "Mathematical Games," *Scientific American*, vol. 10 (1974), pp. 120–125.
[13] Guibas and Odlyzko, op. cit.

11.3 Ergodic Markov Chains

A second important kind of Markov chain we shall study in detail is an *ergodic* Markov chain, defined as follows.

Definition 11.4 A Markov chain is called an *ergodic* chain if it is possible to go from every state to every state (not necessarily in one move). □

In many books, ergodic Markov chains are called *irreducible*.

Definition 11.5 A Markov chain is called a *regular* chain if some power of the transition matrix has only positive elements. □

In other words, for some n, it is possible to go from any state to any state in exactly n steps. It is clear from this definition that every regular chain is ergodic. On the other hand, an ergodic chain is not necessarily regular, as the following examples show.

Example 11.16 Let the transition matrix of a Markov chain be defined by

$$\mathbf{P} = \begin{pmatrix} & 1 & 2 \\ 1 & 0 & 1 \\ 2 & 1 & 0 \end{pmatrix}.$$

Then is clear that it is possible to move from any state to any state, so the chain is ergodic. However, if n is odd, then it is not possible to move from state 0 to state 0 in n steps, and if n is even, then it is not possible to move from state 0 to state 1 in n steps, so the chain is not regular. □

A more interesting example of an ergodic, non-regular Markov chain is provided by the Ehrenfest urn model.

Example 11.17 Recall the Ehrenfest urn model (Example 11.8). The transition matrix for this example is

$$\mathbf{P} = \begin{pmatrix} & 0 & 1 & 2 & 3 & 4 \\ 0 & 0 & 1 & 0 & 0 & 0 \\ 1 & 1/4 & 0 & 3/4 & 0 & 0 \\ 2 & 0 & 1/2 & 0 & 1/2 & 0 \\ 3 & 0 & 0 & 3/4 & 0 & 1/4 \\ 4 & 0 & 0 & 0 & 1 & 0 \end{pmatrix}.$$

In this example, if we start in state 0 we will, after any even number of steps, be in either state 0, 2 or 4, and after any odd number of steps, be in states 1 or 3. Thus this chain is ergodic but not regular. □

Regular Markov Chains

Any transition matrix that has no zeros determines a regular Markov chain. However, it is possible for a regular Markov chain to have a transition matrix that has zeros. The transition matrix of the Land of Oz example of Section 11.1 has $p_{NN} = 0$ but the second power $\mathbf{P}^2$ has no zeros, so this is a regular Markov chain.

An example of a nonregular Markov chain is an absorbing chain. For example, let

$$\mathbf{P} = \begin{pmatrix} 1 & 0 \\ 1/2 & 1/2 \end{pmatrix}$$

be the transition matrix of a Markov chain. Then all powers of $\mathbf{P}$ will have a 0 in the upper right-hand corner.

We shall now discuss two important theorems relating to regular chains.

Theorem 11.7 Let $\mathbf{P}$ be the transition matrix for a regular chain. Then, as $n \to \infty$, the powers $\mathbf{P}^n$ approach a limiting matrix $\mathbf{W}$ with all rows the same vector $\mathbf{w}$. The vector $\mathbf{w}$ is a strictly positive probability vector (i.e., the components are all positive and they sum to one). $\square$

In the next section we give two proofs of this fundamental theorem. We give here the basic idea of the first proof.

We want to show that the powers $\mathbf{P}^n$ of a regular transition matrix tend to a matrix with all rows the same. This is the same as showing that $\mathbf{P}^n$ converges to a matrix with constant columns. Now the jth column of $\mathbf{P}^n$ is $\mathbf{P}^n \mathbf{y}$ where $\mathbf{y}$ is a column vector with 1 in the jth entry and 0 in the other entries. Thus we need only prove that for any column vector $\mathbf{y}$, $\mathbf{P}^n \mathbf{y}$ approaches a constant vector as n tend to infinity.

Since each row of $\mathbf{P}$ is a probability vector, $\mathbf{P}\mathbf{y}$ replaces $\mathbf{y}$ by averages of its components. Here is an example:

$$\begin{pmatrix} 1/2 & 1/4 & 1/4 \\ 1/3 & 1/3 & 1/3 \\ 1/3 & 1/2 & 0 \end{pmatrix} \begin{pmatrix} 1 \\ 2 \\ 3 \end{pmatrix} = \begin{pmatrix} 1/2 \cdot 1 + 1/4 \cdot 2 + 1/4 \cdot 3 \\ 1/3 \cdot 1 + 1/3 \cdot 2 + 1/3 \cdot 3 \\ 1/3 \cdot 1 + 1/2 \cdot 2 + 0 \cdot 3 \end{pmatrix} = \begin{pmatrix} 7/4 \\ 2 \\ 3/2 \end{pmatrix}.$$

The result of the averaging process is to make the components of $\mathbf{P}\mathbf{y}$ more similar than those of $\mathbf{y}$. In particular, the maximum component decreases (from 3 to 2) and the minimum component increases (from 1 to 3/2). Our proof will show that as we do more and more of this averaging to get $\mathbf{P}^n \mathbf{y}$, the difference between the maximum and minimum component will tend to 0 as $n \to \infty$. This means $\mathbf{P}^n \mathbf{y}$ tends to a constant vector. The ijth entry of $\mathbf{P}^n$, $p_{ij}^{(n)}$, is the probability that the process will be in state s_j after n steps if it starts in state s_i. If we denote the common row of $\mathbf{W}$ by $\mathbf{w}$, then Theorem 11.7 states that the probability of being in s_j in the long run is approximately w_j, the jth entry of $\mathbf{w}$, and is independent of the starting state.

11.3. ERGODIC MARKOV CHAINS

Example 11.18 Recall that for the Land of Oz example of Section 11.1, the sixth power of the transition matrix $\mathbf{P}$ is, to three decimal places,

$$\mathbf{P}^6 = \begin{array}{c} \\ \text{R} \\ \text{N} \\ \text{S} \end{array} \begin{pmatrix} \text{R} & \text{N} & \text{S} \\ .4 & .2 & .4 \\ .4 & .2 & .4 \\ .4 & .2 & .4 \end{pmatrix}.$$

Thus, to this degree of accuracy, the probability of rain six days after a rainy day is the same as the probability of rain six days after a nice day, or six days after a snowy day. Theorem 11.7 predicts that, for large n, the rows of $\mathbf{P}$ approach a common vector. It is interesting that this occurs so soon in our example. □

Theorem 11.8 Let $\mathbf{P}$ be a regular transition matrix, let

$$\mathbf{W} = \lim_{n \to \infty} \mathbf{P}^n ,$$

let $\mathbf{w}$ be the common row of $\mathbf{W}$, and let $\mathbf{c}$ be the column vector all of whose components are 1. Then

(a) $\mathbf{wP} = \mathbf{w}$, and any row vector $\mathbf{v}$ such that $\mathbf{vP} = \mathbf{v}$ is a constant multiple of $\mathbf{w}$.

(b) $\mathbf{Pc} = \mathbf{c}$, and any column vector $\mathbf{x}$ such that $\mathbf{Px} = \mathbf{x}$ is a multiple of $\mathbf{c}$.

Proof. To prove part (a), we note that from Theorem 11.7,

$$\mathbf{P}^n \to \mathbf{W} .$$

Thus,

$$\mathbf{P}^{n+1} = \mathbf{P}^n \cdot \mathbf{P} \to \mathbf{WP} .$$

But $\mathbf{P}^{n+1} \to \mathbf{W}$, and so $\mathbf{W} = \mathbf{WP}$, and $\mathbf{w} = \mathbf{wP}$.

Let $\mathbf{v}$ be any vector with $\mathbf{vP} = \mathbf{v}$. Then $\mathbf{v} = \mathbf{vP}^n$, and passing to the limit, $\mathbf{v} = \mathbf{vW}$. Let r be the sum of the components of $\mathbf{v}$. Then it is easily checked that $\mathbf{vW} = r\mathbf{w}$. So, $\mathbf{v} = r\mathbf{w}$.

To prove part (b), assume that $\mathbf{x} = \mathbf{Px}$. Then $\mathbf{x} = \mathbf{P}^n \mathbf{x}$, and again passing to the limit, $\mathbf{x} = \mathbf{Wx}$. Since all rows of $\mathbf{W}$ are the same, the components of $\mathbf{Wx}$ are all equal, so $\mathbf{x}$ is a multiple of $\mathbf{c}$. □

Note that an immediate consequence of Theorem 11.8 is the fact that there is only one probability vector $\mathbf{v}$ such that $\mathbf{vP} = \mathbf{v}$.

Fixed Vectors

Definition 11.6 A row vector $\mathbf{w}$ with the property $\mathbf{wP} = \mathbf{w}$ is called a *fixed row vector* for $\mathbf{P}$. Similarly, a column vector $\mathbf{x}$ such that $\mathbf{Px} = \mathbf{x}$ is called a *fixed column vector* for $\mathbf{P}$. □

Thus, the common row of **W** is the unique vector **w** which is both a fixed row vector for **P** and a probability vector. Theorem 11.8 shows that any fixed row vector for **P** is a multiple of **w** and any fixed column vector for **P** is a constant vector.

One can also state Definition 11.6 in terms of eigenvalues and eigenvectors. A fixed row vector is a left eigenvector of the matrix **P** corresponding to the eigenvalue 1. A similar statement can be made about fixed column vectors.

We will now give several different methods for calculating the fixed row vector **w** for a regular Markov chain.

Example 11.19 By Theorem 11.7 we can find the limiting vector **w** for the Land of Oz from the fact that
$$w_1 + w_2 + w_3 = 1$$
and
$$\begin{pmatrix} w_1 & w_2 & w_3 \end{pmatrix} \begin{pmatrix} 1/2 & 1/4 & 1/4 \\ 1/2 & 0 & 1/2 \\ 1/4 & 1/4 & 1/2 \end{pmatrix} = \begin{pmatrix} w_1 & w_2 & w_3 \end{pmatrix}.$$

These relations lead to the following four equations in three unknowns:
$$\begin{aligned} w_1 + w_2 + w_3 &= 1, \\ (1/2)w_1 + (1/2)w_2 + (1/4)w_3 &= w_1, \\ (1/4)w_1 + (1/4)w_3 &= w_2, \\ (1/4)w_1 + (1/2)w_2 + (1/2)w_3 &= w_3. \end{aligned}$$

Our theorem guarantees that these equations have a unique solution. If the equations are solved, we obtain the solution
$$\mathbf{w} = \begin{pmatrix} .4 & .2 & .4 \end{pmatrix},$$
in agreement with that predicted from $\mathbf{P}^6$, given in Example 11.2. □

To calculate the fixed vector, we can assume that the value at a particular state, say state one, is 1, and then use all but one of the linear equations from $\mathbf{wP} = \mathbf{w}$. This set of equations will have a unique solution and we can obtain **w** from this solution by dividing each of its entries by their sum to give the probability vector **w**. We will now illustrate this idea for the above example.

Example 11.20 (Example 11.19 continued) We set $w_1 = 1$, and then solve the first and second linear equations from $\mathbf{wP} = \mathbf{w}$. We have
$$\begin{aligned} (1/2) + (1/2)w_2 + (1/4)w_3 &= 1, \\ (1/4) + (1/4)w_3 &= w_2. \end{aligned}$$

If we solve these, we obtain
$$\begin{pmatrix} w_1 & w_2 & w_3 \end{pmatrix} = \begin{pmatrix} 1 & 1/2 & 1 \end{pmatrix}.$$

11.3. ERGODIC MARKOV CHAINS

Now we divide this vector by the sum of the components, to obtain the final answer:

$$\mathbf{w} = (\,.4 \quad .2 \quad .4\,)\ .$$

This method can be easily programmed to run on a computer. □

As mentioned above, we can also think of the fixed row vector $\mathbf{w}$ as a left eigenvector of the transition matrix $\mathbf{P}$. Thus, if we write $\mathbf{I}$ to denote the identity matrix, then $\mathbf{w}$ satisfies the matrix equation

$$\mathbf{wP} = \mathbf{wI}\ ,$$

or equivalently,

$$\mathbf{w}(\mathbf{P} - \mathbf{I}) = \mathbf{0}\ .$$

Thus, $\mathbf{w}$ is in the left nullspace of the matrix $\mathbf{P} - \mathbf{I}$. Furthermore, Theorem 11.8 states that this left nullspace has dimension 1. Certain computer programming languages can find nullspaces of matrices. In such languages, one can find the fixed row probability vector for a matrix $\mathbf{P}$ by computing the left nullspace and then normalizing a vector in the nullspace so the sum of its components is 1.

The program **FixedVector** uses one of the above methods (depending upon the language in which it is written) to calculate the fixed row probability vector for regular Markov chains.

So far we have always assumed that we started in a specific state. The following theorem generalizes Theorem 11.7 to the case where the starting state is itself determined by a probability vector.

Theorem 11.9 Let $\mathbf{P}$ be the transition matrix for a regular chain and $\mathbf{v}$ an arbitrary probability vector. Then

$$\lim_{n \to \infty} \mathbf{vP}^n = \mathbf{w}\ ,$$

where $\mathbf{w}$ is the unique fixed probability vector for $\mathbf{P}$.

Proof. By Theorem 11.7,

$$\lim_{n \to \infty} \mathbf{P}^n = \mathbf{W}\ .$$

Hence,

$$\lim_{n \to \infty} \mathbf{vP}^n = \mathbf{vW}\ .$$

But the entries in $\mathbf{v}$ sum to 1, and each row of $\mathbf{W}$ equals $\mathbf{w}$. From these statements, it is easy to check that

$$\mathbf{vW} = \mathbf{w}\ .$$

□

If we start a Markov chain with initial probabilities given by $\mathbf{v}$, then the probability vector $\mathbf{vP}^n$ gives the probabilities of being in the various states after n steps. Theorem 11.9 then establishes the fact that, even in this more general class of processes, the probability of being in s_j approaches w_j.

Equilibrium

We also obtain a new interpretation for **w**. Suppose that our starting vector picks state s_i as a starting state with probability w_i, for all i. Then the probability of being in the various states after n steps is given by $\mathbf{wP}^n = \mathbf{w}$, and is the same on all steps. This method of starting provides us with a process that is called "stationary." The fact that **w** is the only probability vector for which $\mathbf{wP} = \mathbf{w}$ shows that we must have a starting probability vector of exactly the kind described to obtain a stationary process.

Many interesting results concerning regular Markov chains depend only on the fact that the chain has a unique fixed probability vector which is positive. This property holds for all ergodic Markov chains.

Theorem 11.10 For an ergodic Markov chain, there is a unique probability vector **w** such that $\mathbf{wP} = \mathbf{w}$ and **w** is strictly positive. Any row vector such that $\mathbf{vP} = \mathbf{v}$ is a multiple of **w**. Any column vector **x** such that $\mathbf{Px} = \mathbf{x}$ is a constant vector.

Proof. This theorem states that Theorem 11.8 is true for ergodic chains. The result follows easily from the fact that, if **P** is an ergodic transition matrix, then $\bar{\mathbf{P}} = (1/2)\mathbf{I} + (1/2)\mathbf{P}$ is a regular transition matrix with the same fixed vectors (see Exercises 25–28). □

For ergodic chains, the fixed probability vector has a slightly different interpretation. The following two theorems, which we will not prove here, furnish an interpretation for this fixed vector.

Theorem 11.11 Let **P** be the transition matrix for an ergodic chain. Let $\mathbf{A}_n$ be the matrix defined by

$$\mathbf{A}_n = \frac{\mathbf{I} + \mathbf{P} + \mathbf{P}^2 + \cdots + \mathbf{P}^n}{n+1}.$$

Then $\mathbf{A}_n \to \mathbf{W}$, where **W** is a matrix all of whose rows are equal to the unique fixed probability vector **w** for **P**. □

If **P** is the transition matrix of an ergodic chain, then Theorem 11.8 states that there is only one fixed row probability vector for **P**. Thus, we can use the same techniques that were used for regular chains to solve for this fixed vector. In particular, the program **FixedVector** works for ergodic chains.

To interpret Theorem 11.11, let us assume that we have an ergodic chain that starts in state s_i. Let $X^{(m)} = 1$ if the mth step is to state s_j and 0 otherwise. Then the average number of times in state s_j in the first n steps is given by

$$H^{(n)} = \frac{X^{(0)} + X^{(1)} + X^{(2)} + \cdots + X^{(n)}}{n+1}.$$

But $X^{(m)}$ takes on the value 1 with probability $p_{ij}^{(m)}$ and 0 otherwise. Thus $E(X^{(m)}) = p_{ij}^{(m)}$, and the ijth entry of $\mathbf{A}_n$ gives the expected value of $H^{(n)}$, that

11.3. ERGODIC MARKOV CHAINS

is, the expected proportion of times in state s_j in the first n steps if the chain starts in state s_i.

If we call being in state s_j *success* and any other state *failure*, we could ask if a theorem analogous to the law of large numbers for independent trials holds. The answer is yes and is given by the following theorem.

Theorem 11.12 (Law of Large Numbers for Ergodic Markov Chains) Let $H_j^{(n)}$ be the proportion of times in n steps that an ergodic chain is in state s_j. Then for any $\epsilon > 0$,

$$P\left(|H_j^{(n)} - w_j| > \epsilon\right) \to 0 \;,$$

independent of the starting state s_i. $\square$

We have observed that every regular Markov chain is also an ergodic chain. Hence, Theorems 11.11 and 11.12 apply also for regular chains. For example, this gives us a new interpretation for the fixed vector $\mathbf{w} = (.4, .2, .4)$ in the Land of Oz example. Theorem 11.11 predicts that, in the long run, it will rain 40 percent of the time in the Land of Oz, be nice 20 percent of the time, and snow 40 percent of the time.

Simulation

We illustrate Theorem 11.12 by writing a program to simulate the behavior of a Markov chain. **SimulateChain** is such a program.

Example 11.21 In the Land of Oz, there are 525 days in a year. We have simulated the weather for one year in the Land of Oz, using the program **SimulateChain**. The results are shown in Table 11.2.

```
SSRNRNSSSSSSNRSNSSRNSRNSSSNSRRRNSSSNRRSSSSNRSSNSRRRRRRNSSS
SSRRRSNSNRRRRSRSRNSNSRRNRRNRSSNSRNRNSSRRSRNSSSNRSRRSSNRSNR
RNSSSSNSSNSRSRRNSSNSSRNSSRRNRRRSRNRRRNSSSNRNSRNSNRNRSSSRSS
NRSSSNSSSSSNSSSNSNSRRNRNRRRRSRRRSSSSNRRSSSSRSRRRNRRRSSSSR
RNRRRSRSSRRRRSSRNRRRRRNSSRNRSSSNRNSNRRRRNRRRNRSNRRNSRRSNR
RRRSSSRNRRRNSNSSSSSRRRRSRNRSSRRRRSSSRRRNRNRRRSRSRNSNSSRRRR
RNSNRNSNRRNRRRRRSSSNRSSRSNRSSSNSNRNSNSSSNRRSRRRNRRRRNRNRS
SSNSRSNRNRRSNRRNSRSSSRNSRRSSNSRRRNRRSNRRNSSSSSNRNSSSSSSSNR
NSRRRNSSRRRNSSSNRRSRNSSRRNRRNRSNRRRRRRRRRRNSNRRRRRNSRRSSSSN
SNS
```

State	Times	Fraction
R	217	.413
N	109	.208
S	199	.379

Table 11.2: Weather in the Land of Oz.

We note that the simulation gives a proportion of times in each of the states not too different from the long run predictions of .4, .2, and .4 assured by Theorem 11.7. To get better results we have to simulate our chain for a longer time. We do this for 10,000 days without printing out each day's weather. The results are shown in Table 11.3. We see that the results are now quite close to the theoretical values of .4, .2, and .4.

State	Times	Fraction
R	4010	.401
N	1902	.19
S	4088	.409

Table 11.3: Comparison of observed and predicted frequencies for the Land of Oz.

□

Examples of Ergodic Chains

The computation of the fixed vector **w** may be difficult if the transition matrix is very large. It is sometimes useful to guess the fixed vector on purely intuitive grounds. Here is a simple example to illustrate this kind of situation.

Example 11.22 A white rat is put into the maze of Figure 11.4. There are nine compartments with connections between the compartments as indicated. The rat moves through the compartments at random. That is, if there are k ways to leave a compartment, it chooses each of these with equal probability. We can represent the travels of the rat by a Markov chain process with transition matrix given by

$$\mathbf{P} = \begin{array}{c} \\ 1 \\ 2 \\ 3 \\ 4 \\ 5 \\ 6 \\ 7 \\ 8 \\ 9 \end{array} \begin{pmatrix} \begin{array}{ccccccccc} 1 & 2 & 3 & 4 & 5 & 6 & 7 & 8 & 9 \\ 0 & 1/2 & 0 & 0 & 0 & 1/2 & 0 & 0 & 0 \\ 1/3 & 0 & 1/3 & 0 & 1/3 & 0 & 0 & 0 & 0 \\ 0 & 1/2 & 0 & 1/2 & 0 & 0 & 0 & 0 & 0 \\ 0 & 0 & 1/3 & 0 & 1/3 & 0 & 0 & 0 & 1/3 \\ 0 & 1/4 & 0 & 1/4 & 0 & 1/4 & 0 & 1/4 & 0 \\ 1/3 & 0 & 0 & 0 & 1/3 & 0 & 1/3 & 0 & 0 \\ 0 & 0 & 0 & 0 & 0 & 1/2 & 0 & 1/2 & 0 \\ 0 & 0 & 0 & 0 & 1/3 & 0 & 1/3 & 0 & 1/3 \\ 0 & 0 & 0 & 1/2 & 0 & 0 & 0 & 1/2 & 0 \end{array} \end{pmatrix}.$$

That this chain is not regular can be seen as follows: From an odd-numbered state the process can go only to an even-numbered state, and from an even-numbered state it can go only to an odd number. Hence, starting in state i the process will be alternately in even-numbered and odd-numbered states. Therefore, odd powers of **P** will have 0's for the odd-numbered entries in row 1. On the other hand, a glance at the maze shows that it is possible to go from every state to every other state, so that the chain is ergodic.

11.3. ERGODIC MARKOV CHAINS

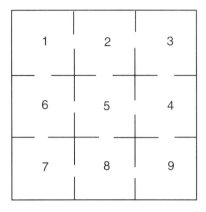

Figure 11.4: The maze problem.

To find the fixed probability vector for this matrix, we would have to solve ten equations in nine unknowns. However, it would seem reasonable that the times spent in each compartment should, in the long run, be proportional to the number of entries to each compartment. Thus, we try the vector whose jth component is the number of entries to the jth compartment:

$$\mathbf{x} = (\,2 \quad 3 \quad 2 \quad 3 \quad 4 \quad 3 \quad 2 \quad 3 \quad 2\,)\;.$$

It is easy to check that this vector is indeed a fixed vector so that the unique probability vector is this vector normalized to have sum 1:

$$\mathbf{w} = (\,\tfrac{1}{12} \quad \tfrac{1}{8} \quad \tfrac{1}{12} \quad \tfrac{1}{8} \quad \tfrac{1}{6} \quad \tfrac{1}{8} \quad \tfrac{1}{12} \quad \tfrac{1}{8} \quad \tfrac{1}{12}\,)\;.$$

□

Example 11.23 (Example 11.8 continued) We recall the Ehrenfest urn model of Example 11.8. The transition matrix for this chain is as follows:

$$\mathbf{P} = \begin{array}{c} \\ 0 \\ 1 \\ 2 \\ 3 \\ 4 \end{array} \begin{array}{c} \begin{array}{ccccc} 0 & 1 & 2 & 3 & 4 \end{array} \\ \left(\begin{array}{ccccc} .000 & 1.000 & .000 & .000 & .000 \\ .250 & .000 & .750 & .000 & .000 \\ .000 & .500 & .000 & .500 & .000 \\ .000 & .000 & .750 & .000 & .250 \\ .000 & .000 & .000 & 1.000 & .000 \end{array} \right) \end{array}.$$

If we run the program **FixedVector** for this chain, we obtain the vector

$$\begin{array}{c} \phantom{\mathbf{w} =} \begin{array}{ccccc} 0 & 1 & 2 & 3 & 4 \end{array} \\ \mathbf{w} = \begin{pmatrix} .0625 & .2500 & .3750 & .2500 & .0625 \end{pmatrix}. \end{array}$$

By Theorem 11.12, we can interpret these values for w_i as the proportion of times the process is in each of the states in the long run. For example, the proportion of times in state 0 is .0625 and the proportion of times in state 1 is .375. The astute reader will note that these numbers are the binomial distribution 1/16, 4/16, 6/16, 4/16, 1/16. We could have guessed this answer as follows: If we consider a particular ball, it simply moves randomly back and forth between the two urns. This suggests that the equilibrium state should be just as if we randomly distributed the four balls in the two urns. If we did this, the probability that there would be exactly j balls in one urn would be given by the binomial distribution $b(n, p, j)$ with $n = 4$ and $p = 1/2$. □

Exercises

1 Which of the following matrices are transition matrices for regular Markov chains?

(a) $\mathbf{P} = \begin{pmatrix} .5 & .5 \\ .5 & .5 \end{pmatrix}$.

(b) $\mathbf{P} = \begin{pmatrix} .5 & .5 \\ 1 & 0 \end{pmatrix}$.

(c) $\mathbf{P} = \begin{pmatrix} 1/3 & 0 & 2/3 \\ 0 & 1 & 0 \\ 0 & 1/5 & 4/5 \end{pmatrix}$.

(d) $\mathbf{P} = \begin{pmatrix} 0 & 1 \\ 1 & 0 \end{pmatrix}$.

(e) $\mathbf{P} = \begin{pmatrix} 1/2 & 1/2 & 0 \\ 0 & 1/2 & 1/2 \\ 1/3 & 1/3 & 1/3 \end{pmatrix}$.

2 Consider the Markov chain with transition matrix

$$\mathbf{P} = \begin{pmatrix} 1/2 & 1/3 & 1/6 \\ 3/4 & 0 & 1/4 \\ 0 & 1 & 0 \end{pmatrix}.$$

(a) Show that this is a regular Markov chain.

(b) The process is started in state 1; find the probability that it is in state 3 after two steps.

(c) Find the limiting probability vector $\mathbf{w}$.

3 Consider the Markov chain with general 2×2 transition matrix

$$\mathbf{P} = \begin{pmatrix} 1-a & a \\ b & 1-b \end{pmatrix}.$$

(a) Under what conditions is $\mathbf{P}$ absorbing?

(b) Under what conditions is $\mathbf{P}$ ergodic but not regular?

(c) Under what conditions is $\mathbf{P}$ regular?

11.3. ERGODIC MARKOV CHAINS

4 Find the fixed probability vector **w** for the matrices in Exercise 3 that are ergodic.

5 Find the fixed probability vector **w** for each of the following regular matrices.

(a) $\mathbf{P} = \begin{pmatrix} .75 & .25 \\ .5 & .5 \end{pmatrix}$.

(b) $\mathbf{P} = \begin{pmatrix} .9 & .1 \\ .1 & .9 \end{pmatrix}$.

(c) $\mathbf{P} = \begin{pmatrix} 3/4 & 1/4 & 0 \\ 0 & 2/3 & 1/3 \\ 1/4 & 1/4 & 1/2 \end{pmatrix}$.

6 Consider the Markov chain with transition matrix in Exercise 3, with $a = b = 1$. Show that this chain is ergodic but not regular. Find the fixed probability vector and interpret it. Show that $\mathbf{P}^n$ does not tend to a limit, but that

$$\mathbf{A}_n = \frac{\mathbf{I} + \mathbf{P} + \mathbf{P}^2 + \cdots + \mathbf{P}^n}{n+1}$$

does.

7 Consider the Markov chain with transition matrix of Exercise 3, with $a = 0$ and $b = 1/2$. Compute directly the unique fixed probability vector, and use your result to prove that the chain is not ergodic.

8 Show that the matrix

$$\mathbf{P} = \begin{pmatrix} 1 & 0 & 0 \\ 1/4 & 1/2 & 1/4 \\ 0 & 0 & 1 \end{pmatrix}$$

has more than one fixed probability vector. Find the matrix that $\mathbf{P}^n$ approaches as $n \to \infty$, and verify that it is not a matrix all of whose rows are the same.

9 Prove that, if a 3-by-3 transition matrix has the property that its *column* sums are 1, then $(1/3, 1/3, 1/3)$ is a fixed probability vector. State a similar result for n-by-n transition matrices. Interpret these results for ergodic chains.

10 Is the Markov chain in Example 11.10 ergodic?

11 Is the Markov chain in Example 11.11 ergodic?

12 Consider Example 11.13 (Drunkard's Walk). Assume that if the walker reaches state 0, he turns around and returns to state 1 on the next step and, similarly, if he reaches 4 he returns on the next step to state 3. Is this new chain ergodic? Is it regular?

13 For Example 11.4 when **P** is ergodic, what is the proportion of people who are told that the President will run? Interpret the fact that this proportion is independent of the starting state.

14 Consider an independent trials process to be a Markov chain whose states are the possible outcomes of the individual trials. What is its fixed probability vector? Is the chain always regular? Illustrate this for Example 11.5.

15 Show that Example 11.8 is an ergodic chain, but not a regular chain. Show that its fixed probability vector **w** is a binomial distribution.

16 Show that Example 11.9 is regular and find the limiting vector.

17 Toss a fair die repeatedly. Let S_n denote the total of the outcomes through the nth toss. Show that there is a limiting value for the proportion of the first n values of S_n that are divisible by 7, and compute the value for this limit. *Hint*: The desired limit is an equilibrium probability vector for an appropriate seven state Markov chain.

18 Let **P** be the transition matrix of a regular Markov chain. Assume that there are r states and let $N(r)$ be the smallest integer n such that **P** is regular if and only if $\mathbf{P}^{N(r)}$ has no zero entries. Find a finite upper bound for $N(r)$. See if you can determine $N(3)$ exactly.

*19 Define $f(r)$ to be the smallest integer n such that for all regular Markov chains with r states, the nth power of the transition matrix has all entries positive. It has been shown,[14] that $f(r) = r^2 - 2r + 2$.

   (a) Define the transition matrix of an r-state Markov chain as follows: For states s_i, with $i = 1, 2, \ldots, r-2$, $\mathbf{P}(i, i+1) = 1$, $\mathbf{P}(r-1, r) = \mathbf{P}(r-1, 1) = 1/2$, and $\mathbf{P}(r, 1) = 1$. Show that this is a regular Markov chain.

   (b) For $r = 3$, verify that the fifth power is the first power that has no zeros.

   (c) Show that, for general r, the smallest n such that $\mathbf{P}^n$ has all entries positive is $n = f(r)$.

20 A discrete time queueing system of capacity n consists of the person being served and those waiting to be served. The queue length x is observed each second. If $0 < x < n$, then with probability p, the queue size is increased by one by an arrival and, inependently, with probability r, it is decreased by one because the person being served finishes service. If $x = 0$, only an arrival (with probability p) is possible. If $x = n$, an arrival will depart without waiting for service, and so only the departure (with probability r) of the person being served is possible. Form a Markov chain with states given by the number of customers in the queue. Modify the program **FixedVector** so that you can input n, p, and r, and the program will construct the transition matrix and compute the fixed vector. The quantity $s = p/r$ is called the *traffic intensity*. Describe the differences in the fixed vectors according as $s < 1$, $s = 1$, or $s > 1$.

---

[14] E. Seneta, *Non-Negative Matrices: An Introduction to Theory and Applications*, Wiley, New York, 1973, pp. 52-54.

11.3. ERGODIC MARKOV CHAINS

21 Write a computer program to simulate the queue in Exercise 20. Have your program keep track of the proportion of the time that the queue length is j for $j = 0, 1, \ldots, n$ and the average queue length. Show that the behavior of the queue length is very different depending upon whether the traffic intensity s has the property $s < 1$, $s = 1$, or $s > 1$.

22 In the queueing problem of Exercise 20, let S be the total service time required by a customer and T the time between arrivals of the customers.

(a) Show that $P(S = j) = (1-r)^{j-1}r$ and $P(T = j) = (1-p)^{j-1}p$, for $j > 0$.

(b) Show that $E(S) = 1/r$ and $E(T) = 1/p$.

(c) Interpret the conditions $s < 1$, $s = 1$ and $s > 1$ in terms of these expected values.

23 In Exercise 20 the service time S has a geometric distribution with $E(S) = 1/r$. Assume that the service time is, instead, a constant time of t seconds. Modify your computer program of Exercise 21 so that it simulates a constant time service distribution. Compare the average queue length for the two types of distributions when they have the same expected service time (i.e., take $t = 1/r$). Which distribution leads to the longer queues on the average?

24 A certain experiment is believed to be described by a two-state Markov chain with the transition matrix $\mathbf{P}$, where

$$\mathbf{P} = \begin{pmatrix} .5 & .5 \\ p & 1-p \end{pmatrix}$$

and the parameter p is not known. When the experiment is performed many times, the chain ends in state one approximately 20 percent of the time and in state two approximately 80 percent of the time. Compute a sensible estimate for the unknown parameter p and explain how you found it.

25 Prove that, in an r-state ergodic chain, it is possible to go from any state to any other state in at most $r - 1$ steps.

26 Let $\mathbf{P}$ be the transition matrix of an r-state ergodic chain. Prove that, if the diagonal entries p_{ii} are positive, then the chain is regular.

27 Prove that if $\mathbf{P}$ is the transition matrix of an ergodic chain, then $(1/2)(\mathbf{I}+\mathbf{P})$ is the transition matrix of a regular chain. *Hint*: Use Exercise 26.

28 Prove that $\mathbf{P}$ and $(1/2)(\mathbf{I}+\mathbf{P})$ have the same fixed vectors.

29 In his book, *Wahrscheinlichkeitsrechnung und Statistik*,[15] A. Engle proposes an algorithm for finding the fixed vector for an ergodic Markov chain when the transition probabilities are rational numbers. Here is his algorithm: For

---
[15] A. Engle, *Wahrscheinlichkeitsrechnung und Statistik*, vol. 2 (Stuttgart: Klett Verlag, 1976).

$$\begin{array}{rrr}
(4 & 2 & 4) \\
(5 & 2 & 3) \\
(8 & 2 & 4) \\
(7 & 3 & 4) \\
(8 & 4 & 4) \\
(8 & 3 & 5) \\
(8 & 4 & 8) \\
(10 & 4 & 6) \\
(12 & 4 & 8) \\
(12 & 5 & 7) \\
(12 & 6 & 8) \\
(13 & 5 & 8) \\
(16 & 6 & 8) \\
(15 & 6 & 9) \\
(16 & 6 & 12) \\
(17 & 7 & 10) \\
(20 & 8 & 12) \\
(20 & 8 & 12)\ .
\end{array}$$

Table 11.4: Distribution of chips.

each state i, let a_i be the least common multiple of the denominators of the non-zero entries in the ith row. Engle describes his algorithm in terms of moving chips around on the states—indeed, for small examples, he recommends implementing the algorithm this way. Start by putting a_i chips on state i for all i. Then, at each state, redistribute the a_i chips, sending $a_i p_{ij}$ to state j. The number of chips at state i after this redistribution need not be a multiple of a_i. For each state i, add just enough chips to bring the number of chips at state i up to a multiple of a_i. Then redistribute the chips in the same manner. This process will eventually reach a point where the number of chips at each state, after the redistribution, is the same as before redistribution. At this point, we have found a fixed vector. Here is an example:

$$\mathbf{P} = \begin{array}{c} \\ 1 \\ 2 \\ 3 \end{array} \begin{array}{c} \begin{array}{ccc} 1 & 2 & 3 \end{array} \\ \left(\begin{array}{ccc} 1/2 & 1/4 & 1/4 \\ 1/2 & 0 & 1/2 \\ 1/2 & 1/4 & 1/4 \end{array} \right) \end{array}.$$

We start with $\mathbf{a} = (4, 2, 4)$. The chips after successive redistributions are shown in Table 11.4.

We find that $\mathbf{a} = (20, 8, 12)$ is a fixed vector.

(a) Write a computer program to implement this algorithm.

(b) Prove that the algorithm will stop. *Hint*: Let $\mathbf{b}$ be a vector with integer components that is a fixed vector for $\mathbf{P}$ and such that each coordinate of

11.4. FUNDAMENTAL LIMIT THEOREM

the starting vector **a** is less than or equal to the corresponding component of **b**. Show that, in the iteration, the components of the vectors are always increasing, and always less than or equal to the corresponding component of **b**.

30 (Coffman, Kaduta, and Shepp[16]) A computing center keeps information on a tape in positions of unit length. During each time unit there is one request to occupy a unit of tape. When this arrives the first free unit is used. Also, during each second, each of the units that are occupied is vacated with probability p. Simulate this process, starting with an empty tape. Estimate the expected number of sites occupied for a given value of p. If p is small, can you choose the tape long enough so that there is a small probability that a new job will have to be turned away (i.e., that all the sites are occupied)? Form a Markov chain with states the number of sites occupied. Modify the program **FixedVector** to compute the fixed vector. Use this to check your conjecture by simulation.

*31 (Alternate proof of Theorem 11.8) Let **P** be the transition matrix of an ergodic Markov chain. Let **x** be any column vector such that $\mathbf{Px} = \mathbf{x}$. Let M be the maximum value of the components of **x**. Assume that $x_i = M$. Show that if $p_{ij} > 0$ then $x_j = M$. Use this to prove that **x** must be a constant vector.

32 Let **P** be the transition matrix of an ergodic Markov chain. Let **w** be a fixed probability vector (i.e., **w** is a row vector with $\mathbf{wP} = \mathbf{w}$). Show that if $w_i = 0$ and $p_{ji} > 0$ then $w_j = 0$. Use this to show that the fixed probability vector for an ergodic chain cannot have any 0 entries.

33 Find a Markov chain that is neither absorbing or ergodic.

11.4 Fundamental Limit Theorem for Regular Chains

The fundamental limit theorem for regular Markov chains states that if **P** is a regular transition matrix then

$$\lim_{n \to \infty} \mathbf{P}^n = \mathbf{W} ,$$

where **W** is a matrix with each row equal to the unique fixed probability row vector **w** for **P**. In this section we shall give two very different proofs of this theorem.

Our first proof is carried out by showing that, for any column vector **y**, $\mathbf{P}^n\mathbf{y}$ tends to a constant vector. As indicated in Section 11.3, this will show that $\mathbf{P}^n$ converges to a matrix with constant columns or, equivalently, to a matrix with all rows the same.

The following lemma says that if an r-by-r transition matrix has no zero entries, and **y** is any column vector with r entries, then the vector **Py** has entries which are "closer together" than the entries are in **y**.

---

[16] E. G. Coffman, J. T. Kaduta, and L. A. Shepp, "On the Asymptotic Optimality of First-Storage Allocation," *IEEE Trans. Software Engineering*, vol. II (1985), pp. 235-239.

Lemma 11.1 Let $\mathbf{P}$ be an r-by-r transition matrix with no zero entries. Let d be the smallest entry of the matrix. Let $\mathbf{y}$ be a column vector with r components, the largest of which is M_0 and the smallest m_0. Let M_1 and m_1 be the largest and smallest component, respectively, of the vector $\mathbf{Py}$. Then

$$M_1 - m_1 \leq (1 - 2d)(M_0 - m_0) .$$

Proof. In the discussion following Theorem 11.7, it was noted that each entry in the vector $\mathbf{Py}$ is a weighted average of the entries in $\mathbf{y}$. The largest weighted average that could be obtained in the present case would occur if all but one of the entries of $\mathbf{y}$ have value M_0 and one entry has value m_0, and this one small entry is weighted by the smallest possible weight, namely d. In this case, the weighted average would equal

$$dm_0 + (1-d)M_0 .$$

Similarly, the smallest possible weighted average equals

$$dM_0 + (1-d)m_0 .$$

Thus,

$$\begin{aligned} M_1 - m_1 &\leq \Big(dm_0 + (1-d)M_0\Big) - \Big(dM_0 + (1-d)m_0\Big) \\ &= (1-2d)(M_0 - m_0) . \end{aligned}$$

This completes the proof of the lemma. $\square$

We turn now to the proof of the fundamental limit theorem for regular Markov chains.

Theorem 11.13 (Fundamental Limit Theorem for Regular Chains) If $\mathbf{P}$ is the transition matrix for a regular Markov chain, then

$$\lim_{n \to \infty} \mathbf{P}^n = \mathbf{W} ,$$

where $\mathbf{W}$ is matrix with all rows equal. Furthermore, all entries in $\mathbf{W}$ are strictly positive.

Proof. We prove this theorem for the special case that $\mathbf{P}$ has no 0 entries. The extension to the general case is indicated in Exercise 5. Let $\mathbf{y}$ be any r-component column vector, where r is the number of states of the chain. We assume that $r > 1$, since otherwise the theorem is trivial. Let M_n and m_n be, respectively, the maximum and minimum components of the vector $\mathbf{P}^n \mathbf{y}$. The vector $\mathbf{P}^n \mathbf{y}$ is obtained from the vector $\mathbf{P}^{n-1}\mathbf{y}$ by multiplying on the left by the matrix $\mathbf{P}$. Hence each component of $\mathbf{P}^n\mathbf{y}$ is an average of the components of $\mathbf{P}^{n-1}\mathbf{y}$. Thus

$$M_0 \geq M_1 \geq M_2 \geq \cdots$$

11.4. FUNDAMENTAL LIMIT THEOREM

and
$$m_0 \leq m_1 \leq m_2 \leq \cdots .$$

Each sequence is monotone and bounded:
$$m_0 \leq m_n \leq M_n \leq M_0 .$$

Hence, each of these sequences will have a limit as n tends to infinity.

Let M be the limit of M_n and m the limit of m_n. We know that $m \leq M$. We shall prove that $M - m = 0$. This will be the case if $M_n - m_n$ tends to 0. Let d be the smallest element of $\mathbf{P}$. Since all entries of $\mathbf{P}$ are strictly positive, we have $d > 0$. By our lemma
$$M_n - m_n \leq (1 - 2d)(M_{n-1} - m_{n-1}) .$$

From this we see that
$$M_n - m_n \leq (1 - 2d)^n (M_0 - m_0) .$$

Since $r \geq 2$, we must have $d \leq 1/2$, so $0 \leq 1 - 2d < 1$, so the difference $M_n - m_n$ tends to 0 as n tends to infinity. Since every component of $\mathbf{P}^n \mathbf{y}$ lies between m_n and M_n, each component must approach the same number $u = M = m$. This shows that
$$\lim_{n \to \infty} \mathbf{P}^n \mathbf{y} = \mathbf{u} ,$$
where $\mathbf{u}$ is a column vector all of whose components equal u.

Now let $\mathbf{y}$ be the vector with jth component equal to 1 and all other components equal to 0. Then $\mathbf{P}^n \mathbf{y}$ is the jth column of $\mathbf{P}^n$. Doing this for each j proves that the columns of $\mathbf{P}^n$ approach constant column vectors. That is, the rows of $\mathbf{P}^n$ approach a common row vector $\mathbf{w}$, or,
$$\lim_{n \to \infty} \mathbf{P}^n = \mathbf{W} .$$

It remains to show that all entries in $\mathbf{W}$ are strictly positive. As before, let $\mathbf{y}$ be the vector with jth component equal to 1 and all other components equal to 0. Then $\mathbf{Py}$ is the jth column of $\mathbf{P}$, and this column has all entries strictly positive. The minimum component of the vector $\mathbf{Py}$ was defined to be m_1, hence $m_1 > 0$. Since $m_1 \leq m$, we have $m > 0$. Note finally that this value of m is just the jth component of $\mathbf{w}$, so all components of $\mathbf{w}$ are strictly positive. $\square$

Doeblin's Proof

We give now a very different proof of the main part of the fundamental limit theorem for regular Markov chains. This proof was first given by Doeblin,[17] a brilliant young mathematician who was killed in his twenties in the Second World War.

---

[17] W. Doeblin, "Exposé de la Théorie des Chaines Simple Constantes de Markov à un Nombre Fini d'Etats," *Rev. Mach. de l'Union Interbalkanique*, vol. 2 (1937), pp. 77–105.

Theorem 11.14 Let **P** be the transition matrix for a regular Markov chain with fixed vector **w**. Then for any initial probability vector **u**, $\mathbf{uP}^n \to \mathbf{w}$ as $n \to \infty$.

Proof. Let X_0, X_1, ... be a Markov chain with transition matrix **P** started in state s_i. Let Y_0, Y_1, ... be a Markov chain with transition probability **P** started with initial probabilities given by **w**. The X and Y processes are run independently of each other.

We consider also a third Markov chain $\mathbf{P}^*$ which consists of watching both the X and Y processes. The states for $\mathbf{P}^*$ are pairs (s_i, s_j). The transition probabilities are given by

$$\mathbf{P}^*[(i,j),(k,l)] = \mathbf{P}(i,j) \cdot \mathbf{P}(k,l) \ .$$

Since **P** is regular there is an N such that $\mathbf{P}^N(i,j) > 0$ for all i and j. Thus for the $\mathbf{P}^*$ chain it is also possible to go from any state (s_i, s_j) to any other state (s_k, s_l) in at most N steps. That is $\mathbf{P}^*$ is also a regular Markov chain.

We know that a regular Markov chain will reach any state in a finite time. Let T be the first time the the chain $\mathbf{P}^*$ is in a state of the form (s_k, s_k). In other words, T is the first time that the X and the Y processes are in the same state. Then we have shown that

$$P[T > n] \to 0 \ \text{as} \ n \to \infty \ .$$

If we watch the X and Y processes after the first time they are in the same state we would not predict any difference in their long range behavior. Since this will happen no matter how we started these two processes, it seems clear that the long range behaviour should not depend upon the starting state. We now show that this is true.

We first note that if $n \geq T$, then since X and Y are both in the same state at time T,

$$P(X_n = j \mid n \geq T) = P(Y_n = j \mid n \geq T) \ .$$

If we multiply both sides of this equation by $P(n \geq T)$, we obtain

$$P(X_n = j, \ n \geq T) = P(Y_n = j, \ n \geq T) \ . \tag{11.1}$$

We know that for all n,

$$P(Y_n = j) = w_j \ .$$

But

$$P(Y_n = j) = P(Y_n = j, \ n \geq T) + P(Y_n = j, \ n < T) \ ,$$

and the second summand on the right-hand side of this equation goes to 0 as n goes to ∞, since $P(n < T)$ goes to 0 as n goes to ∞. So,

$$P(Y_n = j, \ n \geq T) \to w_j \ ,$$

as n goes to ∞. From Equation 11.1, we see that

$$P(X_n = j, \ n \geq T) \to w_j \ ,$$

11.4. FUNDAMENTAL LIMIT THEOREM

as n goes to ∞. But by similar reasoning to that used above, the difference between this last expression and $P(X_n = j)$ goes to 0 as n goes to ∞. Therefore,

$$P(X_n = j) \to w_j ,$$

as n goes to ∞. This completes the proof. $\square$

In the above proof, we have said nothing about the rate at which the distributions of the X_n's approach the fixed distribution $\mathbf{w}$. In fact, it can be shown that[18]

$$\sum_{j=1}^{r} \mid P(X_n = j) - w_j \mid \leq 2P(T > n) .$$

The left-hand side of this inequality can be viewed as the distance between the distribution of the Markov chain after n steps, starting in state s_i, and the limiting distribution $\mathbf{w}$.

Exercises

1 Define $\mathbf{P}$ and $\mathbf{y}$ by

$$\mathbf{P} = \begin{pmatrix} .5 & .5 \\ .25 & .75 \end{pmatrix}, \quad \mathbf{y} = \begin{pmatrix} 1 \\ 0 \end{pmatrix} .$$

Compute $\mathbf{Py}$, $\mathbf{P}^2\mathbf{y}$, and $\mathbf{P}^4\mathbf{y}$ and show that the results are approaching a constant vector. What is this vector?

2 Let $\mathbf{P}$ be a regular $r \times r$ transition matrix and $\mathbf{y}$ any r-component column vector. Show that the value of the limiting constant vector for $\mathbf{P}^n\mathbf{y}$ is $\mathbf{wy}$.

3 Let

$$\mathbf{P} = \begin{pmatrix} 1 & 0 & 0 \\ .25 & 0 & .75 \\ 0 & 0 & 1 \end{pmatrix}$$

be a transition matrix of a Markov chain. Find two fixed vectors of $\mathbf{P}$ that are linearly independent. Does this show that the Markov chain is not regular?

4 Describe the set of all fixed column vectors for the chain given in Exercise 3.

5 The theorem that $\mathbf{P}^n \to \mathbf{W}$ was proved only for the case that $\mathbf{P}$ has no zero entries. Fill in the details of the following extension to the case that $\mathbf{P}$ is regular. Since $\mathbf{P}$ is regular, for some N, $\mathbf{P}^N$ has no zeros. Thus, the proof given shows that $M_{nN} - m_{nN}$ approaches 0 as n tends to infinity. However, the difference $M_n - m_n$ can never increase. (Why?) Hence, if we know that the differences obtained by looking at every Nth time tend to 0, then the entire sequence must also tend to 0.

6 Let $\mathbf{P}$ be a regular transition matrix and let $\mathbf{w}$ be the unique non-zero fixed vector of $\mathbf{P}$. Show that no entry of $\mathbf{w}$ is 0.

---

[18]T. Lindvall, *Lectures on the Coupling Method* (New York: Wiley 1992).

7 Here is a trick to try on your friends. Shuffle a deck of cards and deal them out one at a time. Count the face cards each as ten. Ask your friend to look at one of the first ten cards; if this card is a six, she is to look at the card that turns up six cards later; if this card is a three, she is to look at the card that turns up three cards later, and so forth. Eventually she will reach a point where she is to look at a card that turns up x cards later but there are not x cards left. You then tell her the last card that she looked at even though you did not know her starting point. You tell her you do this by watching her, and she cannot disguise the times that she looks at the cards. In fact you just do the same procedure and, even though you do not start at the same point as she does, you will most likely end at the same point. Why?

8 Write a program to play the game in Exercise 7.

11.5 Mean First Passage Time for Ergodic Chains

In this section we consider two closely related descriptive quantities of interest for ergodic chains: the mean time to return to a state and the mean time to go from one state to another state.

Let $\mathbf{P}$ be the transition matrix of an ergodic chain with states $s_1, s_2, \ldots, s_r$. Let $\mathbf{w} = (w_1, w_2, \ldots, w_r)$ be the unique probability vector such that $\mathbf{wP} = \mathbf{w}$. Then, by the Law of Large Numbers for Markov chains, in the long run the process will spend a fraction w_j of the time in state s_j. Thus, if we start in any state, the chain will eventually reach state s_j; in fact, it will be in state s_j infinitely often.

Another way to see this is the following: Form a new Markov chain by making s_j an absorbing state, that is, define $p_{jj} = 1$. If we start at any state other than s_j, this new process will behave exactly like the original chain up to the first time that state s_j is reached. Since the original chain was an ergodic chain, it was possible to reach s_j from any other state. Thus the new chain is an absorbing chain with a single absorbing state s_j that will eventually be reached. So if we start the original chain at a state s_i with $i \neq j$, we will eventually reach the state s_j.

Let $\mathbf{N}$ be the fundamental matrix for the new chain. The entries of $\mathbf{N}$ give the expected number of times in each state before absorption. In terms of the original chain, these quantities give the expected number of times in each of the states before reaching state s_j for the first time. The ith component of the vector $\mathbf{Nc}$ gives the expected number of steps before absorption in the new chain, starting in state s_i. In terms of the old chain, this is the expected number of steps required to reach state s_j for the first time starting at state s_i.

Mean First Passage Time

Definition 11.7 If an ergodic Markov chain is started in state s_i, the expected number of steps to reach state s_j for the first time is called the *mean first passage time* from s_i to s_j. It is denoted by m_{ij}. By convention $m_{ii} = 0$. □

11.5. MEAN FIRST PASSAGE TIME

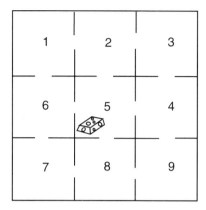

Figure 11.5: The maze problem.

Example 11.24 Let us return to the maze example (Example 11.22). We shall make this ergodic chain into an absorbing chain by making state 5 an absorbing state. For example, we might assume that food is placed in the center of the maze and once the rat finds the food, he stays to enjoy it (see Figure 11.5).

The new transition matrix in canonical form is

$$\mathbf{P} = \begin{array}{c} \\ 1 \\ 2 \\ 3 \\ 4 \\ 6 \\ 7 \\ 8 \\ 9 \\ \hline 5 \end{array} \left(\begin{array}{cccccccc|c} 1 & 2 & 3 & 4 & 6 & 7 & 8 & 9 & 5 \\ 0 & 1/2 & 0 & 0 & 1/2 & 0 & 0 & 0 & 0 \\ 1/3 & 0 & 1/3 & 0 & 0 & 0 & 0 & 0 & 1/3 \\ 0 & 1/2 & 0 & 1/2 & 0 & 0 & 0 & 0 & 0 \\ 0 & 0 & 1/3 & 0 & 0 & 1/3 & 0 & 1/3 & 1/3 \\ 1/3 & 0 & 0 & 0 & 0 & 0 & 0 & 0 & 1/3 \\ 0 & 0 & 0 & 0 & 1/2 & 0 & 1/2 & 0 & 0 \\ 0 & 0 & 0 & 0 & 0 & 1/3 & 0 & 1/3 & 1/3 \\ 0 & 0 & 0 & 1/2 & 0 & 0 & 1/2 & 0 & 0 \\ \hline 0 & 0 & 0 & 0 & 0 & 0 & 0 & 0 & 1 \end{array} \right).$$

If we compute the fundamental matrix $\mathbf{N}$, we obtain

$$\mathbf{N} = \frac{1}{8} \begin{pmatrix} 14 & 9 & 4 & 3 & 9 & 4 & 3 & 2 \\ 6 & 14 & 6 & 4 & 4 & 2 & 2 & 2 \\ 4 & 9 & 14 & 9 & 3 & 2 & 3 & 4 \\ 2 & 4 & 6 & 14 & 2 & 2 & 4 & 6 \\ 6 & 4 & 2 & 2 & 14 & 6 & 4 & 2 \\ 4 & 3 & 2 & 3 & 9 & 14 & 9 & 4 \\ 2 & 2 & 2 & 4 & 4 & 6 & 14 & 6 \\ 2 & 3 & 4 & 9 & 3 & 4 & 9 & 14 \end{pmatrix}.$$

The expected time to absorption for different starting states is given by the vector **Nc**, where

$$\mathbf{Nc} = \begin{pmatrix} 6 \\ 5 \\ 6 \\ 5 \\ 5 \\ 6 \\ 5 \\ 6 \end{pmatrix}.$$

We see that, starting from compartment 1, it will take on the average six steps to reach food. It is clear from symmetry that we should get the same answer for starting at state 3, 7, or 9. It is also clear that it should take one more step, starting at one of these states, than it would starting at 2, 4, 6, or 8. Some of the results obtained from **N** are not so obvious. For instance, we note that the expected number of times in the starting state is 14/8 regardless of the state in which we start. □

Mean Recurrence Time

A quantity that is closely related to the mean first passage time is the *mean recurrence time*, defined as follows. Assume that we start in state s_i; consider the length of time before we return to s_i for the first time. It is clear that we must return, since we either stay at s_i the first step or go to some other state s_j, and from any other state s_j, we will eventually reach s_i because the chain is ergodic.

Definition 11.8 If an ergodic Markov chain is started in state s_i, the expected number of steps to return to s_i for the first time is the *mean recurrence time* for s_i. It is denoted by r_i. □

We need to develop some basic properties of the mean first passage time. Consider the mean first passage time from s_i to s_j; assume that $i \neq j$. This may be computed as follows: take the expected number of steps required given the outcome of the first step, multiply by the probability that this outcome occurs, and add. If the first step is to s_j, the expected number of steps required is 1; if it is to some other state s_k, the expected number of steps required is m_{kj} plus 1 for the step already taken. Thus,

$$m_{ij} = p_{ij} + \sum_{k \neq j} p_{ik}(m_{kj} + 1) ,$$

or, since $\sum_k p_{ik} = 1$,

$$m_{ij} = 1 + \sum_{k \neq j} p_{ik} m_{jk} . \tag{11.2}$$

Similarly, starting in s_i, it must take at least one step to return. Considering all possible first steps gives us

$$r_i = \sum_k p_{ik}(m_{ki} + 1) \tag{11.3}$$

11.5. MEAN FIRST PASSAGE TIME

$$= 1 + \sum_k p_{ik} m_{ki} \; . \qquad (11.4)$$

Mean First Passage Matrix and Mean Recurrence Matrix

Let us now define two matrices **M** and **D**. The ijth entry m_{ij} of **M** is the mean first passage time to go from s_i to s_j if $i \neq j$; the diagonal entries are 0. The matrix **M** is called the *mean first passage matrix*. The matrix **D** is the matrix with all entries 0 except the diagonal entries $d_{ii} = r_i$. The matrix **D** is called the *mean recurrence matrix*. Let **C** be an $r \times r$ matrix with all entries 1. Using Equation 11.2 for the case $i \neq j$ and Equation 11.4 for the case $i = j$, we obtain the matrix equation

$$\mathbf{M} = \mathbf{PM} + \mathbf{C} - \mathbf{D} \; , \qquad (11.5)$$

or

$$(\mathbf{I} - \mathbf{P})\mathbf{M} = \mathbf{C} - \mathbf{D} \; . \qquad (11.6)$$

Equation 11.6 with $m_{ii} = 0$ implies Equations 11.2 and 11.4. We are now in a position to prove our first basic theorem.

Theorem 11.15 For an ergodic Markov chain, the mean recurrence time for state s_i is $r_i = 1/w_i$, where w_i is the ith component of the fixed probability vector for the transition matrix.

Proof. Multiplying both sides of Equation 11.6 by **w** and using the fact that

$$\mathbf{w}(\mathbf{I} - \mathbf{P}) = \mathbf{0}$$

gives

$$\mathbf{wC} - \mathbf{wD} = \mathbf{0} \; .$$

Here **wC** is a row vector with all entries 1 and **wD** is a row vector with ith entry $w_i r_i$. Thus

$$(1, 1, \ldots, 1) = (w_1 r_1, w_2 r_2, \ldots, w_n r_n)$$

and

$$r_i = 1/w_i \; ,$$

as was to be proved. $\square$

Corollary 11.1 For an ergodic Markov chain, the components of the fixed probability vector **w** are strictly positive.

Proof. We know that the values of r_i are finite and so $w_i = 1/r_i$ cannot be 0. $\square$

Example 11.25 In Example 11.22 we found the fixed probability vector for the maze example to be

$$\mathbf{w} = \begin{pmatrix} \frac{1}{12} & \frac{1}{8} & \frac{1}{12} & \frac{1}{8} & \frac{1}{6} & \frac{1}{8} & \frac{1}{12} & \frac{1}{8} & \frac{1}{12} \end{pmatrix}.$$

Hence, the mean recurrence times are given by the reciprocals of these probabilities. That is,

$$\mathbf{r} = \begin{pmatrix} 12 & 8 & 12 & 8 & 6 & 8 & 12 & 8 & 12 \end{pmatrix}.$$

$\Box$

Returning to the Land of Oz, we found that the weather in the Land of Oz could be represented by a Markov chain with states rain, nice, and snow. In Section 11.3 we found that the limiting vector was $\mathbf{w} = (2/5, 1/5, 2/5)$. From this we see that the mean number of days between rainy days is $5/2$, between nice days is 5, and between snowy days is $5/2$.

Fundamental Matrix

We shall now develop a fundamental matrix for ergodic chains that will play a role similar to that of the fundamental matrix $\mathbf{N} = (\mathbf{I} - \mathbf{Q})^{-1}$ for absorbing chains. As was the case with absorbing chains, the fundamental matrix can be used to find a number of interesting quantities involving ergodic chains. Using this matrix, we will give a method for calculating the mean first passage times for ergodic chains that is easier to use than the method given above. In addition, we will state (but not prove) the Central Limit Theorem for Markov Chains, the statement of which uses the fundamental matrix.

We begin by considering the case that $\mathbf{P}$ is the transition matrix of a regular Markov chain. Since there are no absorbing states, we might be tempted to try $\mathbf{Z} = (\mathbf{I} - \mathbf{P})^{-1}$ for a fundamental matrix. But $\mathbf{I} - \mathbf{P}$ does not have an inverse. To see this, recall that a matrix $\mathbf{R}$ has an inverse if and only if $\mathbf{Rx} = \mathbf{0}$ implies $\mathbf{x} = \mathbf{0}$. But since $\mathbf{Pc} = \mathbf{c}$ we have $(\mathbf{I} - \mathbf{P})\mathbf{c} = \mathbf{0}$, and so $\mathbf{I} - \mathbf{P}$ does not have an inverse.

We recall that if we have an absorbing Markov chain, and $\mathbf{Q}$ is the restriction of the transition matrix to the set of transient states, then the fundamental matrix $\mathbf{N}$ could be written as

$$\mathbf{N} = \mathbf{I} + \mathbf{Q} + \mathbf{Q}^2 + \cdots .$$

The reason that this power series converges is that $\mathbf{Q}^n \to \mathbf{0}$, so this series acts like a convergent geometric series.

This idea might prompt one to try to find a similar series for regular chains. Since we know that $\mathbf{P}^n \to \mathbf{W}$, we might consider the series

$$\mathbf{I} + (\mathbf{P} - \mathbf{W}) + (\mathbf{P}^2 - \mathbf{W}) + \cdots . \qquad (11.7)$$

We now use special properties of $\mathbf{P}$ and $\mathbf{W}$ to rewrite this series. The special properties are: 1) $\mathbf{PW} = \mathbf{W}$, and 2) $\mathbf{W}^k = \mathbf{W}$ for all positive integers k. These

11.5. MEAN FIRST PASSAGE TIME

facts are easy to verify, and are left as an exercise (see Exercise 22). Using these facts, we see that

$$
\begin{aligned}
(\mathbf{P} - \mathbf{W})^n &= \sum_{i=0}^{n} (-1)^i \binom{n}{i} \mathbf{P}^{n-i} \mathbf{W}^i \\
&= \mathbf{P}^n + \sum_{i=1}^{n} (-1)^i \binom{n}{i} \mathbf{W}^i \\
&= \mathbf{P}^n + \sum_{i=1}^{n} (-1)^i \binom{n}{i} \mathbf{W} \\
&= \mathbf{P}^n + \left(\sum_{i=1}^{n} (-1)^i \binom{n}{i} \right) \mathbf{W} .
\end{aligned}
$$

If we expand the expression $(1-1)^n$, using the Binomial Theorem, we obtain the expression in parenthesis above, except that we have an extra term (which equals 1). Since $(1-1)^n = 0$, we see that the above expression equals -1. So we have

$$(\mathbf{P} - \mathbf{W})^n = \mathbf{P}^n - \mathbf{W} ,$$

for all $n \geq 1$.

We can now rewrite the series in 11.7 as

$$\mathbf{I} + (\mathbf{P} - \mathbf{W}) + (\mathbf{P} - \mathbf{W})^2 + \cdots .$$

Since the nth term in this series is equal to $\mathbf{P}^n - \mathbf{W}$, the nth term goes to 0 as n goes to infinity. This is sufficient to show that this series converges, and sums to the inverse of the matrix $\mathbf{I} - \mathbf{P} + \mathbf{W}$. We call this inverse the *fundamental matrix* associated with the chain, and we denote it by $\mathbf{Z}$.

In the case that the chain is ergodic, but not regular, it is not true that $\mathbf{P}^n \to \mathbf{W}$ as $n \to \infty$. Nevertheless, the matrix $\mathbf{I} - \mathbf{P} + \mathbf{W}$ still has an inverse, as we will now show.

Proposition 11.1 Let $\mathbf{P}$ be the transition matrix of an ergodic chain, and let $\mathbf{W}$ be the matrix all of whose rows are the fixed probability row vector for $\mathbf{P}$. Then the matrix

$$\mathbf{I} - \mathbf{P} + \mathbf{W}$$

has an inverse.

Proof. Let $\mathbf{x}$ be a column vector such that

$$(\mathbf{I} - \mathbf{P} + \mathbf{W})\mathbf{x} = \mathbf{0} .$$

To prove the proposition, it is sufficient to show that $\mathbf{x}$ must be the zero vector. Multiplying this equation by $\mathbf{w}$ and using the fact that $\mathbf{w}(\mathbf{I} - \mathbf{P}) = \mathbf{0}$ and $\mathbf{w}\mathbf{W} = \mathbf{w}$, we have

$$\mathbf{w}(\mathbf{I} - \mathbf{P} + \mathbf{W})\mathbf{x} = \mathbf{w}\mathbf{x} = 0 .$$

Therefore,
$$(\mathbf{I} - \mathbf{P})\mathbf{x} = \mathbf{0} \ .$$

But this means that $\mathbf{x} = \mathbf{P}\mathbf{x}$ is a fixed column vector for $\mathbf{P}$. By Theorem 11.10, this can only happen if $\mathbf{x}$ is a constant vector. Since $\mathbf{w}\mathbf{x} = 0$, and $\mathbf{w}$ has strictly positive entries, we see that $\mathbf{x} = \mathbf{0}$. This completes the proof. $\square$

As in the regular case, we will call the inverse of the matrix $\mathbf{I} - \mathbf{P} + \mathbf{W}$ the *fundamental matrix* for the ergodic chain with transition matrix $\mathbf{P}$, and we will use $\mathbf{Z}$ to denote this fundamental matrix.

Example 11.26 Let $\mathbf{P}$ be the transition matrix for the weather in the Land of Oz. Then

$$\mathbf{I} - \mathbf{P} + \mathbf{W} = \begin{pmatrix} 1 & 0 & 0 \\ 0 & 1 & 0 \\ 0 & 0 & 1 \end{pmatrix} - \begin{pmatrix} 1/2 & 1/4 & 1/4 \\ 1/2 & 0 & 1/2 \\ 1/4 & 1/4 & 1/2 \end{pmatrix} + \begin{pmatrix} 2/5 & 1/5 & 2/5 \\ 2/5 & 1/5 & 2/5 \\ 2/5 & 1/5 & 2/5 \end{pmatrix}$$
$$= \begin{pmatrix} 9/10 & -1/20 & 3/20 \\ -1/10 & 6/5 & -1/10 \\ 3/20 & -1/20 & 9/10 \end{pmatrix} \ ,$$

so

$$\mathbf{Z} = (\mathbf{I} - \mathbf{P} + \mathbf{W})^{-1} = \begin{pmatrix} 86/75 & 1/25 & -14/75 \\ 2/25 & 21/25 & 2/25 \\ -14/75 & 1/25 & 86/75 \end{pmatrix} \ .$$
$\square$

Using the Fundamental Matrix to Calculate the Mean First Passage Matrix

We shall show how one can obtain the mean first passage matrix $\mathbf{M}$ from the fundamental matrix $\mathbf{Z}$ for an ergodic Markov chain. Before stating the theorem which gives the first passage times, we need a few facts about $\mathbf{Z}$.

Lemma 11.2 Let $\mathbf{Z} = (\mathbf{I} - \mathbf{P} + \mathbf{W})^{-1}$, and let $\mathbf{c}$ be a column vector of all 1's. Then
$$\mathbf{Z}\mathbf{c} = \mathbf{c} \ ,$$
$$\mathbf{w}\mathbf{Z} = \mathbf{w} \ ,$$
and
$$\mathbf{Z}(\mathbf{I} - \mathbf{P}) = \mathbf{I} - \mathbf{W} \ .$$

Proof. Since $\mathbf{P}\mathbf{c} = \mathbf{c}$ and $\mathbf{W}\mathbf{c} = \mathbf{c}$,
$$\mathbf{c} = (\mathbf{I} - \mathbf{P} + \mathbf{W})\mathbf{c} \ .$$

If we multiply both sides of this equation on the left by $\mathbf{Z}$, we obtain
$$\mathbf{Z}\mathbf{c} = \mathbf{c} \ .$$

11.5. MEAN FIRST PASSAGE TIME

Similarly, since $\mathbf{wP} = \mathbf{w}$ and $\mathbf{wW} = \mathbf{w}$,

$$\mathbf{w} = \mathbf{w}(\mathbf{I} - \mathbf{P} + \mathbf{W}) \ .$$

If we multiply both sides of this equation on the right by $\mathbf{Z}$, we obtain

$$\mathbf{wZ} = \mathbf{w} \ .$$

Finally, we have

$$\begin{aligned}(\mathbf{I} - \mathbf{P} + \mathbf{W})(\mathbf{I} - \mathbf{W}) &= \mathbf{I} - \mathbf{W} - \mathbf{P} + \mathbf{W} + \mathbf{W} - \mathbf{W} \\ &= \mathbf{I} - \mathbf{P} \ .\end{aligned}$$

Multiplying on the left by $\mathbf{Z}$, we obtain

$$\mathbf{I} - \mathbf{W} = \mathbf{Z}(\mathbf{I} - \mathbf{P}) \ .$$

This completes the proof. $\square$

The following theorem shows how one can obtain the mean first passage times from the fundamental matrix.

Theorem 11.16 The mean first passage matrix $\mathbf{M}$ for an ergodic chain is determined from the fundamental matrix $\mathbf{Z}$ and the fixed row probability vector $\mathbf{w}$ by

$$m_{ij} = \frac{z_{jj} - z_{ij}}{w_j} \ .$$

Proof. We showed in Equation 11.6 that

$$(\mathbf{I} - \mathbf{P})\mathbf{M} = \mathbf{C} - \mathbf{D} \ .$$

Thus,

$$\mathbf{Z}(\mathbf{I} - \mathbf{P})\mathbf{M} = \mathbf{ZC} - \mathbf{ZD} \ ,$$

and from Lemma 11.2,

$$\mathbf{Z}(\mathbf{I} - \mathbf{P})\mathbf{M} = \mathbf{C} - \mathbf{ZD} \ .$$

Again using Lemma 11.2, we have

$$\mathbf{M} - \mathbf{WM} = \mathbf{C} - \mathbf{ZD}$$

or

$$\mathbf{M} = \mathbf{C} - \mathbf{ZD} + \mathbf{WM} \ .$$

From this equation, we see that

$$m_{ij} = 1 - z_{ij}r_j + (\mathbf{wM})_j \ . \tag{11.8}$$

But $m_{jj} = 0$, and so

$$0 = 1 - z_{jj}r_j + (\mathbf{wM})_j \ ,$$

or
$$(\mathbf{wM})_j = z_{jj}r_j - 1 . \tag{11.9}$$

From Equations 11.8 and 11.9, we have

$$m_{ij} = (z_{jj} - z_{ij}) \cdot r_j .$$

Since $r_j = 1/w_j$,

$$m_{ij} = \frac{z_{jj} - z_{ij}}{w_j} .$$

$\square$

Example 11.27 (Example 11.26 continued) In the Land of Oz example, we find that

$$\mathbf{Z} = (\mathbf{I} - \mathbf{P} + \mathbf{W})^{-1} = \begin{pmatrix} 86/75 & 1/25 & -14/75 \\ 2/25 & 21/25 & 2/25 \\ -14/75 & 1/25 & 86/75 \end{pmatrix} .$$

We have also seen that $\mathbf{w} = (2/5, 1/5, 2/5)$. So, for example,

$$\begin{aligned} m_{12} &= \frac{z_{22} - z_{12}}{w_2} \\ &= \frac{21/25 - 1/25}{1/5} \\ &= 4 , \end{aligned}$$

by Theorem 11.16. Carrying out the calculations for the other entries of $\mathbf{M}$, we obtain

$$\mathbf{M} = \begin{pmatrix} 0 & 4 & 10/3 \\ 8/3 & 0 & 8/3 \\ 10/3 & 4 & 0 \end{pmatrix} .$$

$\square$

Computation

The program **ErgodicChain** calculates the fundamental matrix, the fixed vector, the mean recurrence matrix $\mathbf{D}$, and the mean first passage matrix $\mathbf{M}$. We have run the program for the Ehrenfest urn model (Example 11.8). We obtain:

$$\mathbf{P} = \begin{array}{c} \\ 0 \\ 1 \\ 2 \\ 3 \\ 4 \end{array} \begin{pmatrix} \begin{array}{ccccc} 0 & 1 & 2 & 3 & 4 \end{array} \\ .0000 & 1.0000 & .0000 & .0000 & .0000 \\ .2500 & .0000 & .7500 & .0000 & .0000 \\ .0000 & .5000 & .0000 & .5000 & .0000 \\ .0000 & .0000 & .7500 & .0000 & .2500 \\ .0000 & .0000 & .0000 & 1.0000 & .0000 \end{pmatrix} ;$$

$$\mathbf{w} = \begin{array}{ccccc} 0 & 1 & 2 & 3 & 4 \end{array} \\ \begin{pmatrix} .0625 & .2500 & .3750 & .2500 & .0625 \end{pmatrix} ;$$

11.5. MEAN FIRST PASSAGE TIME

$$\mathbf{r} = \begin{pmatrix} 0 & 1 & 2 & 3 & 4 \\ 16.0000 & 4.0000 & 2.6667 & 4.0000 & 16.0000 \end{pmatrix} ;$$

$$\mathbf{M} = \begin{array}{c} 0 \\ 1 \\ 2 \\ 3 \\ 4 \end{array} \begin{pmatrix} 0 & 1 & 2 & 3 & 4 \\ .0000 & 1.0000 & 2.6667 & 6.3333 & 21.3333 \\ 15.0000 & .0000 & 1.6667 & 5.3333 & 20.3333 \\ 18.6667 & 3.6667 & .0000 & 3.6667 & 18.6667 \\ 20.3333 & 5.3333 & 1.6667 & .0000 & 15.0000 \\ 21.3333 & 6.3333 & 2.6667 & 1.0000 & .0000 \end{pmatrix} .$$

From the mean first passage matrix, we see that the mean time to go from 0 balls in urn 1 to 2 balls in urn 1 is 2.6667 steps while the mean time to go from 2 balls in urn 1 to 0 balls in urn 1 is 18.6667. This reflects the fact that the model exhibits a central tendency. Of course, the physicist is interested in the case of a large number of molecules, or balls, and so we should consider this example for n so large that we cannot compute it even with a computer.

Ehrenfest Model

Example 11.28 (Example 11.23 continued) Let us consider the Ehrenfest model (see Example 11.8) for gas diffusion for the general case of $2n$ balls. Every second, one of the $2n$ balls is chosen at random and moved from the urn it was in to the other urn. If there are i balls in the first urn, then with probability $i/2n$ we take one of them out and put it in the second urn, and with probability $(2n-i)/2n$ we take a ball from the second urn and put it in the first urn. At each second we let the number i of balls in the first urn be the state of the system. Then from state i we can pass only to state $i-1$ and $i+1$, and the transition probabilities are given by

$$p_{ij} = \begin{cases} \frac{i}{2n}, & \text{if } j = i-1, \\ 1 - \frac{i}{2n}, & \text{if } j = i+1, \\ 0, & \text{otherwise.} \end{cases}$$

This defines the transition matrix of an ergodic, non-regular Markov chain (see Exercise 15). Here the physicist is interested in long-term predictions about the state occupied. In Example 11.23, we gave an intuitive reason for expecting that the fixed vector $\mathbf{w}$ is the binomial distribution with parameters $2n$ and $1/2$. It is easy to check that this is correct. So,

$$w_i = \frac{\binom{2n}{i}}{2^{2n}} .$$

Thus the mean recurrence time for state i is

$$r_i = \frac{2^{2n}}{\binom{2n}{i}} .$$

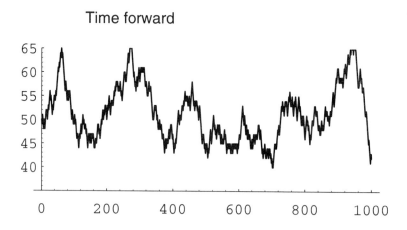

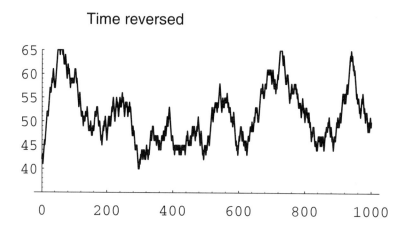

Figure 11.6: Ehrenfest simulation.

Consider in particular the central term $i = n$. We have seen that this term is approximately $1/\sqrt{\pi n}$. Thus we may approximate r_n by $\sqrt{\pi n}$.

This model was used to explain the concept of reversibility in physical systems. Assume that we let our system run until it is in equilibrium. At this point, a movie is made, showing the system's progress. The movie is then shown to you, and you are asked to tell if the movie was shown in the forward or the reverse direction. It would seem that there should always be a tendency to move toward an equal proportion of balls so that the correct order of time should be the one with the most transitions from i to $i-1$ if $i > n$ and i to $i+1$ if $i < n$.

In Figure 11.6 we show the results of simulating the Ehrenfest urn model for the case of $n = 50$ and 1000 time units, using the program **EhrenfestUrn**. The top graph shows these results graphed in the order in which they occurred and the bottom graph shows the same results but with time reversed. There is no apparent difference.

Reversibility

We note that if we had not started in equilibrium, the two graphs would typically look quite different. □

If the Ehrenfest model is started in equilibrium, then the process has no apparent time direction. The reason for this is that this process has a property called *reversibility*. Define X_n to be the number of balls in the left urn at step n. We can calculate, for a general ergodic chain, the reverse transition probability:

$$\begin{aligned} P(X_{n-1} = j | X_n = i) &= \frac{P(X_{n-1} = j, X_n = i)}{P(X_n = i)} \\ &= \frac{P(X_{n-1} = j) P(X_n = i | X_{n-1} = j)}{P(X_n = i)} \\ &= \frac{P(X_{n-1} = j) p_{ji}}{P(X_n = i)} . \end{aligned}$$

In general, this will depend upon n, since $P(X_n = j)$ and also $P(X_{n-1} = j)$ change with n. However, if we start with the vector **w** or wait until equilibrium is reached, this will not be the case. Then we can define

$$p_{ij}^* = \frac{w_j p_{ji}}{w_i}$$

as a transition matrix for the process watched with time reversed.

Let us calculate a typical transition probability for the reverse chain $\mathbf{P}^* = \{p_{ij}^*\}$ in the Ehrenfest model. For example,

$$\begin{aligned} p_{i,i-1}^* &= \frac{w_{i-1} p_{i-1,i}}{w_i} = \frac{\binom{2n}{i-1}}{2^{2n}} \times \frac{2n-i+1}{2n} \times \frac{2^{2n}}{\binom{2n}{i}} \\ &= \frac{(2n)!}{(i-1)!\,(2n-i+1)!} \times \frac{(2n-i+1)\,i!\,(2n-i)!}{2n(2n)!} \\ &= \frac{i}{2n} = p_{i,i-1} . \end{aligned}$$

Similar calculations for the other transition probabilities show that $\mathbf{P}^* = \mathbf{P}$. When this occurs the process is called *reversible*. Clearly, an ergodic chain is reversible if, and only if, for every pair of states s_i and s_j, $w_i p_{ij} = w_j p_{ji}$. In particular, for the Ehrenfest model this means that $w_i p_{i,i-1} = w_{i-1} p_{i-1,i}$. Thus, in equilibrium, the pairs $(i, i-1)$ and $(i-1, i)$ should occur with the same frequency. While many of the Markov chains that occur in applications are reversible, this is a very strong condition. In Exercise 12 you are asked to find an example of a Markov chain which is not reversible.

The Central Limit Theorem for Markov Chains

Suppose that we have an ergodic Markov chain with states $s_1, s_2, \ldots, s_k$. It is natural to consider the distribution of the random variables $S_j^{(n)}$, which denotes the number of times that the chain is in state s_j in the first n steps. The jth component w_j of the fixed probability row vector $\mathbf{w}$ is the proportion of times that the chain is in state s_j in the long run. Hence, it is reasonable to conjecture that the expected value of the random variable $S_j^{(n)}$, as $n \to \infty$, is asymptotic to nw_j, and it is easy to show that this is the case (see Exercise 23).

It is also natural to ask whether there is a limiting distribution of the random variables $S_j^{(n)}$. The answer is yes, and in fact, this limiting distribution is the normal distribution. As in the case of independent trials, one must normalize these random variables. Thus, we must subtract from $S_j^{(n)}$ its expected value, and then divide by its standard deviation. In both cases, we will use the asymptotic values of these quantities, rather than the values themselves. Thus, in the first case, we will use the value nw_j. It is not so clear what we should use in the second case. It turns out that the quantity

$$\sigma_j^2 = 2w_j z_{jj} - w_j - w_j^2 \tag{11.10}$$

represents the asymptotic variance. Armed with these ideas, we can state the following theorem.

Theorem 11.17 (Central Limit Theorem for Markov Chains) For an ergodic chain, for any real numbers $r < s$, we have

$$P\left(r < \frac{S_j^{(n)} - nw_j}{\sqrt{n\sigma_j^2}} < s\right) \to \frac{1}{\sqrt{2\pi}} \int_r^s e^{-x^2/2}\, dx\ ,$$

as $n \to \infty$, for any choice of starting state, where σ_j^2 is the quantity defined in Equation 11.10. $\square$

Historical Remarks

Markov chains were introduced by Andreĭ Andreevich Markov (1856–1922) and were named in his honor. He was a talented undergraduate who received a gold medal for his undergraduate thesis at St. Petersburg University. Besides being an active research mathematician and teacher, he was also active in politics and participated in the liberal movement in Russia at the beginning of the twentieth century. In 1913, when the government celebrated the 300th anniversary of the House of Romanov family, Markov organized a counter-celebration of the 200th anniversary of Bernoulli's discovery of the Law of Large Numbers.

Markov was led to develop Markov chains as a natural extension of sequences of independent random variables. In his first paper, in 1906, he proved that for a Markov chain with positive transition probabilities and numerical states the average of the outcomes converges to the expected value of the limiting distribution (the fixed vector). In a later paper he proved the central limit theorem for such chains. Writing about Markov, A. P. Youschkevitch remarks:

> Markov arrived at his chains starting from the internal needs of probability theory, and he never wrote about their applications to physical

11.5. MEAN FIRST PASSAGE TIME

science. For him the only real examples of the chains were literary texts, where the two states denoted the vowels and consonants.[19]

In a paper written in 1913,[20] Markov chose a sequence of 20,000 letters from Pushkin's *Eugene Onegin* to see if this sequence can be approximately considered a simple chain. He obtained the Markov chain with transition matrix

$$\begin{array}{c} \\ \text{vowel} \\ \text{consonant} \end{array} \begin{pmatrix} \text{vowel} & \text{consonant} \\ .128 & .872 \\ .663 & .337 \end{pmatrix}.$$

The fixed vector for this chain is (.432, .568), indicating that we should expect about 43.2 percent vowels and 56.8 percent consonants in the novel, which was borne out by the actual count.

Claude Shannon considered an interesting extension of this idea in his book *The Mathematical Theory of Communication*,[21] in which he developed the information-theoretic concept of entropy. Shannon considers a series of Markov chain approximations to English prose. He does this first by chains in which the states are letters and then by chains in which the states are words. For example, for the case of words he presents first a simulation where the words are chosen independently but with appropriate frequencies.

> REPRESENTING AND SPEEDILY IS AN GOOD APT OR COME CAN DIFFERENT NATURAL HERE HE THE A IN CAME THE TO OF TO EXPERT GRAY COME TO FURNISHES THE LINE MESSAGE HAD BE THESE.

He then notes the increased resemblence to ordinary English text when the words are chosen as a Markov chain, in which case he obtains

> THE HEAD AND IN FRONTAL ATTACK ON AN ENGLISH WRITER THAT THE CHARACTER OF THIS POINT IS THEREFORE ANOTHER METHOD FOR THE LETTERS THAT THE TIME OF WHO EVER TOLD THE PROBLEM FOR AN UNEXPECTED.

A simulation like the last one is carried out by opening a book and choosing the first word, say it is *the*. Then the book is read until the word *the* appears again and the word after this is chosen as the second word, which turned out to be *head*. The book is then read until the word *head* appears again and the next word, *and*, is chosen, and so on.

Other early examples of the use of Markov chains occurred in Galton's study of the problem of survival of family names in 1889 and in the Markov chain introduced

---

[19]See *Dictionary of Scientific Biography*, ed. C. C. Gillespie (New York: Scribner's Sons, 1970), pp. 124–130.

[20]A. A. Markov, "An Example of Statistical Analysis of the Text of Eugene Onegin Illustrating the Association of Trials into a Chain," *Bulletin de l'Acadamie Imperiale des Sciences de St. Petersburg*, ser. 6, vol. 7 (1913), pp. 153–162.

[21]C. E. Shannon and W. Weaver, *The Mathematical Theory of Communication* (Urbana: Univ. of Illinois Press, 1964).

by P. and T. Ehrenfest in 1907 for diffusion. Poincaré in 1912 dicussed card shuffling in terms of an ergodic Markov chain defined on a permutation group. Brownian motion, a continuous time version of random walk, was introduced in 1900–1901 by L. Bachelier in his study of the stock market, and in 1905–1907 in the works of A. Einstein and M. Smoluchowsky in their study of physical processes.

One of the first systematic studies of finite Markov chains was carried out by M. Frechet.[22] The treatment of Markov chains in terms of the two fundamental matrices that we have used was developed by Kemeny and Snell[23] to avoid the use of eigenvalues that one of these authors found too complex. The fundamental matrix $\mathbf{N}$ occurred also in the work of J. L. Doob and others in studying the connection between Markov processes and classical potential theory. The fundamental matrix $\mathbf{Z}$ for ergodic chains appeared first in the work of Frechet, who used it to find the limiting variance for the central limit theorem for Markov chains.

Exercises

1 Consider the Markov chain with transition matrix

$$\mathbf{P} = \begin{pmatrix} 1/2 & 1/2 \\ 1/4 & 3/4 \end{pmatrix} .$$

Find the fundamental matrix $\mathbf{Z}$ for this chain. Compute the mean first passage matrix using $\mathbf{Z}$.

2 A study of the strengths of Ivy League football teams shows that if a school has a strong team one year it is equally likely to have a strong team or average team next year; if it has an average team, half the time it is average next year, and if it changes it is just as likely to become strong as weak; if it is weak it has 2/3 probability of remaining so and 1/3 of becoming average.

(a) A school has a strong team. On the average, how long will it be before it has another strong team?

(b) A school has a weak team; how long (on the average) must the alumni wait for a strong team?

3 Consider Example 11.4 with $a = .5$ and $b = .75$. Assume that the President says that he or she will run. Find the expected length of time before the first time the answer is passed on incorrectly.

4 Find the mean recurrence time for each state of Example 11.4 for $a = .5$ and $b = .75$. Do the same for general a and b.

5 A die is rolled repeatedly. Show by the results of this section that the mean time between occurrences of a given number is 6.

---

[22]M. Frechet, "Théorie des événements en chaine dans le cas d'un nombre fini d'états possible," in *Recherches théoriques Modernes sur le calcul des probabilités*, vol. 2 (Paris, 1938).

[23]J. G. Kemeny and J. L. Snell, *Finite Markov Chains*.

11.5. MEAN FIRST PASSAGE TIME

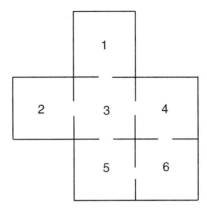

Figure 11.7: Maze for Exercise 7.

6 For the Land of Oz example (Example 11.1), make rain into an absorbing state and find the fundamental matrix **N**. Interpret the results obtained from this chain in terms of the original chain.

7 A rat runs through the maze shown in Figure 11.7. At each step it leaves the room it is in by choosing at random one of the doors out of the room.

(a) Give the transition matrix **P** for this Markov chain.

(b) Show that it is an ergodic chain but not a regular chain.

(c) Find the fixed vector.

(d) Find the expected number of steps before reaching Room 5 for the first time, starting in Room 1.

8 Modify the program **ErgodicChain** so that you can compute the basic quantities for the queueing example of Exercise 11.3.20. Interpret the mean recurrence time for state 0.

9 Consider a random walk on a circle of circumference n. The walker takes one unit step clockwise with probability p and one unit counterclockwise with probability $q = 1 - p$. Modify the program **ErgodicChain** to allow you to input n and p and compute the basic quantities for this chain.

(a) For which values of n is this chain regular? ergodic?

(b) What is the limiting vector **w**?

(c) Find the mean first passage matrix for $n = 5$ and $p = .5$. Verify that $m_{ij} = d(n-d)$, where d is the clockwise distance from i to j.

10 Two players match pennies and have between them a total of 5 pennies. If at any time one player has all of the pennies, to keep the game going, he gives one back to the other player and the game will continue. Show that this game can be formulated as an ergodic chain. Study this chain using the program **ErgodicChain**.

11 Calculate the reverse transition matrix for the Land of Oz example (Example 11.1). Is this chain reversible?

12 Give an example of a three-state ergodic Markov chain that is not reversible.

13 Let $\mathbf{P}$ be the transition matrix of an ergodic Markov chain and $\mathbf{P}^*$ the reverse transition matrix. Show that they have the same fixed probability vector $\mathbf{w}$.

14 If $\mathbf{P}$ is a reversible Markov chain, is it necessarily true that the mean time to go from state i to state j is equal to the mean time to go from state j to state i? *Hint*: Try the Land of Oz example (Example 11.1).

15 Show that any ergodic Markov chain with a symmetric transition matrix (i.e., $p_{ij} = p_{ji}$) is reversible.

16 (Crowell[24]) Let $\mathbf{P}$ be the transition matrix of an ergodic Markov chain. Show that
$$(\mathbf{I} + \mathbf{P} + \cdots + \mathbf{P}^{n-1})(\mathbf{I} - \mathbf{P} + \mathbf{W}) = \mathbf{I} - \mathbf{P}^n + n\mathbf{W},$$
and from this show that
$$\frac{\mathbf{I} + \mathbf{P} + \cdots + \mathbf{P}^{n-1}}{n} \to \mathbf{W},$$
as $n \to \infty$.

17 An ergodic Markov chain is started in equilibrium (i.e., with initial probability vector $\mathbf{w}$). The mean time until the next occurrence of state s_i is $\bar{m}_i = \sum_k w_k m_{ki} + w_i r_i$. Show that $\bar{m}_i = z_{ii}/w_i$, by using the facts that $\mathbf{wZ} = \mathbf{w}$ and $m_{ki} = (z_{ii} - z_{ki})/w_i$.

18 A perpetual craps game goes on at Charley's. Jones comes into Charley's on an evening when there have already been 100 plays. He plans to play until the next time that snake eyes (a pair of ones) are rolled. Jones wonders how many times he will play. On the one hand he realizes that the average time between snake eyes is 36 so he should play about 18 times as he is equally likely to have come in on either side of the halfway point between occurrences of snake eyes. On the other hand, the dice have no memory, and so it would seem that he would have to play for 36 more times no matter what the previous outcomes have been. Which, if either, of Jones's arguments do you believe? Using the result of Exercise 17, calculate the expected to reach snake eyes, in equilibrium, and see if this resolves the apparent paradox. If you are still in doubt, simulate the experiment to decide which argument is correct. Can you give an intuitive argument which explains this result?

19 Show that, for an ergodic Markov chain (see Theorem 11.16),
$$\sum_j m_{ij} w_j = \sum_j z_{jj} - 1 = K.$$

---

[24]Private communication.

11.5. MEAN FIRST PASSAGE TIME

-5 B	20 C
-30 A	15 GO

Figure 11.8: Simplified Monopoly.

The second expression above shows that the number K is independent of i. The number K is called *Kemeny's constant*. A prize was offered to the first person to give an intuitively plausible reason for the above sum to be independent of i. (See also Exercise 24.)

20 Consider a game played as follows: You are given a regular Markov chain with transition matrix $\mathbf{P}$, fixed probability vector $\mathbf{w}$, and a payoff function $\mathbf{f}$ which assigns to each state s_i an amount f_i which may be positive or negative. Assume that $\mathbf{wf} = 0$. You watch this Markov chain as it evolves, and every time you are in state s_i you receive an amount f_i. Show that your expected winning after n steps can be represented by a column vector $\mathbf{g}^{(n)}$, with

$$\mathbf{g}^{(n)} = (\mathbf{I} + \mathbf{P} + \mathbf{P}^2 + \cdots + \mathbf{P}^n)\mathbf{f}.$$

Show that as $n \to \infty$, $\mathbf{g}^{(n)} \to \mathbf{g}$ with $\mathbf{g} = \mathbf{Zf}$.

21 A highly simplified game of "Monopoly" is played on a board with four squares as shown in Figure 11.8. You start at GO. You roll a die and move clockwise around the board a number of squares equal to the number that turns up on the die. You collect or pay an amount indicated on the square on which you land. You then roll the die again and move around the board in the same manner from your last position. Using the result of Exercise 20, estimate the amount you should expect to win in the long run playing this version of Monopoly.

22 Show that if $\mathbf{P}$ is the transition matrix of a regular Markov chain, and $\mathbf{W}$ is the matrix each of whose rows is the fixed probability vector corresponding to $\mathbf{P}$, then $\mathbf{PW} = \mathbf{W}$, and $\mathbf{W}^k = \mathbf{W}$ for all positive integers k.

23 Assume that an ergodic Markov chain has states $s_1, s_2, \ldots, s_k$. Let $S_j^{(n)}$ denote the number of times that the chain is in state s_j in the first n steps. Let $\mathbf{w}$ denote the fixed probability row vector for this chain. Show that, regardless of the starting state, the expected value of $S_j^{(n)}$, divided by n, tends to w_j as $n \to \infty$. *Hint*: If the chain starts in state s_i, then the expected value of $S_j^{(n)}$ is given by the expression

$$\sum_{h=0}^{n} p_{ij}^{(h)}.$$

24 Peter Doyle[25] has suggested the following interpretation for *Kemeny's constant* (see Exercise 19). We are given an ergodic chain and do not know the starting state. However, we would like to start watching it at a time when it can be considered to be in equilibrium (i.e., as if we had started with the fixed vector **w** or as if we had waited a long time). However, we don't know the starting state and we don't want to wait a long time. Peter says to choose a state according to the fixed vector **w**. That is, choose state j with probability w_j using a spinner, for example. Then wait until the time T that this state occurs for the first time. We consider T as our starting time and observe the chain from this time on. Of course the probability that we start in state j is w_j, so we are starting in equilibrium. Kemeny's constant is the expected value of T, and it is independent of the way in which the chain was started. Should Peter have been given the prize?

---

[25]Private communication.

Chapter 12

Random Walks

12.1 Random Walks in Euclidean Space

In the last several chapters, we have studied sums of random variables with the goal being to describe the distribution and density functions of the sum. In this chapter, we shall look at sums of discrete random variables from a different perspective. We shall be concerned with properties which can be associated with the sequence of partial sums, such as the number of sign changes of this sequence, the number of terms in the sequence which equal 0, and the expected size of the maximum term in the sequence.

We begin with the following definition.

Definition 12.1 Let $\{X_k\}_{k=1}^{\infty}$ be a sequence of independent, identically distributed discrete random variables. For each positive integer n, we let S_n denote the sum $X_1 + X_2 + \cdots + X_n$. The sequence $\{S_n\}_{n=1}^{\infty}$ is called a *random walk*. If the common range of the X_k's is $\mathbf{R}^m$, then we say that $\{S_n\}$ is a random walk in $\mathbf{R}^m$. □

We view the sequence of X_k's as being the outcomes of independent experiments. Since the X_k's are independent, the probability of any particular (finite) sequence of outcomes can be obtained by multiplying the probabilities that each X_k takes on the specified value in the sequence. Of course, these individual probabilities are given by the common distribution of the X_k's. We will typically be interested in finding probabilities for events involving the related sequence of S_n's. Such events can be described in terms of the X_k's, so their probabilities can be calculated using the above idea.

There are several ways to visualize a random walk. One can imagine that a particle is placed at the origin in $\mathbf{R}^m$ at time $n = 0$. The sum S_n represents the position of the particle at the end of n seconds. Thus, in the time interval $[n-1, n]$, the particle moves (or jumps) from position S_{n-1} to S_n. The vector representing this motion is just $S_n - S_{n-1}$, which equals X_n. This means that in a random walk, the jumps are independent and identically distributed. If $m = 1$, for example, then one can imagine a particle on the real line that starts at the origin, and at the end of each second, jumps one unit to the right or the left, with probabilities given by the distribution of the X_k's. If $m = 2$, one can visualize the process as taking place in a city in which the streets form square city blocks. A person starts at one corner (i.e., at an intersection of two streets) and goes in one of the four possible directions according to the distribution of the X_k's. If $m = 3$, one might imagine being in a jungle gym, where one is free to move in any one of six directions (left, right, forward, backward, up, and down). Once again, the probabilities of these movements are given by the distribution of the X_k's.

Another model of a random walk (used mostly in the case where the range is $\mathbf{R}^1$) is a game, involving two people, which consists of a sequence of independent, identically distributed moves. The sum S_n represents the score of the first person, say, after n moves, with the assumption that the score of the second person is $-S_n$. For example, two people might be flipping coins, with a match or non-match representing $+1$ or -1, respectively, for the first player. Or, perhaps one coin is being flipped, with a head or tail representing $+1$ or -1, respectively, for the first player.

Random Walks on the Real Line

We shall first consider the simplest non-trivial case of a random walk in $\mathbf{R}^1$, namely the case where the common distribution function of the random variables X_n is given by

$$f_X(x) = \begin{cases} 1/2, & \text{if } x = \pm 1, \\ 0, & \text{otherwise.} \end{cases}$$

This situation corresponds to a fair coin being flipped, with S_n representing the number of heads minus the number of tails which occur in the first n flips. We note that in this situation, all paths of length n have the same probability, namely 2^{-n}.

It is sometimes instructive to represent a random walk as a polygonal line, or path, in the plane, where the horizontal axis represents time and the vertical axis represents the value of S_n. Given a sequence $\{S_n\}$ of partial sums, we first plot the points (n, S_n), and then for each $k < n$, we connect (k, S_k) and $(k+1, S_{k+1})$ with a straight line segment. The *length* of a path is just the difference in the time values of the beginning and ending points on the path. The reader is referred to Figure 12.1. This figure, and the process it illustrates, are identical with the example, given in Chapter 1, of two people playing heads or tails.

Returns and First Returns

We say that an *equalization* has occurred, or there is a *return to the origin* at time n, if $S_n = 0$. We note that this can only occur if n is an even integer. To calculate the probability of an equalization at time $2m$, we need only count the number of paths of length $2m$ which begin and end at the origin. The number of such paths is clearly

$$\binom{2m}{m}.$$

Since each path has probability 2^{-2m}, we have the following theorem.

12.1. RANDOM WALKS IN EUCLIDEAN SPACE

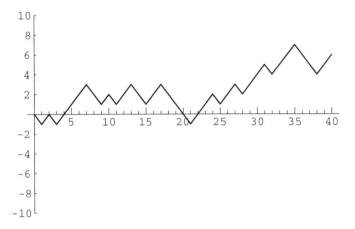

Figure 12.1: A random walk of length 40.

Theorem 12.1 The probability of a return to the origin at time $2m$ is given by

$$u_{2m} = \binom{2m}{m} 2^{-2m} .$$

The probability of a return to the origin at an odd time is 0. □

A random walk is said to have a *first return* to the origin at time $2m$ if $m > 0$, and $S_{2k} \neq 0$ for all $k < m$. In Figure 12.1, the first return occurs at time 2. We define f_{2m} to be the probability of this event. (We also define $f_0 = 0$.) One can think of the expression $f_{2m} 2^{2m}$ as the number of paths of length $2m$ between the points $(0,0)$ and $(2m,0)$ that do not touch the horizontal axis except at the endpoints. Using this idea, it is easy to prove the following theorem.

Theorem 12.2 For $n \geq 1$, the probabilities $\{u_{2k}\}$ and $\{f_{2k}\}$ are related by the equation

$$u_{2n} = f_0 u_{2n} + f_2 u_{2n-2} + \cdots + f_{2n} u_0 .$$

Proof. There are $u_{2n} 2^{2n}$ paths of length $2n$ which have endpoints $(0,0)$ and $(2n,0)$. The collection of such paths can be partitioned into n sets, depending upon the time of the first return to the origin. A path in this collection which has a first return to the origin at time $2k$ consists of an initial segment from $(0,0)$ to $(2k,0)$, in which no interior points are on the horizontal axis, and a terminal segment from $(2k,0)$ to $(2n,0)$, with no further restrictions on this segment. Thus, the number of paths in the collection which have a first return to the origin at time $2k$ is given by

$$f_{2k} 2^{2k} u_{2n-2k} 2^{2n-2k} = f_{2k} u_{2n-2k} 2^{2n} .$$

If we sum over k, we obtain the equation

$$u_{2n} 2^{2n} = f_0 u_{2n} 2^{2n} + f_2 u_{2n-2} 2^{2n} + \cdots + f_{2n} u_0 2^{2n} .$$

Dividing both sides of this equation by 2^{2n} completes the proof. □

The expression in the right-hand side of the above theorem should remind the reader of a sum that appeared in Definition 7.1 of the convolution of two distributions. The convolution of two sequences is defined in a similar manner. The above theorem says that the sequence $\{u_{2n}\}$ is the convolution of itself and the sequence $\{f_{2n}\}$. Thus, if we represent each of these sequences by an ordinary generating function, then we can use the above relationship to determine the value f_{2n}.

Theorem 12.3 For $m \geq 1$, the probability of a first return to the origin at time $2m$ is given by

$$f_{2m} = \frac{u_{2m}}{2m-1} = \frac{\binom{2m}{m}}{(2m-1)2^{2m}} .$$

Proof. We begin by defining the generating functions

$$U(x) = \sum_{m=0}^{\infty} u_{2m} x^m$$

and

$$F(x) = \sum_{m=0}^{\infty} f_{2m} x^m .$$

Theorem 12.2 says that

$$U(x) = 1 + U(x)F(x) . \tag{12.1}$$

(The presence of the 1 on the right-hand side is due to the fact that u_0 is defined to be 1, but Theorem 12.2 only holds for $m \geq 1$.) We note that both generating functions certainly converge on the interval $(-1, 1)$, since all of the coefficients are at most 1 in absolute value. Thus, we can solve the above equation for $F(x)$, obtaining

$$F(x) = \frac{U(x) - 1}{U(x)} .$$

Now, if we can find a closed-form expression for the function $U(x)$, we will also have a closed-form expression for $F(x)$. From Theorem 12.1, we have

$$U(x) = \sum_{m=0}^{\infty} \binom{2m}{m} 2^{-2m} x^m .$$

In Wilf,[1] we find that

$$\frac{1}{\sqrt{1-4x}} = \sum_{m=0}^{\infty} \binom{2m}{m} x^m .$$

The reader is asked to prove this statement in Exercise 1. If we replace x by $x/4$ in the last equation, we see that

$$U(x) = \frac{1}{\sqrt{1-x}} .$$

---

[1] H. S. Wilf, *Generatingfunctionology*, (Boston: Academic Press, 1990), p. 50.

12.1. RANDOM WALKS IN EUCLIDEAN SPACE

Therefore, we have

$$\begin{align} F(x) &= \frac{U(x) - 1}{U(x)} \\ &= \frac{(1-x)^{-1/2} - 1}{(1-x)^{-1/2}} \\ &= 1 - (1-x)^{1/2} \ . \end{align}$$

Although it is possible to compute the value of f_{2m} using the Binomial Theorem, it is easier to note that $F'(x) = U(x)/2$, so that the coefficients f_{2m} can be found by integrating the series for $U(x)$. We obtain, for $m \geq 1$,

$$\begin{align} f_{2m} &= \frac{u_{2m-2}}{2m} \\ &= \frac{\binom{2m-2}{m-1}}{m 2^{2m-1}} \\ &= \frac{\binom{2m}{m}}{(2m-1)2^{2m}} \\ &= \frac{u_{2m}}{2m-1} \ , \end{align}$$

since

$$\binom{2m-2}{m-1} = \frac{m}{2(2m-1)}\binom{2m}{m} \ .$$

This completes the proof of the theorem. $\square$

Probability of Eventual Return

In the symmetric random walk process in $\mathbf{R}^m$, what is the probability that the particle eventually returns to the origin? We first examine this question in the case that $m = 1$, and then we consider the general case. The results in the next two examples are due to Pólya.[2]

Example 12.1 (Eventual Return in $\mathbf{R}^1$) One has to approach the idea of eventual return with some care, since the sample space seems to be the set of all walks of infinite length, and this set is non-denumerable. To avoid difficulties, we will define w_n to be the probability that a first return has occurred no later than time n. Thus, w_n concerns the sample space of all walks of length n, which is a finite set. In terms of the w_n's, it is reasonable to define the probability that the particle eventually returns to the origin to be

$$w_* = \lim_{n \to \infty} w_n \ .$$

This limit clearly exists and is at most one, since the sequence $\{w_n\}_{n=1}^{\infty}$ is an increasing sequence, and all of its terms are at most one.

---
[2]G. Pólya, "Über eine Aufgabe der Wahrscheinlichkeitsrechnung betreffend die Irrfahrt im Strassennetz," Math. Ann., vol. 84 (1921), pp. 149-160.

In terms of the f_n probabilities, we see that

$$w_{2n} = \sum_{i=1}^{n} f_{2i} \ .$$

Thus,

$$w_* = \sum_{i=1}^{\infty} f_{2i} \ .$$

In the proof of Theorem 12.3, the generating function

$$F(x) = \sum_{m=0}^{\infty} f_{2m} x^m$$

was introduced. There it was noted that this series converges for $x \in (-1, 1)$. In fact, it is possible to show that this series also converges for $x = \pm 1$ by using Exercise 4, together with the fact that

$$f_{2m} = \frac{u_{2m}}{2m-1} \ .$$

(This fact was proved in the proof of Theorem 12.3.) Since we also know that

$$F(x) = 1 - (1-x)^{1/2} \ ,$$

we see that

$$w_* = F(1) = 1 \ .$$

Thus, with probability one, the particle returns to the origin.

An alternative proof of the fact that $w_* = 1$ can be obtained by using the results in Exercise 2. □

Example 12.2 (Eventual Return in $\mathbf{R}^m$) We now turn our attention to the case that the random walk takes place in more than one dimension. We define $f_{2n}^{(m)}$ to be the probability that the first return to the origin in $\mathbf{R}^m$ occurs at time $2n$. The quantity $u_{2n}^{(m)}$ is defined in a similar manner. Thus, $f_{2n}^{(1)}$ and $u_{2n}^{(1)}$ equal f_{2n} and u_{2n}, which were defined earlier. If, in addition, we define $u_0^{(m)} = 1$ and $f_0^{(m)} = 0$, then one can mimic the proof of Theorem 12.2, and show that for all $m \geq 1$,

$$u_{2n}^{(m)} = f_0^{(m)} u_{2n}^{(m)} + f_2^{(m)} u_{2n-2}^{(m)} + \cdots + f_{2n}^{(m)} u_0^{(m)} \ . \tag{12.2}$$

We continue to generalize previous work by defining

$$U^{(m)}(x) = \sum_{n=0}^{\infty} u_{2n}^{(m)} x^n$$

and

$$F^{(m)}(x) = \sum_{n=0}^{\infty} f_{2n}^{(m)} x^n \ .$$

12.1. RANDOM WALKS IN EUCLIDEAN SPACE

Then, by using Equation 12.2, we see that

$$U^{(m)}(x) = 1 + U^{(m)}(x)F^{(m)}(x) ,$$

as before. These functions will always converge in the interval $(-1, 1)$, since all of their coefficients are at most one in magnitude. In fact, since

$$w_*^{(m)} = \sum_{n=0}^{\infty} f_{2n}^{(m)} \leq 1$$

for all m, the series for $F^{(m)}(x)$ converges at $x = 1$ as well, and $F^{(m)}(x)$ is left-continuous at $x = 1$, i.e.,

$$\lim_{x \uparrow 1} F^{(m)}(x) = F^{(m)}(1) .$$

Thus, we have

$$w_*^{(m)} = \lim_{x \uparrow 1} F^{(m)}(x) = \lim_{x \uparrow 1} \frac{U^{(m)}(x) - 1}{U^{(m)}(x)} , \qquad (12.3)$$

so to determine $w_*^{(m)}$, it suffices to determine

$$\lim_{x \uparrow 1} U^{(m)}(x) .$$

We let $u^{(m)}$ denote this limit.

We claim that

$$u^{(m)} = \sum_{n=0}^{\infty} u_{2n}^{(m)} .$$

(This claim is reasonable; it says that to find out what happens to the function $U^{(m)}(x)$ at $x = 1$, just let $x = 1$ in the power series for $U^{(m)}(x)$.) To prove the claim, we note that the coefficients $u_{2n}^{(m)}$ are non-negative, so $U^{(m)}(x)$ increases monotonically on the interval $[0, 1)$. Thus, for each K, we have

$$\sum_{n=0}^{K} u_{2n}^{(m)} \leq \lim_{x \uparrow 1} U^{(m)}(x) = u^{(m)} \leq \sum_{n=0}^{\infty} u_{2n}^{(m)} .$$

By letting $K \to \infty$, we see that

$$u^{(m)} = \sum_{2n}^{\infty} u_{2n}^{(m)} .$$

This establishes the claim.

From Equation 12.3, we see that if $u^{(m)} < \infty$, then the probability of an eventual return is

$$\frac{u^{(m)} - 1}{u^{(m)}} ,$$

while if $u^{(m)} = \infty$, then the probability of eventual return is 1.

To complete the example, we must estimate the sum

$$\sum_{n=0}^{\infty} u_{2n}^{(m)} .$$

In Exercise 12, the reader is asked to show that

$$u_{2n}^{(2)} = \frac{1}{4^{2n}} \binom{2n}{n}^2.$$

Using Stirling's Formula, it is easy to show that (see Exercise 13)

$$\binom{2n}{n} \sim \frac{2^{2n}}{\sqrt{\pi n}},$$

so

$$u_{2n}^{(2)} \sim \frac{1}{\pi n}.$$

From this it follows easily that

$$\sum_{n=0}^{\infty} u_{2n}^{(2)}$$

diverges, so $w_*^{(2)} = 1$, i.e., in $\mathbf{R}^2$, the probability of an eventual return is 1.

When $m = 3$, Exercise 12 shows that

$$u_{2n}^{(3)} = \frac{1}{2^{2n}} \binom{2n}{n} \sum_{j,k} \left(\frac{1}{3^n} \frac{n!}{j!k!(n-j-k)!} \right)^2.$$

Let M denote the largest value of

$$\frac{n!}{j!k!(n-j-k)!},$$

over all non-negative values of j and k with $j + k \le n$. It is easy, using Stirling's Formula, to show that

$$M \sim \frac{c}{n},$$

for some constant c. Thus, we have

$$u_{2n}^{(3)} \le \frac{1}{2^{2n}} \binom{2n}{n} \sum_{j,k} \left(\frac{M}{3^n} \frac{n!}{j!k!(n-j-k)!} \right).$$

Using Exercise 14, one can show that the right-hand expression is at most

$$\frac{c'}{n^{3/2}},$$

where c' is a constant. Thus,

$$\sum_{n=0}^{\infty} u_{2n}^{(3)}$$

converges, so $w_*^{(3)}$ is strictly less than one. This means that in $\mathbf{R}^3$, the probability of an eventual return to the origin is strictly less than one (in fact, it is approximately .65).

One may summarize these results by stating that one should not get drunk in more than two dimensions. $\square$

Expected Number of Equalizations

We now give another example of the use of generating functions to find a general formula for terms in a sequence, where the sequence is related by recursion relations to other sequences. Exercise 9 gives still another example.

Example 12.3 (Expected Number of Equalizations) In this example, we will derive a formula for the expected number of equalizations in a random walk of length $2m$. As in the proof of Theorem 12.3, the method has four main parts. First, a recursion is found which relates the mth term in the unknown sequence to earlier terms in the same sequence and to terms in other (known) sequences. An example of such a recursion is given in Theorem 12.2. Second, the recursion is used to derive a functional equation involving the generating functions of the unknown sequence and one or more known sequences. Equation 12.1 is an example of such a functional equation. Third, the functional equation is solved for the unknown generating function. Last, using a device such as the Binomial Theorem, integration, or differentiation, a formula for the mth coefficient of the unknown generating function is found.

We begin by defining g_{2m} to be the number of equalizations among all of the random walks of length $2m$. (For each random walk, we disregard the equalization at time 0.) We define $g_0 = 0$. Since the number of walks of length $2m$ equals 2^{2m}, the expected number of equalizations among all such random walks is $g_{2m}/2^{2m}$. Next, we define the generating function $G(x)$:

$$G(x) = \sum_{k=0}^{\infty} g_{2k} x^k \ .$$

Now we need to find a recursion which relates the sequence $\{g_{2k}\}$ to one or both of the known sequences $\{f_{2k}\}$ and $\{u_{2k}\}$. We consider m to be a fixed positive integer, and consider the set of all paths of length $2m$ as the disjoint union

$$E_2 \cup E_4 \cup \cdots \cup E_{2m} \cup H \ ,$$

where E_{2k} is the set of all paths of length $2m$ with first equalization at time $2k$, and H is the set of all paths of length $2m$ with no equalization. It is easy to show (see Exercise 3) that

$$|E_{2k}| = f_{2k} 2^{2m} \ .$$

We claim that the number of equalizations among all paths belonging to the set E_{2k} is equal to

$$|E_{2k}| + 2^{2k} f_{2k} g_{2m-2k} \ . \tag{12.4}$$

Each path in E_{2k} has one equalization at time $2k$, so the total number of such equalizations is just $|E_{2k}|$. This is the first summand in expression Equation 12.4. There are $2^{2k} f_{2k}$ different initial segments of length $2k$ among the paths in E_{2k}. Each of these initial segments can be augmented to a path of length $2m$ in 2^{2m-2k} ways, by adjoining all possible paths of length $2m-2k$. The number of equalizations obtained by adjoining all of these paths to any one initial segment is g_{2m-2k}, by definition. This gives the second summand in Equation 12.4. Since k can range from 1 to m, we obtain the recursion

$$g_{2m} = \sum_{k=1}^{m} \left(|E_{2k}| + 2^{2k} f_{2k} g_{2m-2k} \right) . \quad (12.5)$$

The second summand in the typical term above should remind the reader of a convolution. In fact, if we multiply the generating function $G(x)$ by the generating function

$$F(4x) = \sum_{k=0}^{\infty} 2^{2k} f_{2k} x^k ,$$

the coefficient of x^m equals

$$\sum_{k=0}^{m} 2^{2k} f_{2k} g_{2m-2k} .$$

Thus, the product $G(x)F(4x)$ is part of the functional equation that we are seeking. The first summand in the typical term in Equation 12.5 gives rise to the sum

$$2^{2m} \sum_{k=1}^{m} f_{2k} .$$

From Exercise 2, we see that this sum is just $(1 - u_{2m})2^{2m}$. Thus, we need to create a generating function whose mth coefficient is this term; this generating function is

$$\sum_{m=0}^{\infty} (1 - u_{2m}) 2^{2m} x^m ,$$

or

$$\sum_{m=0}^{\infty} 2^{2m} x^m + \sum_{m=0}^{\infty} u_{2m} x^m .$$

The first sum is just $(1 - 4x)^{-1}$, and the second sum is $U(4x)$. So, the functional equation which we have been seeking is

$$G(x) = F(4x)G(x) + \frac{1}{1 - 4x} - U(4x) .$$

If we solve this recursion for $G(x)$, and simplify, we obtain

$$G(x) = \frac{1}{(1 - 4x)^{3/2}} - \frac{1}{(1 - 4x)} . \quad (12.6)$$

We now need to find a formula for the coefficient of x^m. The first summand in Equation 12.6 is $(1/2)U'(4x)$, so the coefficient of x^m in this function is

$$u_{2m+2} 2^{2m+1} (m + 1) .$$

The second summand in Equation 12.6 is the sum of a geometric series with common ratio $4x$, so the coefficient of x^m is 2^{2m}. Thus, we obtain

12.1. RANDOM WALKS IN EUCLIDEAN SPACE

$$\begin{aligned} g_{2m} &= u_{2m+2} 2^{2m+1}(m+1) - 2^{2m} \\ &= \frac{1}{2}\binom{2m+2}{m+1}(m+1) - 2^{2m} \ . \end{aligned}$$

We recall that the quotient $g_{2m}/2^{2m}$ is the expected number of equalizations among all paths of length $2m$. Using Exercise 4, it is easy to show that

$$\frac{g_{2m}}{2^{2m}} \sim \sqrt{\frac{2}{\pi}}\sqrt{2m} \ .$$

In particular, this means that the average number of equalizations among all paths of length $4m$ is not twice the average number of equalizations among all paths of length $2m$. In order for the average number of equalizations to double, one must quadruple the lengths of the random walks. □

It is interesting to note that if we define

$$M_n = \max_{0 \le k \le n} S_k \ ,$$

then we have

$$E(M_n) \sim \sqrt{\frac{2}{\pi}}\sqrt{n} \ .$$

This means that the expected number of equalizations and the expected maximum value for random walks of length n are asymptotically equal as $n \to \infty$. (In fact, it can be shown that the two expected values differ by at most $1/2$ for all positive integers n. See Exercise 9.)

Exercises

1 Using the Binomial Theorem, show that

$$\frac{1}{\sqrt{1-4x}} = \sum_{m=0}^{\infty} \binom{2m}{m} x^m \ .$$

What is the interval of convergence of this power series?

2 (a) Show that for $m \ge 1$,

$$f_{2m} = u_{2m-2} - u_{2m} \ .$$

(b) Using part (a), find a closed-form expression for the sum

$$f_2 + f_4 + \cdots + f_{2m} \ .$$

(c) Using part (a), show that

$$\sum_{m=1}^{\infty} f_{2m} = 1 \ .$$

(One can also obtain this statement from the fact that

$$F(x) = 1 - (1-x)^{1/2} \ .)$$

(d) Using Exercise 2, show that the probability of no equalization in the first $2m$ outcomes equals the probability of an equalization at time $2m$.

3 Using the notation of Example 12.3, show that
$$|E_{2k}| = f_{2k} 2^{2m} .$$

4 Using Stirling's Formula, show that
$$u_{2m} \sim \frac{1}{\sqrt{\pi m}} .$$

5 A *lead change* in a random walk occurs at time $2k$ if S_{2k-1} and S_{2k+1} are of opposite sign.

(a) Give a rigorous argument which proves that among all walks of length $2m$ that have an equalization at time $2k$, exactly half have a lead change at time $2k$.

(b) Deduce that the total number of lead changes among all walks of length $2m$ equals
$$\frac{1}{2}(g_{2m} - u_{2m}) .$$

(c) Find an asymptotic expression for the average number of lead changes in a random walk of length $2m$.

6 (a) Show that the probability that a random walk of length $2m$ has a last return to the origin at time $2k$, where $0 \le k \le m$, equals
$$\frac{\binom{2k}{k}\binom{2m-2k}{m-k}}{2^{2m}} = u_{2k} u_{2m-2k} .$$
(The case $k = 0$ consists of all paths that do not return to the origin at any positive time.) *Hint*: A path whose last return to the origin occurs at time $2k$ consists of two paths glued together, one path of which is of length $2k$ and which begins and ends at the origin, and the other path of which is of length $2m - 2k$ and which begins at the origin but never returns to the origin. Both types of paths can be counted using quantities which appear in this section.

(b) Using part (a), show that the probability that a walk of length $2m$ has no equalization in the last m outcomes is equal to $1/2$, regardless of the value of m. *Hint*: The answer to part a) is symmetric in k and $m - k$.

7 Show that the probability of no equalization in a walk of length $2m$ equals u_{2m}.

*8 Show that
$$P(S_1 \ge 0,\ S_2 \ge 0,\ \ldots,\ S_{2m} \ge 0) = u_{2m} .$$

12.1. RANDOM WALKS IN EUCLIDEAN SPACE

Hint: First explain why

$$P(S_1 > 0, S_2 > 0, \ldots, S_{2m} > 0)$$
$$= \frac{1}{2} P(S_1 \neq 0, S_2 \neq 0, \ldots, S_{2m} \neq 0) .$$

Then use Exercise 7, together with the observation that if no equalization occurs in the first $2m$ outcomes, then the path goes through the point $(1, 1)$ and remains on or above the horizontal line $x = 1$.

*9 In Feller,[3] one finds the following theorem: Let M_n be the random variable which gives the maximum value of S_k, for $1 \leq k \leq n$. Define

$$p_{n,r} = \binom{n}{\frac{n+r}{2}} 2^{-n} .$$

If $r \geq 0$, then

$$P(M_n = r) = \begin{cases} p_{n,r}, & \text{if } r \equiv n \pmod{2}, \\ 1, & \text{if } p \geq q, \end{cases}$$

$$P(M_n = r) = \begin{cases} p_{n,r}, & \text{if } r \equiv n \pmod{2}, \\ p_{n,r+1}, & \text{if } r \not\equiv n \pmod{2}. \end{cases}$$

(a) Using this theorem, show that

$$E(M_{2m}) = \frac{1}{2^{2m}} \sum_{k=1}^{m} (4k-1) \binom{2m}{m+k} ,$$

and if $n = 2m + 1$, then

$$E(M_{2m+1}) = \frac{1}{2^{2m+1}} \sum_{k=0}^{m} (4k+1) \binom{2m+1}{m+k+1} .$$

(b) For $m \geq 1$, define

$$r_m = \sum_{k=1}^{m} k \binom{2m}{m+k}$$

and

$$s_m = \sum_{k=1}^{m} k \binom{2m+1}{m+k+1} .$$

By using the identity

$$\binom{n}{k} = \binom{n-1}{k-1} + \binom{n-1}{k} ,$$

show that

$$s_m = 2r_m - \frac{1}{2}\left(2^{2m} - \binom{2m}{m}\right)$$

---

[3]W. Feller, *Introduction to Probability Theory and its Applications*, vol. I, 3rd ed. (New York: John Wiley & Sons, 1968).

and
$$r_m = 2s_{m-1} + \frac{1}{2}2^{2m-1},$$
if $m \geq 2$.

(c) Define the generating functions
$$R(x) = \sum_{k=1}^{\infty} r_k x^k$$
and
$$S(x) = \sum_{k=1}^{\infty} s_k x^k.$$
Show that
$$S(x) = 2R(x) - \frac{1}{2}\left(\frac{1}{1-4x}\right) + \frac{1}{2}\left(\sqrt{1-4x}\right)$$
and
$$R(x) = 2xS(x) + x\left(\frac{1}{1-4x}\right).$$

(d) Show that
$$R(x) = \frac{x}{(1-4x)^{3/2}},$$
and
$$S(x) = \frac{1}{2}\left(\frac{1}{(1-4x)^{3/2}}\right) - \frac{1}{2}\left(\frac{1}{1-4x}\right).$$

(e) Show that
$$r_m = m\binom{2m-1}{m-1},$$
and
$$s_m = \frac{1}{2}(m+1)\binom{2m+1}{m} - \frac{1}{2}(2^{2m}).$$

(f) Show that
$$E(M_{2m}) = \frac{m}{2^{2m-1}}\binom{2m}{m} + \frac{1}{2^{2m+1}}\binom{2m}{m} - \frac{1}{2},$$
and
$$E(M_{2m+1}) = \frac{m+1}{2^{2m+1}}\binom{2m+2}{m+1} - \frac{1}{2}.$$

The reader should compare these formulas with the expression for $g_{2m}/2^{(2m)}$ in Example 12.3.

12.1. RANDOM WALKS IN EUCLIDEAN SPACE

*10 (from K. Levasseur[4]) A parent and his child play the following game. A deck of $2n$ cards, n red and n black, is shuffled. The cards are turned up one at a time. Before each card is turned up, the parent and the child guess whether it will be red or black. Whoever makes more correct guesses wins the game. The child is assumed to guess each color with the same probability, so she will have a score of n, on average. The parent keeps track of how many cards of each color have already been turned up. If more black cards, say, than red cards remain in the deck, then the parent will guess black, while if an equal number of each color remain, then the parent guesses each color with probability 1/2. What is the expected number of correct guesses that will be made by the parent? *Hint*: Each of the $\binom{2n}{n}$ possible orderings of red and black cards corresponds to a random walk of length $2n$ that returns to the origin at time $2n$. Show that between each pair of successive equalizations, the parent will be right exactly once more than he will be wrong. Explain why this means that the average number of correct guesses by the parent is greater than n by exactly one-half the average number of equalizations. Now define the random variable X_i to be 1 if there is an equalization at time $2i$, and 0 otherwise. Then, among all relevant paths, we have

$$E(X_i) = P(X_i = 1) = \frac{\binom{2n-2i}{n-i}\binom{2i}{i}}{\binom{2n}{n}} \ .$$

Thus, the expected number of equalizations equals

$$E\left(\sum_{i=1}^n X_i\right) = \frac{1}{\binom{2n}{n}} \sum_{i=1}^n \binom{2n-2i}{n-i}\binom{2i}{i} \ .$$

One can now use generating functions to find the value of the sum.

It should be noted that in a game such as this, a more interesting question than the one asked above is what is the probability that the parent wins the game? For this game, this question was answered by D. Zagier.[5] He showed that the probability of winning is asymptotic (for large n) to the quantity

$$\frac{1}{2} + \frac{1}{2\sqrt{2}} \ .$$

*11 Prove that

$$u_{2n}^{(2)} = \frac{1}{4^{2n}} \sum_{k=0}^n \frac{(2n)!}{k!k!(n-k)!(n-k)!} \ ,$$

and

$$u_{2n}^{(3)} = \frac{1}{6^{2n}} \sum_{j,k} \frac{(2n)!}{j!j!k!k!(n-j-k)!(n-j-k)!} \ ,$$

---

[4] K. Levasseur, "How to Beat Your Kids at Their Own Game," *Mathematics Magazine* vol. 61, no. 5 (December, 1988), pp. 301-305.

[5] D. Zagier, "How Often Should You Beat Your Kids?" *Mathematics Magazine* vol. 63, no. 2 (April 1990), pp. 89-92.

where the last sum extends over all non-negative j and k with $j+k \le n$. Also show that this last expression may be rewritten as

$$\frac{1}{2^{2n}}\binom{2n}{n}\sum_{j,k}\left(\frac{1}{3^n}\frac{n!}{j!k!(n-j-k)!}\right)^2 .$$

*12 Prove that if $n \ge 0$, then

$$\sum_{k=0}^{n}\binom{n}{k}^2 = \binom{2n}{n} .$$

Hint: Write the sum as

$$\sum_{k=0}^{n}\binom{n}{k}\binom{n}{n-k}$$

and explain why this is a coefficient in the product

$$(1+x)^n(1+x)^n .$$

Use this, together with Exercise 11, to show that

$$u_{2n}^{(2)} = \frac{1}{4^{2n}}\binom{2n}{n}\sum_{k=0}^{n}\binom{n}{k}^2 = \frac{1}{4^{2n}}\binom{2n}{n}^2 .$$

*13 Using Stirling's Formula, prove that

$$\binom{2n}{n} \sim \frac{2^{2n}}{\sqrt{\pi n}} .$$

*14 Prove that

$$\sum_{j,k}\left(\frac{1}{3^n}\frac{n!}{j!k!(n-j-k)!}\right) = 1 ,$$

where the sum extends over all non-negative j and k such that $j+k \le n$. *Hint*: Count how many ways one can place n labelled balls in 3 labelled urns.

*15 Using the result proved for the random walk in $\mathbf{R}^3$ in Example 12.2, explain why the probability of an eventual return in $\mathbf{R}^n$ is strictly less than one, for all $n \ge 3$. *Hint*: Consider a random walk in $\mathbf{R}^n$ and disregard all but the first three coordinates of the particle's position.

12.2 Gambler's Ruin

In the last section, the simplest kind of symmetric random walk in $\mathbf{R}^1$ was studied. In this section, we remove the assumption that the random walk is symmetric. Instead, we assume that p and q are non-negative real numbers with $p+q=1$, and that the common distribution function of the jumps of the random walk is

$$f_X(x) = \begin{cases} p, & \text{if } x=1, \\ q, & \text{if } x=-1. \end{cases}$$

12.2. GAMBLER'S RUIN

One can imagine the random walk as representing a sequence of tosses of a weighted coin, with a head appearing with probability p and a tail appearing with probability q. An alternative formulation of this situation is that of a gambler playing a sequence of games against an adversary (sometimes thought of as another person, sometimes called "the house") where, in each game, the gambler has probability p of winning.

The Gambler's Ruin Problem

The above formulation of this type of random walk leads to a problem known as the Gambler's Ruin problem. This problem was introduced in Exercise 23, but we will give the description of the problem again. A gambler starts with a "stake" of size s. She plays until her capital reaches the value M or the value 0. In the language of Markov chains, these two values correspond to absorbing states. We are interested in studying the probability of occurrence of each of these two outcomes.

One can also assume that the gambler is playing against an "infinitely rich" adversary. In this case, we would say that there is only one absorbing state, namely when the gambler's stake is 0. Under this assumption, one can ask for the probability that the gambler is eventually ruined.

We begin by defining q_k to be the probability that the gambler's stake reaches 0, i.e., she is ruined, before it reaches M, given that the initial stake is k. We note that $q_0 = 1$ and $q_M = 0$. The fundamental relationship among the q_k's is the following:

$$q_k = pq_{k+1} + qq_{k-1} \,,$$

where $1 \le k \le M - 1$. This holds because if her stake equals k, and she plays one game, then her stake becomes $k + 1$ with probability p and $k - 1$ with probability q. In the first case, the probability of eventual ruin is q_{k+1} and in the second case, it is q_{k-1}. We note that since $p + q = 1$, we can write the above equation as

$$p(q_{k+1} - q_k) = q(q_k - q_{k-1}) \,,$$

or

$$q_{k+1} - q_k = \frac{q}{p}(q_k - q_{k-1}) \,.$$

From this equation, it is easy to see that

$$q_{k+1} - q_k = \left(\frac{q}{p}\right)^k (q_1 - q_0) \,. \qquad (12.7)$$

We now use telescoping sums to obtain an equation in which the only unknown is q_1:

$$\begin{aligned}
-1 &= q_M - q_0 \\
&= \sum_{k=0}^{M-1} (q_{k+1} - q_k) \,,
\end{aligned}$$

so

$$-1 = \sum_{k=0}^{M-1}\left(\frac{q}{p}\right)^k (q_1 - q_0)$$
$$= (q_1 - q_0) \sum_{k=0}^{M-1}\left(\frac{q}{p}\right)^k .$$

If $p \neq q$, then the above expression equals

$$(q_1 - q_0)\frac{(q/p)^M - 1}{(q/p) - 1} ,$$

while if $p = q = 1/2$, then we obtain the equation

$$-1 = (q_1 - q_0)M .$$

For the moment we shall assume that $p \neq q$. Then we have

$$q_1 - q_0 = -\frac{(q/p) - 1}{(q/p)^M - 1} .$$

Now, for any z with $1 \leq z \leq M$, we have

$$\begin{aligned}
q_z - q_0 &= \sum_{k=0}^{z-1}(q_{k+1} - q_k) \\
&= (q_1 - q_0)\sum_{k=0}^{z-1}\left(\frac{q}{p}\right)^k \\
&= -(q_1 - q_0)\frac{(q/p)^z - 1}{(q/p) - 1} \\
&= -\frac{(q/p)^z - 1}{(q/p)^M - 1} .
\end{aligned}$$

Therefore,

$$\begin{aligned}
q_z &= 1 - \frac{(q/p)^z - 1}{(q/p)^M - 1} \\
&= \frac{(q/p)^M - (q/p)^z}{(q/p)^M - 1} .
\end{aligned}$$

Finally, if $p = q = 1/2$, it is easy to show that (see Exercise 10)

$$q_z = \frac{M - z}{M} .$$

We note that both of these formulas hold if $z = 0$.

We define, for $0 \leq z \leq M$, the quantity p_z to be the probability that the gambler's stake reaches M without ever having reached 0. Since the game might

12.2. GAMBLER'S RUIN

continue indefinitely, it is not obvious that $p_z + q_z = 1$ for all z. However, one can use the same method as above to show that if $p \neq q$, then

$$q_z = \frac{(q/p)^z - 1}{(q/p)^M - 1} ,$$

and if $p = q = 1/2$, then

$$q_z = \frac{z}{M} .$$

Thus, for all z, it is the case that $p_z + q_z = 1$, so the game ends with probability 1.

Infinitely Rich Adversaries

We now turn to the problem of finding the probability of eventual ruin if the gambler is playing against an infinitely rich adversary. This probability can be obtained by letting M go to ∞ in the expression for q_z calculated above. If $q < p$, then the expression approaches $(q/p)^z$, and if $q > p$, the expression approaches 1. In the case $p = q = 1/2$, we recall that $q_z = 1 - z/M$. Thus, if $M \to \infty$, we see that the probability of eventual ruin tends to 1.

Historical Remarks

In 1711, De Moivre, in his book *De Mesura Sortis*, gave an ingenious derivation of the probability of ruin. The following description of his argument is taken from David.[6] The notation used is as follows: We imagine that there are two players, A and B, and the probabilities that they win a game are p and q, respectively. The players start with a and b counters, respectively.

> Imagine that each player starts with his counters before him in a pile, and that nominal values are assigned to the counters in the following manner. A's bottom counter is given the nominal value q/p; the next is given the nominal value $(q/p)^2$, and so on until his top counter which has the nominal value $(q/p)^a$. B's top counter is valued $(q/p)^{a+1}$, and so on downwards until his bottom counter which is valued $(q/p)^{a+b}$. After each game the loser's top counter is transferred to the top of the winner's pile, and it is always the top counter which is staked for the next game. Then *in terms of the nominal values* B's stake is always q/p times A's, so that at every game each player's nominal expectation is nil. This remains true throughout the play; therefore A's chance of winning all B's counters, multiplied by his nominal gain if he does so, must equal B's chance multiplied by B's nominal gain. Thus,

$$P_a \left(\left(\frac{q}{p}\right)^{a+1} + \cdots + \left(\frac{q}{p}\right)^{a+b} \right) = P_b \left(\left(\frac{q}{p}\right) + \cdots + \left(\frac{q}{p}\right)^a \right) . \quad (12.8)$$

---

[6]F. N. David, *Games, Gods and Gambling* (London: Griffin, 1962).

Using this equation, together with the fact that

$$P_a + P_b = 1,$$

it can easily be shown that

$$P_a = \frac{(q/p)^a - 1}{(q/p)^{a+b} - 1},$$

if $p \neq q$, and

$$P_a = \frac{a}{a+b},$$

if $p = q = 1/2$.

In terms of modern probability theory, de Moivre is changing the values of the counters to make an unfair game into a fair game, which is called a martingale. With the new values, the expected fortune of player A (that is, the sum of the nominal values of his counters) after each play equals his fortune before the play (and similarly for player B). (For a simpler martingale argument, see Exercise 9.) De Moivre then uses the fact that when the game ends, it is still fair, thus Equation 12.8 must be true. This fact requires proof, and is one of the central theorems in the area of martingale theory.

Exercises

1 In the gambler's ruin problem, assume that the gambler initial stake is 1 dollar, and assume that her probability of success on any one game is p. Let T be the number of games until 0 is reached (the gambler is ruined). Show that the generating function for T is

$$h(z) = \frac{1 - \sqrt{1 - 4pqz^2}}{2pz},$$

and that

$$h(1) = \begin{cases} q/p, & \text{if } q \leq p, \\ 1, & \text{if } q \geq p, \end{cases}$$

and

$$h'(1) = \begin{cases} 1/(q-p), & \text{if } q > p, \\ \infty, & \text{if } q = p. \end{cases}$$

Interpret your results in terms of the time T to reach 0. (See also Example 10.7.)

2 Show that the Taylor series expansion for $\sqrt{1-x}$ is

$$\sqrt{1-x} = \sum_{n=0}^{\infty} \binom{1/2}{n} x^n,$$

where the binomial coefficient $\binom{1/2}{n}$ is

$$\binom{1/2}{n} = \frac{(1/2)(1/2 - 1) \cdots (1/2 - n + 1)}{n!}.$$

12.2. GAMBLER'S RUIN

Using this and the result of Exercise 1, show that the probability that the gambler is ruined on the nth step is

$$p_T(n) = \begin{cases} \frac{(-1)^{k-1}}{2p}\binom{1/2}{k}(4pq)^k, & \text{if } n = 2k-1, \\ 0, & \text{if } n = 2k. \end{cases}$$

3 For the gambler's ruin problem, assume that the gambler starts with k dollars. Let T_k be the time to reach 0 for the first time.

(a) Show that the generating function $h_k(t)$ for T_k is the kth power of the generating function for the time T to ruin starting at 1. *Hint*: Let $T_k = U_1 + U_2 + \cdots + U_k$, where U_j is the time for the walk starting at j to reach $j-1$ for the first time.

(b) Find $h_k(1)$ and $h'_k(1)$ and interpret your results.

4 (The next three problems come from Feller.[7]) As in the text, assume that M is a fixed positive integer.

(a) Show that if a gambler starts with an stake of 0 (and is allowed to have a negative amount of money), then the probability that her stake reaches the value of M before it returns to 0 equals $p(1-q_1)$.

(b) Show that if the gambler starts with a stake of M then the probability that her stake reaches 0 before it returns to M equals qq_{M-1}.

5 Suppose that a gambler starts with a stake of 0 dollars.

(a) Show that the probability that her stake never reaches M before returning to 0 equals $1 - p(1-q_1)$.

(b) Show that the probability that her stake reaches the value M exactly k times before returning to 0 equals $p(1-q_1)(1-qq_{M-1})^{k-1}(qq_{M-1})$. *Hint*: Use Exercise 4.

6 In the text, it was shown that if $q < p$, there is a positive probability that a gambler, starting with a stake of 0 dollars, will never return to the origin. Thus, we will now assume that $q \geq p$. Using Exercise 5, show that if a gambler starts with a stake of 0 dollars, then the expected number of times her stake equals M before returning to 0 equals $(p/q)^M$, if $q > p$ and 1, if $q = p$. (We quote from Feller: "The truly amazing implications of this result appear best in the language of fair games. A perfect coin is tossed until the first equalization of the accumulated numbers of heads and tails. The gambler receives one penny for every time that the accumulated number of heads exceeds the accumulated number of tails by m. The 'fair entrance fee' equals 1 independent of m.")

---

[7]W. Feller, op. cit., pg. 367.

7 In the game in Exercise 6, let $p = q = 1/2$ and $M = 10$. What is the probability that the gambler's stake equals M at least 20 times before it returns to 0?

8 Write a computer program which simulates the game in Exercise 6 for the case $p = q = 1/2$, and $M = 10$.

9 In de Moivre's description of the game, we can modify the definition of player A's fortune in such a way that the game is still a martingale (and the calculations are simpler). We do this by assigning nominal values to the counters in the same way as de Moivre, but each player's current fortune is defined to be just the value of the counter which is being wagered on the next game. So, if player A has a counters, then his current fortune is $(q/p)^a$ (we stipulate this to be true even if $a = 0$). Show that under this definition, player A's expected fortune after one play equals his fortune before the play, if $p \neq q$. Then, as de Moivre does, write an equation which expresses the fact that player A's expected final fortune equals his initial fortune. Use this equation to find the probability of ruin of player A.

10 Assume in the gambler's ruin problem that $p = q = 1/2$.

(a) Using Equation 12.7, together with the facts that $q_0 = 1$ and $q_M = 0$, show that for $0 \leq z \leq M$,

$$q_z = \frac{M - z}{M}.$$

(b) In Equation 12.8, let $p \to 1/2$ (and since $q = 1 - p$, $q \to 1/2$ as well). Show that in the limit,

$$q_z = \frac{M - z}{M}.$$

Hint: Replace q by $1 - p$, and use L'Hopital's rule.

11 In American casinos, the roulette wheels have the integers between 1 and 36, together with 0 and 00. Half of the non-zero numbers are red, the other half are black, and 0 and 00 are green. A common bet in this game is to bet a dollar on red. If a red number comes up, the bettor gets her dollar back, and also gets another dollar. If a black or green number comes up, she loses her dollar.

(a) Suppose that someone starts with 40 dollars, and continues to bet on red until either her fortune reaches 50 or 0. Find the probability that her fortune reaches 50 dollars.

(b) How much money would she have to start with, in order for her to have a 95% chance of winning 10 dollars before going broke?

(c) A casino owner was once heard to remark that "If we took 0 and 00 off of the roulette wheel, we would still make lots of money, because people would continue to come in and play until they lost all of their money." Do you think that such a casino would stay in business?

12.3 Arc Sine Laws

In Exercise 12.1.6, the distribution of the time of the last equalization in the symmetric random walk was determined. If we let $\alpha_{2k,2m}$ denote the probability that a random walk of length $2m$ has its last equalization at time $2k$, then we have

$$\alpha_{2k,2m} = u_{2k} u_{2m-2k} \ .$$

We shall now show how one can approximate the distribution of the α's with a simple function. We recall that

$$u_{2k} \sim \frac{1}{\sqrt{\pi k}} \ .$$

Therefore, as both k and m go to ∞, we have

$$\alpha_{2k,2m} \sim \frac{1}{\pi \sqrt{k(m-k)}} \ .$$

This last expression can be written as

$$\frac{1}{\pi m \sqrt{(k/m)(1-k/m)}} \ .$$

Thus, if we define

$$f(x) = \frac{1}{\pi \sqrt{x(1-x)}} \ ,$$

for $0 < x < 1$, then we have

$$\alpha_{2k,2m} \approx \frac{1}{m} f\left(\frac{k}{m}\right) \ .$$

The reason for the $\approx$ sign is that we no longer require that k get large. This means that we can replace the discrete $\alpha_{2k,2m}$ distribution by the continuous density $f(x)$ on the interval $[0,1]$ and obtain a good approximation. In particular, if x is a fixed real number between 0 and 1, then we have

$$\sum_{k<xm} \alpha_{2k,2m} \approx \int_0^x f(t)\,dt \ .$$

It turns out that $f(x)$ has a nice antiderivative, so we can write

$$\sum_{k<xm} \alpha_{2k,2m} \approx \frac{2}{\pi} \arcsin \sqrt{x} \ .$$

One can see from the graph of this last function that it has a minimum at $x = 1/2$ and is symmetric about that point. As noted in the exercise, this implies that half of the walks of length $2m$ have no equalizations after time m, a fact which probably would not be guessed.

It turns out that the arc sine density comes up in the answers to many other questions concerning random walks on the line. Recall that in Section 12.1, a random walk could be viewed as a polygonal line connecting $(0,0)$ with (m, S_m). Under this interpretation, we define $b_{2k,2m}$ to be the probability that a random walk of length $2m$ has exactly $2k$ of its $2m$ polygonal line segments above the t-axis.

The probability $b_{2k,2m}$ is frequently interpreted in terms of a two-player game. (The reader will recall the game Heads or Tails, in Example 1.4.) Player A is said to be in the lead at time n if the random walk is above the t-axis at that time, or if the random walk is on the t-axis at time n but above the t-axis at time $n-1$. (At time 0, neither player is in the lead.) One can ask what is the most probable number of times that player A is in the lead, in a game of length $2m$. Most people will say that the answer to this question is m. However, the following theorem says that m is the least likely number of times that player A is in the lead, and the most likely number of times in the lead is 0 or $2m$.

Theorem 12.4 If Peter and Paul play a game of Heads or Tails of length $2m$, the probability that Peter will be in the lead exactly $2k$ times is equal to

$$\alpha_{2k,2m} .$$

Proof. To prove the theorem, we need to show that

$$b_{2k,2m} = \alpha_{2k,2m} . \tag{12.9}$$

Exercise 12.1.7 shows that $b_{2m,2m} = u_{2m}$ and $b_{0,2m} = u_{2m}$, so we only need to prove that Equation 12.9 holds for $1 \leq k \leq m-1$. We can obtain a recursion involving the b's and the f's (defined in Section 12.1) by counting the number of paths of length $2m$ that have exactly $2k$ of their segments above the t-axis, where $1 \leq k \leq m-1$. To count this collection of paths, we assume that the first return occurs at time $2j$, where $1 \leq j \leq m-1$. There are two cases to consider. Either during the first $2j$ outcomes the path is above the t-axis or below the t-axis. In the first case, it must be true that the path has exactly $(2k-2j)$ line segments above the t-axis, between $t = 2j$ and $t = 2m$. In the second case, it must be true that the path has exactly $2k$ line segments above the t-axis, between $t = 2j$ and $t = 2m$.

We now count the number of paths of the various types described above. The number of paths of length $2j$ all of whose line segments lie above the t-axis and which return to the origin for the first time at time $2j$ equals $(1/2)2^{2j}f_{2j}$. This also equals the number of paths of length $2j$ all of whose line segments lie below the t-axis and which return to the origin for the first time at time $2j$. The number of paths of length $(2m-2j)$ which have exactly $(2k-2j)$ line segments above the t-axis is $b_{2k-2j,2m-2j}$. Finally, the number of paths of length $(2m-2j)$ which have exactly $2k$ line segments above the t-axis is $b_{2k,2m-2j}$. Therefore, we have

$$b_{2k,2m} = \frac{1}{2}\sum_{j=1}^{k} f_{2j}b_{2k-2j,2m-2j} + \frac{1}{2}\sum_{j=1}^{m-k} f_{2j}b_{2k,2m-2j} .$$

We now assume that Equation 12.9 is true for $m < n$. Then we have

12.3. ARC SINE LAWS

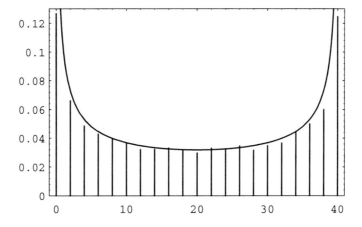

Figure 12.2: Times in the lead.

$$\begin{aligned}
b_{2k,2n} &= \frac{1}{2}\sum_{j=1}^{k} f_{2j}\alpha_{2k-2j,2m-2j} + \frac{1}{2}\sum_{j=1}^{m-k} f_{2j}\alpha_{2k,2m-2j} \\
&= \frac{1}{2}\sum_{j=1}^{k} f_{2j}u_{2k-2j}u_{2m-2k} + \frac{1}{2}\sum_{j=1}^{m-k} f_{2j}u_{2k}u_{2m-2j-2k} \\
&= \frac{1}{2}u_{2m-2k}\sum_{j=1}^{k} f_{2j}u_{2k-2j} + \frac{1}{2}u_{2k}\sum_{j=1}^{m-k} f_{2j}u_{2m-2j-2k} \\
&= \frac{1}{2}u_{2m-2k}u_{2k} + \frac{1}{2}u_{2k}u_{2m-2k} \ ,
\end{aligned}$$

where the last equality follows from Theorem 12.2. Thus, we have

$$b_{2k,2n} = \alpha_{2k,2n} \ ,$$

which completes the proof. □

We illustrate the above theorem by simulating 10,000 games of Heads or Tails, with each game consisting of 40 tosses. The distribution of the number of times that Peter is in the lead is given in Figure 12.2, together with the arc sine density.

We end this section by stating two other results in which the arc sine density appears. Proofs of these results may be found in Feller.[8]

Theorem 12.5 Let J be the random variable which, for a given random walk of length $2m$, gives the smallest subscript j such that $S_j = S_{2m}$. (Such a subscript j must be even, by parity considerations.) Let $\gamma_{2k,2m}$ be the probability that $J = 2k$. Then we have

$$\gamma_{2k,2m} = \alpha_{2k,2m} \ .$$

□

---

[8]W. Feller, op. cit., pp. 93–94.

The next theorem says that the arc sine density is applicable to a wide range of situations. A continuous distribution function $F(x)$ is said to be *symmetric* if $F(x) = 1 - F(-x)$. (If X is a continuous random variable with a symmetric distribution function, then for any real x, we have $P(X \le x) = P(X \ge -x)$.) We imagine that we have a random walk of length n in which each summand has the distribution $F(x)$, where F is continuous and symmetric. The subscript of the *first maximum* of such a walk is the unique subscript k such that

$$S_k > S_0, \ldots, S_k > S_{k-1}, S_k \ge S_{k+1}, \ldots, S_k \ge S_n \ .$$

We define the random variable K_n to be the subscript of the first maximum. We can now state the following theorem concerning the random variable K_n.

Theorem 12.6 Let F be a symmetric continuous distribution function, and let α be a fixed real number strictly between 0 and 1. Then as $n \to \infty$, we have

$$P(K_n < n\alpha) \to \frac{2}{\pi} \arcsin \sqrt{\alpha} \ .$$

$\square$

A version of this theorem that holds for a symmetric random walk can also be found in Feller.

Exercises

1 For a random walk of length $2m$, define ϵ_k to equal 1 if $S_k > 0$, or if $S_{k-1} = 1$ and $S_k = 0$. Define ϵ_k to equal -1 in all other cases. Thus, ϵ_k gives the side of the t-axis that the random walk is on during the time interval $[k-1, k]$. A "law of large numbers" for the sequence $\{\epsilon_k\}$ would say that for any $\delta > 0$, we would have

$$P\left(-\delta < \frac{\epsilon_1 + \epsilon_2 + \cdots + \epsilon_n}{n} < \delta\right) \to 1$$

as $n \to \infty$. Even though the ϵ's are not independent, the above assertion certainly appears reasonable. Using Theorem 12.4, show that if $-1 \le x \le 1$, then

$$\lim_{n \to \infty} P\left(\frac{\epsilon_1 + \epsilon_2 + \cdots + \epsilon_n}{n} < x\right) = \frac{2}{\pi} \arcsin \sqrt{\frac{1+x}{2}} \ .$$

2 Given a random walk W of length m, with summands

$$\{X_1, X_2, \ldots, X_m\} \ ,$$

define the *reversed* random walk to be the walk W^* with summands

$$\{X_m, X_{m-1}, \ldots, X_1\} \ .$$

(a) Show that the kth partial sum S_k^* satisfies the equation

$$S_k^* = S_m - S_{n-k} \ ,$$

where S_k is the kth partial sum for the random walk W.

12.3. ARC SINE LAWS

(b) Explain the geometric relationship between the graphs of a random walk and its reversal. (It is not in general true that one graph is obtained from the other by reflecting in a vertical line.)

(c) Use parts (a) and (b) to prove Theorem 12.5.

Appendix A
Normal distribution table

NA (0,d) = area of shaded region

	.00	.01	.02	.03	.04	.05	.06	.07	.08	.09
0.0	.0000	.0040	.0080	.0120	.0160	.0199	.0239	.0279	.0319	.0359
0.1	.0398	.0438	.0478	.0517	.0557	.0596	.0636	.0675	.0714	.0753
0.2	.0793	.0832	.0871	.0910	.0948	.0987	.1026	.1064	.1103	.1141
0.3	.1179	.1217	.1255	.1293	.1331	.1368	.1406	.1443	.1480	.1517
0.4	.1554	.1591	.1628	.1664	.1700	.1736	.1772	.1808	.1844	.1879
0.5	.1915	.1950	.1985	.2019	.2054	.2088	.2123	.2157	.2190	.2224
0.6	.2257	.2291	.2324	.2357	.2389	.2422	.2454	.2486	.2517	.2549
0.7	.2580	.2611	.2642	.2673	.2704	.2734	.2764	.2794	.2823	.2852
0.8	.2881	.2910	.2939	.2967	.2995	.3023	.3051	.3078	.3106	.3133
0.9	.3159	.3186	.3212	.3238	.3264	.3289	.3315	.3340	.3365	.3389
1.0	.3413	.3438	.3461	.3485	.3508	.3531	.3554	.3577	.3599	.3621
1.1	.3643	.3665	.3686	.3708	.3729	.3749	.3770	.3790	.3810	.3830
1.2	.3849	.3869	.3888	.3907	.3925	.3944	.3962	.3980	.3997	.4015
1.3	.4032	.4049	.4066	.4082	.4099	.4115	.4131	.4147	.4162	.4177
1.4	.4192	.4207	.4222	.4236	.4251	.4265	.4279	.4292	.4306	.4319
1.5	.4332	.4345	.4357	.4370	.4382	.4394	.4406	.4418	.4429	.4441
1.6	.4452	.4463	.4474	.4484	.4495	.4505	.4515	.4525	.4535	.4545
1.7	.4554	.4564	.4573	.4582	.4591	.4599	.4608	.4616	.4625	.4633
1.8	.4641	.4649	.4656	.4664	.4671	.4678	.4686	.4693	.4699	.4706
1.9	.4713	.4719	.4726	.4732	.4738	.4744	.4750	.4756	.4761	.4767
2.0	.4772	.4778	.4783	.4788	.4793	.4798	.4803	.4808	.4812	.4817
2.1	.4821	.4826	.4830	.4834	.4838	.4842	.4846	.4850	.4854	.4857
2.2	.4861	.4864	.4868	.4871	.4875	.4878	.4881	.4884	.4887	.4890
2.3	.4893	.4896	.4898	.4901	.4904	.4906	.4909	.4911	.4913	.4916
2.4	.4918	.4920	.4922	.4925	.4927	.4929	.4931	.4932	.4934	.4936
2.5	.4938	.4940	.4941	.4943	.4945	.4946	.4948	.4949	.4951	.4952
2.6	.4953	.4955	.4956	.4957	.4959	.4960	.4961	.4962	.4963	.4964
2.7	.4965	.4966	.4967	.4968	.4969	.4970	.4971	.4972	.4973	.4974
2.8	.4974	.4975	.4976	.4977	.4977	.4978	.4979	.4979	.4980	.4981
2.9	.4981	.4982	.4982	.4983	.4984	.4984	.4985	.4985	.4986	.4986
3.0	.4987	.4987	.4987	.4988	.4988	.4989	.4989	.4989	.4990	.4990
3.1	.4990	.4991	.4991	.4991	.4992	.4992	.4992	.4992	.4993	.4993
3.2	.4993	.4993	.4994	.4994	.4994	.4994	.4994	.4995	.4995	.4995
3.3	.4995	.4995	.4995	.4996	.4996	.4996	.4996	.4996	.4996	.4997
3.4	.4997	.4997	.4997	.4997	.4997	.4997	.4997	.4997	.4997	.4998
3.5	.4998	.4998	.4998	.4998	.4998	.4998	.4998	.4998	.4998	.4998
3.6	.4998	.4998	.4999	.4999	.4999	.4999	.4999	.4999	.4999	.4999
3.7	.4999	.4999	.4999	.4999	.4999	.4999	.4999	.4999	.4999	.4999
3.8	.4999	.4999	.4999	.4999	.4999	.4999	.4999	.4999	.4999	.4999
3.9	.5000	.5000	.5000	.5000	.5000	.5000	.5000	.5000	.5000	.5000

Appendix B

Number of adult children of various statures born of 205 mid-parents of various statures.
(All female heights have been multiplied by 1.08)

Heights of the Mid-parents in inches.	Heights of the adult children.														Total number of		
	Below 62.2	63.2	64.2	65.2	66.2	67.2	68.2	69.2	70.2	71.2	72.2	73.2	Above	Adult children.	Mid-parents.	Medians
Above	..	..	..	..	..	..	..	..	..	..	1	3	..	4	5	..
72.5	..	..	..	..	..	..	1	2	1	2	7	2	4	19	6	72.2
71.5	..	..	..	1	3	4	3	5	10	4	9	2	2	43	11	69.9
70.5	1	..	1	1	1	3	12	18	14	7	4	3	3	68	22	69.5
69.5	..	1	16	4	17	27	20	33	25	20	11	4	5	183	41	68.9
68.5	1	7	11	16	25	31	34	48	21	18	4	3	..	219	49	68.2
67.5	..	3	5	14	36	38	28	38	19	11	4	..	..	211	33	67.6
66.5	..	3	3	5	2	17	17	14	13	4	..	..	..	78	20	67.2
65.5	1	..	9	5	7	11	11	7	7	5	2	..	..	66	12	66.7
64.5	1	1	4	4	1	5	5	2	..	..	..	..	..	23	5	65.8
Below	1	..	2	4	1	2	2	1	..	..	..	..	..	14	1	..
Totals	5	7	32	59	48	117	138	120	167	99	64	41	17	14	928	205	..
Medians	..	..	66.3	67.8	67.9	67.7	67.9	68.3	68.5	69.0	69.0	70.0	..	..	..	..

Note.— In calculating the Medians, the entries have been taken as referring to the middle of the squares in which they stand. The reason why the headings run 62.2, 63.2, &c., instead of 62.5, 63.5, &c., is that the observations are unequally distributed between 62 and 63, 63 and 64, &c., there being a strong bias in favour of integral inches. After careful consideration, I concluded that the headings, as adopted, best satisfied the conditions. This inequality was not apparent in the case of the Mid-parents. Source: F. Galton, "Regression towards Mediocrity in Hereditary Stature", *Royal Anthropological Institute of Great Britain and Ireland*, vol.15 (1885), p.248.

Appendix C

Life Table

Number of survivors at single years of Age, out of 100,000 Born Alive, by Race and Sex: United States, 1990.

	All races				All races		
Age	Both sexes	Male	Female	Age	Both sexes	Male	Female
0	100000	100000	100000	43	94707	92840	96626
1	99073	98969	99183	44	94453	92505	96455
2	99008	98894	99128	45	94179	92147	96266
3	98959	98840	99085	46	93882	91764	96057
4	98921	98799	99051	47	93560	91352	95827
5	98890	98765	99023	48	93211	90908	95573
6	98863	98735	99000	49	92832	90429	95294
7	98839	98707	98980	50	92420	89912	94987
8	98817	98680	98962	51	91971	89352	94650
9	98797	98657	98946	52	91483	88745	94281
10	98780	98638	98931	53	90950	88084	93877
11	98765	98623	98917	54	90369	87363	93436
12	98750	98608	98902	55	89735	86576	92955
13	98730	98586	98884	56	89045	85719	92432
14	98699	98547	98862	57	88296	84788	91864
15	98653	98485	98833	58	87482	83777	91246
16	98590	98397	98797	59	86596	82678	90571
17	98512	98285	98753	60	85634	81485	89835
18	98421	98154	98704	61	84590	80194	89033
19	98323	98011	98654	62	83462	78803	88162
20	98223	97863	98604	63	82252	77314	87223
21	98120	97710	98555	64	80961	75729	86216
22	98015	97551	98506	65	79590	74051	85141
23	97907	97388	98456	66	78139	72280	83995
24	97797	97221	98405	67	76603	70414	82772
25	97684	97052	98351	68	74975	68445	81465
26	97569	96881	98294	69	73244	66364	80064
27	97452	96707	98235	70	71404	64164	78562
28	97332	96530	98173	71	69453	61847	76953
29	97207	96348	98107	72	67392	59419	75234
30	97077	96159	98038	73	65221	56885	73400
31	96941	95962	97965	74	62942	54249	71499
32	96800	95785	97887	75	60557	51519	69376
33	96652	95545	97804	76	58069	48704	67178
34	96497	95322	97717	77	55482	45816	64851
35	96334	95089	97624	78	52799	42867	62391
36	96161	94843	97525	79	50026	39872	59796
37	95978	94585	97419	80	47168	36848	57062
38	95787	94316	97306	81	44232	33811	54186
39	95588	94038	97187	82	41227	30782	51167
40	95382	93753	97061	83	38161	27782	48002
41	95168	93460	96926	84	35046	24834	44690
42	94944	93157	96782	85	31892	21962	41230

Index

$n!$, 80
π, estimation of, 43–46

absorbing Markov chain, 415
absorbing state, 416
AbsorbingChain (program), 421
absorption probabilities, 420
Ace, Mr., 241
Ali, 178
alleles, 348
AllPermutations (program), 84
ANDERSON, C. L., 157
annuity, 246
    life, 247
    terminal, 247
arc sine laws, 493
area, estimation of, 42
Areabargraph (program), 46
asymptotically equal, 81

Baba, 178
babies, 14, 250
Banach's Matchbox, 255
BAR-HILLEL, M., 176
BARNES, B., 175
BARNHART, R., 11
BAYER, D., 120
Bayes (program), 147
Bayes probability, 136
Bayes' formula, 146
BAYES, T., 149
beard, 153
bell-shaped, 47
Benford distribution, 195
BENKOSKI, S., 40
Bernoulli trials process, 96
BERNOULLI, D., 227

BERNOULLI, J., 113, 149, 310–312
Bertrand's paradox, 47–50
BERTRAND, J., 49, 181
BertrandsParadox (program), 49
beta density, 168
BIENAYMÉ, I., 310, 378
BIGGS, N. L., 85
binary expansion, 69
binomial coefficient, 93
binomial distribution, 99, 184
    approximating a, 329
Binomial Theorem, 103
BinomialPlot (program), 99
BinomialProbabilities (program), 98
Birthday (program), 78
birthday problem, 77
blackjack, 247, 253
blood test, 254
Bose-Einstein statistics, 107
Box paradox, 181
BOX, G. E. P., 213
boxcars, 27
BRAMS, S., 179, 182
Branch (program), 382
branching process, 377
    customer, 394
BranchingSimulation (program), 388
bridge, 181, 182, 199, 203, 287
BROWN, B. H., 38
BROWN, E., 425
Buffon's needle, 44–46, 51–53
BUFFON, G. L., 9, 44, 50–51
BuffonsNeedle (program), 45
bus paradox, 164

calendar, 38
cancer, 147

canonical form of an absorbing
    Markov chain, 417
car, 137
CARDANO, G., 30–31, 110, 249
cars on a highway, 66
CASANOVA, G., 11
Cauchy density, 218, 401
cells, 347
Central Limit Theorem, 325
    for Bernoulli Trials, 330
    for Binomial Distributions, 328
    for continuous independent trials
        process, 358
    for discrete independent random
        variables, 345
    for discrete independent trials
        process, 343
    for Markov Chains, 464
    proof of, 398
chain letter, 389
characteristic function, 398
Chebyshev Inequality, 305, 316
CHEBYSHEV, P. L., 313
chi-squared density, 216, 296
Chicago World's Fair, 52
chord, random, 47, 54
chromosomes, 348
CHU, S.-C., 110
CHUNG, K. L., 153
Circle of Gold, 389
Clinton, Bill, 196
clover-leaf interchange, 39
CLTBernoulliGlobal, 332
CLTBernoulliLocal (program), 329
CLTBernoulliPlot (program), 327
CLTGeneral (program), 345
CLTIndTrialsLocal (program), 342
CLTIndTrialsPlot (program), 341
COATES, R. M., 305
CoinTosses (program), 3
Collins, People v., 153, 202
color-blindness, 424
conditional density, 162
conditional distribution, 134
conditional expectation, 239

conditional probability, 133
CONDORCET, Le Marquis de, 12
confidence interval, 334, 360
conjunction fallacy, 38
continuum, 41
convolution, 286, 291
    of binomial distributions, 289
    of Cauchy densities, 294
    of exponential densities, 292, 300
    of geometric distributions, 289
    of normal densities, 294
    of standard normal densities, 299
    of uniform densities, 292, 299
CONWAY, J., 432
CRAMER, G., 227
craps, 235, 240, 468
Craps (program), 235
CROSSEN, C., 161
CROWELL, R., 468
cumulative distribution function, 61
    joint, 165
customer branching process, 394
cut, 120

Dartmouth, 27
darts, 56, 57, 59, 60, 64, 71, 163, 164
Darts (program), 58
DAVID, F. N., 86, 337, 489
DAVID, F. N., 32
de MOIVRE, A., 37, 88, 148, 336, 489
de MONTMORT, P. R., 85
de MÉRÉ, CHEVALIER, 4, 31, 37
degrees of freedom, 217
DeMere1 (program), 4
DeMere2 (program), 4
density function, 56, 59
    beta, 168
    Cauchy, 218, 401
    chi-squared, 216, 296
    conditional, 162
    exponential, 53, 66, 163, 205
    gamma, 207
    joint, 165
    log normal, 224
    Maxwell, 215

INDEX

    normal, 212
    Rayleigh, 215, 295
    t-, 360
    uniform, 60, 205
derangement, 85
DIACONIS, P., 120, 251
Die (program), 225
DieTest (program), 297
distribution function, 1, 19
    properties of, 22
    Benford, 195
    binomial, 184
    geometric, 184
    hypergeometric, 193
    joint, 142
    marginal, 143
    negative binomial, 186
    Poisson, 187
    uniform, 183
DNA, 348
DOEBLIN, W., 449
DOYLE, P. G., 87, 470
Drunkard's Walk example, 416, 419–421, 423, 427, 443
Dry Gulch, 279

EDWARDS, A. W. F., 108
Egypt, 30
Ehrenfest model, 410, 433, 441, 460, 461
EHRENFEST, P., 410
EHRENFEST, T, 410
EhrenfestUrn (program), 462
EISENBERG, B., 160
elevator, 89, 116
Emile's restaurant, 75
ENGLE, A., 445
envelopes, 179, 180
EPSTEIN, R., 287
equalization, 472
equalizations
    expected number of, 479
ergodic Markov chain, 433
ESP, 250, 251
EUCLID, 85

Euler's formula, 202
Eulerian number, 127
event, 18
events
    attraction of, 160
    independent, 139, 164
    repulsion of, 160
existence of God, 245
expected value, 226, 268
exponential density, 53, 66, 163, 205
extinction, problem of, 379

factorial, 80
fair game, 241
FALK, R., 161, 176
fall, 131
fallacy, 38
FELLER, W., 11, 106, 107, 191, 201, 218, 254, 344
FERMAT, P., 4, 32–35, 112–113, 156
Fermi-Dirac statistics, 107
figurate numbers, 108
financial records
    suspicious, 196
finite additivity property, 23
FINN, J., 178
First Fundamental Mystery of Probability, 232
first maximum of a random walk, 496
first return to the origin, 473
Fisher's Exact Test, 193
FISHER, R. A., 252
fixed column vector, 435
fixed points, 82
fixed row vector, 435
FixedPoints (program), 82
FixedVector (program), 437
flying bombs, 191, 201
Fourier transform, 398
FRECHET, M., 466
frequency concept of probability, 70
frustration solitaire, 86
Fundamental Limit Theorem for Regular Markov Chains, 448
fundamental matrix, 419

for a regular Markov chain, 457
for an ergodic Markov chain, 458

GALAMBOS, J., 303
GALILEO, G., 12
Gallup Poll, 14, 336
Galton board, 99, 351
GALTON, F., 281, 345, 350, 377
GaltonBoard (program), 99
Gambler's Ruin, 426, 486, 487
gambling systems, 241
gamma density, 207
GARDNER, M., 181
gas diffusion
    Ehrenfest model of, 410, 433, 441, 460, 461
GELLER, S., 176
GeneralSimulation (program), 9
generating function
    for continuous density, 394
    moment, 366, 395
    ordinary, 370
genes, 348, 411
genetics, 345
genotypes, 348
geometric distribution, 184
geometric series, 29
GHOSH, B. K., 160
goat, 137
GONSHOR, H., 425
GOSSET, W. S., 360
grade point average, 343
GRAHAM, R., 251
GRANBERG, D., 161
GRAUNT, J., 246
Greece, 30
GRIDGEMAN, N. T., 51, 181
GRINSTEAD, C. M., 87
GUDDER, S., 160

HACKING, I., 30, 148
HAMMING, R. W., 283, 284
HANES data, 345
Hangtown, 279
Hanover Inn, 65

hard drive, Warp 9, 66
Hardy-Weinberg Law, 349
harmonic function, 428
Harvard, 27
hat check problem, 82, 85, 105
heights
    distribution of, 345
helium, 107
HEYDE, C., 378
HILL, T., 196
Holmes, Sherlock, 91
HorseRace (program), 6
hospital, 14, 250
HOWARD, R. A., 406
HTSimulation (program), 6
HUDDE, J., 148
HUIZINGA, F., 389
HUYGENS, C., 147, 243–245
hypergeometric distribution, 193
hypotheses, 145
hypothesis testing, 101

Inclusion-Exclusion Principle, 104
independence of events, 139, 164
    mutual, 141
independence of random variables
    mutual, 143, 165
independence of random
    variables, 143, 165
independent trials process, 144, 168
interarrival time, average, 208
interleaving, 120
irreducible Markov chain, 433
Isle Royale, 202

JAYNES, E. T., 49
JOHNSONBOUGH, R., 153
joint cumulative distribution
    function, 165
joint density function, 165
joint distribution function, 142
joint random variable, 142

KAHNEMAN, D., 38
Kemeny's constant, 469, 470
KEMENY, J. G., 200, 406, 466

INDEX

KENDALL, D. G., 378
KEYFITZ, N., 383
KILGOUR, D. M., 179, 182
KINGSTON, J. G., 157
KONOLD, C., 161
KOZELKA, R. M., 344

Labouchere betting system, 12, 13
LABOUCHERE, H. du P., 12
LAMPERTI, J., 267, 324
LAPLACE, P. S., 51, 53, 350
last return to the origin, 482
Law (program), 310
Law of Averages, 70
Law of Large Numbers, 307, 316
    for Ergodic Markov Chains, 439
    Strong, 70
LawContinuous (program), 318
lead change, 482
LEONARD, B., 256
LEONTIEF, W. W., 426
LEVASSEUR, K., 485
library problem, 82
life table, 39
light bulb, 66, 72, 172
Linda problem, 38
LINDEBERG, J. W., 344
LIPSON, A., 161
Little's law for queues, 276
Lockhorn, Mr. and Mrs., 65
log normal density, 224
lottery
    Powerball, 204
LUCAS, E., 119

MAISTROV, L., 150, 310
MANN, B., 120
margin of error, 335
marginal distribution function, 143
Markov chain, 405
    absorbing, 415
    ergodic, 433
    irreducible, 433
    regular, 433
Markov Chains
    Central Limit Theorem for, 464
    Fundamental Limit Theorem for
        Regular, 448
MARKOV, A. A., 464
martingale, 241, 242, 428
    origin of word, 11
martingale betting system, 11, 14, 248
matrix
    fundamental, 419
MatrixPowers (program), 407
maximum likelihood
    estimate, 198, 202
Maximum Likelihood
    Principle, 91, 117
Maxwell density, 215
maze, 440, 453
McCRACKEN, D., 10
mean, 226
mean first passage matrix, 455
mean first passage time, 452
mean recurrence matrix, 455
mean recurrence time, 454
memoryless property, 68, 164, 206
milk, 252
modular arithmetic, 10
moment generating function, 366, 395
moment problem, 368, 398
moments, 365, 394
Monopoly, 469
MonteCarlo (program), 42
Monty Hall problem, 136, 161
moose, 202
mortality table, 246
mule kicks, 201
MULLER, M. E., 213
multiple-gene hypothesis, 348
mustache, 153
mutually independent events, 141
mutually independent random
    variables, 143

negative binomial distribution, 186
NEGRINI, M., 196
New York Times, 340
New York Yankees, 118, 253

New-Age Solitaire, 130
NEWCOMB, S., 196
NFoldConvolution (program), 287
normal density, 47, 212
NormalArea (program), 322
nursery rhyme, 84

odds, 27
ordering, random, 127
ordinary generating function, 370
ORE, O., 30, 31
outcome, 18
Oz, Land of, 406, 439

Pascal's triangle, 94, 103, 108
PASCAL, B., 4, 32–35, 107, 112–113, 156, 242, 245
paternity suit, 222
PEARSON, K., 9, 351
PENNEY, W., 432
People v. Collins, 153, 202
PERLMAN, M. D., 45
permutation, 79
    fixed points of, 82
Philadelphia 76ers, 15
photons, 106
Pickwick, Mr., 153
Pilsdorff Beer Company, 279
PITTEL, B., 256
point count, 287
Poisson approximation to the binomial distribution, 189
Poisson distribution, 187
    variance of, 262
poker, 95
polls, 333
Polya urn model, 152, 174
ponytail, 153
posterior probabilities, 145
Powerball lottery, 204
PowerCurve (program), 102
Presidential election, 336
PRICE, C., 86
prior probabilities, 145
probability
    Bayes, 136
    conditional, 133
    frequency concept of, 2
    of an event, 19
    transition, 406
    vector, 407
problem of points, 32, 112, 147, 156
process, random, 128
PROPP, J., 256
PROSSER, R., 200
protons, 106
PÓLYA, G., 15, 17, 475

quadratic equation, roots of, 73
quantum mechanics, 107
QUETELET, A., 350
Queue (program), 208
queues, 186, 208, 275
quincunx, 351

RABELAIS, F., 12
racquetball, 157
radioactive isotope, 66, 71
RAND Corporation, 10
random integer, 39
random number generator, 2
random ordering, 127
random process, 128
random variable, 1, 18
    continuous, 58
    discrete, 18
    functions of a, 210
    joint, 142
random variables
    independence of, 143
    mutual independence of, 143
random walk, 471
    in n dimensions, 17
RandomNumbers (program), 3
RandomPermutation (program), 82
rank event, 160
raquetball, 13
rat, 440, 453
Rayleigh density, 215, 295
records, 83, 234

INDEX

Records (program), 84
regression on the mean, 282
regression to the mean, 345, 352
regular Markov chain, 433
reliability of a system, 154
restricted choice, principle of, 182
return to the origin, 472
    first, 473
    last, 482
    probability of eventual, 475
reversibility, 463
reversion, 352
riffle shuffle, 120
RIORDAN, J., 86
rising sequence, 120
rnd, 42
ROBERTS, F., 426
Rome, 30
ROSS, S., 270, 276
roulette, 13, 237, 432
run, 229
RÉNYI, A., 167

SAGAN, H., 237
sample, 333
sample mean, 265
sample space, 18
    continuous, 58
    countably infinite, 28
    infinite, 28
sample standard deviation, 265
sample variance, 265
SAWYER, S., 412
SCHULTZ, H., 255
SENETA, E., 378, 444
service time, average, 208
SHANNON, C. E., 465
SHOLANDER, M., 39
shuffling, 120
SHULTZ, H., 256
SimulateChain (program), 439
simulating a random variable, 211
snakeeyes, 27
SNELL, J. L., 87, 175, 406, 466
snowfall in Hanover, 83

spike graph, 6
Spikegraph (program), 6
spinner, 41, 55, 59, 162
spread, 266
St. Ives, 84
St. Petersburg Paradox, 227
standard deviation, 257
standard normal random
    variable, 213
standardized random variable, 264
standardized sum, 326
state
    absorbing, 416
    of a Markov chain, 405
    transient, 416
statistics
    applications of the Central Limit
        Theorem to, 333
stepping stones, 412
SteppingStone (program), 412
stick of unit length, 73
STIFEL, M., 110
STIGLER, S., 350
Stirling's formula, 81
STIRLING, J., 88
StirlingApproximations
    (program), 81
stock prices, 241
StockSystem (program), 241
Strong Law of Large
    Numbers, 70, 314
suit event, 160
SUTHERLAND, E., 182

t-density, 360
TARTAGLIA, N., 110
tax returns, 196
tea, 252
telephone books, 256
tennis, 157, 424
tetrahedral numbers, 108
THACKERAY, W. M., 14
THOMPSON, G. L., 406
THORP, E., 247, 253
time to absorption, 419

TIPPETT, L. H. C., 10
traits, independence of, 216
transient state, 416
transition matrix, 406
transition probability, 406
tree diagram, 24, 76
    infinite binary, 69
Treize, 85
triangle
    acute, 73
triangular numbers, 108
trout, 198
true-false exam, 267
Tunbridge, 154
TVERSKY, A., 14, 38
Two aces problem, 181
two-armed bandit, 170
TwoArm (program), 171
type 1 error, 101
type 2 error, 101
typesetter, 189

ULAM, S., 11
unbiased estimator, 266
uniform density, 205
uniform density function, 60
uniform distribution, 25, 183
uniform random variables
    sum of two continuous, 63
unshuffle, 122
USPENSKY, J. B., 299
utility function, 227

VANDERBEI, R., 175
Vandermonde determinant, 369
variance, 257, 271
    calculation of, 258
variation distance, 128
VariationList (program), 128
volleyball, 158
von BORTKIEWICZ, L., 201
von MISES, R., 87
von NEUMANN, J., 10, 11
vos SAVANT, M., 40, 86, 136, 176, 181

Wall Street Journal, 161
watches, counterfeit, 91
WATSON, H. W., 378
WEAVER, W., 465
Weierstrass Approximation Theorem, 315
WELDON, W. F. R., 9
Wheaties, 118, 253
WHITAKER, C., 136
WHITEHEAD, J. H. C., 181
WICHURA, M. J., 45
WILF, H. S., 91, 474
WOLF, R., 9
WOLFORD, G., 159
Woodstock, 154

Yang, 130
Yin, 130

ZAGIER, D., 485
Zorg, planet of, 90